PETROLEUM AND THE CONTINENTAL SHELF OF NORTH-WEST EUROPE

Volume 1

GEOLOGY

Proceedings of a conference organized jointly by the Geological Society of London, The Institute of Geological Sciences, The Institute of Petroleum and the Petroleum Exploration Society of Great Britain. The following were also associated: The United Kingdom Offshore Operators' Association, The American Association of Petroleum Geologists, Deutsche Gesellschaft für Mineralölwissenschaft und Kohle-chemie, L'Association Française des Techniciens du Pétrole; Norsk Petroleums Institutt.

PETROLEUM AND THE CONTINENTAL SHELF OF NORTH-WEST EUROPE

Volume 1

GEOLOGY

EDITED BY

AUSTIN W. WOODLAND
(Institute of Geological Sciences, London)

A HALSTED PRESS BOOK

JOHN WILEY & SONS
NEW YORK—TORONTO

PUBLISHED IN THE U.S.A. AND CANADA BY
HALSTED PRESS
A DIVISION OF JOHN WILEY & SONS, INC., NEW YORK

Library of Congress Cataloging in Publication Data
Main entry under title:

Petroleum and the continental shelf of north-west
    Europe.

    "A Halsted Press book."
    "Proceedings of a conference organized jointly
by the Geological Society of London . . . [et al.]"
    CONTENTS: v. 1. Geology.
    1. Petroleum in submerged lands—Europe—Congresses.
2. Continental shelf—Europe—Congresses. I. Woodland,
Austin W. II. Geological Society of London.
TN874.A1P47   553'.28'0916336   75-14329
ISBN 0-470-95993-2

WITH 299 ILLUSTRATIONS

© APPLIED SCIENCE PUBLISHERS LTD 1975

Printed in Great Britain by Galliard (Printers) Ltd, Great Yarmouth

ORGANIZING COMMITTEE

C. J. O. Moorhouse *(Chairman)*, Institute of Petroleum
Dr J. M. Bowen, Petroleum Exploration Society of Great Britain
R. Dyk, United Kingdom Offshore Operators' Association
E. G. Everett, United Kingdom Offshore Operators' Association
Dr L. V. Illing, Petroleum Exploration Society of Great Britain
K. A. D. Inglis, Institute of Petroleum
D. C. Ion, Institute of Petroleum
H. Jagger, Institute of Petroleum
C. H. Maynard *(Secretary)*, Institute of Petroleum
Dr A. J. Smith, Geological Society of London
Dr A. W. Woodland *(Editor)*, Institute of Geological Sciences

Contents

Foreword

By JAMES MOORHOUSE

(Chairman, Conference Organizing Committee)

IN the words of a leading geologist who attended the Conference at which the papers published in this book were presented: "This was the most important and significant geological conference ever presented in the European area, and possibly in the world, in view of the great importance of the North Sea development in the total world energy picture". A bold claim you may think but certainly the Conference led to a frank exchange of information which was rather unique at such an early stage in the development of a new petroleum province. Furthermore it was the first joint meeting in the United Kingdom between geologists and environmental specialists from industry, universities, government and agencies from around the world.

The Conference was jointly organized by The Geological Society of London, The Institute of Geological Sciences, The Institute of Petroleum, and The Petroleum Exploration Society of Great Britain. The United Kingdom Offshore Operators' Association was also closely associated with the Conference as were the American Association of Petroleum Geologists, Deutsche Gesellschaft für Mineralölwissenschaft und Kohlechemie, L'Association Française des Techniciens du Pétrole and Norsk Petroleums Institutt. It also enjoyed the backing of The Royal Society in London and after the Conference its distinguished Foreign Secretary, Sir Kingsley Dunham, F.R.S., remarked that this was "One of the only scientific conferences I have attended where almost everything was new!" But judge for yourself. Within the covers of this book there are 38 papers on the geology of the Continental Shelf of north-west Europe which were presented; the major emphasis, however, was on the various North Sea oil and gas producing basins. The geological presentations were unique, firstly in that the majority of the papers were from commercial (as opposed to academic) sources, mainly oil companies, and secondly that most of the material had never been previously publicly presented.

On the last two days of this week-long Conference, attention was switched from the geology *per se* to potential environmental problems associated with exploration, production, treatment and transportation of crude oil and with the operational and regulatory measures to minimize these. The papers covering these aspects are, however, the subject of a separate volume being published almost simultaneously under the title "Petroleum and the Continental Shelf of North West Europe: Volume 2 Environmental Protection". They also cover the possible effects of offshore operations on local amenities and on marine life as seen through the eyes of leading researchers in this field.

To pick out the name of anyone in particular who played a notable part in the proceedings of a Conference such as this is always difficult but there is one person especially who I think should be singled out for mention, namely Dr A. W. Woodland of the Institute of Geological Sciences, who gallantly undertook to edit each and every one of the papers that appear in this volume. I think you will agree with me that he has done an admirable task. Finally, I should like to record my warmest thanks to the many others who ensured the success of this Conference, including the members of the Conference Organizing Committee, our distinguished chairmen, and the presenters and joint authors of the papers, and the staff of the Institute of Petroleum who acted as the organizers.

Opening Address

By Mr JOHN SMITH

(Parliamentary Under-Secretary of State for Energy)

As Britain's Minister with special responsibility for North Sea oil and gas it gives me very real pleasure to open a conference which brings together so many specialists in the world of offshore exploration.

I am told that there are around 1000 experts here, 300 of you from abroad. And that there has never before been such a concentration of geologists and geophysicists to discuss the North Sea bed.

By the end of the week, therefore, we cannot fail to have a better understanding of the geology and resources of the continental shelf of this corner of the world, and of the geology from which further oil and gas could be discovered.

Few people could have suspected nine years ago, when gas was first discovered on the UK Continental Shelf, that there were such vast reserves of both oil and gas.

Nor that by 1980 we could be virtually self-sufficient in these resources.

But by early next year we expect our first oil to come ashore—between 3m and 5m tons. And in 1980 oil production could be in the range of 100m to 140m tons a year.

Oil was discovered on the UK Continental Shelf only five years ago. Now, as my Secretary of State told Parliament earlier this month, proven and probable reserves from commercial oil fields total 1160 million tons.

It is of course too early yet to assess the probable reserves of other promising finds including the possible extension into UK waters of the Norwegian Statfjord field.

A great deal more exploration work, even under existing licences, still remains to be done and I am sure further discoveries will be made. It is to be hoped that some of these discoveries will eventually prove to be of the same order as those finds we have made so far. And, of course, there are large tracts which have not yet been licensed, nor even designated.

As you know, there is great debate going on at the moment about the possible size of reserves which may be eventually discovered, but in Government we cannot base policy on speculation.

The development of the UK Continental Shelf has therefore been extremely rapid. In the British sector of the North Sea some 630 wells have been drilled—four times the number that have been drilled in any other sector—and 24 fields have been discovered. The companies involved deserve very real praise for this remarkable achievement.

Some of the companies are now reaching the stage, as recent announcements have shown, where they are prepared to release information on some of their fields to other licensees, instead of waiting until they are obliged to.

We in government have been receiving companies' seismic and well information all along, under the terms of the licences that have been granted. But this information has been confidential, and the geology of the North Sea has therefore been under wraps as far as other companies, and interested outsiders such as universities, have been concerned.

The Government, through the Institute of Geological Sciences, is already building up a geological picture of the Continental Shelf as a whole—mostly from the information supplied by licensees, but also from work carried out by the IGS itself on sub-surface and sea bed geology—to enable a full assessment of the UK potential resources to be properly evaluated.

But the individual companies are able to look at what is under the North Sea only through their own "window". Until they are in a position to inter-relate all the windows, and to compare different interpretations of similar systems, there will be little incentive to explore areas which have so far not attracted much attention. There will be little encouragement to look at different sorts of plays. Smaller companies are less likely to come in to pick up second generation finds.

The likelihood therefore of exchange of information that has previously been commercially confidential—no doubt some of this new information will be released here this week—is refreshing. In fact such exchange is vital if we are to build up quickly a public model of the geology of the different areas and formations in which oil and gas are found.

So far we do know that the North Sea has undergone very complex geology which has led to oil and gas

1

being found in almost every geological system—from the Carboniferous to the Tertiary. One of the most exciting finds has been the occurrence of volcanic ashes in the north of the North Sea. This has given geologists considerable food for thought.

We cannot of course talk about the discovery and the development of such reserves without talking about their implications for the environment. And I know that this Conference will be spending a great deal of time on this aspect.

Wherever large quantities of oil are being handled pollution will always be an obvious, and very real, cause for concern. The UK Government is very concerned about this and is determined to get on top of, and keep on top of, this issue.

In Britain we already have an Offshore Pollution Liability Agreement, a compensation scheme entered into voluntarily by United Kingdom offshore operators. This scheme is an interim measure.

Next year the British Government will host a conference at which we hope a convention on pollution liability will be signed by North Sea governments. This will be one of the results of the work started by the London Conference on Safety and Pollution Safeguards which we sponsored in March last year.

But the best way by far is to prevent, rather than cure, pollution. We know from our close liaison with both platform manufacturers and the oil companies that substantial safety margins are being incorporated into the designs now being built. Our new regulations, which require permanent offshore structures in the UK sector of the North Sea have to have a certificate of fitness by 31 August, 1975, pay close attention to the need to prevent accident and to ensure a high standard of operation.

In fact we are the first of the countries undertaking operations in the difficult conditions prevailing in the North Sea to introduce comprehensive legislation governing the safety of operations and of the personnel and equipment employed in them. Other governments have already shown interest in our legislation.

The successful exploitation of the North Sea hydrocarbon resources, with the new data and technological innovation it will produce, will lead, we know, to world-wide development of offshore petroleum in waters that have until now seemed impossible to work.

Continued international cooperation, between governments and between the kind of organisations gathered here today, will play a large part in this in the years ahead, especially as we move into ever deeper waters.

For this reason, as well as many others, I wish this Conference, as an example of international cooperation, every success.

The Tectonic Development of Great Britain and the Surrounding Seas

By P. E. KENT

(President of the Geological Society of London)

SUMMARY

This paper reviews the tectonic development of the British Isles and of the surrounding continental shelf as a co-ordinated area. The subject is essentially the post-Caledonian history, with the main emphasis on the evidence provided by post-Carboniferous formations.

Except for Cornubia the main positive units are fundamentally eroded Caledonian massifs, clearly defined and well advanced in peneplanation by the Lower Carboniferous. Of these only the Pennine Blocks appear to have been deeply buried by later Carboniferous sediments. Except for the Scottish Highlands the positive areas all appear to have remained near sea-level during the Carboniferous, subject to intermittent periods of sedimentation and erosion. The Scottish Highlands region appears to have been the dominant sediment source within the British area.

After the Hercynian orogeny and a major erosion phase, the Permo-Triassic basins were developed by a combination of contemporary large scale down-warping and faulting, producing graben-like features in the centres of the main offshore basins, but the positive areas of the Carboniferous retained their identities, and in consequence made only minor contributions to the sedimentary fill of adjoining basins. The same general regime continued in the Jurassic, with contemporaneous faulting documented in the northern North Sea.

Sedimentation controlled by this structural pattern continued into the Lower Cretaceous, ended by widespread faulting, uplift and erosion at various times from the early Aptian to Albian or Cenomanian. On the edge of the Viking Graben sagging continued well into the Upper Cretaceous, but this case appears to be exceptional. Taphrogenic control ended with this episode: the last major dislocation of British basinal areas. The Upper Cretaceous and Tertiary were deposited in mostly unfaulted relatively simple basins, the former widely transgressive, the latter in more restricted areas and associated with broad folding, *e.g.* the rise of the Weald from Lower Tertiary onwards, the relative uplift of the Pennines and other positive units probably extending over a large part of the Tertiary period. Onshore the East Anglian Massif remained depressed and extensive Palaeozoic/Mesozoic high areas in the mid and northern North Sea and elsewhere likewise lost their separate identities and sank with the base of the Tertiary/Cretaceous basin.

INTRODUCTION

It is a somewhat difficult assignment to write a review-type paper for a symposium which includes three other reviews, and in which a great part of the data is due to be released for the first time later in the proceedings. Under these circumstances it seems most useful to attempt a review of the development of the north-west European Continental Shelf as a co-ordinated unit, leaning heavily on the 150 years of geological knowledge of the land areas for background, but relating this to the offshore which now provides a major amount of data relevant to broader trends, and which continues the history in much more detail for the post-Cretaceous.

The seas around modern Britain largely occupy the successor basins to those of the Permo-Triassic, and it is the edges of these basins, with attenuated sequences, which are most obviously sensitive to regional movement. These two factors dictate much of the relationships within modern land and sea areas, for the stratigraphical breaks are more clearly defined in what are now the land areas, whereas the sequence overall is more complete in the submerged modern basins. Understanding of both is necessary for an adequate regional analysis.

The writer has to express his deep indebtedness to the geological and geophysical staff of the British Petroleum Company for providing much of the background from which this paper grew, and to the Directors of the Company for authorizing its use and for providing the necessary supporting facilities. In the final stages he has had valuable comments on the draft from P. J. Walmsley, A. M. Spencer, B. G. Williams and G. G. Leckie, which have added materially to accuracy and completeness of the presentation.

STRUCTURAL GEOGRAPHY OF THE REGION

Although it is sufficiently obvious that the geography of any area changes through geological time, it is necessary to make the point that in consequence structural–geographical names (however generalized) must carry a time-connotation. As an example may be quoted the Wales–Brabant Massif (Midland Barrier, London–Brabant Platform, St Georges Land) which was essentially a late Palaeozoic feature; it was interrupted by Mesozoic downwarps in St Georges Channel, in the West Midlands and

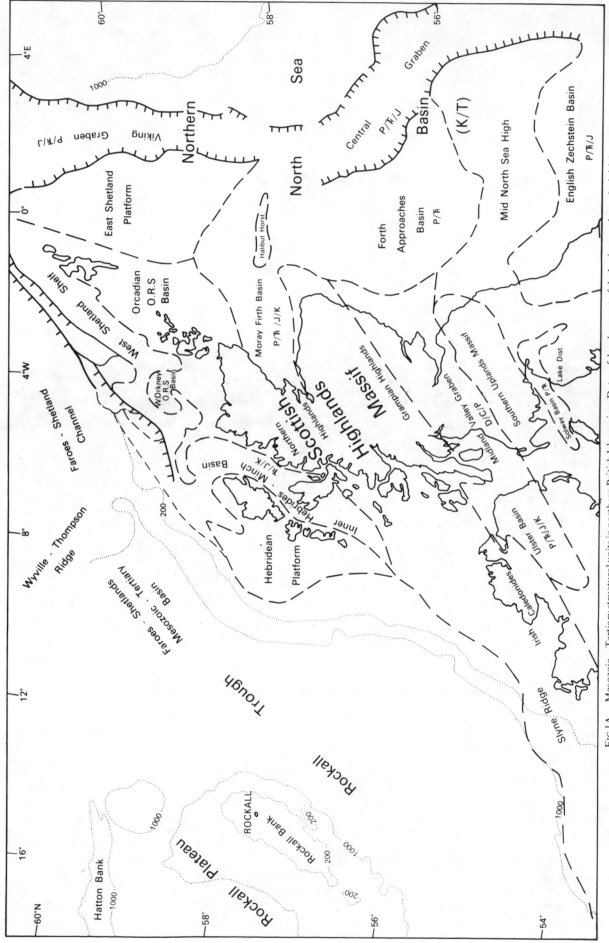

FIG 1A. Mesozoic — Tertiary structural units in northern British Isles region. Dates of development of the basins are shown by initials.

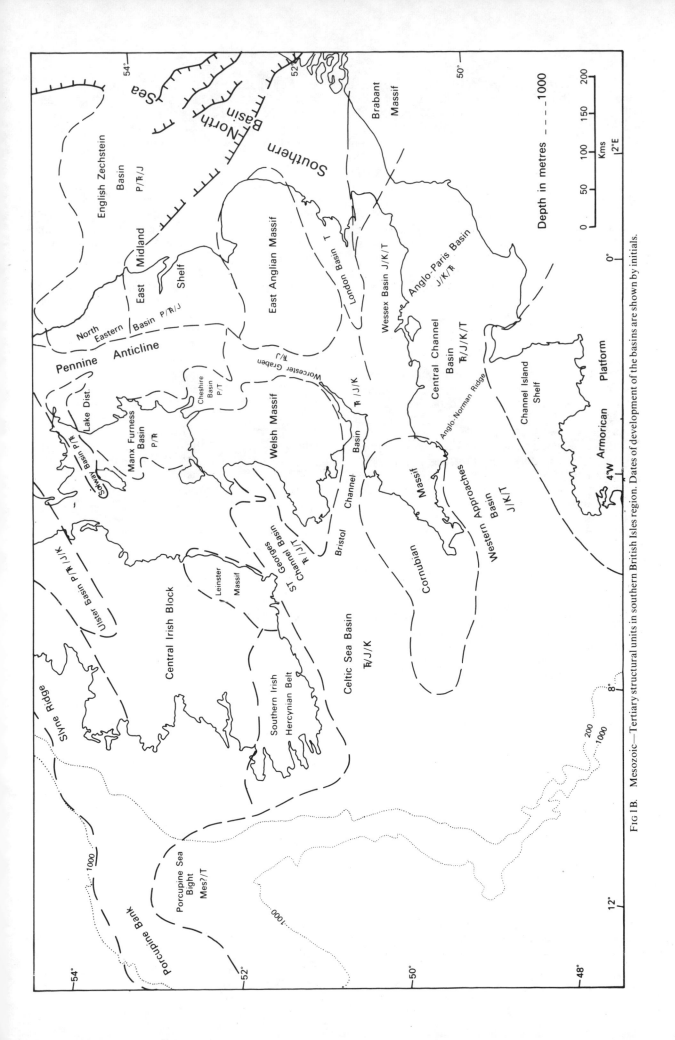

FIG 1B. Mesozoic—Tertiary structural units in southern British Isles region. Dates of development of the basins are shown by initials.

(partially) in the southern North Sea into separate units—South-eastern Ireland, the Welsh Massif, the East Anglian Massif and Brabant— so that the original broad name is inapplicable after the Permian. Unless otherwise qualified the regional structures are here defined in terms of their Mesozoic and Tertiary expression, and the accompanying map (Fig 1A & B) shows features of that period.

The British Isles are an emergent part of the north-west European Continental Shelf including a group of structures which have had a positive tendency over a long period of time. Most of the Palaeozoic outcrops include a foundation of rocks folded, metamorphosed and granite-intruded during the Caledonian orogeny. The Cornubian peninsula and southern Ireland are the only emergent sections of the Hercynian orogenic belt; deformation further north at that time was limited to relatively gentle foreland folding.

Uplift and associated erosion in connection with the Hercynian orogeny stripped much of the thin and probably discontinuous Carboniferous cover from the Wales–Brabant Massif and brought the Pennines into existence as an axial upwarp, linking former separated massifs (Alston, Askrigg and probably Derbyshire). East of the Pennines the Permian was deposited on a structural shelf in the North-eastern Basin which was marginal to the main Southern North Sea Basin, and progressive subsidence during the Trias spread eastwards and southwards across the English Midlands. In western England a series of cratonic partly faulted basins developed. The Cheshire Basin, with 3000–3500m of Permo-Trias, connected southwards with the Worcester half-graben and westwards with the Irish Sea basins, which in turn linked southwards with the downwarps of St George's Channel (a further breakdown of the Wales–Brabant Massif) and northwards with the basins of northern Ireland, with the Scottish Midland Valley subsidence and with the later Hebridean Basin.

Permos-Triassic rocks lap onto the irregular surfaces of the Palaeozoic/Precambrian highs of Cornubia, Wales, the East Anglian Massif and such other remnants of the Wales–Brabant Massif as Charnwood; the Lake District and the Southern Uplands, further accentuated the distinctiveness of the early Carboniferous positive areas. Jurassic basin development followed the Permo-Triassic trends.

At this time also the axial north–south graben system of the mid-North Sea began to develop, with Permian subsidence (partly fault controlled) cutting across the Westphalian basin and extending progressively northwards as the Trias/Jurassic Viking Graben of the north. This rift linked with a series of subsidiary basins: the Sole Pit Trough in the south with an abnormal thickness of Jurassic (analogous to Jurassic depocentres in Germany; the Forth Approaches or South Scotland Basin, apparently mainly Permo-Triassic; and the Moray Firth Basin, much faulted, developed in the Permian in the central part of the much broader Orcadian basin of Devonian date, which continued to subside through the Trias, Jurassic and Lower Cretaceous.

Subsequently, in the Eastern (Mid) Channel and southern England Wessex Basin, in the south Midlands and north-eastern England, Jurassic and then Lower Cretaceous basin-development largely followed the pattern established in the Trias, and this was probably true of north-western England although evidence is now very scanty.

Upper Cretaceous basin development was broadly coincident with the earlier Mesozoic more complex troughs and basins, but extended widely onto adjoining shelves, and was no longer subject to the fault control which earlier characterised the central parts of the larger basins. Locally, inversion resulted in increased Upper Cretaceous subsidence outside the Jurassic troughs.

During the Mesozoic the shallow Hebridean Basin, which became the site of the Tertiary volcanic centres, developed west of the Highlands Massif, lapping onto a broad Outer Hebrides shelf which extends to the continental edge. It was presumably linked with the Ulster Basin farther south, which contains Permian as well as thick Trias but only a remnant of Jurassic beneath the Cretaceous Chalk and Tertiary lavas. West of Ireland the Porcupine Bight, flanked seawards by the Porcupine Bank believed to be a subsided continental segment, contains a sedimentary column, still undrilled, presumed to include both Mesozoic and Tertiary rocks—the latter perhaps dominating, as in the Outer Channel Basin (southern Celtic Sea).

Except for the major protracted vulcanicity in the Midland Valley of Scotland (Devonian/Carboniferous), volcanic centres were much less widespread than the rifting with which they are perhaps genetically associated. Permian faulting was associated with vulcanicity only in the Midland Valley of Scotland and in Devonshire. The mid-North Sea graben system led to Jurassic vulcanicity in the Forties area, as in the north-west Netherlands, extending from the oil-field northwards, east of the Moray Firth. The opening of the north Atlantic was preceded by a long period of fault-controlled subsidence, but the major volcanic centres of the Western Isles of Scotland and of northern Ireland are localized in space and time, whether due to passage over a "hot spot" or for other reasons. Volcanoes have thus been a frequent but not continuous feature of the development of the region.

THE SEQUENCE OF TECTONIC EVENTS

There is a world-wide tendency to classify events in shelf areas in terms of orogenic phases elsewhere, although precise contemporaneity is often doubtful and genetic relationship more tenuous still. In the Carboniferous in particular there has been application to Britain of the orogenic place names devised by Stille on mainland Europe, and there is a comparable procedure in the Mesozoic of referring to Early Cimmerian (Kimmerian), Late Cimmerian, Laramide and even Nevadian phases. Use of these names may be a convenience but is, in the writer's view, undesirable, carrying as they do an implication of orogenic (compressional) movement for events which are not of proved orogenic origin and which in many cases are probably largely due to eustatic control.

In the British Carboniferous, Rayner (1953, p. 279) concluded that most of the evidence for orogeny in the Lower Carboniferous was decidedly slim; that only the Sudetic phase could be correlated with a widespread unconformity (base of Millstone Grit/Bowland Shales) but that this was essentially a matter of differential movement between blocks and gulfs. Correlation of the earlier Carboniferous in Britain by Ramsbottom (1973) has now shown that the Dinantian sequence is essentially cyclic, the main unconformities being due to this feature combined with regional transgression and at least some of these cycles can be identified in Belgium and the USA, implying probability of a eustatic origin. In the Mesozoic, a great deal of the movement classified as early or late Cimmerian is associated with halokinetic displacement in Germany and in the North Sea, and again is not orogenic in type, or necessarily in fundamental control. No detectable shortening of the underlying floor is associated with its development, and the equivalent displacements in Britain (the "Jurassic axes") involve tilts only of a degree or less.

Deltaic conditions in the Upper Carboniferous, with deposition largely close to sea-level, provided a fine index for cyclic control, as also did the evaporitic deposition in the Permian (Zechstein cycles). Cyclic control is recognizable also in the Bunter (Wills, 1970) although it is too complex for wide correlation. In the Keuper and Jurassic also cyclic movements continued but were subordinate to other factors.

The contemporaneity of events over wide areas, whether epeirogenic or eustatic, is nevertheless important in analysis of the history of development of offshore structures. The main Mesozoic and Tertiary interruptions in deposition may be listed as follows, with the further comment that during periods when water was generally shallow, deposition was particularly sensitive to interruption by extraneous factors, and that shoreline areas (such as the Mendips) show additional but localized unconformities.

Upper Permian, Zechstein: Contemporaneous graben formation in western Britain and the North Sea, probably continuing during Triassic.

Mid-Triassic: Hardegsen unconformity (stratigraphic break usually without measurable angular discordance).

Late and End-Triassic: (i) Inception of halokinetic movements in North Sea salt basins; (ii) Rhaetian instability (minor breaks and rapidly varying depositional conditions; institution of new positive areas).

End Lower Jurassic: Low angle unconformity at beginning of Bajocian (Aalenian) coinciding with regional shallowing and inception of deltaic and estuarine conditions in northern Britain and most of North Sea.

Mid-Middle Jurassic, early Bathonian (Vesulian): Widespread low-angle unconformity and stratigraphic breaks, associated with faulting offshore.

Beginning Upper Jurassic, Lower Callovian: Wide spread of clastics possibly due to cyclic control.

Mid-Upper Jurassic, Upper Oxfordian: Minor stratigraphic breaks and regression (cyclic?).

Late-Upper Jurassic/early Neocomian: Major faulting in North Sea Basin; local erosion of Jurassic in southern basin; followed by widespread transgression.

Early Aptian: Folding, local faulting, uplift and erosion particularly in southern England.

Early-Middle Albian: Widespread uplift erosion and angular unconformity.

Cenomanian: Onlap and local unconformity.

Early Palaeocene: Widespread gentle warping, uplift and regression.

Early Eocene (Ypresian): Local uplift (*e.g.* of Wealden area) followed by transgression.

Middle and Upper Eocene: 2–3 phases of cyclic regression/transgression.

Oligocene: Uplift and regression.

Miocene (imprecisely dated): Development of large monoclinal folds in southern England and Channel.

This list is derived mainly from evidence in eastern and southern England. Data are less complete for the marine areas (in particular in the North Sea) but indicate a broadly parallel history, allowing for the circumstance that in centres of basins deposition would not be interrupted as frequently as in the marginal areas. Some major movements—in particular the development of growth faults along the western side of the West Shetland Shelf—at present remain undated except that by analogy they are thought to be at least as old as Triassic.

This summary with the history traced in the

following pages, is a strong warning against the common assumption in British geological writing that post-Hercynian movements were mainly Alpine. The largest regional episodes of vertical movements were in fact Permo-Triassic and immediately pre-Albian, with mid-Tertiary one of the less active periods.

FIG 2. Lower Carboniferous structural units; offshore distribution is largely hypothetical.

ANCIENT POSITIVE AREAS

The history of the old positive areas of Britain is important not only as part of the general tectonic framework, but also because they provide a potential source of detrital sediments, notably hydrocarbon-bearing source and reservoir rocks for the surrounding subsiding basins.

With one exception they originated as high-standing eroded remnants of Lower Palaeozoic and older rocks deformed and metamorphosed during the multiple phases of the Caledonian orogeny (Fig 2). The exception is Cornubia, the south-western peninsula, which lay within the Hercynian orogenic belt and has, for the most part, a later origin. Farther north, Hercynian and later movements were essentially of foreland folding type or entirely epeirogenic in character.

Description here starts with the East Anglian Massif, the London Platform of many authors, which unlike its exposed homologues has remained buried, with its envelope of Upper Palaeozoic,

Mesozoic and later sediments available to demonstrate the later Phanerozoic history.

East Anglian Stable Massif

Along routes from Coventry, Leicester or Norwich to London it is a remarkable fact that the Palaeozoic platform is little more than a field's length below the modern surface. Attenuated Mesozoic rocks in fact overlie a remarkably flat plateau bevelled across the pre-Mesozoic rocks, most appropriately called the East Anglian Massif, alternatively the London Platform. (London is peripherally situated; the feature extends to the Wash.) This is the Mesozoic expression of a high block which at other times was continuous with a Welsh highland, extending westwards into Ireland (St George's Land), and into Belgium (Anglo-Belgian ridge or London–Brabant Platform) (Wills, 1951). Across it the Upper Cretaceous truncates Jurassic to rest directly on Palaeozoic rocks.

This relationship was not due to a simple uplift, but represents relative stability between surrounding troughs through the whole of Permian and Mesozoic times (Kent, 1947).

The general nature of the floor of older Palaeozoic rocks was discussed by Stubblefield and Bullerwell (1966). The dips in Lower Palaeozoic rocks are almost invariably high (40–80°) which could imply strong folding or particularly dense faulting—more probably the former. In discussion of their paper Brunstrom pointed out that magnetic data indicated a depth to the basement crystalline floor of 7000–10 000m in Huntingdonshire and Cambridgeshire.

Seismic surveys in the broad belt of country south of the Charnwood–Wash crystalline ridge confirm this concept: records for example across the basin parallel to the Norfolk coast show only a characterless "mush" of indefinite thickness below the main unconformity. Together with dominance of argillaceous rocks in the borehole sections this has been taken to indicate a very thick Lower Palaeozoic geosynclinal section, strongly deformed and postulated to be a branch of the Caledonian system sweeping from the Ardennes to the western Pennines (Kent, 1974). This mass is however not appreciably metamorphosed, and it is debatable whether it could be part of the Caledonian orogenic belt. There is an additional structural problem presented by the remarkably consistent Tremadocian date of rocks below the main unconformity in the central and western Midlands (Bulman and Ruston, 1973).

In terms of Mesozoic development this rigid block is a discrete entity, margined on the west by a trough through the Cotswolds and the Severn Basin, on the south by the Wessex–Weald basin, and on the east and north by the North Sea basin with its landward continuation.

Throughout Mesozoic times it is clear that this massif has remained close to sea-level, with a

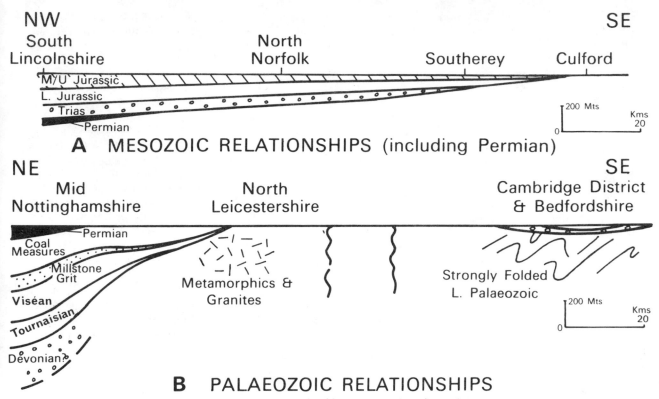

FIG 3. Sections illustrating stratigraphic relationships on the northern flank of the East Anglia Massif.

relative oscillation measured in a few hundreds of feet, sometimes suffering mild erosion, at other times subsiding sufficiently for the deposition of attenuated representatives of the various formations better developed down flank (Fig 3A). It shows no signs of having been a highland area, subject to large scale denudation, since at least the early Permian, except for a short but important interval in the Lower Cretaceous (Berriasian–Barremian) when it provided a source for the Wealden sediments to the south (Allen, 1967).

This same type of history can in fact be traced further back in time (Fig 3B). Coal Measures, Millstone Grit and Carboniferous Limestone all thin out from the basin of northern England against the variously called Midland Barrier or Wales–Brabant Massif, and thicken again south of it into the basin extending from South Wales across to the Kent coalfield. Over the block, thin remnants of Lower Carboniferous which survive —as at Cambridge and Northampton—show that through that long period also the massif was near sea-level, with a delicate balance between deposition and erosion.

Old Red Sandstone with a marine Devonian element is more widespread—rather more so than Wills' 1973 map indicated. It has been reached in a number of borings but is not usually penetrated; scanty seismic data and general relationships suggest that it too is a relatively thin veneer on older Palaeozoic rocks. It shows no evidence of strong folding in the East Anglian area, but

transgresses across steeply dipping, sheared Lower Palaeozoic rocks.

In summary, the East Anglian Massif is an area folded and compacted by Caledonian movements which—viewed on a continental scale—has remained unfolded and non-subsiding since the Silurian or early Devonian, a remarkable record of stability. Its history has been outlined here at some length, because as a massif which has retained its sedimentary cover and hence its historical record of movement, it provides an important standard for comparison with other positive areas in Britain.

Cornubia and the Welsh Highland

The present surface rocks of the Cornubian peninsula are essentially a product of the Armorican geosyncline and of the subsequent orogeny, terminating with granite intrusions. Continuation of this province offshore towards the south-west is indicated by geophysical surveys along the topographic ridge extending to the Scilly Isles and Haig Fras. Along the strike to the east almost nothing is known of corresponding structures: there must be complex thrusting in the concealed Palaeozoic rocks continuing that known near Boulogne across the Straits of Dover, but this remains to be elucidated.

The Cornubian peninsula itself was broadly blocked out in Permo-Triassic times. Triassic rocks lap onto the old rocks along the south side of the Bristol Channel, which is a faulted syncline containing a fairly complete sequence of Jurassic rocks (Lloyd *et al.*, 1973). A small patch of Trias survives near Bideford as further witness to the antiquity of the west coast. The eastern and southern sides of the block are defined by the edge of the Mesozoic Wessex Basin on land, and by the edge of the comparable Trias-Jurassic basins in the English Channel (Curry, Hamilton and Smith, 1970). On the east (Dorset) in particular it is clear that the various Mesozoic formations are thinning towards a basin edge in the direction of this massif.

It is known that the older Palaeozoic outcrops were unroofed in Permian times, and a fjord-like valley filled with Permo-Triassic red beds extends half way across the the peninsula near Crediton. It may consequently be supposed that the greater part of the erosion of this Armorican block was completed by the early Mesozoic and that it survived as a low-lying massif, transgressed to a variable extent by the more extensive marine incursions.

Wales is a more difficult problem. This extensive area of metamorphic Lower Palaeozoic provides little direct clue on post-Caledonian history. There has been much speculation on the dating of the main erosion phase, with uplift estimated at various dates from Permo-Trias to Miocene by different authors.

Wills (1973) has shown that peneplanation of the Lower Palaeozoic rocks almost certainly reached a very advanced stage by Lower Carboniferous if not in Devonian times, the massif having remained since then as an essentially unfolded low-lying block between subsiding basins (Fig. 4). The subdued topography is indicated by two lines of evidence: by the convergence of unconformities onto the block with the survival of small outliers of Old Red Sandstone and Carboniferous rocks on the block (Carboniferous Limestone at Corwen, Old Red Sandstone near Clun and in Carmarthenshire); and independently by the almost total absence of any detritus supply to the surrounding Lower Carboniferous sea (Rayner, 1967, p. 172).

By analogy with other massifs, and in line with the general absence of Welsh detritus in the Mesozoic basins, it may be presumed that the peneplaned Welsh block, once formed, continued near sea-level throughout the Mesozoic. In contrast to Cornubia, the East Anglian Massif, the Pennines and the Southern Uplands, there are however no relics of late Palaeozoic, Trias or later Mesozoic rocks in minor fault angles or downwarps in the central area of Wales (although there are suspected Lower Tertiary relics in south Wales according to T. N. George), suggesting that the early "summit plane" which was dissected by late Tertiary erosion was already an important distance (perhaps 1000m) beneath the late Palaeozoic to Mesozoic surface.

It may thus be presumed that the broad domed peneplain, from which the Welsh mountains were dissected in post-Pliocene times by radial river

systems, had been developed by the late Devonian and Carboniferous: that the massif lay near sea-level, truncated by mild erosion or temporarily buried during intermittent transgressions over a long period before the final (late Tertiary?) uplift.

The results of the Mochras boring are directly relevant to this problem. This borehole located two miles (3km) from the exposed Lower Cambrian of Harlech, penetrated 602m of Pleistocene and Oligocene, 896m of Lower Jurassic and ended in Trias. The Lias is entirely fine-grained and Trias clastics can be matched with westerly sources (Lleyn peninsula), leading to a deduction that at least North Wales was deeply buried beneath Jurassic (Wood and Woodland, 1971). It is nevertheless possible that the bounding Mochras fault was submarine and developed contemporaneously in the Jurassic; this would not conflict with the faunal evidence. The greater part of the movement on the fault (totalling 4500m according to Wood and Woodland) was however pre-Oligocene and—from Cardigan Bay evidence—mainly pre-Upper Cretaceous. The later phase was not post-Oligocene, as has been indicated, but is intra-Oligocene on the seismic evidence of wedging of the Tertiary beds (Fig. 7). Thus assuming conjugate movement, the Welsh massif may well have been stripped of any earlier Mesozoic sediments by the Upper Cretaceous, and suffered a late phase of intra-Oligocene uplift.

There could be a close analogy between the Welsh Highland area and the East Anglian massif, similarly a region just "awash" from the Devonian to the Lower Cretaceous; the difference is that the post-Cretaceous tilting left East Anglia near sea-level but elevated the Welsh massif by perhaps 2000m.

The Pennines and the Lake District

The Pennines consist fundamentally of three blocks in north–south alignment, each comprising a relatively unfolded expanse of Lower Carboniferous, which are separated by contemporaneous Lower Carboniferous troughs or "gulfs". The northerly two massifs—the Alston and Askrigg blocks—are known to have shallow foundations of mildly metamorphosed Lower Palaeozoic sediments with Caledonian granite intrusions, and it is thought that this relationship may apply also to the southernmost block, the Derbyshire Dome, although direct evidence is at present lacking.

The post-Caledonian landscape of the northerly blocks was reduced to a surface of low relief by early Lower Carboniferous times, and was effectively submerged early in the Viséan. The blocks were then the site of shelf-type deposition, but in the intervening troughs much greater thicknesses of predominantly basinal type sediments were deposited. Subsequently the whole area was submerged beneath the deltaic Millstone Grit and Coal Measures, the latter spreading with

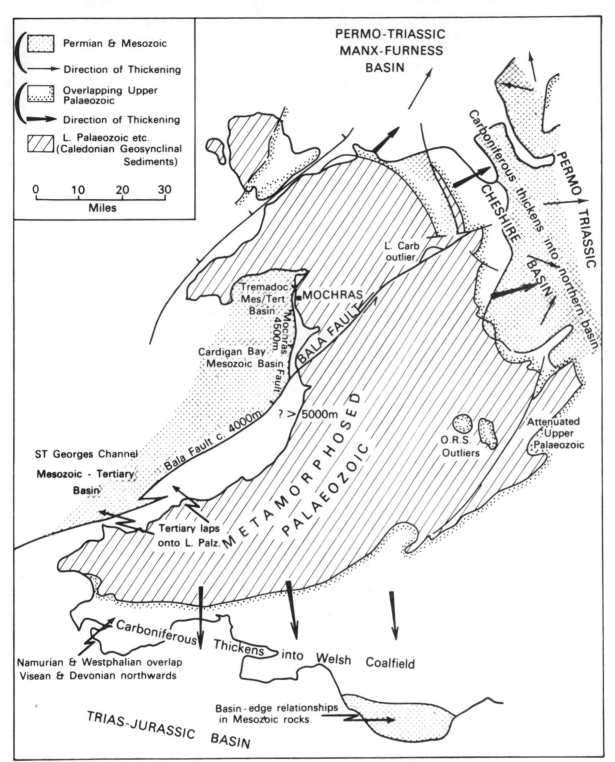

FIG 4.   Structural relationships of Welsh Massif to Carboniferous and Mesozoic basins.

only locally known differentiation over block and basin alike. The area was remote from the main Hercynian orogenic belt, but the relatively incompetent rocks between the blocks were compressed into broad open folds in late or post-Westphalian times, and the area was subject to major tension-faulting before deposition of the Permian. Major faults in the block-edges were active at this time and an extremely complex pattern of minor faulting was developed in the basinal segments.

The Permian and later history is a matter of deduction. On the east the Carboniferous rocks are truncated (at low angles) by a well developed sub-Permian peneplain, which must have originally continued westwards for a number of miles across the now exposed blocks and basins, together with some part of the succeeding Mesozoic.

Dolomitization of the Carboniferous Limestone is taken to be evidence that the Permian, or Permo-Triassic sea, extended a long way onto the Derbyshire block and by analogy probably onto the northerly blocks also. Coarse sands and gravels in sink-holes in Derbyshire are now dated as Pliocene on palynological evidence, but they are believed to be largely derived from a former Trias cover and indicate that this block, perhaps with the others, was partly submerged by early Mesozoic sediments.

Sedimentary units of the eastern basin tend to thin westwards from the North Sea coast towards outcrop, and the Pennines probably bore only an attenuated succession of Mesozoic, deposition being controlled by a hinge-line at or near the western sides of the Pennine blocks which were in each case faulted. Considerable physiological relief is indicated in the westerly part of the Pennines by Permian and Triassic rocks which fill deep valleys incised into Carboniferous in Staffordshire and Lancashire.

The Lake District block is structurally now an appendage of the Pennines, separated therefrom by a Permo-Triassic fault basin, but it had a somewhat different earlier history. It represents a section of the Caledonian geosyncline, filled with tens of thousands of metres of metamorphosed sediments and volcanic rocks, which was intensely folded, bevelled and subsequently transgressed by Lower Carboniferous. The Carboniferous was itself gently domed and truncated by Permo-Trias, both dipping peripherally as a result of subsequent tilting. Older Palaeozoic rocks provided only a modest detrital contribution to these formations. Lias of normal fine-grained lithology is present in the Carlisle basin to the north. Again we have a massif of Caledonian rocks transgressed by Carboniferous and maintaining a positive position in the Mesozoic; again, however, it does not appear to have formed a highland capable of supplying any large quantity of detritus to the surrounding basins.

Southern Uplands of Scotland

To a considerable degree the post-Armorican history of the Southern Uplands is analogous to that of the massifs south of the border. The isoclinally folded and metamorphosed Lower Palaeozoic rocks were planed off and then tension-faulted during Lower Old Red Sandstone times; this formation survives in growth-fault and fold basins marginal to the Midland Valley. The area was partly transgressed by Upper Old Red Sandstone, Lower and Upper Carboniferous in conformable series (George, 1965, pp. 22–5). Progressive submergence of the area during the Lower Carboniferous has been deduced (George, 1960, Fig 15), but it is not known whether a complete cover was achieved by the Coal Measures (Frances, 1965, p. 350). The Upper Palaeozoic sediments however formed only a thin veneer on the planed-off massif, perhaps a fiftieth of their thickness in the adjoining basins. Thick Permo-Triassic sandstones (red beds) occur in structural basins with a north–south trend (Craig, 1965, p. 397). Their wide occurrence would be consistent with maintenance of the Southern Uplands generally near base level through New Red Sandstone times, with locally accentuated deposition associated with contemporaneous faulting.

It would be a logical (although currently unsupported) assumption that the Southern Uplands remained near sea-level through Jurassic times, submerged by the most powerful transgressions (? the Cretaceous Sea). They apparently contributed little sediment to the developing sedimentary basins.

The Scottish Highlands

The Scottish Highlands differ from all the more southerly Caledonian massif areas, having remained topographically high, subject to erosion and acting as a major source of detritus for the Old Red Sandstone, the Carboniferous and the Mesozoic sediments of the surrounding basins.

The process of denudation was well advanced during deposition of the Lower Old Red Sandstone, in which conglomerates demonstrate first the erosion of metamorphic Ordovician and then of the Dalradian floor (Waterston, 1965, p. 275). Enormous quantities of sediment were derived from the Highlands, so that in the Midland Valley, the Lower Old Red alone measures (with its lavas) 6000m in Kincardineshire, and in north-east Scotland the Middle Old Red comprises 6000m of sediments deposited at or near water-level or sub-aerially. Major faulting—amounting to 2500–3000m on the Highland Boundary Fault—preceded deposition of the Upper Old Red Sandstone, which transgressed across the earlier division onto the metamorphic rocks and itself measures many thousands of metres.

In the Midland Valley earlier Carboniferous (Tournaisian) may be represented by part of the Upper Old Red Sandstone (Waterston, 1965, p. 302), and the Viséan of Scotland and Northumbria contrasts with that of the rest of Britain in the large proportion of clastic sediments, with a tendency to fresh-water and deltaic facies related to a northern source. The source area included almost the whole of the Highlands (Francis, 1965, p. 349) and perhaps also a broad belt to the east, subsequently bevelled. The sedimentary section of the Lower Carboniferous (dominantly clastic) measures some 1000–1500m, the Millstone Grit and Coal Measures a further 600–900m. The last mentioned are not particularly large figures, but the measure of uplift and erosion of the Highland source has to be related to the whole Coal Measure delta, extending southwards into the English Midlands.

The extent of the Devonian–Carboniferous northern highland area is not defined by existing data, but in the Permo-Triassic deposition (marked by relatively small outcropping remnants) outlined the present Scottish Highlands; the Moray Firth Basin was well developed and the Minch Basin was probably undergoing its major subsidence.

As demonstrated by George (1965, pp. 30–31), the Highland Massif persisted as a source-area through the Jurassic, with clear evidence of littoral or paralic conditions on both east and west coasts. The thick sandstones of the Lias, Bajocian, Bathonian, Upper Jurassic and Lower Cretaceous, are symptomatic of hinterland relief. Differential movement between massif and basins was associated with fault movement—in the Kimmeridge on the east coast (part of the Great Glen fault system), and in the late Jurassic or early Cretaceous in Skye (the 600-m Camasunary fault).

Uplift of the Scottish Highlands was thus a very long term, secular effect, with movement in the Tertiary only the latest part of the history.

Movement of the Great Glen fault is an important element in the development of the Highlands and of the adjoining basins. Originally it had been assumed to be essentially a vertical fracture, with a mainly vertical throw, but in an elegant exposition Kennedy (1946) demonstrated strong evidence that it is a sinistral transcurrent fault with a shift of 65 miles (105km). Later authors have both supported and disagreed with the direction of shift. The continuation of the fault towards the north-east has also led to differences of opinion. It is most usually taken to swing parallel to the Caithness coast and to pass either through or close to the Shetlands; thus Flinn (1969) and others have identified it with the Walls Boundary Fault there. Vogt (1954) has however suggested that it could be identified in northern Norway near Kristiansund.

In relation to development of the North Sea Basin it is however important to recognize that the fault splays east of Inverness, with branches following the Caithness coast towards the Shetlands, and spreading out in the Moray Firth (Fig 9). Which of these is the main fault is of lesser importance, for any member of the system may be transcurrent in origin. Although sharp changes of sedimentary thickness occur across the faults in the Moray Firth these are not necessarily due to vertical movement. A complex and very long-term history is likely for all the branches of the system.

Irish Massif

Ireland has been regarded as a single positive unit in relation to basin subsidence (*e.g.* by Armstrong, 1972, p. 469), but this concept needs qualification.

Northern Ireland is aligned with the Scottish Midland Valley graben, and although the structure is not simple it is an area which subsided extensively in the Permo-Triassic, so that an initial invasion of the Zechstein sea was followed by Triassic deposition, including thick Keuper with salt beds. Subsidence continued into the Lias but above this the sequence is defective, with transgressive Upper Cretaceous Greensand and Chalk followed by Eocene volcanics.

The Central Irish plain appears to be more nearly analogous to the Pennine blocks, with a wide extent of nearly flatlying but fairly thick Lower Carboniferous carbonates. It is unclear whether the limestone was succeeded by thick Upper Carboniferous, or whether it was extensively transgressed by Mesozoic rocks. The absence of relics of such incidents suggests that, like the Pennine blocks, any Mesozoic (at least) was attenuated.

South-eastern Ireland includes a Lower Palaeozoic tract and is clearly analogous in its earlier history to Wales, being part of the Caledonian orogenic belt, while the south is crossed by the Hercynian orogen with its sharply folded Devonian and Carboniferous. A single relic of Upper Cretaceous Chalk has been found on land (Walsh, 1966), but in an area so poorly mapped others may well occur. The inference is that southern Ireland at least was a positive area in the late Mesozoic, since only this highly transgressive deposit is present. Chalk has been found flooring the surrounding seas, particularly in St George's Channel, where it blankets a considerable area of earlier Cretaceous and Jurassic. Whether the southern Irish land-area was continuously a Mesozoic high, or whether it was elevated and lost an earlier cover during Lower Cretaceous movement remains a matter of inference, with a long-term positive history perhaps the more probable.

REGIONAL HISTORY:
PERMIAN–MESOZOIC–TERTIARY BASINS

The system of Mesozoic and Tertiary basins originated in the Permian—which is tectonically "Mesozoic" in the present context—and illustrates well the accidental location of the modern coastline in relation to structural negative units. Thus the Jurassic/Tertiary Wessex Basin of southern England is continuous with the Paris Basin, the London Basin is the tip of the major Tertiary downwarp of the North Sea and the English North-eastern Basin is stratigraphically and structurally part of the Southern North Sea Basin. The dominantly Permo-Triassic fault basins of western England (Cheshire Basin and Worcester Graben) are closely analogous to the northern Irish Sea, Cardigan Bay and smaller fault troughs of the west.

Description is from the south anticlockwise around the nucleus of positive features which characterize the British Isles.

The Wessex Basin and English Channel Basins

The Wessex Basin is a major Mesozoic–Tertiary downwarp extending from the Cornubian

Highland across southern England, limited on the north by the East Anglian Massif but continuous towards the south with the basin of the Eastern Channel and the Paris Basin.

Mesozoic subsidence took a fairly simple form despite the complexity of the Hercynian orogenic belt which it conceals. Little detail is known of the Permo-Triassic history: it is presumed that a major downwarp developed east of Cornubia, linking the outcropping development of east Devonshire with the Trias of the Worcester basin. During the Jurassic there was nearly continuous unbroken deposition in the central parts of the basin (Lees and Taitt, 1945), with a complementary history on the margins of intermittent deposition and minor uplift which can be correlated with events over a much wider area (Arkell, 1933, p. 87, etc.). Minor uplift and erosion took place in southern England as early as the Portlandian (Taitt and Kent, 1958) and more strongly in the Aptian. Within the Wealden basin Allen (1967, Fig 7) has shown that Jurassic detritus was entering from both south and north, implying Neocomian uplift of the margins. The overall subsidence came to an end a little later, the Albian transgressing across a planed-off surface after an episode of folding and faulting. The unconformity is well documented in Dorset (Arkell, 1947).

As elsewhere in Britain and the British seas, the Chalk was widely transgressive and essentially unaffected by structural complications in the underlying rocks; it was followed by a regression and Tertiary deposition was much more limited, centred on the Hampshire and London Basins (the latter an invasion of the southern part of the East Anglian Massif). The Lower Tertiary survives over a far smaller area than the Jurassic, and was presumably thin or absent over the Weald and Straits of Dover area.

The main Tertiary deformation of the Wessex Basin and Eastern Channel was post-Oligocene, and took the form of inversion structures flanked by highly asymmetric folds or monoclines—the Purbeck Anticline in Dorset, the Isle of Wight Monocline, and two analogous structures in mid-Channel south of the Isle of Wight (Smith and Curry, 1974). The landward cases were originally taken to be compressional Alpine folds, but it is now thought that vertical displacement is dominant over lateral compression and that these structures may be related to movement of deep-seated fault-blocks activated at about the time of the mid-Tertiary orogeny. Thus the Purbeck and Isle of Wight monoclines appear to have developed over post-Jurassic, pre-Albian normal faults formerly displaced in the opposite sense (down to south).

The North-eastern Basin of England and the Southern North Sea

The North-eastern Basin lies east of the Pennines and north of the East Anglian Massif; it is continuous with the basin of the south-western North Sea, and it has a Mesozoic link with the west and south.

At surface the structure is particularly simple, with generally north–south strikes and only minor interruptions in the general easterly dip. At depth it is rather more complex. The Carboniferous rocks apparently overlie a shelf-type Lower Palaeozoic development; the Carboniferous basin itself had its centre west of the Permian–Mesozoic outcrops and began with a diversification into shelf and trough deposition reflecting epeirogenic control in the Viséan (possibly also in the Tournaisian) (Rayner, 1953, 1967; Kent, 1966, 1967). This pattern of block and basin probably continues offshore, with thick Namurian indicating a basinal development, for example off the north Yorkshire coast. The main Upper Carboniferous basin was centred in the western Pennines. Both the Millstone Grit and Coal Measures are attenuated south-east and east into Lincolnshire—subsidence of the delta basin being limited southwards by the hinge effect of the Wales–Brabant Massif (Wills, 1951, Pl. ix).

Along the Pennines flank the Permian transgressed across the planed-off Carboniferous epeirogenic structures and foreland folding. In the succeeding Trias progressive subsidence of the marginal parts of the East Anglian Massif permitted transgression some 30 to 50 miles (48–80km) farther south, developed an east–west link across the south of the Pennines to the deep subsidences of western England, and also widened and deepened a late Coal Measures connection south through the Cotswolds and Worcestershire to the Wessex Basin. This structural pattern was maintained through the Jurassic with minor changes due to succeeding transgressions and regressions, subsidence being related to the hinge effect of the East Anglian Massif margins and, inferentially, a corresponding structural control provided by the Pennine swell.

The shelf-area formed a single subsiding unit through the Trias, but became differentiated at the beginning of the Jurassic by development of the Market Weighton Block in south-east Yorkshire; this was essentially a slowly subsiding tilted block some 12 miles (19km) across which acted as a hinge between a stable shelf in Lincolnshire and an unstable shelf in the Yorkshire Basin (Kent, 1955, 1974). It has no traceable existence before the Rhaetic; it controlled thicknesses of nearly all the Jurassic and Lower Cretaceous formations, but was finally transgressed with modest prior erosion by the Upper Cretaceous, in a manner closely similar to relations over the East Anglian Massif.

The Jurassic stable shelf—geographically coincident with the East Midlands Shelf (Kent, 1967)—extends some 50 miles (80km) beyond the coast, being broadly coincident with the post-Cretaceous Wolds Syncline of Donovan (1963). It is limited on the east by the English Zechstein basin (Cook, 1965). The Yorkshire Basin (unstable shelf) similarly continues offshore, with a development of broad post-Cretaceous folds which have been mapped by Dingle (1971).

The North-eastern Basin shared the history of the southern North Sea in the Permian and Triassic; it was essentially an undifferentiated subsiding basin, in which the formations are continuously correlatable from Germany across to eastern England (Geiger and Hopping, 1968). Instability developed late in the Trias—the first movements of the halokinetic structures offshore are dated as end-Triassic (Brunstrom and Walmsley, 1969). On land the subdivision of the North-eastern Basin began in the Rhaetic, and continued through much of the Jurassic.

Apart from the eustatically controlled major variations of water depth, there was on land an interruption of deposition with an angular discordance at the end of the Lias (pre-Dogger in the original English sense) and again in the late Bajocian/early Bathonian. In each case differential movements were measured in tens of metres. No comparable movement took place at the end of the Middle Jurassic in eastern England (in contrast to mainland Europe), but during the shallow-water phase of the Upper Oxfordian (Corallian), deposition in Yorkshire was sufficiently sensitive to tectonic control to show minor movements. Offshore most of the evidence of Jurassic development has been lost following local uplift off the Yorkshire coast, by inversion of the Sole Pit Trough which had subsided differentially in the

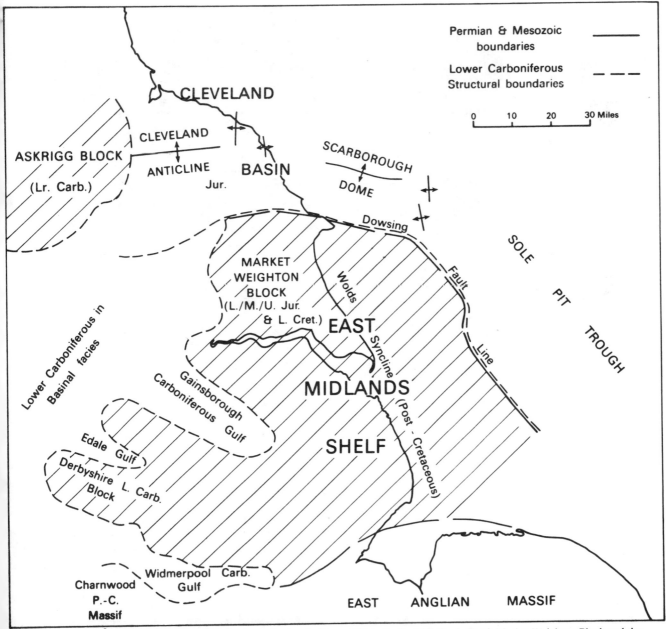

FIG 5.   Structural units in North-eastern England. During the Permo-Trias, the East Midlands Shelf, Market Weighton Block and the Cleveland area formed a single depositional area continuous with the English Zechstein Basin of the Southern North Sea.

Lias, and by pre-Cretaceous erosion of Jurassic over a wide area beyond this.

The late Aptian or Albian movement was the main Mesozoic tectonic event in eastern England, leading to truncation of the attenuated Jurassic formations transgressing onto the East Anglian Massif, and of most of them similarly in south Yorkshire (Market Weighton). The same phase of movement was widespread in the southern North Sea, so that Chalk (with variable amounts of Lower Cretaceous) transgressed widely and came to rest on Trias. The unconformity is accentuated over halokinetic structures but is essentially a much broader feature.

Post-Cretaceous movement involved inversion of the Jurassic Cleveland basin of north-east Yorkshire to form the complex Cleveland Anticline, a case closely analogous to the Weald, the basin-floor here also probably remaining synclinal (Kent, 1974) and also inversion on a larger scale of the Sole Pit Trough on the edge of the English Evaporite Basin (Brunstrom and Walmsley, 1969; Ziegler, 1974).

Also partly post-Cretaceous is a major fault-trend separating the Cleveland and Sole Pit basins from the stable shelf to the south and south-west; this produced a crush zone in the Chalk at Flamborough Head; it heads thence eastwards into the North Sea as the Dowsing Fault which swings south-eastwards towards the Rhine Graben complex (Fig 5). This structural line has a long history, extending back through the Jurassic and possibly pre-Permian.

These features were subsidiary to the secular down-warping of the main North Sea Tertiary basin, its axis in these latitudes near the Netherlands coast. Local perturbations due to continued halokinesis of Permian salt-plugs reached a maximum in the Tertiary north of latitude 54° N, and continued into the Quaternary. In the salt-basin the movements of salt were balanced by sedimentation between the plugs, so that as Brunstrom and Walmsley commented (1969, pp. 881–2) the base of the Quaternary is almost parallel to the base of the Permian, although separated by up to 3000m of largely deformed strata.

Western Fault Basins—Cheshire, Worcester and Irish Sea Basins

In contrast to eastern England, where the Permo-Triassic rocks were deposited in conventional down-warps and thin gently and progressively towards the surrounding contemporary highs, deposition in the west was largely related to deep troughs bounded by faults or sharp monoclines. On land these troughs developed contemporaneously with deposition from the Lower Permian into the Trias. Thick Upper Triassic salt deposits are present in several of them. The later history is not well known, but the minor amounts of Jurassic rocks surviving indicate that

subsidence of the land basins was never later comparably large.

The southernmost trough is the Worcester Basin—a one-sided rift with Trias faulted against the Palaeozoic and Precambrian of the Malvern Hills, the beds thinning gently eastwards to the flank of the East Anglian Massif. A Triassic fill of at least 1000m was deduced on stratigraphic evidence; gravity data have since indicated a figure of 2500m (Cook and Thirlaway, 1956, with discussion).

The Cheshire Basin is becoming progressively better documented. A gravity survey by the Anglo–American Oil Company first demonstrated the graben-like form (White, 1949) and an estimate of 2000m of Permo-Triassic fill was made from stratigraphical evidence (Kent, 1949). Seismic surveys confirmed by drilling now indicate a depth to the pre-Permian floor in excess of 3000m in the central and southern parts of the basin.

On the northern margin of the basin, near Manchester, contemporaneous movement of faults was recorded in Upper Permian times, with sharp changes of thickness affecting the basal sands. Sedimentary thicknesses round the basin margins add up to considerably less than those in the central areas and contemporary subsidence of the trough must be accepted. The Trias is cut by large faults, in the past assumed to be mid-Tertiary but by analogy with offshore basins more likely to be pre-Tertiary and perhaps pre-Upper Cretaceous.

A Triassic basin of comparable size occupies the eastern Irish Sea, its eastern margin facing the coastal plains of West Lancashire. In part the eastern boundary is formed by major faults (as off the north-western Lake District and south of Fleetwood); elsewhere (Furness district) there is an onlapping relationship to the older rocks of the Lake District dome. In the Formby area movement contemporary with Permian and Triassic deposition is indicated by drilling and seismic results.

Bott (1968) has summarised gravity and magnetic evidence for two other analogous but smaller basins: these are the Stranraer Basin, a Permo-Triassic depression of 1.3km depth crossing the Scottish coast, fault-bounded on the east; and the Solway Firth Basin, estimated to have about 4km of fill. Two other gravity features—the Peel Basin immediately west of the Isle of Man and the Kish Bank Basin off the Dublin coast—are perhaps Carboniferous depressions.

The East Irish Sea Basin is flanked by thick Lower Carboniferous rocks on the east, by thinner Upper and Lower Carboniferous resting on the Lower Palaeozoic along the North Wales coast—with an Old Red Sandstone representative in Anglesey—and Lower Carboniferous rocks form the northern and southern tips of the Isle of Man. Subsidence of this Irish Sea Basin was thus initiated at least as early as the Carboniferous. Bott (1968, p. 14) has in fact suggested that the northern

Irish Sea basin system dates from the end of the Caledonian orogeny rather than the Triassic, but this hypothesis is so far unconfirmed. The earlier tectonic pattern may have been less simple than that of the Permo-Trias.

The southern Irish Sea includes a complex of Carboniferous to Tertiary subsidences (Fig 6). In the western part the north-east–south-west Central Irish Sea Basin has a Carboniferous fill, extending from Anglesey to the south-east Irish coast (Al-Shaikh, 1970) with a thickness estimated at 2500m. On the south-east this syncline is separated by the Irish Sea geanticline from a Mesozoic–Tertiary basin system extending through Tremadoc and Cardigan Bays to St George's Channel and the Celtic Sea; this area also may contain Upper Palaeozoic, but evidence is so far lacking.

Close to the Harlech coast the Tremadoc Bay Basin is a trapdoor-type subsidence faulted in the east; the main subsidence was post-Lias and pre-Oligocene with a subsidiary intra-Oligocene movement discussed previously (Fig 7) (Bullerwell and McQuillan, 1969, Fig 2). The bounding Mochras fault is estimated to measure at least 4500m, with 3000m of post-Palaeozoic sediments present on the down-thrown side (Wood and Woodland, 1971). This basin is continuous with that of Cardigan Bay, also fault-bounded on its eastern side, which has a shallow connection south-eastwards (across the St Tudwals Arch) to the main

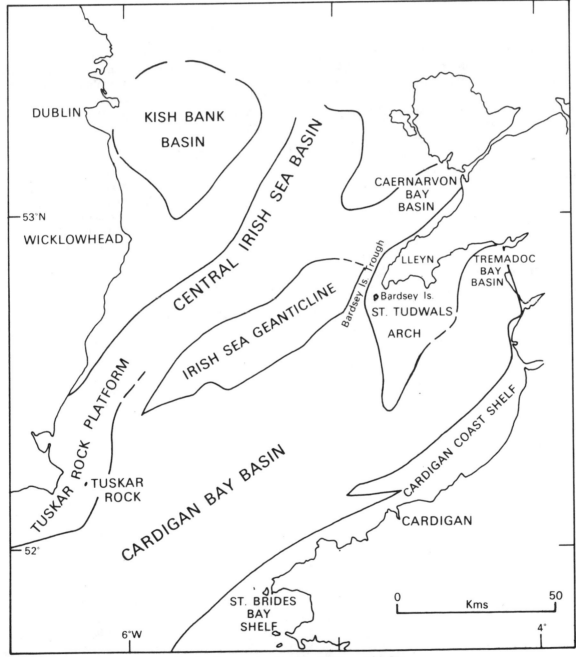

FIG 6. Structural units of the Southern Irish Sea.

West

East

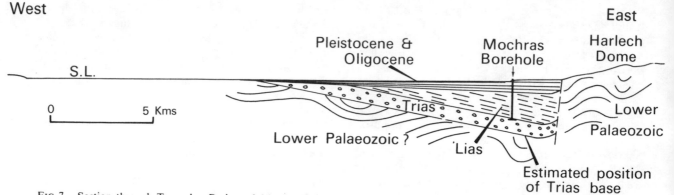

FIG 7. Section through Tremadoc Basin and Mochras boring. Truncation of Jurassic shows that the main subsidence was pre-Oligocene. Wedging of the Oligocene members shows the final movement to be contemporary with deposition. (Based on Bullerwell and McQuillan, 1969, Wright *et al.*, 1969.)

St George's Channel Basin (Dobson *et al.*, 1973). Both the Cardigan Bay and St George's Channel Basins are bounded on the south-east by the Bala Fault system, comparable in throw to the Mochras Fault (Fig 8).

Subsidence of this basin-system began at least as early as the Permo-Trias, with a deep evaporite sequence (undated) capable of producing diapirs in a setting comparable to those of the North Sea. The Permo-Triassic and Jurassic column has been estimated as 4km thick (Blundell *et al.*, 1971). These formations were strongly faulted, tilted and planed off in pre-Upper Cretaceous times, and their present distribution reflects fault-control which is related to deep Caledonoid trends. Blundell *et al.* (1971) emphasized the contrast between these complex fault-structures and the simple Tertiary basin with sediments transgressing onto older Palaeozoic; later work (Dobson *et al.*, 1973) shows that the main faulting is likely to

predate also the Chalk and some Cretaceous clastics (? Aptian–Albian—compare the southern North Sea). Some faulting is clearly post-Cretaceous and probably post-Palaeogene, but the broad structural contrast between pre-Albian taphrogenic control and post-Albian down-warping remains valid.

Mid North Sea High and Ringkøbing-Fyn High

The Mid North Sea High is structurally a depressed eastwards-plunging continuation of the Southern Uplands. Presumably its earlier history was very similar; its present form is essentially a legacy of Hercynian movements, forming a buried high between the seaward continuations of Upper Carboniferous basins of the Midland Valley of Scotland and Northumberland (Armstrong, 1972, Fig 6). The Ringkøbing-Fyn High is in line with it, separated from it by the Central Graben (East Dogger Bank Graben of Sorgenfrei, 1969) part of

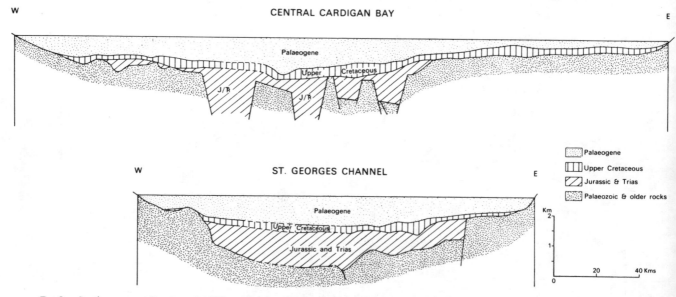

FIG 8. Sections across Southern Irish Sea, based on Blundell *et al.*, 1971. Later work indicates greater complexity but stratigraphical relationships are not basically different.

the mid North Sea graben system which here contains Permian evaporites, dating the origin of the main tensional feature.

The Mid North Sea High formed the northern boundary of the westerly evaporite basin (English Zechstein Basin) of the southern North Sea. The Zechstein extends over the south-westerly part of it in an essentially dolomite facies; it is succeeded by relatively undisturbed thin Trias, directly overlain by Cretaceous. It was thus an area of stable shelf type, its stratigraphy analogous to that of the East Anglian Massif, standing at a generally lower level than the Southern Uplands continuation, and finally subject to gentle downwarping as a marginal part of the Tertiary basin.

The Ringkøbing-Fyn High extending west from mid-Denmark had a somewhat similar history. It lacks the Permian and formed a boundary to the north-west German Rotliegendes and Zechstein salt basins, but was transgressed by Trias (the sequence is defective and thin compared with basins to north and south), by Jurassic (although Upper Jurassic is very poorly developed), and by Upper Cretaceous, before partaking in the broad Tertiary subsidence of the main basin.

The Midland Valley of Scotland and the Forth Approaches Basin

The Midland Valley of Scotland has had a long and somewhat complicated history. On land the story of its development ends with the Permian volcanics, but it is possible to continue it a little further from the results of North Sea exploration.

The northern fault of the graben, the Highland Boundary Fault, is believed to have originated in the Ordovician and to have affected Silurian deposition. The rift, as such, dates from early Devonian times when the Southern Uplands Fault was also developed. Lower Old Red Sandstone measures up to 6000m in the deeper part of the negative area, and beneath Upper Old Red Sandstone a 3000m cut-out is estimated across the Highland Boundary Fault near Loch Lomond with corresponding relationships in the south (George, 1965, p. 22). George ascribes this phase to terminal Caledonian movements.

During the Carboniferous the Highland Boundary Fault appears to have been quiescent, while the Southern Uplands Fault was "almost continuously active" (loc. cit.). Up to 3500m of Carboniferous rocks accumulated in the Midland Valley, very little on the Southern Uplands block. Volcanic rocks make up a large part of both Old Red Sandstone and Lower Carboniferous in the graben, continuing on a minor scale in the Upper Carboniferous and Permian. Some Permo-Trias remains in the Mauchline Basin.

The Midland Valley so far appears to be unique in Britain, as a rift originating early in the Devonian and continuing development into the Permo-Trias. Evidence on the origin of the other Permo-Triassic fault basins of western Britain is inconclusive in relation to such a long-term history, and more deep drilling is required to show whether they are analogous.

Within the landward part of the Midland Valley any Permian and later sediments have been removed by erosion, but seismic survey shows that the Midland Valley continues a moderate distance east-north-east into the North Sea, becoming progressively more of a broad depression than a rift, with an ill-defined eastern limit. Some 50 to 80 miles (80–129km) north-east of Edinburgh the depression is occupied by a Permo-Triassic sedimentary basin, with a salt-plug field related to Permian (presumably Zechstein) evaporites, emplaced in Keuper times. The Permo-Triassic basin cuts across the lines of the Midland Valley boundary-faults before they die out, showing that their development was mainly pre-Zechstein.

The Moray Firth Basin

The Moray Firth is flanked on the west by Triassic and Jurassic residuals, resting unconformably on Old Red Sandstone beyond the line of the Great Glen Fault. Triassic sandstone occurs to the south at Elgin, and a development of arenaceous Lower Cretaceous in the submerged area was deduced from rafted masses in the Drift. The Jurassic transgressed in a series of episodes onto coastal platforms cut into the faulted Highlands margin (George, 1965, p. 31).

The Firth coincides with a gravity low, initially interpreted as due to a granite but now recognized as due to a Mesozoic basin (Fig. 9), which has been mapped by the IGS (Chesher et al., 1972). Industrial work has covered the area with a close net of seismic lines and three deep wells have been drilled, but the results have not been released.

The Mesozoic downwarp lies within the Old Red Sandstone Orcadian Basin. Carboniferous rocks have not been found, and post-Devonian breakdown of the area appears to date from the Permian, which is represented by marine Zechstein deposited in an arm of the North Sea basin (Armstrong, 1972, p. 471; Ziegler, 1974, Fig 6). The Mesozoic succession is much as might be deduced from land outcrops, with a predominantly sandy Trias, thin and shallow water Lower, Middle and early Upper Jurassic, deeper water Kimmeridge shale and predominantly sandy Lower Cretaceous. The pre-Upper-Cretaceous column totals about 2000m.

This sequence is cut by numerous large faults, coinciding with sharp differences in formational thicknesses, which on sea-floor mapping evidence are mainly pre-Upper Cretaceous, and are believed to include both end-Jurassic and Aptian movements. If these were normal structures there would be an implication of penecontemporaneous growth, but they may be dominantly transcurrent fractures with a history linked to that of the Great Glen Fault itself.

In the area where the Forth Approaches Basin,

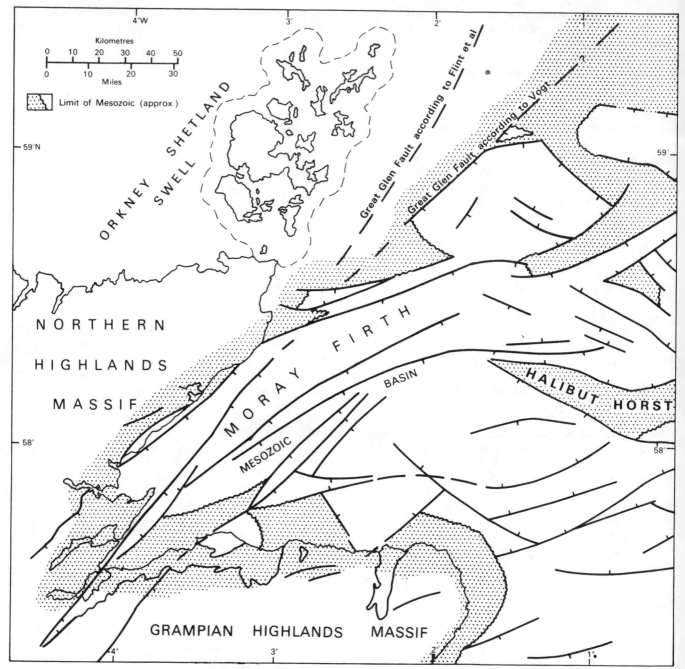

FIG 9. Post-Jurassic fault trends in the Moray Firth, based on seismic surveys by Hamilton Bros. and British Petroleum Development Co.

Central Graben and Viking Graben meet, east–west fault trends such as the Halibut Horst (Fig. 9) are aligned with the southern side of the Moray Firth Basin, and the gentle east–west Tertiary fold of the Forties oil-field (Thomas *et al.*, 1974, p. 396) is perhaps related to a comparable basement feature. The small degree of offsetting of the Central Graben and Viking Graben at this latitude could be related to later transcurrent movement of a branch of the Great Glen Fault system, discussed above. It is notable that this area of crossing trends is the only part of the north-west European shelf where Jurassic volcanics are known (Howitt *et al.*, Paper 28).

East Shetland Platform and Viking Graben

The East Shetland Platform is structurally a stable shelf continuous with the Orkney–Shetland Platform and beyond that with the similar West Shetland Shelf. Its tectonic history however is best considered in relation to the subsiding trough which flanks it on the east for 150 miles (241km) along the northern North Sea midline. As noted above, the boundary is heavily but not continuously faulted, and can be regarded as a faulted hinge (Fig. 10).

Most boreholes and most of the seismic work have been located in or close to the graben, and the great majority of penetrations have ended in the

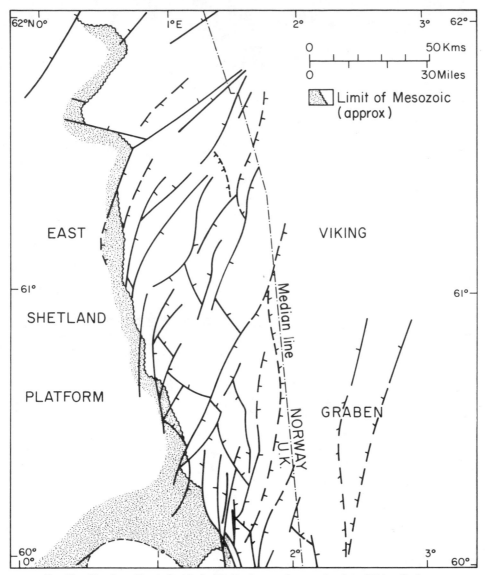

FIG 10. Northern North Sea fault distribution on the margin of the Viking Graben.

Mesozoic fill, so that information on the origin of this main structural line is limited. It is however known that the East Shetland Platform includes metamorphic rocks at least locally, and that both Upper Palaeozoic sediments and metamorphic rocks occur in the floor of the trough. The latter carry the implication that the East Shetland Trough with the Viking Graben, the main tensional trough of the North Sea, was not produced by pull-apart of the continental crust in association with early Atlantic opening but is a subsided segment.

The initial date of subsidence in these latitudes is not accurately established. The geometry of the graben fill indicates that Jurassic, and in all probability also Triassic rocks, were deposited in banked (onlap) relationship to the fault scarp. A belt of Permian evaporite crosses the line of the Mid North Sea High, separating it from the Ringkøbing-Fyn High (the East Dogger Bank Graben of Sorgenfrei, 1969) indicating that rifting there dates at least from the late Zechstein. Further

north Permian occurs on the western flank as well as in the trough on the latitude of Aberdeen (extending westwards with the Moray Firth Basin), and it is known in the Viking rift system further north (Howitt, 1974) (Fig. 11).

There is thus no need to assume that the Mid North Sea rift-system began in the south and spread northwards; it appears to be at least as old as Upper Permian. Whether the origin goes back into Old Red Sandstone times remains unknown; analogy with the Midland Valley of Scotland suggests this as a possibility, and the rather different but sub-parallel Great Glen Fault system is also mainly Devonian in origin.

Subsidence of the rift areas continued throughout the Jurassic into Lower Cretaceous, when transgression across fault lines became marked. It was partly or largely contemporaneous with deposition, but present information on lithology does not indicate any major widespread contribution of coarse sediment from the adjoining

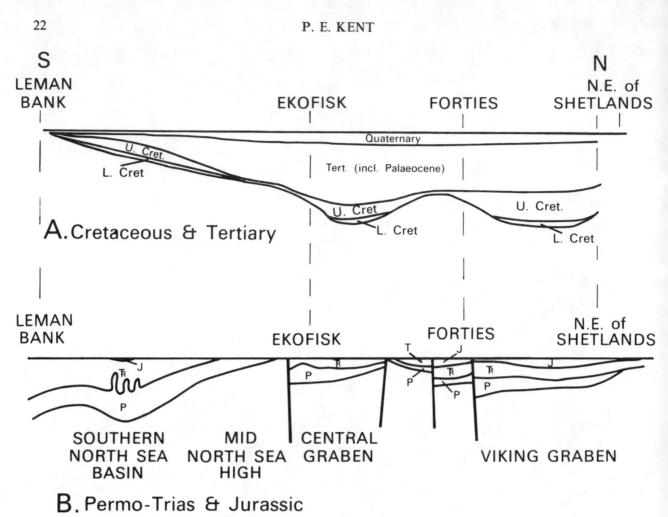

FIG 11.   Longitudinal section of North Sea Basin showing relationships of (A) Cretaceous and Tertiary, deposited in intra-cratonic downwarps and (B) Permian to Jurassic, controlled by fault basins. Based on Howitt (1974).

East Shetland Platform edge, and this may have remained mostly at or a little below sea level through this time interval. The whole system, in fact, seems to have been subject to a shallow water regime (unstable shelf) until the Upper Cretaceous, when northward passage of Chalk into a thick argillaceous development is interpreted as due to incoming of a relatively deep water facies, related to the development of much greater depths in the adjoining newly formed north-east Atlantic.

The trough itself remained structurally complex. Apart from the varying fault pattern along its edge (likely to be associated with variation in detailed history of movement) it included large tilted blocks which underwent movement contemporary with Jurassic sedimentation. The history of these movements is still being worked out.

Gentle folding of the East Shetlands Platform post-dated the Upper Cretaceous (which extended nearly to the modern coast, mostly as a thin veneer but a little thicker in small basinal areas) and pre-dated early Tertiary deposition. From that time onwards the northern North Sea formed a simple, broad synclinal down-warping area, its north–south axis coinciding almost exactly with the midline, characterized by continuous and fairly

even subsidence through the whole Tertiary and Quaternary periods.

Orkney–Shetland High and North-western Basins

The Orkney–Shetlands region is largely occupied by a broad high in structural continuity with the Northern Highlands. It is complex, with gently folded Palaeozoic basins rimmed with shallow or outcropping crystalline basement—the Orcadian Old Red Sandstone Basin continuing from north-east Scotland, and a group of Old Red Sandstone and/or Carboniferous basins west of the Orkneys (Bott and Watts, 1970) (Fig 12). This central area of crystalline rocks and their Upper Palaeozoic basins appears to have remained rigid and unfolded in post-Palaeozoic times, but was displaced by large faults, the Great Glen splays on the east (with possibly transcurrent shift) and very large faults with vertical throw on the west (Fig. 13).

Of the latter, the West Shetland Boundary Fault now separates the West Shetland Shelf from a Mesozoic–Tertiary trapdoor type basin, and the Rona Fault Belt (a complex of ten or more fractures of varying size) bounds the main Faroe–Shetland Mesozoic–Tertiary Basin of the

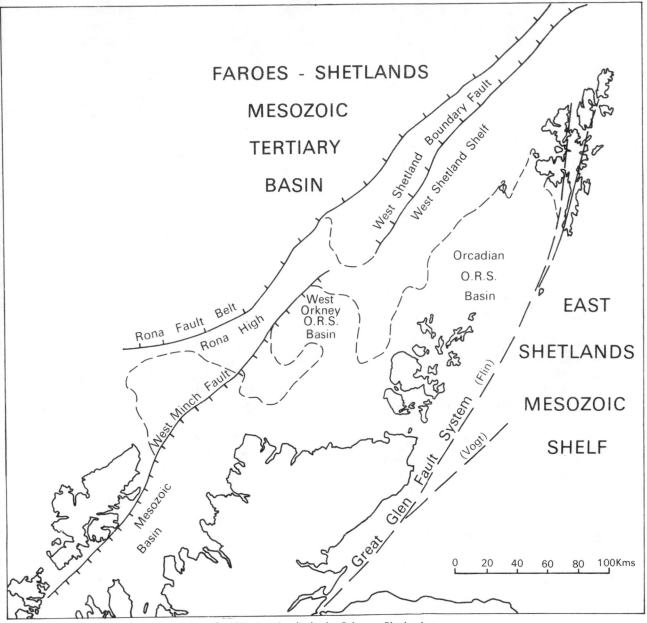

FIG 12. Structural units in the Orkney—Shetland area.

North Atlantic. These involve very large vertical displacements—the West Shetland Boundary Fault at least 3.5km; the Rona Fault Belt perhaps twice as much in total.

The West Shetland Basin, between these two fault lines, has a wedge-shaped fill reaching its maximum thickness close to the Boundary Fault, and thinning north-westwards onto the Rona High and its plunging continuation. The date of the sediments is as yet unconfirmed, but they appear likely to include both Mesozoic and Tertiary. The structure is analogous, in mirror image, to the contemporaneous fault-block development in East Greenland, which began to develop in the Permian (Collinson, 1972).

The main outer fault system, the Rona Fault Belt limiting the continental shelf, includes smaller

inward tilted blocks similarly faulted down towards the ocean. This is a common feature of Atlantic margins, as in the eastern USA (Burk, 1968). The faulting here was mainly or entirely pre-Tertiary and may well have been pre-Upper Cretaceous, as in the North Sea, a matter which borehole control will settle.

Minch and Inner Hebrides Basins

The Rona Basement High swings westwards across the general north-north-east trend of the basins, forming a bar between the West Shetland Basin and the Minch depression. North of Skye this (the Northern Minch) is a single basin with small outcrops of New Red Sandstone resting on Basement crystalline rocks on the surrounding coasts, and with a fill comprising thin Jurassic on

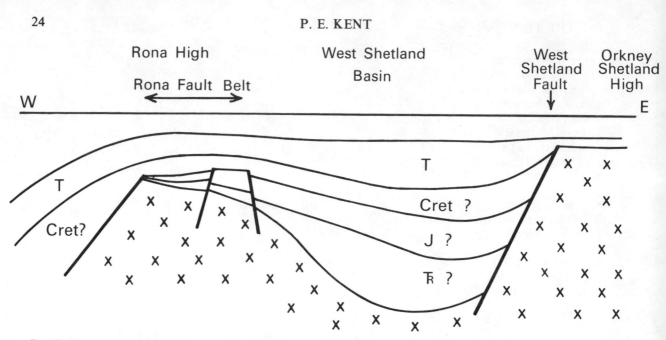

FIG 13. West Shetland Basin; diagrammatic section illustrating sedimentary relationships. Stratigraphic dating is based on seismic velocities only.

rather more than 3km of Trias. The floor is partly of Torridonian, which forms a ridge across the basin (Chesher *et al.*, 1972) but non-metamorphic (probably Upper Palaeozoic) sediments are locally present beneath the Permo-Triassic sequence.

From Skye southwards a double basin is developed (Binns *et al.*, 1973). The westerly depression (Southern Minch Basin) is bounded on the west by the Minch Fault system; it is assymetrical with the deepest part (more than 1.4km of sediments) on the north-west side near the fault. The easterly depression (unnamed; here referred to as the Inner Hebrides Basin) is bounded on the north-west by the Camasunary Fault, which is known to have a Jurassic/Cretaceous history of movement (Fig 14).

The bedded Mesozoic rocks are widely overlain by Tertiary volcanic rocks in the southerly areas, blanketing them particularly on the islands around the contemporary plutonic centres. Upper Cretaceous overlaps the earlier Mesozoic to rest on

metamorphic Basement on the Morvern mainland (George, 1965, p. 32) and is itself overlapped by the volcanics. The volcanics rest on basement on the central ridge in southern Skye, and approach it very closely along the line of the Minch Fault, probably resting on Lower Jurassic and Permian west of the fault near Benbecula and transgressing the fault-line itself east of Uist. It is evident that the Mesozoic basin development is pre-Tertiary and that the two large controlling faults are also mainly pre-Tertiary. The Upper Cretaceous is also transgressive, at least locally, but it is not yet clear whether it post-dates the main fault-basin development.

The Hebridean Platform

The wide continental shelf extending from the Outer Hebrides to the 100-fathom line constitutes an important section of the north-west European continental shelf. Regretfully, air magnetometer surveys checked by seismic reflection lines have

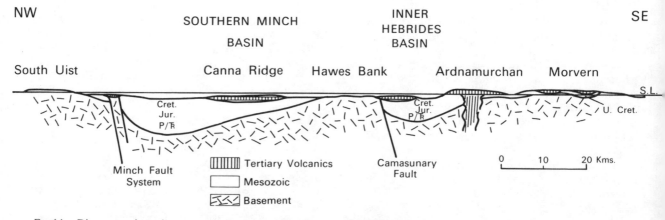

FIG 14. Diagrammatic section across the Sea of the Hebrides, showing late Mesozoic faulting and overlapping relations of Cretaceous and Tertiary rocks.

shown that it consists almost entirely of shallow basement crystalline rocks (presumably largely Lewisian) with only minor occurrences of sedimentary cover, ascribed to the Mesozoic by Roberts (in press).

Its eastern boundary, founded by the Minch fault, seems likely to have had a Mesozoic history, and by analogy it may be supposed to have separated the Minch Permo-Mesozoic (a half-graben in which sediments thickened westwards to the fault as in the southerly Hebrides) from a Hebridean platform with a history comparable to that of the East Shetland Platform.

Porcupine Bight and Bank

South of the Hebrides the oceanic shelf-break deepens from 200 to 1000m, so that the continental shelf (structurally defined) follows the northern side of the Slyne Ridge and then swings southwards to include the Porcupine Bank and its plunging continuation, the Porcupine Ridge. This topographic arc curves round the Porcupine Seabight Trough (Fig. 15).

The Slyne Ridge continues the Caledonian trend of southern Scotland and northern Ireland and may be supposed to be autochthonous. Stride *et al.* (1969, p. 63) have suggested that the Seabight Trough opened by lateral transit of the edge of the European continental plate rather than vertical subsidence of a continental segment; this would be in line with a limited sequence of sediments in the trough. Gravity data can be interpreted as indicating thin crust beneath the Seabight Trough (Gray and Stacey, 1970) and relation to magnetic anomaly 32 would give a maximum age of 80my (Basal Campanian). Clarke *et al.* (1971, p. 76)

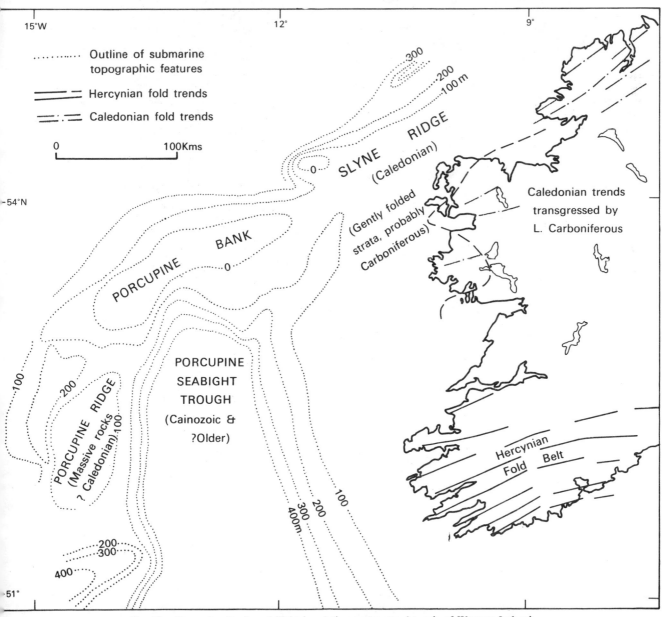

FIG 15. Porcupine Bank and Bight in relation to structural trends of Western Ireland.

however suspect that older rocks (perhaps including Upper Palaeozoic) may be present in the trough and that this is a relatively old feature. More detailed seismic surveys followed by drilling will be necessary to resolve this problem; for the time being the area appears to be a subsided western section of Ireland, the time and mechanism of the down-warp remaining uncertain.

Bristol Channel and Celtic Basin

Lloyd *et al.* (1973) have shown that the outer Bristol Channel is occupied by a faulted east–west synclinal outcrop of Mesozoic rocks, extending in time from the Trias to late Jurassic (late Kimmeridgian or Portlandian) with only minor stratigraphic breaks. The Jurassic thickness aggregates nearly 1400m, one of the thicker British developments. Interruptions in deposition correspond with general indications of periodical shallowing or lack of sediments, but intra-formational folding or faulting is recognized. The main disturbance of the beds was deduced to be "Lower Cretaceous or mid-Tertiary" (*loc. cit.*). Analogy with structures along the strike in Wiltshire would suggest that the former date is more likely.

On the northern edge of the syncline Lower Lias overlaps Trias onto flat-lying Carboniferous rocks in Glamorgan, and on the southern flank almost achieves a similar cut-out at the north-western end of the Quantock Hills. This is a normal edge-of-basin relationship and indicates that at least in the Lower Mesozoic (and in all probability later) the basinal downwarp was developing *pari passu* with deposition. Except for an Upper Oxfordian lignitic sand (believed to be derived from the south) there is however no coarse detritus identified as from the adjoining Palaeozoic outcrops, which (as noted above) may well have been topographically low.

George (1971, p. 372) has made the suggestion that residual pipeclay deposits in South Wales and southern Ireland are likely to be Lower Tertiary, an occurrence also in line with the basin-edge concept of stratigraphic relations.

Westwards the Bristol Channel basin extends into the northerly Celtic Sea, much of which is floored with Palaeocene and Upper Cretaceous Chalk. Seismic work has indicated that beneath a widespread unconformity Lower Cretaceous, Jurassic and probably Triassic rocks are present to considerable thickness in basins and fault troughs. Ziegler (1974) indicates a Jurassic depocentre in this area. Faulting appears to be dominant over folding, with a predominant strike of approximately west-south-west–east-north-east, as in the Bristol Channel, presumably controlled by the Hercynian grain of the older rocks. The structural history seems likely to be analogous to that of Wiltshire and Dorset, with major movements essentially of epeirogenic type and dominantly of middle Cretaceous (Aptian/Albian) date. More specific information will emerge as borehole control becomes available.

Western Approaches Basin

The Western Approaches Basin trends east-north-east–west-south-west parallel to the Hercynian trends, and originated as a trough soon after the Hercynian–Variscan orogeny (Curry *et al.*, 1970). Devonian and Carboniferous rocks are present in synclines both north-west and south of the Cornubian peninsula, and rocks strongly resembling the British Permo-Triassic rest on an eroded surface of older rocks in mid-Channel.

Subsidence in the central belt continued in the Jurassic (1000m of finer-grained rocks than is usual on land) and in the Lower Cretaceous (mainly of Wealden facies). As in Dorset, and in the Celtic Sea basin south of Ireland, the area was strongly faulted before an Upper Cretaceous transgression which extended onto "basement" beyond the limits of the earlier trough (Fig. 16).

Gentle warping preceded deposition of the Eocene, and the Tertiary is dominantly of deep

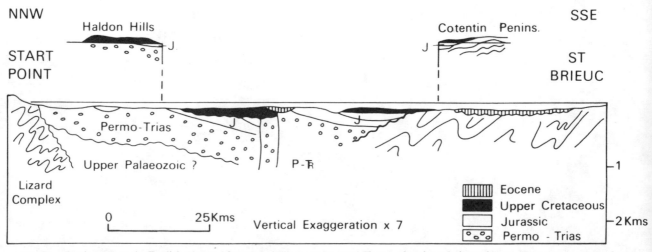

FIG 16.   Section across the English Channel from South Devon to Western France, showing relations of Upper Cretaceous and Tertiary to Permian—Mesozoic graben structures and basin edges.

shelf limestones except for a possible Oligocene emergence. The alternation of unconformities and transgressions has been ascribed to a variable rate of opening of the North Atlantic (*loc. cit.*) but their coincidence with world-wide incidents alternatively suggests a broader (eustatic) element of control.

IMPLICATIONS FOR CONTINENTAL SHELF DEVELOPMENT

The history of the development of the north-west European continental shelf is fully compatible with the broad terms of the plate-tectonics hypothesis in its implications for the late opening of the north Atlantic. An excellent review has recently been published by Ziegler (1974), and it would be invidious to restate his conclusions.

A notable feature emerging from the present review is the occurrence in other major basins of the structural pattern found in the Northern North Sea—that is, a more or less median fault trough system developed contemporaneously with Permo-Triassic to Lower Cretaceous deposition, blanketed by a much wider basin fill of mainly unfaulted Cretaceous and Tertiary. Major intra-Jurassic faulting is so far documented only in the northern North Sea; it is suspected but not yet established elsewhere.

This general relationship is found in each of the larger basins where the full Mesozoic history is known (northern North Sea, Cardigan Bay, Western English Channel). It is true of the southern North Sea if allowance is made for the complicating effects of halokinesis; it may be true of the Bristol Channel/Northern Celtic Sea, and, in modified form, probably of relationships on the edge of the Faeroes–Shetland basin. Thus there was a long history of taphrogenic control, with at least intermittent faulting contributing to subsidence throughout much of the Permian to Lower Cretaceous (Aptian) interval. The trends were mostly (but not entirely) parallel to the later Atlantic margin, and are symptomatic of very long term tension, reasonably explained as due to plastic thinning of the deep crust. This process ended when the magnetically-documented plate movement began; the lateral movement was then accomplished by a different mechanism as the ocean proper developed.

It has been emphasized elsewhere that this is a history characterizing other Atlantic-type coasts (notably Greenland, East Africa and western Australia (Kent, 1973)), where breakdown of the continental margin began in late Carboniferous to early Permian. Australia even shows the post-Jurassic regional unconformity which developed when fault control of subsidence ended.

The change of the world machinery requires more explanation than it has received so far. The nearly simultaneous world-wide inception of epeirogenic movement, its large scale and long-term persistence (much longer than the subsequent plate movement phase) establish it as a dominant element in continental development away from the Pacific region.

Also in contrast to the Pacific region, the occurrence of volcanic episodes in space and time is irregular and localized. On this and on other Atlantic margins eruptions occurred within the region from the Devonian onwards, and (except for a possible peaceful interval in Upper Permian and Trias) occur in nearly all parts of the stratigraphic column. They are however often limited to one area (as during Lower/Middle Jurassic) or to a minor proportion of the region (as during Lower Permian) despite the presumed uniformity of stress conditions. This localization perhaps most likely reflects the heterogeneity of the crust, which is a factor not usually emphasized in broad-brush geology. It provides a warning against too facile generalizations from regional concepts.

REFERENCES

Allen, P., 1967. Origin of the Hastings facies in north-western Europe. *Proc. Geol. Ass.*, **78**, 27–105.

Al-Shaikh, Z. D. 1970. The geological structure of part of the central Irish Sea. *Geophys. Jl. R. astra. Soc.*, **20**, 233–7.

Arkell, W. J. 1933. *The Jurassic System in Great Britain.* Oxford.

Arkell, W. J. 1947. The geology of the country around Weymouth, Swanage, Corfe and Lulworth. *Mem. geol. Surv. Gt. Br.*

Armstrong, G. 1972. Review of the geology of the British Continental Shelf. (Presented at Joint Meeting *Inst. Min. Metall./Inst. Min. Eng.*, Lond., 28/1/72.)

Binns, P. E., McQuillin, R. and Kenolty, N. 1973. The geology of the Sea of the Hebrides. *Rep. Inst. Geol. Sci.*, **73/14**, 43 pp.

Blundell, D. J., Davey, F. J. and Graves, L. J. 1971. Geophysical surveys over the South Irish Sea and Nymphe Bank. *Jl. geol. Soc. Lond.*, **127**, 339–75.

Bott, M. H. P. 1968. *The Geological Structure of the Irish Sea Basin.* Oliver and Boyd, Edinburgh.

Bott, M. H. P. and Watts, A. B. 1970. Geophysical investigations on the continental shelf and slope around the Hebrides, Orkneys and Shetlands (abstr., with discussion). *Proc. geol. Soc. Lond.*, **1662**, 80–2.

Brunstrom, R. G. W. and Walmsley, P. J. 1969. Permian evaporites in North Sea Basin. *Bull. Am. Ass. Petrol. Geol.*, **53**, 870–83.

Bullerwell, W. and McQuillin, R. 1969. Preliminary report on a seismic reflection survey in the Southern Irish Sea, July 1968. *Rep. Inst. Geol. Sci.*, **69/2**, 7 pp.

Bulman, O. M. B. and Rushton, A. W. A. 1973. Tremadoc faunas from boreholes in Central England. *Bull. geol. Surv. Gt. Br.*, **43**, 1–40.

Burk, J. A. 1968. Buried ridges within continental margins. *Trans. N.Y. Acad. Sci., Ser. 2*, **30**, 397–409.

Chesher, J. A., Deegan, C. E., Ardus, D. A., Binns, P. E. and Fannin, N. G. T. 1972. IGS marine drilling with m.v. Whitethorn in Scottish waters 1970–71. *Rep. Inst. geol. Sci.*, **72/10**, 25 pp.

Clarke, R. H., Bailey, R. J. and Taylor Smith, D. 1971. Seismic reflection profiles of the continental margin west of Ireland. *In* F. M. Delany (Ed.) ICSU/SCOR Working Party 31 symposium Cambridge 1970: The geology of the East Atlantic continental margin 2. Europe. *Rep. Inst. Geol. Sci.*, **70/14**, 67–76.

Collette, B. J. 1960. The gravity field of the North Sea, *In* G. J.

Bruins (Ed.) Gravity Expeditions 1948–58. *Netherlands Geodetic Commission Delft*, 47–96.

Collinson, J. D. 1972. The Røde ø Conglomerate of inner Scoresby Sund and the Carboniferous(?) and Permian rocks west of the Schuchert Flod. *Meddr. Grønland*, **192**, 47.

Cook, A. H. and Thirlaway, H. I. S. 1955. The geological results of measurements of gravity in the Welsh borders. *Q. Jl. geol. Soc. Lond.*, **111**, 47–70.

Cook, E. E. 1965. Geophysical operations in the North Sea. *Geophysics*, **30**, 495–510.

Craig, G. Y. 1965. Permian and Triassic. *In: The Geology of Scotland. G. Y. Craig (Ed.)*, Oliver & Boyd, Edinburgh, 383–400.

Curry, D., Hamilton, D. and Smith, A. J. 1970. Geological evolution of the western English Channel and its relation to the nearby continental margin. *In:* F. M. Delany (Ed.) ICSU/SCOR Working Party symposium Cambridge 1970: The geology of the East Atlantic continental margin 2. Europe. *Rep. Inst. geol. Sci.*, **70/14**, 129–42.

Dingle, R. V. 1971. A marine geological survey off the north-east coast of England (Western North Sea). *Jl. geol. Soc. Lond.*, **127**, 303–38.

Dobson, M. R., Evans, W. E. and Whittington, R. 1973. The geology of the South Irish Sea. *Rep. Inst. geol. Sci.*, **73/11**, 35 pp.

Donovan, D. T. 1963. *The geology of British Seas*. Univ. of Hull.

Dunham, K. C. and Poole, E. G. 1974. The Oxfordian Coalfield. *Jl. geol. Soc. Lond.*, **130**, 387–91.

Flinn, D. 1969. A geological interpretation of the aeromagnetic maps of the Continental Shelf around Orkney and Shetland. *Geol. Jl.*, **6**, 279–92.

Francis, E. H. 1965. Carboniferous. *In: The Geology of Scotland. G. Y. Craig (Ed.)*, Oliver & Boyd, Edinburgh, 309–57.

Geiger, M. E. and Hopping, C. A. 1968. Triassic stratigraphy of the southern North Sea Basin. *Phil. Trans. R. Soc., Ser. B.*, **254** (790), 1–36.

George, T. N. 1960. The stratigraphical evolution of the Midland Valley. *Trans. geol. Soc. Glasg.*, **24**, 32–107.

George, T. N. 1965. The geological growth of Scotland. *In: G. Y. Craig (Ed.), The Geology of Scotland*, Oliver & Boyd, Edinburgh, 1–48.

George, T. N. 1971. Discussion (p. 372) of Blundell, D. J. *et al.* Geophysical Surveys over South Irish Sea and Nymphe Bank. *Jl. geol. Soc. Lond.*, **127**, 337–75.

Gray, F. and Stacey, A. P. 1970. Gravity and magnetic interpretation of Porcupine Bank and Porcupine Bight. *Deep Sea Res.*, **17**, 467.

Howitt, F. 1974. North Sea oil in a world context. *Nature, Lond.*, **249**, 700–3.

Kennedy, W. Q. 1946. The Great Glen Fault. *Q. Jl. geol. Soc. Lond.*, **102**, 41–76.

Kent, P. E. 1947. A deep boring at North Creake, Norfolk. *Geol. Mag.*, **84**, 2–18.

Kent, P. E. 1949. A structure contour map of the surface of the buried Pre-Permian rocks of England and Wales. *Proc. Geol. Ass.*, **60**, 87–103.

Kent, P. E. 1955. The Market Weighton Structure. *Proc. Yorks. geol. Soc.*, **30**, 197–224.

Kent, P. E. 1966. The structure of the concealed Carboniferous Rocks of N.E. England. *Proc. Yorks. geol. Soc.*, **35**, 323–52.

Kent, P. E. 1967. A contour map of the sub-Carboniferous surface in the north-east Midlands. *Proc. Yorks. geol. Soc.*, **36**, 127–33.

Kent, P. E. 1973. Geology and geophysics in the discovery of giant oil fields. *APEA Jl.*, **13**, 3–8.

Kent, P. E. 1974. Structural History. *In: D. H. Rayner and J. E. Hemmingway (Eds.), The Geology and Mineral Resources of Yorkshire*. Yorks. Geol. Soc., 13–28.

Lees, G. M. and Cox, P. T. 1937. The geological results of the present search for oil in Great Britain by the D'Arcy Exploration Company Ltd., *Q. Jl. geol. Soc. Lond.*, **93**, 156–94.

Lees, G. M. and Taitt, A. H. 1945. The geological results of the search for oilfields in Great Britain. *Q. Jl. geol. Soc. Lond.*, **101**, 255–317.

Lloyd, A. J., Savage, R. J. G., Stride, A. H. and Donovan, D. T. 1973. The geology of the Bristol Channel floor. *Phil. Trans. R. Soc., Ser. A.*, **274**, 595–626.

Ramsbottom, W. H. C. 1973. Transgressions and regressions in the Dinantian: a new synthesis of British Dinantian stratigraphy. *Proc. Yorks. geol. Soc.*, **39**, 567–607.

Rayner, D. H. 1953. The Lower Carboniferous Rocks in the north of England: A Review. *Proc. Yorks. geol. Soc.*, **28**, 279.

Rayner, D. H. 1967. *The Stratigraphy of the British Isles*. Cambridge University Press. 453 pp.

Roberts, D. G. (*In press*). Structural development of the British Isles continental margin and the Rockall Plateau. *In: C. A. Burk and C. L. Drake (Eds.), Geology of Continental Margins*. Springer-Verlag.

Smith, A. J. and Curry, D. 1974. The structure and geological evolution of the English Channel. *Proc. R. Soc. (In press)*.

Sorgenfrei, T. 1969. Geological perspectives in the North Sea areas. *Meddr. dansk. geol. Foren.*, **19**, 160–96.

Stride, A. H., Curray, J. R., Moore, D. G. and Belderson, R. H. 1969. Marine geology of the Atlantic Continental margin of Europe. *Phil. Trans. R. Soc., Ser. A.*, **264**, 31–75.

Stubblefield, J. and Bullerwell, W. 1966. Some results of a recent geological survey boring in Huntingdonshire. *Proc. geol. Soc.*, **1637**, 35–40.

Taitt, A. H. and Kent, P. E. 1958. Deep boreholes at Portsdown and Henfield. *BP Technical Pub. Lond.*

Thomas, A. N., Walmsley, P. J. and Jenkins, D. A. L. 1974. Forties Field, North Sea. *Bull. Am. Ass. Petrol. Geol.*, **58**, 396–406.

Vogt, T. 1954. The lateral compression in Norway and the Great Glen Fault in Scotland I and II. *Forh. Kon. Norske Vidensk. Selsk.*, **27**, 42–53.

Walsh, P. T. 1966. Cretaceous outliers in South-West Ireland and their implications for Cretaceous palaeogeography. *Q. Jl. geol. Soc. Lond.*, **122**, 63–84.

Waterston, C. D. 1965. Old Red Sandstone. *In: The Geology of Scotland. G. Y. Craig (Ed.)*, Oliver & Boyd, Edinburgh, 269–308.

White, P. H. N. 1949. Gravity data obtained in Great Britain by the Anglo–American Oil Co. Ltd. *Q. Jl. geol. Soc. Lond.*, **104**, 339–64.

Wills, L. J. 1951. *A Palaeogeographical Atlas of the British Isles and adjacent parts of Europe*. Blackie & Son, Ltd. Lond.

Wills, L. J. 1970. The Triassic succession in the central Midlands in its regional setting. In Symposium: Triassic rocks of the British Isles. *Q. Jl. geol. Soc. Lond.*, **126**, 225–83.

Wills, L. J. 1973. A palaeogeographical map of the Palaeozoic floor below the Permian and Mesozoic formations in England and Wales. *Mem. Geol. Soc. Lond.*, **7**.

Wood, A. and Woodland, A. W. 1971. Introduction to the Llanbedr (Mochras Farm) Borehole. *In:* A. W. Woodland (Ed.) 1971. The Llanbedr (Mochras Farm) Borehole. *Rep. Inst. geol. Sci.*, **71/18**, 1–9.

Ziegler, P. A. (*In press*). Bergen Conference December 1973.

2

The Seismic Structure of the Western Approaches and the South Armorican Continental Shelf and its Geological Interpretation

By F. AVEDIK

(Centre National pour L'Exploitation des Océans)

SUMMARY

Based on all available seismic refraction data, the seismic structure of the Western Approaches to the English Channel and the South Armorican continental shelf has been established. The correlation of seismic velocities showed a three-fold structure, related to geological layers as follows:
1. High-velocity, deep sequence (4.4–6.6km/sec): metamorphic basement and Lower to Middle Palaeozoic.
2. Intermediate sequence (3.6–4.3km/sec): Permo-Triassic (and Jurassic?) and possibly also Lower Cretaceous on the margin of the Western Approaches.
3. Upper sequence (1.9–3.6km/sec): Cretaceous and Cenozoic sediments.

An acoustic unconformity appears to be present between the Upper sequence and the High velocity and Intermediate sequences.

Different geomorphology characterizes these layers: the Palaeozoic is tectonized and generally follows the Hercynian structural trends; mainly Permo-Triassic sediments fill the depressions of the Palaeozoic floor, but only in the Western Approaches; eastwards-transgressive Cretaceous and Cenozoic sediments cover discordantly the preceding layers.

As inferred from the seismic structure, three major sedimentary phases characterize the geology of the Western Approaches and the South Armorican shelf:
1. a Palaeozoic sequence, ending with the Hercynian orogeny;
2. a continental-type sequence, from the Hercynian orogeny until the opening of the Bay of Biscay (only in the Western Approaches);
3. a mainly marine-type sequence, starting with the opening of the Bay of Biscay and continuing to present times.

INTRODUCTION

The Western Approaches to the English Channel is an approximately triangle-shaped region bordered by the two Palaeozoic land areas of Cornwall and of the Armorican Massif and, on the side of the open ocean, by the edge of the continental shelf. The sea covering the Western Approaches shallows eastwards from depths of about 200m to 60m. The bathymetry of the South Armorican margin shows the same tendency and the sea shallows from about 200m on the margin landwards.

Although the exploration of the seabed here began in the last century, only limited information on the surface geology of the Western Approaches and the adjacent shelf areas was available until the end of World War II (Dangeard, 1971). Since then, thanks to the shallow-penetration seismic profiling and core-sampling done by universities and other institutions, a remarkably detailed knowledge has been acquired of the morphology and structure of the mainly Mesozoic and Cenozoic strata, but only as far as about 7° W. It has been postulated on the basis of the data available that a rather deep, sediment-filled trough extends under the axial part of the floor of the Western Approaches (Andreieff *et al.*, 1971; Boillot *et al.*, 1971; Larsonneur, 1971;

Vanney *et al.*, 1971). As a result of seismic refraction surveys, carried out mainly by British surveyors and by the Bureau de Recherches Géologiques et Minières (BRGM) between 1955 and 1968, more information on the deeper geological structure has been obtained (Bunce *et al.*, 1964; Day *et al.*, 1956; Frappa and Horn, 1971; Hill and King, 1953). The seismic data confirmed the presence of the subsurface axial trough and indicated that it was filled with about 3.5km of sediment underlain by a high-velocity substratum.

The Western Approaches trough gained the interest of the petroleum companies in the nineteen-sixties. The extensive seismic reflexion surveys which they have carried out since 1970, have led to a more precise definition of the geometry of this structure. To complement the reflexion data, a large scale seismic refraction survey was initiated in 1971 and completed by the Centre Océanologique de Bretagne in the fall of 1973. The surveyed area, which includes 28 refraction stations, extends from 47° 30′ to 49° 30′ N, and about 4° to 8° W.

The sonobuoy technique for the detection of seismic signals had to be adapted to cope with the high tidal currents in the surveyed area. Essentially, we used hydrophones or geophones positioned on the sea floor; the seismic signal from the bottom

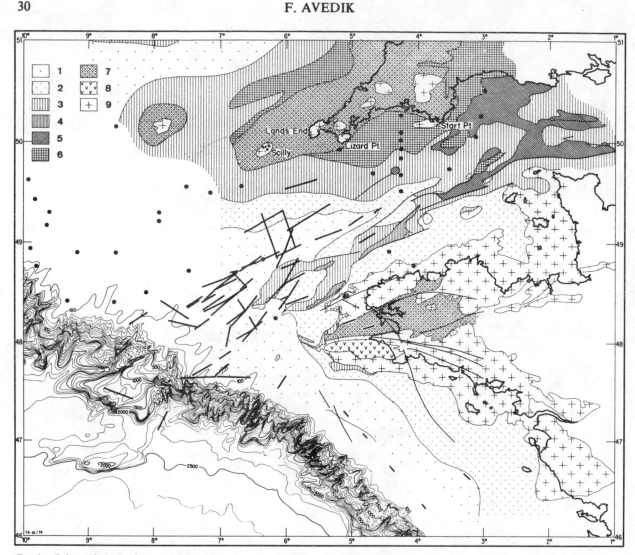

Fig 1. Schematic geological map of the Western Approaches and adjacent areas, showing the positions of the seismic refraction stations. 1—2: Cenozoic; 3: Upper Cretaceous; 4: Lower Cretaceous; 5: Jurassic; 6: Permo-Triassic; 7: Devonian; 8: "greenstones"; 9: granites. Dots and short lines: previous seismic refraction stations. Long lines: profiles by Centre Océanologique de Bretagne, Brest, in cruises 1971—73.

detector was fed through a conductor line to a radio transmitter on the sea-surface, and then received and recorded on the ship. This method ensured drift-free positioning of the detectors and led to a considerably better signal-to-noise ratio than that achieved by other systems of detection (Avedik and Renard, 1973).

The general trend of magnetic anomalies, which is assumed to reflect the structural trend, is mainly north-east to south-west in the Western Approaches and, close to the continental shelf edge it curves in to the north-west (Hill and Vine, 1965). Therefore, to minimize errors in the recorded velocities due to abrupt changes in dip, the chosen orientation of the profiles was mainly north-east to south-west, that is along the strike of the apparent structural trend (Fig 1). Generally the profiles were not reversed, or were run in two directions around a central receiving point. Reverse profiles were restricted to areas of presumed subsurface basins. This procedure seemed to be the best compromise

for obtaining the maximum structural information and the least disturbed velocity records. The length of the profiles varied from about 20 to 50km and shots were fired about every 100m. The seismic source was a 360cu in air-gun. For about half of the profiles, two guns were used and shot simultaneously. Navigation was based on Decca and satellite fixes and the error between horizontal distances computed from navigational data and those deduced from direct arrivals ranged in most cases from 1 to 3 percent.

It is general practice to establish the seismic structure of an area on the basis of lateral correlation of similar velocities. It is then assumed that similar velocities represent the same geological layer. Clearly, this is not necessarily justified. Lateral velocity variations may occur within a layer due to lithological changes; high velocity carbonates could easily be confused with both metamorphosed and crystalline rocks because they generally fall roughly in the same velocity class;

there may also be layers geologically different with the same seismic velocities. Another potential difficulty is the "hidden-layer" problem of a layer resting under a higher velocity one. In this case the lower velocity surface does not constitute a refracting interface and the lower velocity layer generally remains undetected. Finally, in areas without deep boreholes, direct correlation between the subsurface geology and the seismic horizons is obviously impossible. In such areas (and these include the Western Approaches and the South Armorican shelf) only the geology of the surrounding land and the surface geology of the sea-floor yield guidelines for the interpretation of the seismic data. In spite of these difficulties and ambiguities, however, seismic refraction remains an indispensable tool for subsurface exploration.

GEOLOGICAL BACKGROUND

In order to provide a "geological guide" for the interpretation of the seismic structure, we briefly summarize the history of the area, as deduced from observations on the surrounding land areas and from what is known of the geology of the seabed (refer to Fig 1).

The oldest, north-east to south-west, structural trend recognized in Cambrian and Precambrian strata in north-eastern Brittany relates to the so-called Cadomian orogeny (550–600my). Ordovician, Silurian and Devonian sediments were deposited during subsequent Lower Palaeozoic transgressive phases. In Cornwall, the oldest sediments at surface belong to the Devonian. Interfingering of Devonian marine facies with continental Old Red Sandstone deposits in Cornwall, and also in Brittany, suggests that the southern shore-line of the "Old Red Sandstone Continent" (North Atlantic Continent) stretched across this area.

The Hercynian orogeny (~350my), accompanied by extensive metamorphic and intrusive activity, considerably remodelled the whole region. It involved the development of large east–west, north-west–south-east and, later again, roughly east–west trending arches through Brittany, Cornwall and Southern Ireland, and it seems likely that the creation of these arches reactivated older tectonic zones of weakness. In southern Brittany, the South Armorican shear-zone developed into a north-east to west curved arch whereas the main tectonic directions related to this orogeny in northern Brittany are north-east–south-west and north-west–south-east. After the Hercynian orogeny Cornwall and Brittany remained emerged, and mainly in Permian and possibly also in Triassic times these areas underwent extensive erosion. The geological evidence of this denudation are the continental-type New Red Sandstone deposits which are in direct contact with Devonian strata to the south of Cornwall.

A glance at the geological map of France and Great Britain shows that in lower Mesozoic times, there must have been some structural barrier between Start Point and Cotentin which protected the Western Approaches from an overall, westwards transgression, although channels may have breached this ridge between the present Western Approaches and the Paris Basin. There is presently no geological evidence of any important Lower Mesozoic transgression eastwards. However, Upper Cretaceous sediments, indicating a major eastwards transgression cover large areas of the present day Western Approaches and lie discordantly on the Devonian and Permo-Triassic deposits. The eastwards transgressive tendency was probably related to the opening of the North Atlantic and the Bay of Biscay.

The transgressive Upper Mesozoic deposits crossed the Start Point–Cotentin barrier and joined the Upper Mesozoic of the Paris Basin. With intermissions, the eastwards transgression continued into the Cenozoic. Approximately north-east–south-west trending faults, accompanied by linear swells of the post-Palaeozoic sediments from Cotentin to the western limit of the explored area have been detected on the seabed.

THE SEISMIC STRUCTURE

A preliminary interpretation based on the correlation of velocities found beneath a series of

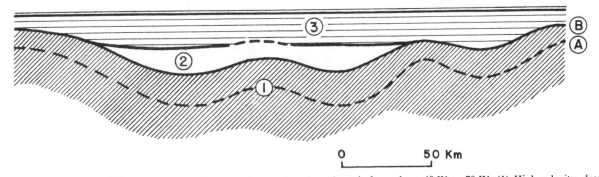

FIG 2. Schematic seismic structure of the Western Approaches along its axis from about 4° W to 7° W. (1) High velocity, deep sequence (4.4—6.6km/sec). (A) and (B) indicate layers "A" and "B". (2) Intermediate sequence (4.0—4.3km/sec). (3) Upper sequence (1.9—3.6km/sec).

stations aligned along a profile following the north-east–south-west axis of the Western Approaches was presented in February 1974 at the discussion meeting on the "Geology of the English Channel" (Avedik, 1974). This profile between approximately 4° and 7° W, revealed the following seismic structure (Fig 2):

1. a high-velocity, deep sequence, characterized by tectonized, conformable layers (velocity range from 4.4 to 6.6km/sec);
2. an intermediate sequence, filling depressions in the deep tectonized sequence (4.1–4.3km/sec);
3. an upper, low to medium velocity sequence, with almost horizontal layers (1.9–3.6 km/sec).

An acoustic unconformity appears to be present between the high-velocity, tectonized, deep sequence and the upper, horizontally layered sequence. This unconformity is correlated with the geological unconformity between the Palaeozoic and Cretaceous strata in the Western Approaches.

All available seismic refraction data have been compiled in this work to present an overall seismic structure of the Western Approaches and the South Armorican continental shelf.

Throughout the area the seismic structure is characterized by the same velocity groups as noted above, except for one additional velocity class (3.6–3.9km/sec). In this paper the three-fold classification of velocities is maintained and as a first approach, the 3.6–3.9km/sec velocity class is included with the intermediate sequence.

High Velocity Sequence (4.4–6.6km/sec). The velocities within the sequence may be split into two groups (Fig 2): (a) 5.3–6.6km/sec (layer "A"); (b) 4.4–5.0km/sec (layer "B").

(a) Layer "A" was detected beneath almost all stations in the Western Approaches and the South Armorican continental shelf. The isobath map (Fig 3) shows the rapid deepening of this high-velocity surface towards the axis of the Western Approaches, where it reaches its maximum depth of about 4.0–4.5km and forms an approximately

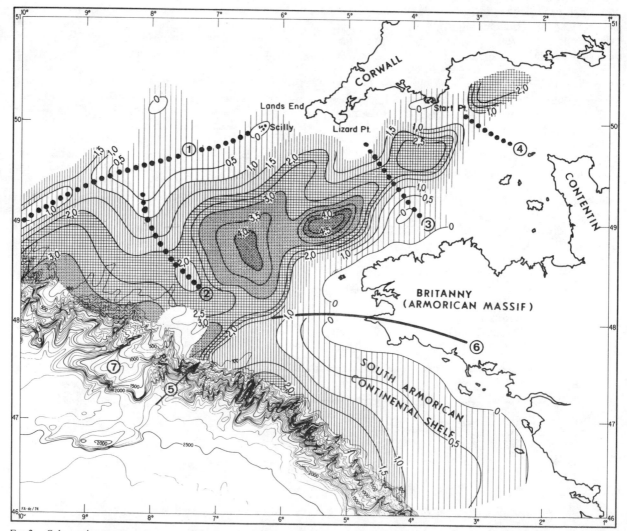

FIG 3.   Schematic structure-contours of the metamorphic basement (depth in km). Structures: (1) Lands End—Scilly axis; (2) North-west—south-east basement high; (3) Lizard—Brittany ridge; (4) Start—Cotentin ridge; (5) North Armorican Fault; (6) South Armorican Shear Zone; (7) Meriadzek terrace.

triangle-shaped depression oriented north-east–south-west. West of Brittany, and on the South Armorican continental shelf, layer "A" slopes rather gently to the south-west and reaches the continental shelf edge at a subsurface depth of about 1.5–2.0km.

To infer the nature of the high-velocity layer "A", the correspondence between the observed velocities and those measured on outcrops or strata of known geological setting has to be established. The metamorphic and igneous rocks on the surrounding land areas have generally velocities in the 5 to 7km/sec range, which agree well with those obtained for the layer "A". The various stages of metamorphism displayed by these rocks in Brittany and the frequent acid or basic intrusives associated with them, explain the relatively large scatter of observed values in this velocity class. Possibly a similar geology characterizes the areas under the sea, in which case it seems unlikely that one could differentiate the different geological units solely on the basis of their seismic velocities. We shall refer therefore to layer "A" in a global way, designating it as the "metamorphic basement" comprised of rocks in various stages of metamorphism, often associated with acid and basic intrusives.

There is no difficulty in assigning the observed layer "A", which is found in a relatively shallow depth around the Armorican Massif, to the "metamorphic basement". The same is true close to the Lizard and the Scilly Islands, where basic intrusives and Hercynian granites crop out. Here layer "A" forms a shallow, southwards dipping surface. Away from land and outcrops, the layer plunges to great depth and it becomes increasingly difficult to decide whether its apparent continuity is indeed a continuation of the "metamorphic basement". In particular, the presence of high velocity carbonates, which may mask the "true" basement is possible. However, the depth of the numerous magnetic sources, computed from the magnetic anomalies, shows such close correspondence to the depth of layer "A" throughout the mapped area (Segoufin, 1974) that it seems reasonable to assume that the high-velocity surface of layer "A" corresponds closely to the "metamorphic basement".

The observed north-east–south-west, east–west and north-west–south-east structural features of the "metamorphic basement" in the Western Approaches and the South Armorican continental shelf can be logically related to the structural directions observed in Brittany and Cornwall. One

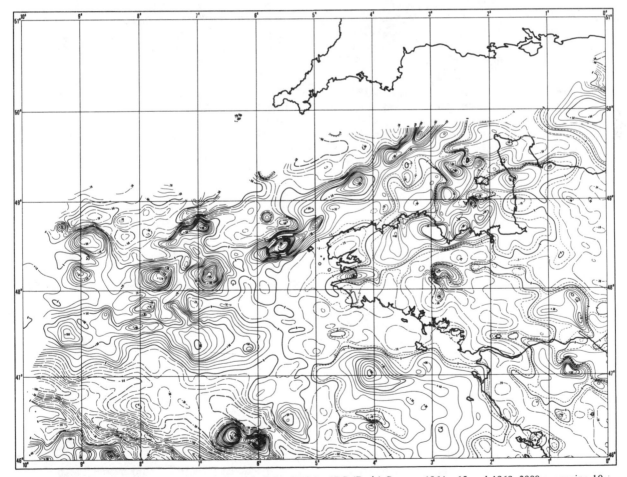

FIG 4. Magnetic anomaly map of the total magnetic field. CNRS—IPG (Paris) Surveys 1964—65 and 1969, 3000m, spacing 10γ.

of the outstanding structural features is the system of north-east–south-west faults (Fig 3). This system appears to cross the continental shelf edge south of the Meriadzek complex and seems to be accompanied by a sedimentary swell of similar orientation as evidenced by the surface geology of the seabed. In the forthcoming discussion I shall refer to this fault system as the North Armorican Fault.

The alignment of the strong magnetic anomalies associated with the North Armorican Fault, after a sharp south-eastwards turn in the vicinity of the Isle of Wight, appears to be linked to the north-west–south-east oriented anomalous magnetic zone in the Paris Basin (Fig 4). The lateral offset observed in the North Armorican Fault west of Brittany is also the locus of a strong magnetic anomaly. The computed source depths indicate the presence of basic material, possibly basalt, intruded at a relatively shallow depth in this fractured zone (around 1 to 2km) (Segoufin, 1974). Further, the seismic interface associated with the Conrad discontinuity which is detected at a depth of about 15km beneath the southern part of a north-west–south-east oriented refraction profile from Lands End to Morlaix in Brittany shows a large bulge (Holder and Bott, 1971), several kilometres high beneath the North Armorican Fault which might have been the source of the basic intrusions rising to much higher levels. If one accepts this two-level intrusive structure as typical for the North Armorican Fault, an alternative interpretation may also be considered, which regards these two stages as time-separated geological events. The first event, associated with the generalized extensive phases of the early Mesozoic produced a large swell of the mantle intruding into the dislocations. The second one occurred during a later reactivation of this weak zone of the crust and produced smaller intrusive bodies unwarping and invading sedimentary layers. I shall add some more comments on this feature in the third section of this chapter.

The North Armorican Fault, considering its north-east–south-west orientation, may be the remnant evidence of an old, possibly Cadomian (Caledonian?) phase of tectonic activity, reactivated successively during later events.

The structure of the basement high which separates the two deep central depressions is probably complex. It seems to be related as well to the north Armorican granite trend, characterized by low gravity anomalies (Fig 5) as to the north-east–south-west oriented basement high. The rather positive gravity anomalies appear to follow the trend of the latter basement high and possibly signify material of basic origin (Allan, 1961).

The north-west–south-east oriented ridges in the eastern area which induce a step-like, easterly shallowing of the basement and link the Lizard and Start Point complexes to Brittany and the Cotentin peninsula respectively, follow very closely the observed north-west–south-east Hercynian trend and probably originated during the Hercynian orogeny. Also, the high gravity values and their trend (Fig 5) clearly support the seismic interpretation. The ridges must have played a considerable role as eastern barriers during the post-Hercynian sedimentary history of the Western Approaches.

The basement high from Lands End to the continental margin—the Lands End–Scilly axis—is formed by a succession of Hercynian granite batholiths; on the surface at Lands End and the Scilly Islands, they plunge gently towards the continental margin. These deeper batholiths are also reflected in the north-east–south-west trend of low gravity values (Fig. 5) which are indicators of Hercynian granites in this area (Bott and Smithson, 1967). The very high seismic velocities (>6.2km/sec) and rather strong gravity anomalies south of this axis are strong evidence of high grade metamorphism associated with the batholiths. The nature of the relative basement high, which slopes and curves to the south-south-east, and its relation to the Lands End–Scilly axis is still unknown. The relatively high gravity and strong magnetic anomalies partially associated with this structure suggest the presence of material of basic origin.

Southwards to the North Armorican Fault, around Brittany and beneath the South Armorican continental shelf, the metamorphic basement remains at a shallow depth, suggesting that it is an extension of the Armorican Massif as far as the continental margin. In this area the South Armorican Shear Zone constitutes a major Hercynian tectonic feature (Fig 3). The shear zone forms a large, north-east to west incurved arch in southern Brittany and geological observations indicate that it extends under the sea as far as 5° W. Considerable lateral displacements of the northern and southern sectors of the Armorican Massif are supposed to have taken place along the shear zone, but the sense of movement is controversial (Debelmas, 1974, pp. 125–38, 150–4). The shear zone, if its trend is extrapolated, seems to "butt" against the North Armorican Fault. The basement forms a large, elongated feature south to this hypothetical extension of the shear zone, which geometry suggests that the southern sector of the Armorican Massif may have been subjected to a left-lateral displacement. This sense of motion would fit the pattern of gravity anomalies on both sides of the shear zone (Sibuet, 1972). On the South Armorican continental shelf, close to 46° N, one has to note the westwards incurved isobaths which appear to indicate the northern slope of a westwards trending basement high. This basement high is probably a major structure which extends far to the south. One may regard this structural high as the separation of the South Armorican Continental Shelf and the Aquitanian domain.

It should be mentioned that beneath some longer profiles (>30km), mainly in the area of the large

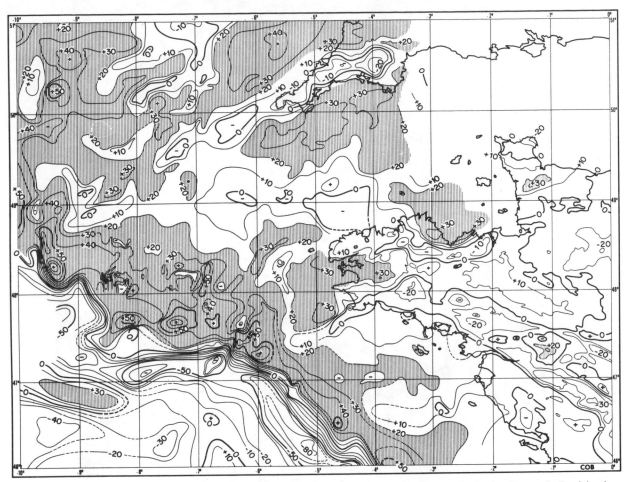

FIG 5. Free-air gravity anomalies (Bouguer anomalies on continents). Published by Bureau de Recherches et de Participation Minières, 1963, and Davey (1970).

central basin of the Western Approaches, secondary arrivals yielded velocities in the 6.5–7.3km/sec range. However, as these arrivals appeared at horizontal distances which were close to the maximum radio range of the emitters, it could not be investigated whether these events represent refracted waves or wide-angle reflections. The time–distance graph of the station at 48° 42′ N and 06° 43′ W (for the profile oriented north-east–south-west) shows an arrival which seems to correspond to a reflection with vertical two-way travel time of 3.6 seconds. On the basis of the available data it seems possible that a seismic interface is present in the central area of the Western Approaches at a depth of about 6–8km. The positions of stations featuring such deep arrivals, the measured apparent velocities and the corresponding depths of this seismic discontinuity are shown in the Appendix.

(b) Layer "B" (4.4–5.0km/sec) (Fig 2), is the second refracting interface in the high-velocity sequence and it is taken to represent a layer which covers the metamorphic basement. It is present over the major part of the mapped area. This material constitutes a north-north-westwards thinning sedimentary wedge which strikes north-east–south-west and shoulders against the North

Armorican Fault, where it attains its maximum development of about 2km (Fig 6). The refracting surface of layer "B" is deepest in the central areas (about 2.5km) and follows the step-like shallowing of the "metamorphic basement" to the east. To the north, towards the Lands End–Scilly axis, layer "B" rises gradually to a shallow depth and appears to be linked to the large Devonian outcrop found around Lands End.

Southwards to the North Armorican Fault, around Brittany, and on the South Armorican continental shelf, layer "B" forms a rather thin sheet and wedges out toward the present land areas and also toward the basement high, which seems to separate the South Armorican and the Aquitanian continental shelves.

Two remarkable morphological features of layer "B" should be noted. One is its rather uniform thickness in the north-east–south-westwards trending depressions, and the other is its thinning over the observed Hercynian uplands (the Lizard–Brittany and Start–Cotentin ridges, and the Lands End–Scilly axis). The almost uniform thickness of the layer in a north-east–south-west direction which contrasts markedly with the observed westwards increase in the thickness of Mesozoic strata and the most probable erosion of

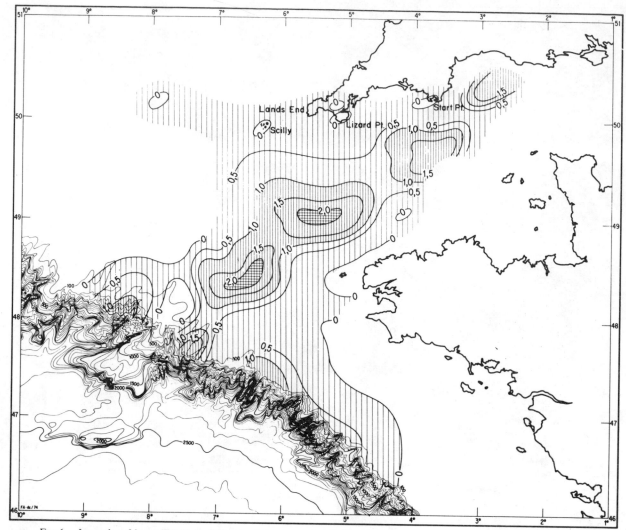

FIG 6.   Isopachs of layer "B" (4.4—5.0km/sec), interpreted as Lower and Middle Palaeozoic. Contour interval 0.5km.

the Hercynian uplands may be evidence of the existence of an extensive sedimentary sequence before the Hercynian orogeny, that is in Lower and Middle Palaeozoic times. In addition, the seismic velocities measured on both continental and submarine outcrops fall in the same velocity class. One is thus led to the conclusion that layer "B" corresponds to Palaeozoic strata, and its refracting surface represents closely the "Palaeozoic floor".

The thinning of the Palaeozoic layer is not evident over the north-west–south-east basement high near the continental shelf edge of the Western Approaches, and this might signify that the structure, if only partially, originated in a later geological event.

The apparent absence of the Palaeozoic west of the previously mentioned basement high near the continental margin is dubious. Velocity determinations from sonobuoy data obtained on the Meriadzek terrace, beyond the continental slope, indicate the presence of a layer with a seismic velocity of 4.7km/sec. The correlation of this layer with the strata in the Western Approaches, which have a similar velocity and supposed to be Palaeozoic, is hypothetical. The station is remote

from the continental margin and lies in a domain generally considered as a transition zone. On the other hand, the morphology of the Meriadzek terrace (Smith and Van Riessen, 1973), a unique feature on this portion of the continental margin, suggests gravity sliding, perhaps resulting from the collapse of the continental margin. Thus, the presence of Palaeozoic strata on the Meriadzek terrace appears quite possible and, if this interpretation is correct, it could explain the partial removal of this layer from the continental shelf edge. The collapse of the continental margin may have occurred with the rotation of Spain (Iberian craton) and possibly followed lines of weakness along the continental margin resulting from the combined effect of tectonics associated with the North Armorican Fault and the South Armorican Shear Zone.

Intermediate Sequence (3.6–4.3km/sec). The material characterized by velocities in the 3.6–4.3km/sec range (intermediate sequence) was detected under the Western Approaches only between the North Armorican Fault and what I may describe as the southern slope of the Lands End–Scilly axis. The sedimentary layers belonging

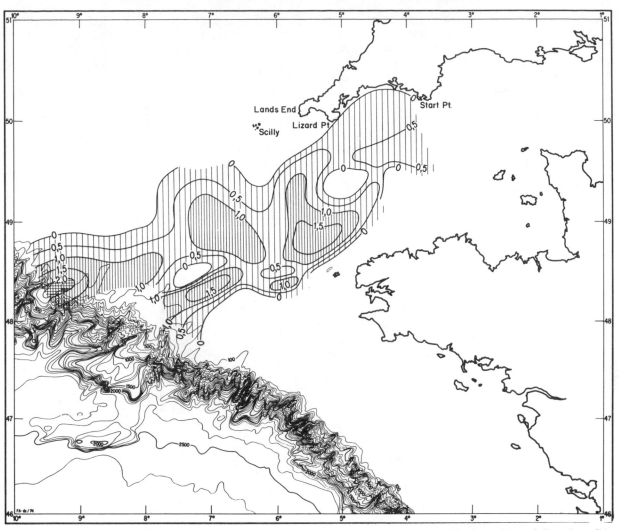

FIG 7.   Isopachs of the intermediate seismic sequence (3.6—4.3km/sec), associated with Permo-Triassic (and Jurassic?), and possibly Lower-Cretaceous on the margin. Contour interval 0.5km.

to the intermediate seismic sequence appear to fill depressions in the Palaeozoic floor between the continental margin and the Lizard–Brittany and Start Point–Cotentin ridges, the latter forming a major structural barrier which seems to have separated the present Western Approaches and Paris Basin during the deposition of these layers (Fig 7).

The analysis of velocities within the layer shows a predominance of higher velocities (4.0–4.3km/sec) in the central areas. In contrast, the somewhat lower velocities (3.6–3.9km/sec) characterize material deposited mainly on the eastern and western periphery, that is in the eastern channels and in the east-north-east–west-south-west oriented, eastwards-thinning sedimentary wedges on the continental margin. Here the 3.6–3.9km/sec material partially overlies the basement. The observed differentiation of velocities in the intermediate sequence is probably the expression of its variable lithological composition which, in turn, must reflect deposition under changing conditions or during different geological times.

Similar geological conclusions can be drawn from the acoustic discordance observed between the seismically different structures of the tectonized high-velocity sequence (Palaeozoic), whose depressions are filled by sediments represented by the intermediate sequence, and the upper, low- to medium-velocity sequence (see above; Fig 2), whose seismic structure shows almost horizontal layering. This acoustic unconformity appears to be similar to the geological unconformity observed in the Western Approaches between Cretaceous and Palaeozoic strata. The similarity between the acoustic and geological unconformities lead one to the conclusion that the sediments represented by the intermediate seismic sequence were likely to have been deposited during Devonian (Carboniferous?) to Cretaceous times. One of the sedimentary components of the sediment-fill may well be Permo-Triassic, mainly continental deposits that were a product of the extensive Permo-Triassic erosional phase. Such Permo-Triassic sediments are known to crop out south of Cornwall.

The opening of the North Atlantic and the

particular position of Spain are clearly pertinent to understanding the late Palaeozoic and Lower Mesozoic sedimentary history of the Western Approaches and that of the South Armorican continental shelf. Various investigators (Bard *et al.*, 1971; Le Pichon *et al.*, 1971; Sibuet *et al.*, 1971; Stauffer and Tarling, 1971) place the northern limit of Spain, or the Iberian craton, in the vicinity of the Meriadzek terrace and have inferred that its separation from the present continental margin took place only in the post-Jurassic times. It seems possible that the Western Approaches may have been protected from widespread Lower Mesozoic transgression from the west because Spain blocked off the major part of the western outlets. In this case the deposition of Lower Mesozoic sediments (possibly Triassic marine and Jurassic Aquitanian or South Portugal type) would have been prevented. The two Hercynian uplands, the Lands End–Scilly axis and the Armorican Massif, may have formed barriers in the north and south respectively at this time.

The relatively thick sediment fill in the central areas wedges out in the eastern areas to a thin sheet of deposits or is lacking over the Lizard–Brittany and the Start–Cotentin Hercynian ridges. The thinning out is not an erosional feature only; it suggests that communication between the Western Approaches and the Paris Basin was considerably restricted in Lower Mesozoic times by these Hercynian structural highs. This does not exclude the possibility of a Lower Mesozoic "trough" linking the western, possibly oceanic areas, through the Western Approaches to the Paris Basin, but it greatly reduces its importance. Furthermore, the most westerly Jurassic outcrop so far found in the Western Approaches (about 4° 30′ W, slightly to the east of the Lizard–Brittany ridge) (Fig 1) has been dated as Lias and the younger Jurassic sediments clearly show an eastwards regressive tendency (Larsonneur, 1971).

Velocity measurements in the English Permo-Triassic show two velocity classes (Wyrobek, 1959). The continental-type deposits, similar to those observed south of Cornwall, are in the 3.0km/sec velocity range and rarely exceed 3.6km/sec even at greater depth. Evaporitic material in this formation raises the seismic velocity to about 4.1–4.3km/sec. The seismic velocity of Jurassic deposits in the Paris Basin is generally in the 3.6–3.9km/sec range and only exceptionally reaches about 4.2km/sec. The 3.8km/sec measured on Jurassic outcrops in the eastern part of the Western Approaches agrees well with velocities obtained in the Paris Basin. The Aquitanian or South-Portugal-type carbonates fall generally in the high-velocity class (about 4.4–4.6km/sec) and often show values which are comparable to those of metamorphic and igneous rocks. If the seismic velocities of these significant samples are compared with those comprising the intermediate sequence (3.6–3.9 and 4.0–

4.3km/sec), one notices a rather close agreement between the velocity ranges, except for the class of high-velocity carbonates, and it seems a relatively simple matter to deduce the geological nature of this seismic sequence. But if we take into consideration the possible existence of sediments with a source lying to the west and deposited during eastwards transgressive phases, then the extrapolation of velocities (measured, for example, in the eastern areas and belonging to a different geological domain) becomes questionable. Indeed, we lack geological evidence on the existence of the Lower Mesozoic transgressional phases originating in the west and on the westwards extension of sedimentary sequences derived from the east. Moreover, because we have no borehole data from the western areas of the Approaches, it is not possible to formulate any really sure conclusions on the geological nature of the intermediate sequence.

It can be postulated that Permo-Triassic continental-type deposits partially fill the central depressions and the eastern "channels" (Fig 7). These deposits, due to their lower velocities (3.0–3.6km/sec), are possibly masked ("hidden layers") in the central depressions by the higher velocity cover (4.0–4.3km/sec), which might be the expression of an evaporitic layer or the remnants of an eastwards transgressive Triassic or Jurassic phase. In the eastern "channels" the Permo-Triassic deposits may be present under thin, higher-velocity (3.8km/sec) Jurassic material linked to the Paris Basin facies.

The east-north-east-west-south-west oriented wedges on the continental margin (Fig 7) are filled with sediments whose seismic velocity falls in the 3.6–3.9km/sec class and correspond closely to the velocities measured on Jurassic outcrops on the eastern periphery of the Western Approaches. However, because of the eastwards transgressive aspect of these wedges, association of this material with the Jurassic found in the eastern areas is questionable. In addition, the velocity of these sediments does not match the generally higher ones of the Jurassic of the South Portugal or Aquitanian type. It therefore seems advisable to link the origin of this material to a post-Jurassic geological event.

To summarize evidence on the geological nature of the intermediate seismic sequence representing the infilling of depressions of the Palaeozoic floor, it appears that the sequence in the central depressions is composed of Permo-Triassic continental-type deposits, overlain by an evaporitic layer or remnants of Triassic or Jurassic sediments originating from an eastwards transgressive phase. This transgression may have found its way through the previously mentioned structural barriers formed by the Lands End–Scilly axis, the Iberian craton and the Armorican Massif, respectively in the north, west and south. In the eastern channels, except for the "Permian basin" south of Cornwall, the Permo-Triassic appears to be covered by the westwards transgressive Jurassic phase linked to the Paris Basin, with this westwards extension limited by the Lizard–Brittany and Start–Cotentin ridges. The sedimentary wedges on the margin are possibly post-Jurassic.

Upper Sequence (1.9–3.6km/sec). The seismic structure of the upper, low- to medium-velocity sequence indicates almost horizontally layered strata discordantly covering the high-velocity and intermediate sequences, that is, both the Palaeozoic floor and the fill of its depressions. This "acoustic unconformity" located at the base of the upper sequence is interpreted as being the geological unconformity observed in the Western Approaches between Cretaceous layers and Palaeozoic strata. Therefore, one should seek correlations between the seismic layers of this sequence and the Cretaceous and Cenozoic strata present on the seabed over the major part of the Western Approaches and the South Armorican continental shelf.

Three distinct velocity groups have been detected in the Upper sequence: (a) 3.2–3.6km/sec, average 3.4km/sec, layer C_1; (b) 2.6–2.9km/sec, average 2.8km/sec, layer C_2; and (c) 1.9–2.4km/sec, layer C_3.

(a) Layer C_1 ($\sim$3.4km/sec) appears to cover discordantly the high-velocity and the intermediate sequences, that is the Palaeozoic and the fill of its depressions, which is interpreted as Permo-Triassic (and Jurassic?) deposits. Layer C_1 is present on the South Armorican continental shelf (maximum thickness about 400–600m) as well as under the Western Approaches, mainly between the North Armorican Fault and the southern slope of the Lands End–Scilly axis. In both areas it is characterized by an eastwards thinning sedimentary wedge, whose thickness reaches about 1.5–2km on the margin of the Approaches decreasing to about 500–600m in the central areas and even less towards the eastern Lizard–Brittany and Start Point–Cotentin ridges which seem to block its eastwards extension. Also it seems to thin out and to disappear on the southern slope of the Lands End–Scilly axis as well as south to the North Armorican Fault in the eastern and central Western Approaches. The wedging out of this layer is also observed towards the basement high interpreted as the separation of the South Armorican shelf and the Aquitanian basin. In the eastern areas its discrimination from the Permo-Triassic is not evident because of only slight velocity contrast. Based on the fact that velocities in the 3.2–3.6km/sec range appeared as the first discernible velocities on the time–distance graphs close to the few Lower Cretaceous outcrops, it seems probable that Layer C_1 represents, at least partially, the Lower Cretaceous. The somewhat higher velocity observed (3.6–3.9km/sec) in the marginal wedges of the intermediate sequence, associated with the Lower Cretaceous on the basis of geological inferences, can possibly be due to a different facies of this geological layer.

(b) Layer C_2 ($\sim$2.8km/sec) represents a fairly good refracting surface and appears to cover a much larger surface than layer C_1 both on the South Armorican shelf and in the Western Approaches. However its structure is closely similar to that of layer C_1, that is an eastwards thinning sedimentary wedge, whose thickness decreases from a maximum of about 2km on the margin of the Approaches to about 600m in the central and 300m in the eastern areas, but breaches over the eastern ridges towards the Paris basin. On the South Armorican shelf, layer C_2 reaches only about 300m and wedges out rapidly towards land.

Velocity measurements on outcropping Upper Cretaceous strata fall generally in the 2.6–2.9km/sec range, thus it was deduced that Layer C_2 is representative of Upper Cretaceous sediments.

The isopachs (Fig 8) show the combined thickness of Layers C_1 and C_2, that is the total Cretaceous in the Western Approaches and the South Armorican continental shelf.

In the Western Approaches the Cretaceous represents an approximately triangle-shaped, eastwards thinning sedimentary wedge, extending mainly between the southern slope of the Lands End–Scilly axis and the North Armorican Fault. In the eastern channel, over the ridges, only the Upper Cretaceous appears to be present.

On the South Armorican shelf, southwards to the North Armorican Fault, only a limited thickness of Cretaceous sediments seems to have accumulated. Nevertheless, its eastwards transgressive tendency is similar to the one observed in the Western Approaches.

The clearly eastwards transgressive character of the Cretaceous layers signifies a major subsidence phase which generally follows the rifting and break up of the continent at its margin (Beck, 1972). This phenomenon is most likely a long lasting event, possibly occurring in phases. The physical process which takes place after the separation of continents, and which causes the subsidence of marginal areas is as yet not well understood, but it is observed on the margins of all the continents. It is likely an isostatic process accompanied by the migration of basic mantle material towards the continent from beneath the growing basin beyond the continental slope.

The transgressive tendency of the Lower and Upper Cretaceous in the Western Approaches and South Armorican shelf is thus most probably related to the break up of the Iberian craton from the Armorican Massif and adjacent areas, which was possibly initiated in late Jurassic times. The timing of this event, deduced from purely seismic data agrees well with the results of palaeomagnetic interpretation on samples and other data from the Bay of Biscay (Sibuet, 1972).

An interesting feature is the partial removal of

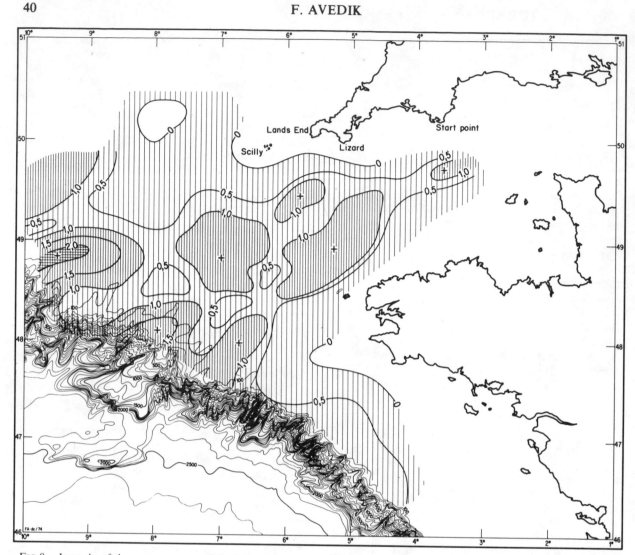

FIG 8.   Isopachs of the upper sequence (1.9—3.6km/sec) representing the Lower and Upper Cretaceous. Contour interval 0.5km.

the Upper Cretaceous sediments from the sedimentary swells over parts of the north-west–south-east basement high near the continental margin and the linear swell following the trend of the North Armorican Fault. Both features relate to strong magnetic and positive gravity anomalies, suggesting basic intrusions. In attempting to explain and date these features one must take account of the considerable bulging of the deep seismic interface beneath the North Armorican Fault and the relative shallow depth of smaller intrusive bodies (see discussion on the high-velocity sequence, layer "A"). In short, it appears very difficult to associate the height and width of the deep intrusion with the relatively small sedimentary swell, the height of which seems to be only some hundred metres. Thus two-stage intrusive activity seems probable, the first and deep phase associated with the extensions of the early Mesozoic which led to the break up of continents, the second with a later, possibly Eocene phase, related to the Alpine movements.

(c) Layer C_3 (1.9–2.4km/sec) the reliability of velocity determinations is often dubious because of the limited vertical extent of these sediments. Layer C_3 apparently contains Cenozoic material, whose thickness reaches about 300–500m in the central areas of the Western Approaches and about 600–800m on the continental margins. Where thicker Cenozoic is present, a distinctive refracting surface with velocities in the 2.2–2.4km/sec range seems to belong to the Eocene or Palaeocene.

CONCLUDING REMARKS ON THE POST-PALAEOZOIC SEDIMENTARY HISTORY, INCLUDING THE PERMO-TRIASSIC

It has been shown that the sedimentary history of the Western Approaches and the South Armorican continental shelf involved the deposition of at least three distinctive sedimentary sequences related to different, major geological events: (1) a Palaeozoic sequence, ending with the Hercynian orogeny; (2) a mainly continental-type sedimentary sequence, from the Hercynian orogeny until the opening of the Bay of Biscay; and (3) the mainly marine-type,

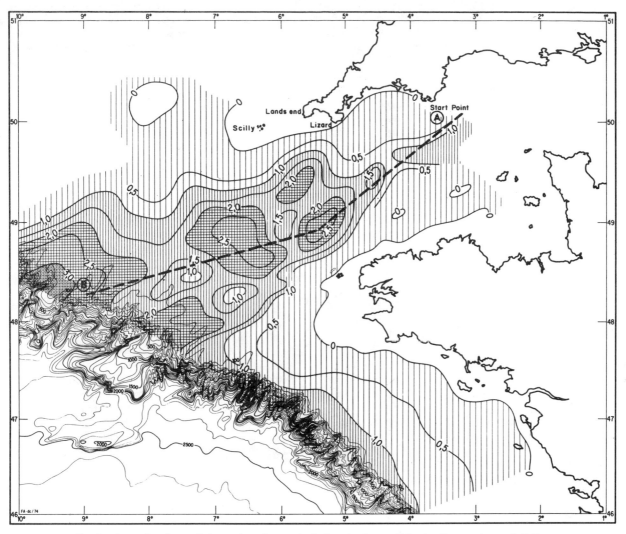

FIG 9. Isopachs of post-Palaeozoic sediments including the Permo-Triassic. Contour intervals 0.5km.

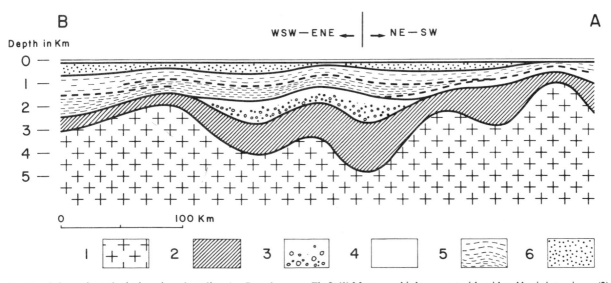

FIG 10. Schematic geological section along line A—B as shown on Fig 9. (1) Metamorphic basement with acid and basic intrusions; (2) Lower and Middle Palaeozoic; (3) Permo-Triassic (and Jurassic?); (4) Jurassic (and Permo-Triassic?); (5) Lower and Upper Cretaceous; (6) Cenozoic.

Upper Mesozoic and Cenozoic sequence, which represents sedimentation starting with the opening of the Bay of Biscay and continuing to present times.

The Palaeozoic geosyncline was considerably remodelled during the Hercynian orogeny when uplands and depressions were formed along east–west, north-east–south-west and north-west–south-east structural trends. At the end of the orogeny, the Western Approaches formed a closed area bordered by the Hercynian uplands—the Lands End–Scilly axis in the north and the Lizard–Brittany and Start–Cotentin ridges to the east. The southern and western structural borders of the area were respectively the Armorican craton, whose northern limit seems to be the North Armorican Fault, and the Iberian craton. The sediments of the mainly continental-type post-Hercynian sequence were laid down in the north-east–south-west oriented basin, floored by depressions and smaller ridges. At the end of this sedimentary phase some evaporitic material may have been laid down in the central depressions.

Alternatively, early Mesozoic material may have accumulated on the top of continental-type deposits prior to the large transgressive phase, possibly following the opening of the Bay of Biscay. Whatever the nature of this layer, it appears that its deposition represented the end of the accumulation of mainly continental-type sedimentary sequences in the Western Approaches. The area was a peneplain with the depressions of the Palaeozoic floor filled in. The resulting total thickness of the deposits in the central depressions reached about 1.0–1.5km.

The opening of the Bay of Biscay, possibly in early Cretaceous times, moved the Iberian craton, the western barriers of the Western Approaches and the South Armorican continental shelf towards the North Atlantic. The eastwards transgression flooded the area and deposition of the marine Cretaceous and Cenozoic sedimentary sequence began. Because of the rapid subsidence of the margin between the Lands End–Scilly axis and the North Armorican Fault, approximately 2–3km of sediments accumulated. Towards the east, however, the thickness of the deposits decreased to about 1.0km or less. Southwards to the North Armorican Fault, a geologically different domain is present—the Armorican craton. Due to a lower subsidence rate of its peripheries, a sedimentary cover of only about 1.0–1.5km accumulated on the South Armorican continental shelf, wedging out landwards. This mainly marine sedimentary sequence is still in process of accumulation in the Western Approaches and on the South Armorican continental shelf.

Figure 9 shows the thickness of post-Palaeozoic sediments, including the Permo-Triassic. The total thickness in the central depressions of the Western Approaches reaches about 2.0–2.5km in which the probable Permo-Triassic (and Jurassic?) represent about 1.0–1.5km. On the margin about 2–3km, mainly Cretaceous and Cenozoic sediments have been accumulated. On the South Armorican continental shelf only about 1.0–1.3km of Cretaceous and Cenozoic cover is present and this wedges out towards land.

The geological section (Fig 10) along line A–B (of Fig. 9) clearly shows the different sedimentary sequences: the tectonized Palaeozoic sediments, the fill-in of depressions of the Palaeozoic floor and the transgressive Cretaceous and Cenozoic sedimentary phase.

In conclusion it is emphasized that the sedimentary history which I have outlined was deduced to a great extent from indirect geophysical data of limited extent. The interpretation of such data, by its nature, is controversial and will remain so until enough borehole data are available.

ACKNOWLEDGEMENTS

The author gratefully acknowledges the assistance given through critical comments and discussion of the manuscript by Dr J. Francheteau, Dr D. Needham, Dr G. Pautot, Dr V. Renard and expresses thanks to Mrs Y. Potard for typing the manuscript and Mr D. Carré for the drafting of the diagrams.

APPENDIX

Position of stations	Apparent horizontal velocity in km/sec	Depth in km
48° 42′ N 06° 43′ W	6.6	6.8
48° 49′ N 06° 24′ W	6.5	6.1
48° 38′ N 06° 59′ W	6.8	5.1
48° 33′ N 06° 36′ W	7.3	7.7
48° 35′ N 06° 02′ W	7.2	7.9
48° 51′ N 06° 10′ W	7.0	11.3 (?)
49° 19′ N 06° 24′ W	7.1	7.1
48° 22′ N 05° 59′ W	6.6	4.4

REFERENCES

Allan, T. D. 1961. A magnetic survey in the Western English Channel. *Q. Jl. geol. Soc. Lond.*, **117**, 157–71.

Andreieff, P., Bouysse, Ph., Horn, R. and Moncardini, Ch. 1971. Contribution à l'étude géologique des Approches Occidentales de la Manche. Colloque sur la géologie de la Manche. *Mem. Bur. Rech. geol. Min.*, **79**, 31–58.

Avedik, F. and Renard, V. 1973. Seismic refraction on the continental shelves with detectors on the sea-floor. *Geophys. Prospect.*, **21**, 220–8.

Avedik, F. 1974. Seismic refraction survey in the Western Approaches. Preliminary results. Discussion meeting on the geology of the English Channel. February, 1974. London (*In press*).

Bard, J. P., Capdevila, R. and Matte, Ph. 1971. La structure de la chaine hercynienne de la Meseta ibérique: comparison avec les segments voisins. *In: J. Debyser, X. Le Pichon and L. Montadert (Eds.), Histoire structurale du Golfe de Gascogne. Technip. Paris. Tome 1, I*, 4–68.

Beck, R. H. 1972. The Oceans, the new frontier in exploration. *APEA Journal*, **12** (2), 5–28.

Boillot, G., Horn, R. and Lefort, J. P. 1971. Evolution structurale de la Manche occidentale au Secondaire et au Tertiaire. Colloque sur la geologie de la Manche. *Mem. Bur. Rech. geol. Min.*, **79**, 79–86.

Bott, M. H. P. and Smithson, S. B. 1967. Gravity investigations of subsurface shape and mass distributions of granite batholithes. *Bull. Geol. Soc. Am.*, **78**, 859–78.

Bunce, E. T., Crampin, S., Hersey, J. B. and Hill, M. N. 1964. Seismic refraction observations on the continental boundary West of Britain. *Jl. Geophys. Res.*, **69**, 3853–63.

Bur. Rech. Geol. Min. (Eds.) 1974. Geological map of the English Channel 1:1 000 000 scale. Established by Boillot, G., Lefort, J. P., Cressard, A. and Musellec, P., Service Geologique Nationale.

Dangeard, L. 1971. Historique des premières recherches de géologie sous-marines dans la Manche. Colloque sur la géologie de la Manche. *Mem. Bur. Rech. geol. Min.*, **79**, 13–15.

Day, A. A., Hill, M. N., Laughton, A. S. and Swallow, J. C. 1956. Seismic prospecting in the Western Approaches of the English Channel. *Q. Jl. geol. Soc. Lond.*, **112**, 15–44.

Debelmas, J. 1974. *Géologie de la France. Vol. 1, Vieux massifs et Grandes Bassins sedimentaires*. Doin, Paris, 294 pp.

Frappa, M. and Horn, R. 1971. Etude par sismique refraction du plateau continental au large d'Ouessant. *Bull. Inst. Geol. Bassin Aquitaine*, **11**, 401–10.

Hill, M. N. and King, W. B. R. 1953. Seismic prospecting in the English Channel and its geological interpretation. *Q. Jl. geol. Soc. Lond.*, **109**, 1–19.

Hill, M. N. and Vine, F. J. 1965. A preliminary magnetic survey of the Western Approaches to the English Channel. *Q. Jl. geol. Soc. Lond.*, **121**, 463–75.

Holder, A. P. and Bott, M. H. P. 1971. Crustal structure in the vicinity of South-west England. *Geophys. Jl. R. astr. Soc.*, **23**, 465–89.

Larsonneur, C. 1971. Données sur l'évolution post-hercynienne de la Manche. Colloque sur la géologie de la Manche. *Mem. Bur. Rech. geol. Min.*, **79**, 203–14.

Le Pichon, X., Bonnin, J., Francheteau, J. and Sibuet, J. C. 1971. Une hypothèse d'évolution tectonique du Golfe de Gascogne. *In: J. Debyser, X. Le Pichon and L. Montadert (Eds.), Histoire structurale du Golfe de Gascogne. Technip. Paris. Tome 2. VI.* 11, 44 pp.

Revoy, M. 1968. Etude d'un profil de seismic refraction en Bretagne. (*Thesis*), *Laboratoire de Physique de l'Ecole Normale Superieure*.

Segoufin, J. 1974. Structure du plateau continental Armoricain. Discussion meeting on the geology of the English Channel, February 1974. London (*in press*).

Sibuet, J. C. 1972. Contribution de la gravimétrie à l'étude de la Bretagne et du plateau continental adjacent. *C. r. Somm. Seanc. Soc. geol. Fr.*, Pt 3, 124–32.

Sibuet, J. C., Pautot, G. and Le Pichon, X. 1971. Interpretation structurale du Golfe de Gascogne à partir des Profils de sismique. *In: J. Debyser, X. Le Pichon and L. Montadert (Eds.), Histoire structurale du Golfe de Gascogne. Technip. Paris. Tome 2, VI*, 10, 31 pp.

Smith, S. G. and Van Riessen, E. D. 1973. The Meriadzek terrace: a marginal plateau. *Mar. Geophys. Res.*, **2**, 83–94.

Stauffer, K. W. and Tarling, D. H. 1971. Age of the Bay of Biscay. New paleomagnetic evidence. *In: J. Debyser, X. Le Pichon and L. Montadert (Eds.), Histoire structurale du Golfe de Gascogne. Technip. Paris. Tome 1, II*, 2, 18 pp.

Stride, A. H., Curray, J. R., Moore, D. G. and Belderson, R. H. 1969. Marine geology of the Atlantic Continental Margin of Europe. *Phil. Trans. R. Soc. Lond.*, Series A, **264**, 31–75.

Vanney, J. R., Scolari, G., Lapierre, F., Martin, G. and Dieucho, O. 1971. Carte géologique provisoire de la plate-forme continentale armoricaine (decembre 1970). *In: J. Debyser, X. Le Pichon and L. Montadert (Eds.), Histoire structurale de Golfe de Gascogne. Technip. Paris. Tome 1, III*, 20 pp.

Walcott, R. I. 1970. An isostatic origin for basement uplifts. *Canadian Jl. Earth Sci.*, **7**, 931–37.

Wyrobek, S. M. 1959. Well velocity determination in the English Trias, Permian and Carboniferous. *Geophys. Prospect.*, **7**, 218–30.

3

Geology and Petroleum Possibilities West of the United Kingdom

By D. R. WHITBREAD

(Cluff Oil Limited)

SUMMARY

The geological succession and structure of the basins west of the UK are described from scanty published data and inference. These basins are the West Shetland complex, West Orkney, North Minch, South Minch, Lough Foyle–Islay, Lough Neagh–Kintyre, Irish Sea complex, Kish Bank, Cardigan Bay, Celtic Sea–Western Approaches–Bristol Channel complex and Porcupine Seabight Basin (west of Ireland). A sedimentary wedge thickens off the continental shelf west of Scotland and Ireland and its age is related to the opening of the north-east Atlantic. The concept of a West British graben (from 62° N to 50° N) analogous to the possible graben system in the North Sea is rejected as an hypothesis to integrate the development of the West Britain basins. Instead, the evolution of these basins is related to reactivation of basement fault trends and differential Mesozoic and Tertiary subsidence. The relation of north-east Atlantic evolution to the West Shetland complex is discussed.

INTRODUCTION

This paper gives a brief summary of the known and inferred stratigraphy of the basins shown in Fig.1. The number of deep tests drilled by oil companies in these basins is extremely limited and the results of these boreholes are, of course, confidential. However, from scout information on these wells (West Shetland area—8 drilled, 1 drilling; Irish Sea—3 drilled, 1 drilling; Cardigan Bay—1 drilled; Celtic Sea (Irish Waters)—13 drilled, 1 drilling; Celtic Sea (British Waters)—2 drilled, together with outcrop data and geophysical information, an adequate idea of the geology of the basins west of the UK can be assembled. Figure 1 shows in general terms the location of the basins which are shown in more detail in Figs 3–6 inclusive. The origin and evolution of these basins appears to be related to reactivation of pre-existing faults, with differential down-warping controlling the degree of asymmetry within individual basins.

It is not possible to link the evolution of these basins with any deep crustal event such as incipient rifting and spreading (Naylor *et al.*, 1974). In Figs 1–7 differentiation of basinal areas is often arbitrary and generalized and in many cases thin sediment cover (often late Tertiary and/or Quaternary) exists over areas notated as outcropping "Economic Basement". This concept of "Economic Basement" is commonly employed by geologists engaged in the search for hydrocarbons. It is a valuable concept whereby all rocks deemed incapable of containing economic quantities of hydrocarbons are treated as a unit—in the present area all rocks of Carboniferous and pre-Carboniferous age.

At least two important exceptions within this broad classification must be given, which would allow significant accumulations of hydrocarbons both within and derived from Economic Basement. Firstly, the Devonian of the Orcadian province, which is provisionally placed within the "Economic Basement", may well have acted both as partial source rock and primary reservoir for an unknown quantity of hydrocarbons. Secondly, Upper Carboniferous (Westphalian) Coal Measures appear to have acted as a source (by deep burial and gasification of coals) for most, if not all, of the dry gas found to date in the southern North Sea, north Holland and Germany area. Hence, the "Economic Basement" has here acted as source for a vast quantity of hydrocarbons. This same unit has probably acted as source rock for hydrocarbons in the Irish Sea complex and the northerly extent of Coal Measures and coals of the Calciferous Sandstone Measures west of Britain is clearly a vital point to be established (Machrihanish Coalfield, approximately 55° 25′ N and 5° 38′ W, is the present known northern limit of outcropping source-rock Coal Measures).

Migration of hydrocarbons from sources within "Economic Basement" to overlying sediments could account for a significant percentage of present recoverable reserves in the North Sea. Allocation of age within the "Economic Basement" from interval velocity determinations derived from deep seismic profiling is usually erroneous since these deeper velocities are usually derived from spurious reflections and there are no reliable guidelines for such age allocation from deeply buried "Economic Basement" rocks. To complicate matters, many Mesozoic and Permian units (both clastic and carbonate) show what might have been considered anomalously high interval velocities from well evidence east of the UK.

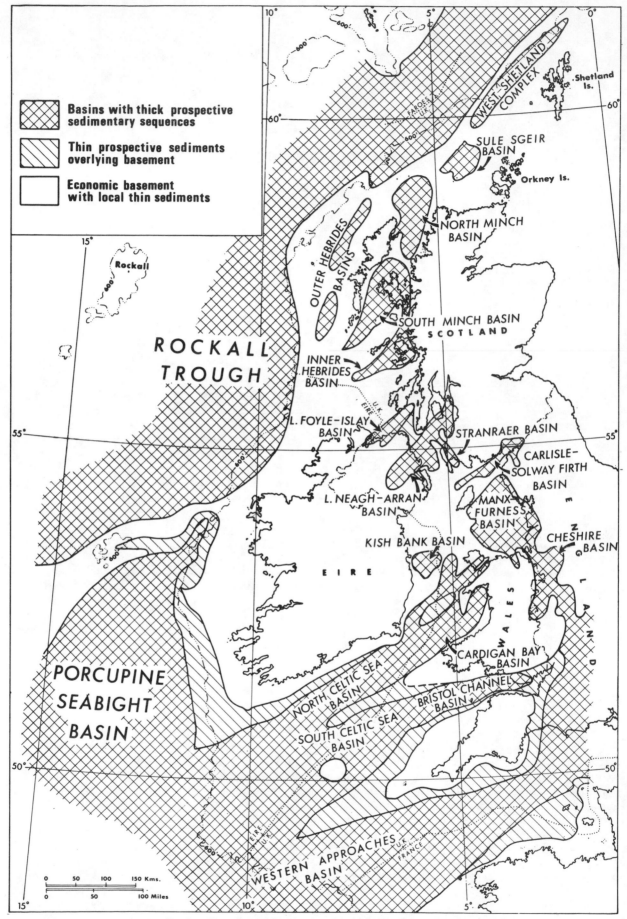

Legend:

- Basins with thick prospective sedimentary sequences
- Thin prospective sediments overlying basement
- Economic basement with local thin sediments

ROCKALL TROUGH

PORCUPINE SEÁBIGHT BASIN

Rockall

WEST SHETLAND COMPLEX

Shetland Is.

SULE SGEIR BASIN

Orkney Is.

NORTH MINCH BASIN

OUTER HEBRIDES BASINS

SOUTH MINCH BASIN

SCOTLAND

INNER HEBRIDES BASIN

L. FOYLE-ISLAY BASIN

STRANRAER BASIN

CARLISLE-SOLWAY FIRTH BASIN

L. NEAGH-ARRAN BASIN

MANX-FURNESS BASIN

KISH BANK BASIN

CHESHIRE BASIN

EIRE

WALES

CARDIGAN BAY BASIN

NORTH CELTIC SEA BASIN

BRISTOL CHANNEL BASIN

SOUTH CELTIC SEA BASIN

ENGLAND

WESTERN APPROACHES BASIN

0 50 100 150 Kms.

0 50 100 Miles

FIG 1. General map showing location of basins in west Britain.

STRATIGRAPHY

General

Most, if not all, of the stratigraphic units referred to herein are represented by often attenuated and highly faulted outcrops scattered through west Britain and Ireland. These outcrops present a highly biased sample of the sequences to be expected offshore (Kent, 1975). Experience of facies changes (from land outcrop into the offshore) derived from the North Sea makes it possible for controlled speculation about the nature and thickness of offshore sequences west of the UK.

Permo-Triassic

Permo-Triassic sediments crop out through the Inner Hebrides, probably the Outer Hebrides (Steel, 1974), Northern Ireland, around the Irish Sea and in south-west England. In the Hebrides area alluvial fan, flood plain and playa environments have been recognized (Steel, 1971, 1974; Bruck et al., 1967), which have their equivalents in the southern North Sea. The great thickness and unusual facies of the sequence cropping out near Stornoway has led to doubt about the age of these rocks, but recent palynological work has indicated that the rocks are no older than Permo-Triassic. Poor preservation of palynomorphs prevents more specific age attribution. In south-west Scotland the classic dune sands exposed at Mauchline, Dumfries, Lochmaben and Arran (Craig, 1965) indicate that desert environments were also certainly developed well away from the southern North Sea and there are indications that similar dune sands are present beneath the Irish Sea and Lough Neagh–Arran basins. These dune sands interdigitate with piedmont fault-scarp breccias in the Galloway basins. In passing, it is worth mentioning the very early Permian volcanic rocks (dominantly quartz-dolerites) of Scotland. An interesting analogy with the early Permian of Norway, Germany and the North Sea can be drawn here.

In Northern Ireland the Magnesian Limestone has been encountered in the deeper parts of the Lough Neagh–Arran basin in boreholes and also at outcrop. It is a matter of conjecture whether this facies of the Permian extends north and west of this basin into the Minch and West Shetland basins. Around the margins of the Irish Sea the Permo-Triassic is thickly developed and Well 110/8-2 evidently encountered a full succession of Keuper, Bunter and Permian including a thick evaporite sequence. The recent gas discovery 110/2-1, apparently produced gas from a relatively shallow horizon (T.D. approximately 4100ft (1250m)) probably from the Bunter Sandstone. Oil was produced in the Formby area from Keuper Waterstones on the east margin of the Irish Sea. This shallow field appears to be a secondary accumulation from a deeper source. The Kish Bank Basin probably consists of a Permo-Triassic sequence—no wells have yet tested this sequence however. In the Cardigan Bay Basin a thick Permo-Triassic sequence exists and 106/24-1 apparently bottomed in Triassic without reaching prognosed total depth.

Further south still in the Celtic Sea, thick Permo-Triassic sequences are known to occur and the outcrops in south-west England and offshore around this peninsula are in a marginal facies. In south Devon and Dorset some 8000ft (2438m) of Permian and 1600ft (488m) of Triassic crop out (Audley-Charles, 1970; Laming, 1966). The succession offshore from this marginal sequence is little known but the models of Permian and Triassic facies change known from the southern North Sea may be applied (with caution) to the Celtic Sea. Palaeowind direction is unknown but may be expected to be from the south and east (cf. Dawlish Sand palaeowinds) with consequent disposition of dune sand facies west of the marginal facies referred to above. In the Puriton borehole in Somerset (51° 10′ N and 03° 00′ W) halite interbedded with gypsum and sandstones of Keuper age indicates the probability of substantial halites at least in the North and South Celtic Sea basins. Thick salt is seen on seismic profiles in these basins but its age is as yet unpublished though often assumed to be Triassic. There is no reason to believe that the Permo-Trias facies west of Britain are fundamentally different to facies in the North Sea and certainly no reason to expect marine Permian (as developed in, for example, Spitzbergen), as development of oceanic crust and incipient rifting of the Atlantic is believed to have started in the Jurassic. More critical for the occurrence of Permo-Trias hydrocarbons seems to be the disposition of Westphalian Coal Measures source rocks.

Jurassic

Jurassic rocks are exposed both in the Hebridean area and in south-west England. Between these two extremes there are but scattered outcrops in northern England and Northern Ireland. However, in Mochras 1 (Fig 5) some 4300ft (1311m) of Liassic were cored beneath unconformable Tertiary and above conformable Triassic (Wood and Woodland, 1968). It is also known that Jurassic rocks are extensively developed in the Cardigan Bay Basin and in the Celtic Sea. In the North Celtic Sea Basin, thick Upper Jurassic with undetermined thicknesses of Lower and Middle Jurassic have been encountered in many of the Marathon wells (Fig 6).

On Skye and Raasay the Jurassic totals more than 3100ft (945m) while in the Mull–Morvern region it is more than 1000ft (305m) thick. The questionable limits of the Upper Jurassic in both areas make precise thickness determination impossible but it is likely that there are changes of facies and thickness between north and south Skye

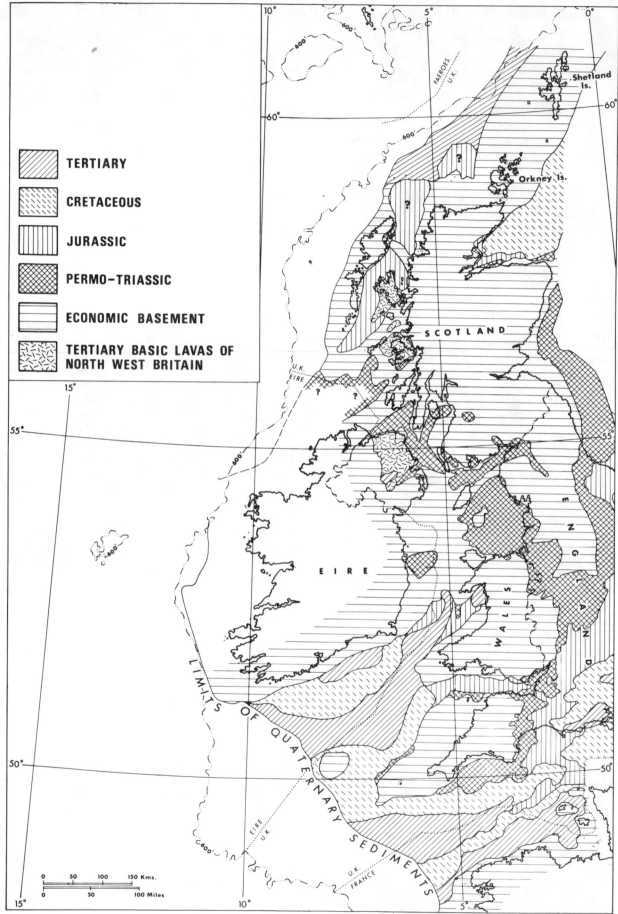

FIG 2.   Generalized outcrop map of west Britain.

(Sykes, *in press*) with a corresponding marked increase in sand percentage in the thicker succession in south Skye. The sandstones of the Middle Jurassic, so called "Great Estuarine Series", are in fact more properly attributed to deltaic deposition (Hudson, 1964). They were apparently derived from the Scottish Highlands and are underlain by a thin oil shale on Raasay and east Skye which was used for distillation and production of limited quantities of crude oil during the Second World War. The exploration of the sub-surface extension of the Jurassic into the North Minch and West Shetland basins is perhaps the most interesting objective on the north-west European shelf in the next few years in the light of the important Jurassic discoveries in the Viking Graben of the North Sea. It must be presumed that the Jurassic is a prime objective for most, if not all, wells drilled to date west of Shetland. Well 205/21-1 encountered "indications of hydrocarbons" and it is interesting to speculate on the age of the reservoir. Well 205/30-1 was apparently abandoned below the Jurassic at a location close to the south-east boundary fault of the West Shetland Basin.

The Liassic exposed in Northern Ireland, near Carlisle and also drilled in Mochras 1 probably represents the erosional remnants of a once much more extensive sequence including Middle and possibly Upper Jurassic, which can be expected to be present in the deeper parts of the Cardigan Bay Basin. In the Bristol Channel Basin it is estimated (Lloyd *et al.*, 1973) that 5250ft (1600m) of Lower, Middle and Upper Jurassic overlie at least 700ft (213m) of Trias. The Jurassic here is dominantly clastic with only thin carbonates—a possible guide to facies in the North Celtic Sea Basin. Certainly the Upper Jurassic drilled in parts of the North Celtic Sea Basin is anomalously thick compared with that at outcrop in south-west England, with a thick Purbeckian–?Portlandian overlying Kimmeridgian. In south-west England the Jurassic aggregates about 4000ft (1219m). Thick sandstones are known in the Middle and Upper Lias and "Corallian", and carbonates are present throughout the sequence. Potential source rocks also occur through the sequence, though they are dominant in the Oxfordian and Kimmeridgian. The recent British Gas discovery at Wytch Farm in Dorset is believed to be from an equivalent of the Upper Liassic Bridport Sand. The offshore extension of Jurassic into the Celtic Sea (*sensu lato*) is unknown but Jurassic objectives must be prime targets through much of this area.

Before leaving the Jurassic, the vexed question of the age and significance of the Cimmerian unconformity may be mentioned. There are a number of unconformities within the Jurassic–Lower Cretaceous. Identification, from seismic sections, of one of these unconformities as *the* Cimmerian is a grossly oversimplified generalization. The exact age of the unconformities in this part of the column clearly reveals a large number of stratigraphic breaks. Equation of any one of these geologically established breaks with the seismically identifiable "Cimmerian" shrouds the problem in greater darkness. It is of passing interest that the name Cimmerian (which is, of course, nothing to do with the Kimmeridgian) is derived from a tribe believed by the ancients to live in perpetual darkness; hence the adjective is now used to denote dense darkness. Perhaps results of drilling and careful stratigraphic analysis will shed light on this densely dark unconformity.

Cretaceous

Upper Cretaceous rocks, often in a marginal sandy chalk facies, are widely but sparsely distributed from the Inner Hebrides through to Northern Ireland. The Upper Cretaceous is next found at outcrop onshore in south-west England where it rests with marked angular unconformity on rocks down to and including Carboniferous. However, the chalk is known extensively from sea-floor sampling in the North Celtic Sea and Western Approaches basin (Curry *et al*, 1970) and is inferred in the Cardigan Bay Basin (Dobson *et al*, 1973) and in the subsurface of both Celtic Sea basins. There is some indication from analysis of interval velocity data from the West Shetland Basin that the Upper Cretaceous, presumed to be present in this area, is in shaly facies rather than the well known chalk facies known in the North Sea. Such a facies change would tie in with the known northward facies change occurring in the northern North Sea from pure chalk (south) to shales (north). Hence the sandy chalks present in the Hebridean outcrops may represent the northerly limit of chalk facies on the west side of the UK.

The Lower Cretaceous is absent at outcrop everywhere except in southern England and the eastern part of the Western Approaches Basin. At the base of the Upper Cretaceous there is everywhere in the area of this paper a major unconformity and it appears that only in the deeper part of the basins are Lower Cretaceous sediments exposed. In the Marathon Kinsale Head gas discovery, it appears that the reservoir rocks (48/25-1 payzone from 2700ft to 3000ft (823 to 914m)) are deltaic sandstones of Lower Cretaceous age similar to those in the Weald area of southern England. The presence of deltaic Lower Cretaceous in the Celtic Sea should be no surprise to those familiar with the stratigraphy of Portugal, northern Spain and the Scotian Shelf of Eastern Canada. It is probable that a marine embayment existed in Lower Cretaceous to the south of the Celtic Sea and the Iberian margin (in a prerotational position). Lower Cretaceous rocks are not definitely proved anywhere else in the area of this paper but are most probably present beneath Upper Cretaceous cover in the deeper parts of some basins, notably the West Shetland Basin. There are scout reports of Lower Cretaceous sandstones present in several West Shetland wells.

Tertiary and Quaternary

The early Tertiary was a time of intense igneous activity in western Scotland and Northern Ireland and the Tertiary volcanic centres have long been studied in detail (Fig 2). Much work by the Institute of Geological Sciences and other bodies has led to a more precise knowledge of the offshore extent of these volcanics and of newly recognized centres such as the Blackstones Plutonic Centre (Binns *et al.*, 1974). Apart from these well-known Tertiary volcanics and associated plutonics, there are thin terrigenous sedimentary layers between sills in Scotland and Ireland. However, Tertiary rocks are notably represented in the Mochras Borehole where some 1740ft (530m) of non-marine Tertiary were encountered. Around Lough Neagh up to 1150ft (351m) of non-marine post volcanic clays and sands of Tertiary age have been drilled in the Washing Bay Borehole.

By contrast, the Tertiary of the Celtic Sea and Western Approaches basins is marine and dominantly carbonate (Curry *et al.*, 1970). Palaeocene (Danian) carbonates are overlain unconformably in the western part of the Celtic Sea basins by Eocene carbonates. Eocene carbonates are overlain in turn unconformably by Miocene carbonates. Oligocene appears to be missing at the sea-floor outcrop (although one possibly Oligocene sample has been recovered). The Oligocene was the period of deposition of the onshore Bovey Tracey clays, the Mochras lignitic

clays and the Wessex basin non-marine limestones, clays and sands. The Plio-Pleistocene and Holocene thicken markedly west of the limit shown on Fig 2 and this makes further construction of a sea-floor outcrop map difficult. On the shelf-edge Eocene and Miocene rocks have been recovered (Curry *et al.*, 1962) and a thick Tertiary sequence, cut by an unconformity of probable Oligocene age, is clearly visible on seismic data in the Western Approaches.

In the West Shetland area there is little evidence from dredge or core of Tertiary rocks although a thick Tertiary sequence can be seen clearly on the seismic sections covering the West Shetland Basin and it is also possible to link seismically with the known Tertiary of the North Sea. Finally, the wells drilled to date in this area have proved a thick Tertiary sequence.

BASIN SUMMARY

The West Shetland complex (Fig 3) comprises the West Shetland Basin *sensu stricto* flanked on its seaward side by a north-east plunging complex basement ridge, on the flanks of which most wells to date have been drilled, and the shelf-edge basin thickening into the Faeroes Trough on the seaward side of the ridge. Published seismic sections across this complex (Whitbread, 1972) indicate the abrupt south-east fault margin to the West Shetland Basin with (to the north-west) thick Tertiary overlying a

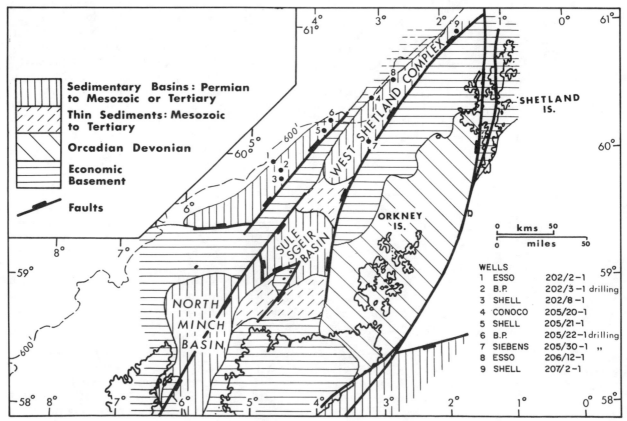

Fig 3. West Orkney and Shetland Basins.

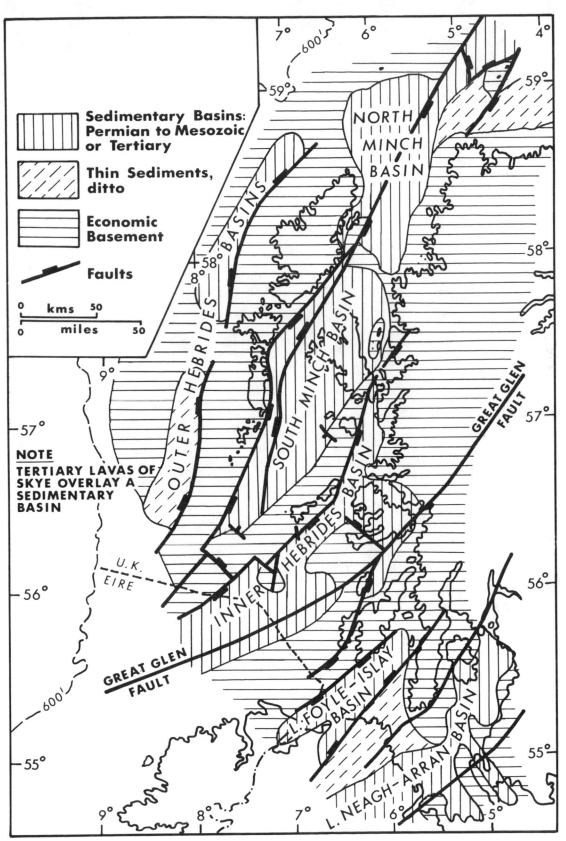

FIG 4.   West Scottish basins.

thick Mesozoic succession to a total thickness probably exceeding 25 000ft (7620m). The sections published to date do not extend seawards north-west across the basement ridge into the deep water of the Faeroes Channel.

South of these linear north-north-east-south-south-west trending basins is the fault-bounded Sule Sgeir Basin with a relatively thin Tertiary sequence overlying thick Mesozoic. To the south-east of this basin, west of the Orkneys, there is a series of north-north-east-south-south-west trending faults bounding small basins with probably a Mesozoic sequence overlying Orcadian Devonian. West of this area lies the North Minch Basin. This basin is transected by the north-north-east-south-south-west trending Minch Fault complex. This fault may have been active as a wrench fault in pre-Mesozoic times but it is clear that it, and its splays, displayed dominantly vertical movement in Permian-Recent times. The sense of movement of the fault has been the subject of much conjecture (Steel, 1971), mainly arguing from the sediment transport directions of the Stornoway beds. The North Minch Basin shallows markedly south of a line joining the Butt of Lewis and Cape Wrath, south to the outcropping ridge of Torridonian which separates this basin from the South Minch Basin. This last basin, together with the Inner Hebrides and Lough Foyle-Islay basins is well described (Binns et al., 1974) from geophysical data and shallow coring (Fig 4). North and west of Ireland there is a lack of knowledge of the distribution of Mesozoic basins apart from the work summarized by Dobson and Evans (1974). From deep seismic data it is clear that a very rapidly seawards-thickening sequence is encountered at or close to the edge of the present continental shelf into the Rockall Trough (for seismic sections of Rockall Trough see Whitbread, 1972). The same seawards-thickening wedge is encountered west of the Outer Hebrides. Over most of the continental shelf west of the Hebrides seismic data shows little or no sediment cover (apart from the shallow Outer Hebrides basins) until the edge of the continental shelf, where Tertiary and probably also Mesozoic sediments thicken rapidly into the Rockall Trough-Faeroes Channel. This situation is in marked contrast to the West Shetland area where very thick sedimentary basins exist on the shelf.

On Fig 5, the Lough Neagh-Arran Basin appears very similar to the Lough Foyle-Islay Basin immediately north-west of it, but the Dalradian of the Kintyre peninsula and its south-westwards continuation in Northern Ireland may well have acted as a partial or even complete barrier during Permian and later time. The Manx-Furness Basin is now relatively well known from IGS and University work (Wright et al., 1971; Bott and Young, 1971). This basin contains a thick Permian and Triassic sequence and the recently drilled 110/2-1 well has proved gas-productive, probably

from the Bunter sandstone. The north-west portion of the basin is fault-controlled and some faulting also appears to separate this basin from the Solway Firth Basin to the north, although a thin Permo-Triassic cover probably exists over the north-east-south-west trending Ramsay-Whitehaven ridge. The basin is also strongly fault-controlled on the east and south margin with a faulted connection to the Cheshire Basin to the south-east. There is no connection south-westwards to either the Kish Bank Basin or the Cardigan Bay Basin. Kish Bank appears to be strongly fault-controlled on the northern and south-western flanks and to have a Permo-Triassic infill. The Cardigan Bay Basin has been described in detail (Dobson et al., 1973) and is strongly fault-controlled on all margins except where it passes south-westwards in to the North Celtic Sea Basin. The infill of the Cardigan Bay Basin ranges from presumed Permian to Tertiary with strong pre-Upper Cretaceous and pre-Neogene unconformities visible on seismic data. There is also strong evidence of presumed Triassic salt activity (Dobson et al., 1973).

In Fig 6, the Celtic Sea—extending from the south Irish coast to Brittany—is subdivided into the North and South Celtic Sea basins (the latter extending eastwards into the Bristol Channel Basin) and the Western Approaches Basin. In Fig 2, the superficial geology of this area is shown out to the limit of Quaternary cover. The North Celtic Sea Basin is best known, from drilling, and here a sequence from Permian to Neogene is well established. So far, there has been no deep drilling at all in the Western Approaches Basins. To the west of all three of these basins the Tertiary thickens considerably and the Palaeocene-Danian appears to be in a similar facies to that in the Ekofisk complex of Norway (Curry et al., 1970). There is extensive faulting on the margins of all these basins trending east-north-east-west-south-west, probably along reactivated ancient lines of weakness first developed in the Armorican orogeny. The west limit of the Pembroke ridge is not known but the feature appears effectively to separate the North and South Celtic Sea basins in the Upper Cretaceous and Tertiary, if not before. A seismic profile crossing the west part of the North Celtic Sea Basin in a north-west-south-east direction has been published (Whitbread, 1972).

EVOLUTION OF THE BASINS TO THE WEST OF BRITAIN

This brief and generalized summary of the basins from West Shetland in the north to the Western Approaches in the south shows the diversity of basin form and fill that exists. The key role of pre-existing fault trends in determining the shape of these basins must be emphasized. Close examination of the geological and geophysical data

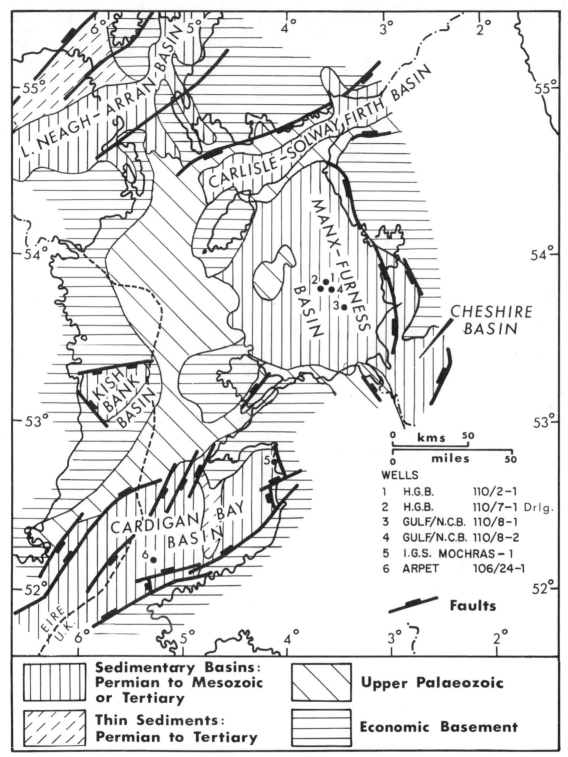

FIG 5.   Irish Sea and Cardigan Bay basins.

covering the basins makes the unification of the basins as previously proposed (Naylor *et al.*, 1974) untenable. There is no evidence for a line of more or less continuous grabenal basins such as may exist beneath the North Sea (which has been attributed to incipient oceanic rifting and basinal collapse during Mesozoic time) down the west coast of Britain. This is not to deny the possible presence of Tertiary *rrr* junctions beneath some of the volcanic/plutonic centres of western Britain, but the present form of the western basins is believed to be a product of intense erosion and active fault movement at several periods during their evolution—notably pre-Upper Cretaceous, post-Upper Cretaceous, pre-Miocene–post-Eocene, and probably also Recent. It is probable that before these periods of erosion Mesozoic sediments were much more widespread over areas

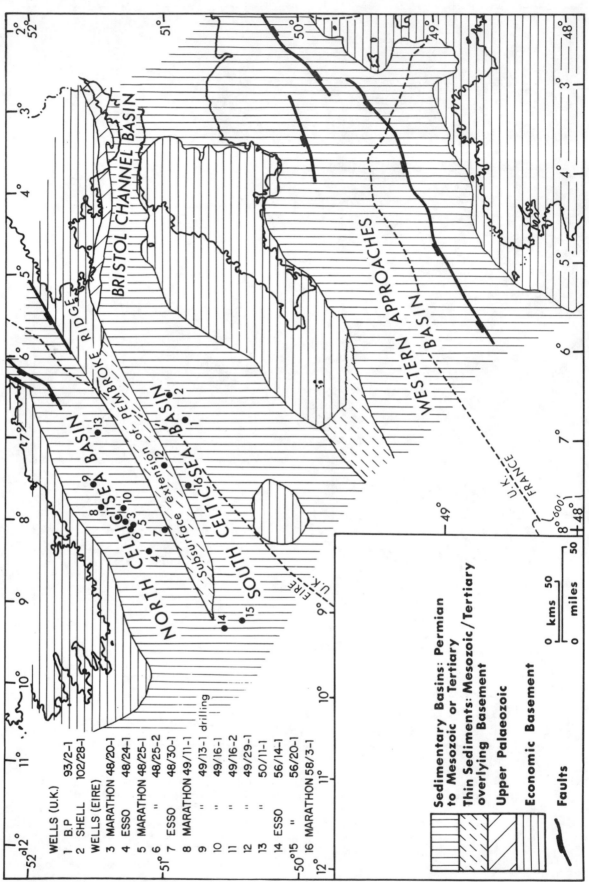

FIG 6. Celtic Sea Basins and Western Approaches.

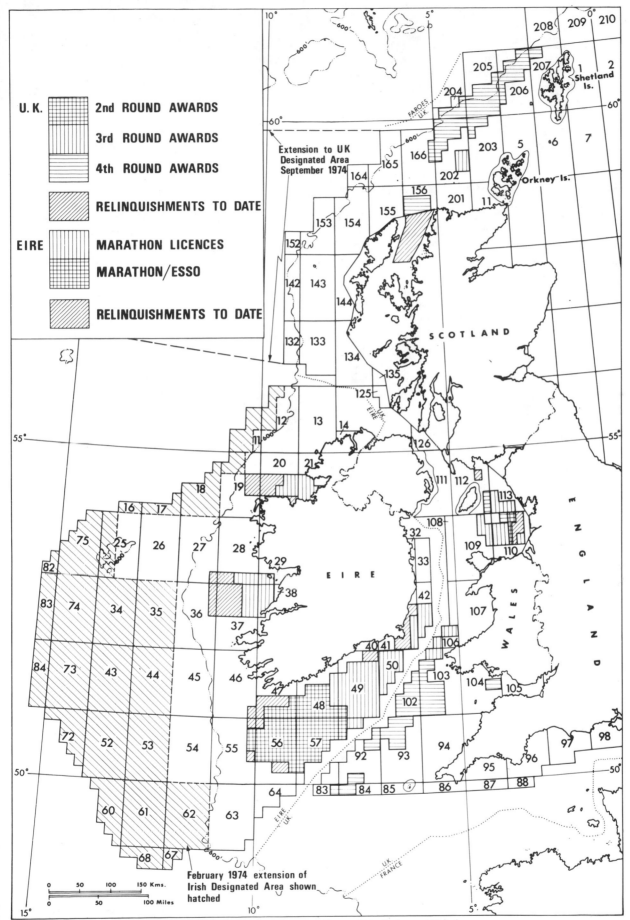

FIG 7. Location of current licensed blocks in west Britain.

now denoted in Fig 1 as "Economic Basement". The Upper Cretaceous remnant preserved at Ballydeanlea in Co. Kerry, Ireland, as a karst breccia is perhaps a pointer to further discoveries of a similar type. A feature common to the continental margin from 62° to 48° N, including the area west of Ireland not discussed here, is the seaward-thickening wedge of Tertiary and probably Mesozoic sediment going into the deeper waters of the Faeroes Channel–Rockall Trough–Porcupine Seabight and (far) Western Approaches.

The evolution of the Rockall Trough will be mentioned elsewhere during this meeting but it appears that oceanic crust is first dateable (from magnetic anomalies) as appearing at the base of the Upper Cretaceous (Bailey *et al.*, 1974). The appearance of dateable magnetic striping must be preceded by a period of influx of truly oceanic waters which probably occurred in this area in the lower part of the Jurassic and this in turn is presumably preceded by regional uplift and probably a period of evaporite formation such as occurred in many of the basins west of Britain during the Triassic. It is salutary in discussing the probable age of separation of continental masses to mention the published results of drilling on the east Canadian shelf (Austin and Howie, 1973; Austin, 1973) where the stratigraphy is relatively well established. It seems that the Faeroes Channel–Rockall Trough opened well before the area west of the Western Approaches. On this basis a marked difference in the ages and facies of the lowest sedimentary fill in the basins here discussed probably exists—that is, between the West Shetland and Western Approaches basins respectively. The role of Porcupine Bank and any possible incipient rifting on the site of Porcupine Seabight is not discussed here—suffice it to say that a thick Mesozoic and Tertiary section exists in the Seabight (Whitbread, 1972) with mobile evaporites which are possibly the eastern counterpart of the Jurassic Argo salt known on the Scotia Shelf (McIver, 1973). The major faulting and evolution of the West Shetland complex seems to be due to relief of stress related to opening of Rockall Trough which may have begun to separate as early as the Triassic.

In summary, no single system of rifting accounts for the present disposition of the basins west of Britain. These basins are related to two or more phases of development of the north-east Atlantic, to extensive faulting and subsidence along pre-existing basement fault trends and to widespread erosion during the Mesozoic, Tertiary and Recent.

HYDROCARBON POSSIBILITIES

Figure 7 shows the present acreage holdings in both British and Irish waters. Because of the highly competitive situation in both the UK and Ireland,

little can be said about the age or type of rocks believed to be most interesting for the petroleum geologist.

In the Kinsale Head discovery, Marathon have established the first economic accumulation of hydrocarbons in the area of this summary. Estimates of recoverable reserves of gas vary from 1–3×10^{12}scf. The field is believed to produce from relatively shallow Lower Cretaceous sandstones which could give rise to production problems in this water depth—about 300ft (91m). British Gas have very recently announced successful testing of gas in the Irish Sea on 110/2-1 but no reserve estimates are possible at this stage. In the West Shetlands, Shell have announced that 205/21-1 found "indications of hydrocarbons". In the Celtic Sea, Esso (on farm-out from Marathon) also found oil indications in 48/24-1. This last report was wildly exaggerated in the Irish press and was re-reported as the discovery of a vast oilfield—a salutary reminder of the power, if not the responsibility, of the fourth estate.

Clearly the reservoirs found productive to date warrant, and will involve, further detailed exploration. In addition many horizons in the Jurassic are most promising objectives and at Wytch Farm (outside the scope of this summary) Liassic sands are believed to be the productive horizon in the British Gas oil discovery. In the light of North Sea success, the whole Jurassic is clearly to be regarded as most interesting, both as reservoir and source. Additional plays exist where presumed source-type Jurassic is thought to be in fault juxtaposition with presumed reservoir Permo-Triassic or perhaps, in some areas, yet older rocks. Finally, but not least in importance, the various objectives in the Lower Tertiary—particularly in the Celtic Sea and West Shetlands, remain to be explored. It is worth remembering that it was Tertiary discoveries in the North Sea which sparked off the present spate of oil discoveries in that area.

ACKNOWLEDGEMENTS

In conclusion, I wish to thank the Chairman and Directors of Cluff Oil for permission to present this summary and to Mr M. B. Scott for critically reading the manuscript.

REFERENCES

Amoco Canada Petroleum Co. & Imperial Oil. 1973. Regional Geology of the Grand Banks. *Bull. Can. Petrol. Geol.*, **21**, 479–503.

Audley-Charles, M. G. 1970. Stratigraphical correlation of the Triassic rocks of the British Isles. *Q. Jl. geol. Soc. Lond.*, **126**, 19–48.

Austin, G. H. 1973. Regional geology of eastern Canada offshore. *Bull. Am. Ass. Petrol. Geol.*, **57**, 1250–1275.

Austin, G. H. and Howie, R. D. 1973. *In:* Hood, P. S., McMillan, N. J. and Pelletier, B. R. (Eds). Earth science symposium on offshore eastern Canada. Regional geology of offshore Canada. *Geol. Surv. Canada Paper*, 71/23, 73–107.

Bailey, R. J., Grzywacz, J. M. and Buckley, J. S. 1974. Seismic reflection profiles of the continental margin bordering the Rockall Trough. *Jl. geol. Soc. Lond.*, **130**, 55–69.

Binns, P. E., McQuillin, R. and Kenolty, N. 1974. The geology of the sea of the Hebrides. *Rep. Inst. Geol. Sci.*, 73/14, 43 pp.

Bott, M. H. P. and Young, D. G. G. 1971. Gravity measurements in the north Irish Sea. *Jl. geol. Soc. Lond.*, **126**, 413–434.

Bruck, P. M., Dedman, R. E. and Wilson, R. C. L. 1967. The New Red Sandstone of Raasay and Scalpay, Inner Hebrides. *Scott. Jl. Geol.*, **3**, 168–80.

Craig, G. Y. 1965. Permian and Triassic. *In: G. Y. Craig. (Ed.), The Geology of Scotland*, Oliver & Boyd, Edinburgh, 383–400.

Curry, D., Martini, E., Smith, A. J. and Whittard, W. F. 1962. The geology of the western approaches of the English Channel; 1. Chalky rocks from the upper reaches of the continental slope. *Phil. Trans. R. Soc. Lond., Series B*, **245**, 267–90.

Curry, D., Hamilton, D. and Smith, A. J. 1970. Geological evolution of the western English Channel and its relation to the nearby continental margin. *In:* Delany, F. M. (Ed.), ICSU/SCOR Working Party symposium, Cambridge. The geology of the east Atlantic continental margin, 2, Europe. *Rep. Inst. Geol. Sci.*, 70/14, 129–42.

Dobson, M. R., Evans, W. E. and Whittington, R. 1973. The geology of the south Irish Sea. *Rep. Inst. Geol. Sci.*, 73/11, 35 pp.

Dobson, M. R. and Evans, D. 1974. The geological structure of the Malin Sea. *Jl. geol. Soc. Lond.*, **130**, 475–78.

Hudson, J. D. 1964. The petrology of the sandstones of the Great Estuarine Series and the Jurassic palaeogeography of Scotland. *Proc. geol. Ass.*, **75**, 499–527.

Kent, P. E. 1975. The tectonic development of Great Britain and the surrounding seas. *Paper 1, this volume.*

Laming, D. J. C. 1966. Imbrication, paleocurrents and other sedimentary features in the Lower New Red Sandstone, Devonshire, England. *Jl. Sedim. Pet.*, **36**, 940–59.

Lee, G. W. and Pringle, J. 1932. A synopsis of the Mesozoic rocks of Scotland. *Trans. geol. Soc. Glasg.*, **19**, 158–224.

Lloyd, A. J., Savage, R. J. G., Stride, A. H. and Donovan, D. T. 1973. The geology of the Bristol Channel floor. *Phil. Trans. R. Soc. Lond., Series A*, **274**, 595–626.

McIver, N. L. 1972. Cenozoic and Mesozoic stratigraphy of the Nova Scotia shelf. *Can. J. Earth Sci.*, **9**, 54–70.

McQuillin, R. and Bacon, M. 1974. Preliminary report on seismic reflection surveys in sea areas around Scotland, 1969–73. *Rep. Inst. Geol. Sci.*, **74/12**, 7 pp.

Naylor, D., Pegrum, R. M., Rees, G. and Whiteman, A. J. 1974. Nordsjøbassengene. The North Sea trough system. *Noroil*, **2**, 17–22.

Steel, R. J. 1971. The New Red Sandstone movement on the Minch Fault. *Nature Phys. Sci.*, **234**, 158–9.

Steel, R. J. 1974. New Red Sandstone flood plain and piedmont sedimentation in the Hebridean province, Scotland. *Jl. Sedim. Pet.*, **44**, 336–57.

Sykes, R. *Scott. Jl. Geol., In press.*

Whitbread, D. R. 1972. *In:* North Sea conferences 1 & 2. The hydrocarbon potential of western Britain and Ireland. *Fin. Times Conference Proceedings*, 81–4.

Wood, A. and Woodland, A. W. 1968. Borehole at Mochras, west of Llanbedr, Merionethshire. *Nature, Lond.*, **219** (5161), 1352–54.

Woodland, A. W. (Ed.). 1971. The Llanbedr (Mochras Farm) Borehole. *Rep. Inst. Geol. Sci.*, **71/18**, 115 pp.

Wright, J. E., Hull, J. H., McQuillin, R. and Arnold, S. E. 1971. Irish Sea investigations 1969–70. *Rep. Inst. Geol. Sci.*, **71/19**, 55 pp.

DISCUSSION

D. N. Holt (Freeman, Fox and Partners): (1) Are any of the possibly productive basins we have been hearing about on topographic lows on the Continental Shelf in water depths exceeding 100–150m?

(2) What basins if any run across the shelf onto the Continental slope and beyond?

(3) What is the outer limit beyond which everything is below the economic basement? Does this, in accordance with plate tectonics and Continental Drift theory coincide with the edge of the Continental Shelf?

The reasons for these questions is that such environments require new and different technology of exploitation such as Excess Buoyancy or Tension-leg production-platforms, and speaking as one of those who is currently involved in the development of such projects I would like to get some idea what the future demands for such equipment is likely to be.

Dr D. R. Whitbread: (1) Although many of the basins described west of Britain are represented in part by bathymetric lows on the continental shelf, there is no straightforward relationship between bathymetry and basin outline. The complications of Quaternary sedimentation superimposed on these Permian-Mesozoic-Tertiary basins make any straightforward correlation between bathymetric low and structural basin impossible. For the record, it is of interest to note that for the Persian Gulf there is little or no correlation between bathymetric features and hydrocarbon prospective structures. There too, present-day bathymetry largely reflects Quaternary sedimentation although, with hindsight only, it is sometimes possible to identify hydrocarbon occurrences with bathymetric highs.

(2) West of Britain, three basin complexes appear to cross the continental shelf and to sub-crop on the continental slope. These are, from north to south, (i) the westerly basin of the west Shetland complex, (ii) the fault-bounded basins that appear to be associated with the extension of the Great Glen Fault west of northern Ireland, and (iii) the western extension of the Celtic Sea and Western Approaches Basin.

(3) This question is difficult to understand and appears to show a misuse of the concept of "economic basement". West of Britain, in the Faeroe Channel, the Rockall Trough and the south-west and south extension of this Trough, a thick sequence of Tertiary and possibly also of late Mesozoic rocks is shown on seismic reflection records. "Economic Basement", in the sense of rocks deemed incapable of containing economic quantities of hydrocarbons, appears to lie at great depths below this Tertiary and ?Mesozoic cover. Perhaps part of the problem is that west of Britain, "Economic Basement", in the sense in which this term is used in the North Sea, may well be of a different (and younger) age than the Carboniferous that is taken as "Economic Basement" in the North Sea.

Areas such as the Anton Dohrn Seamount show a thin sedimentary cover over what is assumed to be rocks non-prospective for hydrocarbon accumulation.

D. G. Roberts: Water depths of 100–150m occur locally in the Irish Sea. Practically the whole of the Sea of the Hebrides Basin south of Skye lies under water depths greater than 150m, and depths over most of the North Minch Basin exceed 100m.

Although the less resistant rocks of the sedimentary basins have been eroded to form hollows in the base-Quaternary surface, the irregular distribution of Quaternary sediment, which ranges from 0 to 250m, is an important secondary control on water depth.

P. E. Binns: The Mesozoic basins on the shelf west of the Outer Hebrides, the Sea of the Hebrides and Malin Sea are associated with topographic lows due in part to glacial effects in water depths exceeding 100–150m. The relationship of these basins to the shelf and slope is not entirely clear, though the geophysical and geological evidence shows a series of small (?Mesozoic) basins developed along a splay of the Great Glen Fault which can be traced to the shelf-edge. Bearing in mind the probable Middle Jurassic-Lower Cretaceous age of the rift movements that structured the margin, it is possible that down-faulted remnants of these basins may underlie the slope. One example may be to the west of Shetland as observed by Dr Whitbread. It is difficult to define the outer limit beyond which everything is below "Economic Basement". The continent–ocean boundary in the Rockall Trough is effectively marked by the 2.0-second isopach which may correspond to such a limit.

4

Recent Developments in the Geology of the Irish Sea and Cheshire Basins

By V. S. COLTER and K. W. BARR

(British Gas Corporation)

SUMMARY

Recent exploratory drilling in the Irish Sea has confirmed the existence of a Permo-Triassic basin analogous to the Cheshire Basin.

Offshore wells have found a Triassic marl/salt/sandstone sequence of over 2400m, overlaying a Permian sequence of some 500m, comprising an upper shale/evaporite series equivalent to the Manchester Marl and a lower shale/sandstone Collyhurst Sandstone equivalent. These beds rest on Carboniferous rocks.

Sparse microfloras do not allow precise chronostratigraphical correlations to be made. Lithostratigraphical correlations between offshore wells and the outcrops of Cumbria and Lancashire are possible, but difficulties arise in extending some correlations into the central and southern Cheshire Basin. Specifically, well evidence suggests a rapid sanding-out of the Manchester Marl away from the outcrop, but whether or not the absence of distinct Collyhurst Sandstone and Manchester Marl at the southern end of the basin is due to this process, or to onlap by younger beds, remains unclear.

INTRODUCTION

In view of the close geological and geographical relationships of the Irish Sea and Cheshire basins it is proposed to treat them as a single geological province and to discuss their stratigraphies and structures concurrently. Both are primarily Permo-Triassic basins, and it is with the Permo-Triassic formations and their subsequent geological history that this account is principally concerned.

For the present purpose the Irish Sea is defined as the region lying between south-western Scotland (the Southern Uplands), the coast of Cumbria and Lancashire, the north coast of Wales, and a line between Anglesey and the Isle of Man. The Cheshire Basin is well known and needs no further definition at this stage.

In the Irish Sea, geophysical surveys, sea-floor sampling and four recent offshore exploratory wells have confirmed the existence of a Permo-Triassic basin, somewhat analogous to the adjacent Cheshire Basin, where seismic surveys and recent deep drilling have also shown the presence of a thick Permo-Triassic section.

ACKNOWLEDGEMENT

The work of the Institute of Geological Sciences, and several universities geophysical studies, especially those of M. H. P. Bott (1964, 1965, 1968), have contributed to the delineation of the Irish Sea basin, and this published information has been incorporated in the present review. The permission to make use of the exploratory well data supplied by British Petroleum, British Gas Corporation, Gulf/National Coal Board and Trend Exploration Ltd is gratefully acknowledged.

REGIONAL GEOLOGICAL FRAMEWORK

The Irish Sea and Cheshire basins may be regarded as two Permo-Triassic basins connected by a relatively shallow sill between the Welsh Palaeozoic massif and the Carboniferous uplift of the Lancashire Coalfields. During the Permo-Triassic these two basins behaved in much the same way and had comparable histories.

The regional setting of the Irish Sea and Cheshire Basins is shown in Fig 1. The northern margin of the Irish Sea Basin is formed by the Lower Palaeozoic rocks of the Southern Uplands of Scotland. To the east lie the Carboniferous uplands of the western Pennines and the lower Palaeozoic uplift of the Lake District. The southern margin is formed by the Lower Palaeozoic uplift of North Wales, with fringing Carboniferous, and further west lies the Precambrian complex of Anglesey. The western margin is controlled by the Isle of Man–Anglesey structure, which sea-floor sampling has shown to be composed of Carboniferous, Lower Palaeozoic and Precambrian rocks (Wright *et al.*, 1971). Geophysical evidence (Bott, 1964, 1965, 1968; Wright *et al.*, 1971) indicates the existence of a buried Palaeozoic ridge connecting the Isle of Man to the Whitehaven area at the north-west of the Lake District Uplift. This feature effectively subdivides the Irish Sea Basin proper from a northern sub-basin dominated by an extension of the Solway–Carlisle syncline.

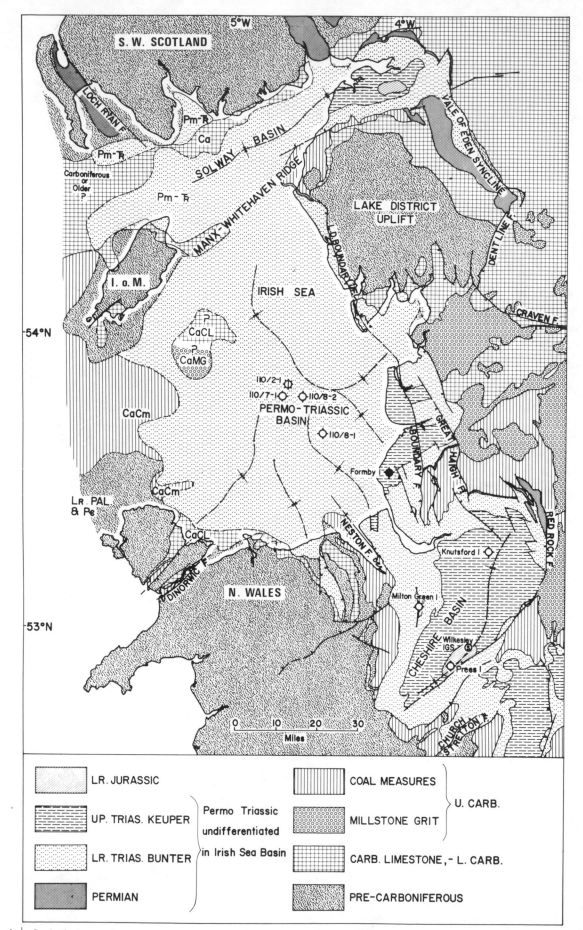

FIG 1. Geological map of the Irish Sea—Cheshire Basin region. (Based on IGS "Ten-mile" Geological Map of Great Britain, and Irish Sea data from Wright *et al.*, 1971.) It should be noted that the sub-divisions of the Permo-Triassic as mapped are not entirely in accord with the stratigraphy adopted in this paper as presented in Figs 3 and 4.

The related Cheshire Basin is similarly flanked by Carboniferous and Lower Palaeozoic rocks, comprising the Lancashire Coalfield to the north, the Welsh–Shropshire uplands to the west and south, and the Pennines to the east. In common with other Permo-Triassic basins in and around the British Isles, the Irish Sea and Cheshire basins owe their inception to the block faulting, uplift and down-warping which marked the later stages of the Variscan orogeny. During the earlier stages of this orogeny the Carboniferous and underlying Palaeozoic rocks of this region were folded along both Caledonoid (east-north-east–west-south-west) and Pennine (north–south) axes (Fig 1). Superimposed on these is a system of north-north-west–south-south-east faults (Charnian trend), probably initiated in pre-Permian times and reactivated later, the most important result of which was the formation of several graben structures, such as the Vale of Eden and the Vale of Clwyd. Moreover, it appears probable that the Irish Sea Basin itself owes its formation to graben-faulting along a series of major faults having a similar trend, some representatives of which have been later reactivated as the marginal faults along the boundary of the Carboniferous uplifts of Lancashire (Boundary Fault, Great Haigh Fault, etc.) and the Lake District Boundary Fault. Other such faults within the basin are the Neston Fault along the Dee estuary, and further north those controlling the parallel alignment of the Permian outcrops of Loch Ryan and Luce Bay. North of the Manx–Whitehaven Palaeozoic ridge the basin morphology is, however, dominated by the Solway sub-basin. On the western margin between the Isle of Man and Anglesey faulting appears to play only a minor role, and the basin edge is characterized by an on-lap of Permo-Triassic on to older rocks.

The Cheshire Basin is in essence a half-graben, having an inherited Caledonoid (north-east–south-west) alignment, in which basin formation was controlled by subsidence along the Red Rock Fault and its south-western extension. The latter is *en echelon* with, and may be related to, the Church Stretton system. Again, as in the Irish Sea Basin, the western margin is one of sedimentary onlap.

STRATIGRAPHY

The stratigraphy of the Cheshire Basin established on the basis of outcrop and shallow borehole data can now be extended into the subsurface, following the drilling of the two deep wells Knutsford No. 1 and Prees No. 1 by the British Petroleum–British Gas group and Trend Exploration Ltd respectively.

In the case of the Irish Sea some stratigraphic control was already provided round the rim of the basin by outcrops, shallow borehole data and by the BP Formby wells. This has now been supplemented by the results of the offshore wells drilled by Hydrocarbons GB Ltd, a subsidiary of British Gas Corporation (110/2-1 and 110/7-1), and the Gulf/NCB group (110/8-1 and 110/8-2), which have revealed the presence of a Permo-Triassic succession comparable in many ways to that of the Cheshire Basin. The stratigraphy and correlation of the Irish Sea, Cheshire Basin and related areas are given in Fig 3. A correlation diagram of the well sections in the Irish Sea and Cheshire Basin is shown in Fig 4.

The problems of Permo-Triassic stratigraphic nomenclature are as yet not fully resolved (Kent, 1970; Ager, 1970; Pattison et al., 1973; Smith et al., 1974), particularly those involving long-range correlations and the recognition of time–stratigraphic units. These problems are inherent in the prevailing continental regime in which the sediments accumulated. Palynological studies on samples from the new wells have been particularly disappointing. For these reasons, the traditional rock stratigraphic names are used herein, and in the discussion no attempt is made to treat Permian sediments separately from those of the Trias. The Permo-Trias boundary is here put conventionally at the base of the Bunter Pebble Beds, but in the absence of any internationally agreed boundary in the European marine successions, the choice of this horizon, even if it is a time-line, is merely a convenience. In those sections in which the Pebble Beds are absent, even this expedient is impossible. In correlating the wells, use has largely been made of the gamma-ray/sonic logs, which permit the recognition of certain lithological and diagenetic zones in the sandstones, as well as a wide-ranging gamma-ray marker, which is here correlated with the base of the Keuper Sandstone.

Pre-Permo-Trias

Figure 1 shows the outcrops of pre-Permo-Triassic rocks around the two basins and Fig 2 shows an interpretation of the Pre-Permian subcrop. Of the deep wells drilled, the Gulf/NCB 110/8-1 well found beneath the Permian sediments 7m of barren red beds overlying shales and fine- to medium-grained sandstones whose possible age ranges from Namurian to Westphalian B. The well 110/8-2 found reddened beds of uncertain age, but of Namurian or Westphalian aspect. Knutsford No. 1 in the Cheshire Basin found shales, sandstones and coals of Westphalian age, and Prees No. 1 at the southern end of the same basin found 161m of Upper Westphalian reddened shales overlying Lower Palaeozoic (Ordovician/Silurian).

At the outcrop, the lowest unit of the Permo-Trias, the Collyhurst Sandstone, rests on Carboniferous beds ranging in age from Namurian (Millstone Grit) to late Westphalian (Ardwick group) which testifies to the amount of uplift and erosion that took place prior to the initiation of the Permo-Triassic sedimentation.

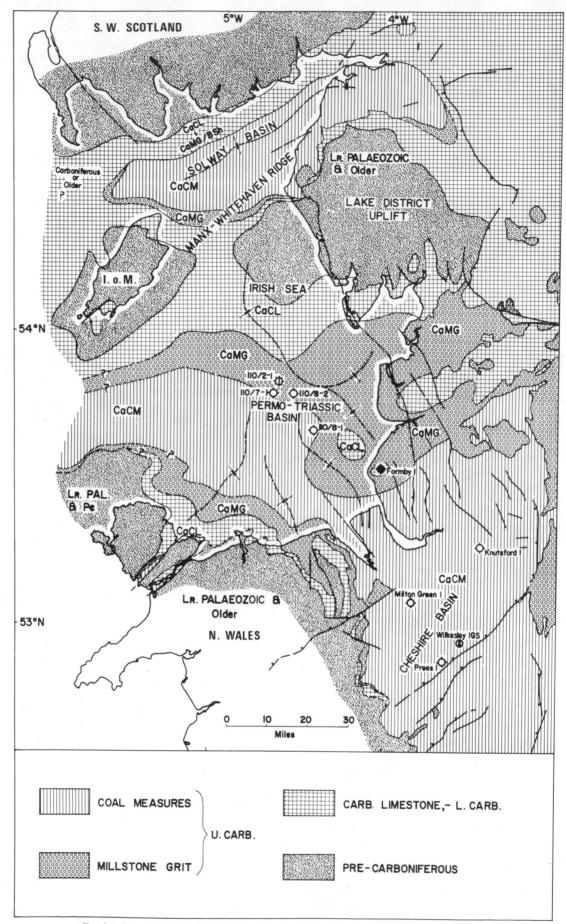

FIG 2. Subcrop map of pre-Permo-Triassic rocks, Irish Sea—Cheshire Basin region.

New Red Beds Beneath the Keuper Sandstone

The Permo-Triassic rocks of the Cheshire and Irish Sea Basins beneath the Keuper Sandstone can be regarded as being composed of a thick body of red sandstones, sub-divided by: (a) beds laid down during a marine incursion from the north west, comprising the Manchester Marl/St. Bees Shale with associated and correlative carbonates and evaporites; (b) by higher conglomerates originating from the south, the Bunter Pebble Beds (Fig 3).

These two formations are not both present throughout the two basins. The presence or absence of these beds has resulted in the use of differing names for the associated and equivalent

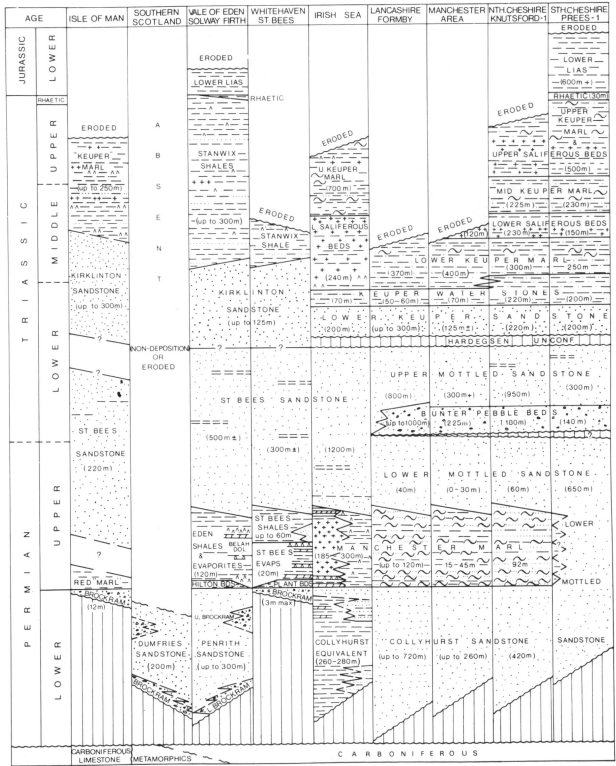

FIG 3.　Stratigraphic correlation chart; Irish Sea, Cheshire Basin and adjacent areas.

sandy beds in various places. Thus, in areas where the Manchester Marl is present, a lower sandstone formation—the Collyhurst Sandstone—can be separated from the Lower Mottled Sandstone, whereas in areas where the Manchester Marl is absent, the term "Lower Mottled Sandstone" includes all sandstone between the Bunter Pebble Beds and the Carboniferous. Similarly, in the presence of the Bunter Pebble Beds, the upper sandstone is called the "Upper Mottled Sandstone", whereas in the absence of the Pebble Beds, the sandstones overlying the Manchester Marl become the undifferentiated "St Bees Sandstone" of the northern outcrop and Irish Sea (Fig 3).

Collyhurst, Lower Mottled and Penrith Sandstones

These sandstones, occurring above, below and lateral to the Manchester Marl and its facies equivalents, and below the Bunter Pebble Beds, are a continental series and appear often to be characterized by intervals of dune bedding and "millet-seed" sandstone (Wills, 1951, plate xii, B).

The Collyhurst Sandstone crops out in places around the northern rim of the Cheshire Basin (between Manchester and St Helens), along the eastern rim of the Irish Sea Basin (north-west side of the Lancashire Coal Field), and is known in coal borings and in the Formby wells. It rests unconformably on rocks ranging from Upper Coal Measures to Millstone Grit, and is variable in thickness, ranging from a few metres to over 250m in the Manchester area (Tonks et al., 1931; Poole and Whiteman, 1955; Magraw, 1966), and from 0 to 720m in the Formby wells (Kent, 1948; Falcon and Kent, 1960). The variable thickness has been attributed to deposition over a deeply dissected topography, and to deposition associated with active fault scarps. The sandstones are generally light-coloured, often with well-rounded, frosted grains ("millet seed sandstone") and dune-bedding. Conglomerates and pebbly sandstones are present locally. Similar sandstones, to judge from drill cuttings, were penetrated in the Knutsford No. 1 well.

No separate Manchester Marl and basal Permo-Triassic sandstone are distinguishable at the southern outcrop of the Cheshire Basin or in the Prees No. 1 well. Correlations based on certain well log and lithological criteria in the Upper Mottled Sandstones and Bunter Pebble Beds between Prees No. 1 and Knutsford No. 1 suggest that in the former the Lower Mottled Sandstone includes both a part of the Collyhurst Sandstone and an arenaceous equivalent of the Manchester Marl (Fig 4). Thompson (1970a, Fig 2) shows that west of Warrington and into the Wirral Peninsula, the Manchester Marl is also absent and consequently an undifferentiated Lower Mottled Sandstone lies between the Bunter Pebble Beds and the Coal Measures.

In the Gulf/NCB Irish Sea well 110/8-1, the Collyhurst Sandstone interval is represented by an alternating sandstone and shale section, whilst the equivalent beds in the nearby well 110/8-2 consist entirely of shale and siltstone. This shaling out of the Collyhurst Sandstone towards the centre of the basin is reminiscent, albeit on a smaller scale, of that in the Rotliegendes of the Southern North Sea Basin and north-west Germany. Hence it is inferred that the central part of the basin is predominantly shale, more or less surrounded by a sandstone facies around the uplifted margins (Fig 5a). On the north-eastern margin of the Irish Sea Basin, the basal Permian clastics are represented by the thin breccias of the Brockram seen at Whitehaven, which consists of locally derived Lower and Upper Palaeozoic fragments. In the Vale of Eden, a Lower Brockram breccia is overlain by the aeolian Penrith Sandstones, which include an Upper Brockram. In some instances, in the Penrith Sandstones, silica cement in optical continuity with the grains is present (Eastwood, 1953). Farther north, in Dumfriesshire and Kirkcudbrightshire, comparable aeolian sandstones are found, which in some cases overlie basalt lavas and which may range down to Westphalian "E" in age (Mykura, 1965, Fig 3).

The regional distribution of the various lithologies is summarized in the palaeofacies map Fig 5a.

Manchester Marl and St Bees Shales and Evaporites

The Manchester Marl of the outcrop follows the Collyhurst Sandstone with apparent conformity, and consists of shales, marl, calcareous siltstones and thin calcareous sandstone layers. Some of the calcareous beds are fossilferous, with an impoverished Upper Permian Bakevellia fauna (Pattison et al., 1973). In thickness this formation varies from 16 to 75m in the Manchester area, and to over 215m in the Formby wells, where it is quite sandy.

In the Cheshire Basin the marl of the outcrop has apparently been rapidly replaced basinwards by sandy beds with minor shale layers, penetrated in Knutsford No. 1. The absence of the Manchester Marl between Warrington and the Wirral Peninsula (Thompson, 1970b, Fig 2) may be due to a complete change from shale to sandstone. As mentioned above, the equivalents, if any, of the Manchester Marl in Prees No. 1 and at the southern outcrop are not discernible. However, the view that the Lower Mottled Sandstone of the southern end of the Cheshire Basin includes the equivalents of the Collyhurst Sandstone and the Manchester Marl is supported by subsurface correlations. Parallelism between the top of the Carboniferous, the base and top of the Bunter Pebble Beds, and the top of a silicified zone in the Upper Mottled Sandstone in the Knutsford No. 1 and Prees No. 1 wells suggests

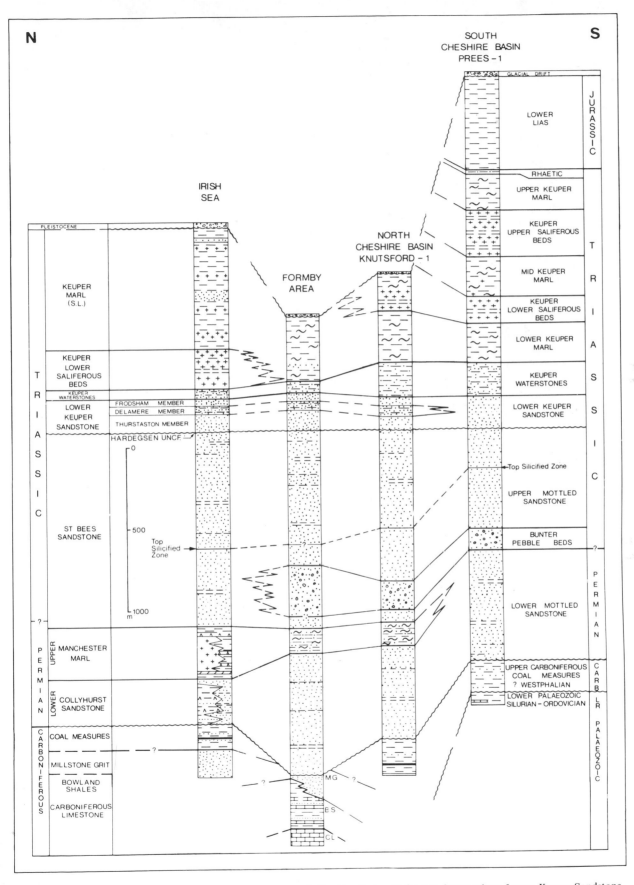

FIG 4.   Correlation diagram based on well data; Irish Sea and Cheshire Basins. Reference datum = base Lower Keuper Sandstone
(Hardegsen unconformity).

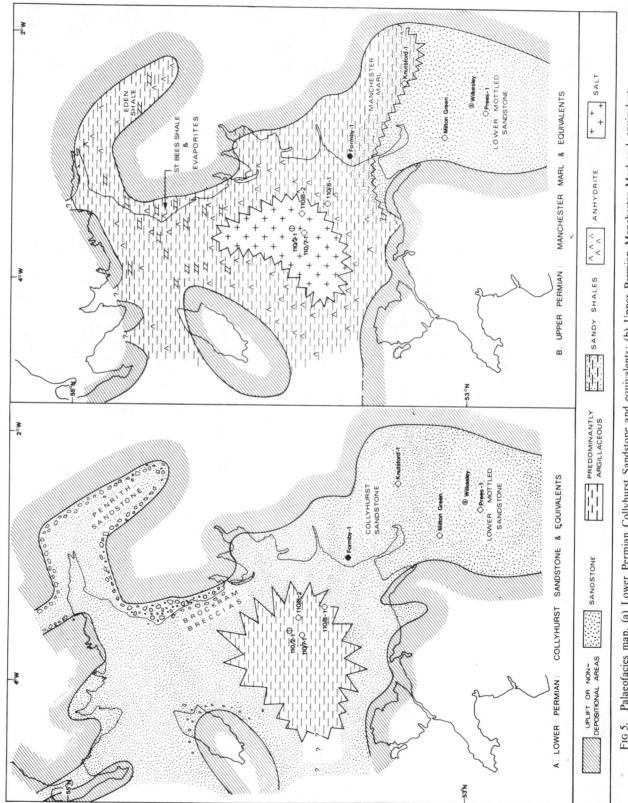

Fig 5. Palaeofacies map. (a) Lower Permian Collyhurst Sandstone and equivalents; (b) Upper Permian Manchester Marl and equivalents.

that, in the latter, little of the section beneath the Bunter Pebble Beds has been lost either at the base by southward onlap onto an old land area, or at the top by erosion at the base of the base of the Pebble Beds (Fig 4).

In the Irish Sea Basin, predominantly shaly sections occur again in the well 110/8-1 and in northern Isle of Man. Between these occurrences, in well 110/8-2 the equivalent section is largely replaced by salt. No transitional carbonate/sulphate facies between the shale and salt is known from the subsurface in the immediate area, but the Whitehaven boreholes described by Arthurton and Hemingway (1972) indicate the existence on the rim of the basin of an area of inter- to supra-tidal carbonate and anhydrite sediments, comprising the St Bees Evaporites. These carbonates and evaporites are followed by the fine clastics of the St Bees Shales.

In the Vale of Eden, a comparable succession, the Eden Shales (Arthurton, 1971), overlies the Penrith Sandstone. These beds have at their base the Hilton Plant Beds, and contain frequent anhydrite and gypsum interbeds, as well as a 6m-thick carbonate, the Belah Dolomite, in the upper part.

The palaeofacies map (Fig 5b) shows the interpreted distribution of the lithological units and emphasizes the relatively rapid transition from sandstone to shale in the Cheshire Basin and from shales to predominant salt in the centre of the Irish Sea Basin.

Bunter Pebble Beds

The Bunter Pebble Beds rest unconformably on the questionably Permian Bridgnorth Sandstone south of the Cheshire Basin, and in places onlap onto Permian breccias and older rocks (Pattison *et al.*, 1973). These authors tentatively extend this unconformity into South Lancashire and North Cheshire (*op. cit.*, Table 2).

This formation has also been described in Cheshire (Fitch *et al.*, 1966) as having been deposited as "braided river lag gravel, channel-bar and swale-fill sediments of a large subaerial alluvial fan". Lithologies represented include sandstones, sandstones with shattered pebbles and massive conglomerates. Current directions and grain-size distribution indicate derivation from the south and south-east. Among the characteristic components are the well-known pebbles of liver-coloured quartzites, whilst some pebbles have been identified as having come from the Brittany area.

In the subsurface of the Cheshire Basin, in Knutsford No. 1 and Prees No. 1, precise identification of the upper and lower limits of this formation is hampered by the difficulty of recognizing fragments of scattered pebbles in drill cuttings. In both wells, the base of the unit is picked at an abrupt decrease in porosity above the Lower Mottled Sandstone, as seen on sonic logs; the top is picked at the last pebbles seen in the samples. Liver-coloured and milky quartzite pebbles have been seen in Knutsford No. 1 and Prees No. 1. In both of these wells, the Bunter Pebble Beds seems to have been affected by silicification, which also affects the lower part of the overlying Upper Mottled Sandstone.

Pebble beds have also been recognized in the Formby Wells (Kent, 1948), but not in the 110/8-1 and 2 wells of the Irish Sea. The HGB wells 110/2-1 and 110/7-1 were not drilled deep enough to investigate possible Bunter Pebble Beds equivalents. Pebble beds are also missing at Garstang in Lancashire, at a similar latitude to the Gulf wells, and from the Solway and Eden Basins, which is consistent with the inferred southern origin of these conglomeratic beds.

Upper Mottled Sandstone and St Bees Sandstone

These formations include all sandstones beneath the Keuper Sandstone and above the Bunter Pebble Beds (Upper Mottled Sandstone), or above the Manchester Marl in the absence of the Pebble Beds (St Bees Sandstone).

Evidence regarding the environment of deposition of the sandstones is reviewed by Audley-Charles (1970b). In general, at the outcrop, although cross-bedding is present, evidence of aeolian deposition is rare. Fluviatile and lacustrine conditions were postulated by Fitch *et al.*(1966) in Cheshire, although these authors also detected aeolian influence near the top of the Upper Mottled Sandstone. No information on bedding characteristics is available from the Formby wells, Knutsford No. 1 or Prees No. 1, but the bulk of sand grains are angular to subrounded. The occurrence, however, of well-rounded, frosted sand grains, correlated with zones of better porosity suggests some aeolian influence here also.

The St Bees Sandstone of the 110/8-2 well appears, from the gamma-ray log, to be rather more shaly in general and to have more distinct shale laminae than that of the Knutsford well. This may possibly be related to distance from the source of sediment, as in the case of the shaling out of the Collyhurst Sandstone, and the replacement of the Manchester Marl by salt in the Irish Sea Basin, although the equivalent section in Prees No. 1 also appears somewhat shaly.

A feature of the Upper Mottled Sandstone of the Formby wells, Knutsford No. 1 and Prees No. 1, and the St Bees Sandstone of the 110/8-2 well is a zone of silicification, which is seen as a decrease to uniformly low values in porosity on the sonic logs. This zone, which includes the Pebble Beds where present, maintains a striking uniformity of thickness, being 505m in Prees No. 1, 459m in Knutsford No. 1 and 480m in 110/8-2 (Fig 4). It is not present in the structurally higher well 110/8-1. Approximately 305m of silicified sandstone was reported by well-site geologists in the lower part of the section in Formby No. 6 and silicification was reported from Formby No. 4 (BP confidential report), but no sonic logs exist with which to correlate these units with those in the other wells.

The uniformity in thickness of this silicified zone in wells so far apart, and its absence in the shallow section of well 110/8-1, suggests the effect of overburden on some originally susceptible depositional unit. The particular susceptibility is implied by the absence of silicification in the underlying Lower Mottled and Collyhurst Sandstones of Knutsford. In Prees No. 1, the Lower Mottled Sandstone immediately beneath the silicified Bunter Pebble Beds shows higher porosities, but a gradual decrease in porosity is seen in this unit, until at the base the values are similar to those in the Pebble Beds and Upper Mottled Sandstone. This also suggests the effect of overburden.

The top of the Upper Mottled Sandstone has been correlated with the widespread intra-Bunter event, the Hardegsen unconformity (Warrington, 1970; Fig 3). Some support is given to this view by the correlations from the top of the Carboniferous to the base of the Keuper Sandstone in Fig. 4, which suggest that most of the additional section in Knutsford No. 1, compared to Prees No. 1, occurs between the top of the silicified zone, or certainly the top of the Pebble Beds, and the base of the Keuper Sandstone. However unlikely it may be that the lithological markers represent time planes, their parallelism is impressive. Wherever the thinning in the Prees succession may have taken place, it is evident that the depocentre during the accumulation of the succession from base Collyhurst Sandstone to top Upper Mottled Sandstone does not coincide with the present structural low in the Prees area.

Keuper Series

The upper part of the Triassic sequence, broadly grouped as the "Keuper", seems by its greater lateral persistence and uniformity to reflect more stable conditions throughout the region; it represents the final stages of continental degradation prior to the Jurassic marine transgression. With the exception of the initial Keuper Sandstone, the sediments are mainly fine-grained clastics representing end-cycle sedimentation. In essence they consist of continental-type sandstones, including aeolian dune sands, siltstones, mudstones and shales, locally with evaporitic intercalations, and with two well-defined evaporitic phases. The succession is shown in Fig 3. Following Warrington (1970b), the base of the "Middle" Trias is taken at about the top of the Keuper Waterstones Formation, and the base of the "Upper" Trias at the base of the Upper Saliferous Formation. As Warrington stresses, these Triassic lithofacies units are probably diachronous and their correlations with time stratigraphic stages are hazardous.

The base of the series is now regarded as being the Hardegsen unconformity (Warrington, 1970b) although the evidence for this episode is not very strong in and around the Cheshire Basin, still less in the Irish Sea. In places, however, the basal beds of the Keuper Sandstone are coarse and occasionally conglomeratic (Edwards and Trotter, 1954), suggesting some rejuvenation of the source areas.

Keuper Sandstone

The Keuper Sandstone of the Cheshire Basin has been described in detail by Thompson (1970a, 1970b) and his sub-divisions and lithofacies interpretations are followed.

Thompson recognizes three basic lithofacies units: (a) a lower, partly aeolian, partly fluviatile unit, the *Thurstaston Member*, whose general facies is comparable to that of the underlying Bunter Upper Mottled Sandstone; (b) a middle fluviatile unit characterized by red pebbly sandstones with interbedded silts and shales, the *Delamere Member*; and (c) an uppermost predominantly aeolian sandstone unit, the *Frodsham Member*, which includes some spectacular dune-bedded sandstones. In some of these sandstones, current-bedding and other sedimentary features are indicative of fluviatile deposition, probably as outwash fans or braided flood-plain deposits. Periodically, during the earlier stages, and more or less completely during the later stages, fluviatile conditions were replaced by widespread dune fields.

In the wells Prees No. 1, Knutsford No. 1, Formby No. 6 and 110/8-2, the base of the Keuper Sandstone is picked by the writers at a gamma-ray log marker which separates the Upper Mottled Sandstone from an overlying less radioactive unit. In the North Sea, the Hardegsen unconformity (top Bunter Sandstone of Geiger and Hopping, 1969) is often marked by the presence immediately above of a conspicuous gamma-ray marker. This marker is also seen in the onshore wells in eastern England figures by Balchin and Ridd (1970; Fig 3). The top of the Keuper Sandstone as shown in Fig 5 is picked at another gamma-ray change to the more argillaceous Waterstones. Gamma-ray/sonic logs through the interval thus identified as Keuper Sandstone in Knutsford No. 1, Formby No. 6 and 110/8-2 show a threefold division namely: (a) a lower cleaner sandstone, (b) a more argillaceous sand with shale bands and (c) an upper cleaner sandstone. Correlation of these units with Thompson's Thurstaston, Delamere and Frodsham Members is supported by both the sample descriptions and sonic/porosity logs from Knutsford No. 1, which show that more porous sandstone with aeolian grains are confined to the upper and lower sandstones, whereas the middle member is characterized by lower porosity. It should be noted, however, that these subdivisions are less apparent in wells 110/2-1 and 110/7-1. In Prees No. 1 to the south, this threefold division of the Keuper Sandstone is not recognizable, which is consistent with Thompson's observation that the

Delamere Member is absent at the southern end of the Cheshire Basin (Thompson, 1970b; Fig 6).

Farther north in Cumbria and the Isle of Man, the Keuper Sandstone, as well as the overlying Waterstones and Lower Keuper Marl, apparently pass laterally into a continuous sandstone sequence—the Kirklinton Sandstone, which also has some aeolian ("millet seed") bands.

Keuper Waterstones

The overlying Keuper Waterstones follow transitionally, and as a whole seem to represent a passage series into the succeeding finer clastics of the Keuper Marl. The Waterstones, as typically developed, consist of a flaggy alternation of shales, mudstones and fine sandstone layers, red, purplish to chocolate-brown in colour. At outcrop they range in thickness from 70 to 220m. They are probably mainly lagoonal deposits, possibly with some intertidal influence (Fitch et al., 1966), and may thus be a feeble reflection of the Muschelkalk transgression of the North Sea–Germanic Basin.

In wells Prees No. 1, Knutsford No. 1, Formby No. 6, and 110/8-2, the Waterstones unit is identified by a marked increase in gamma-radiation above the Keuper Sandstone. The top is somewhat arbitrarily picked, as the formation is gradational to the overlying beds—the Lower Keuper Marl in the Cheshire basin and the Lower Saliferous Beds in the Irish Sea. In the Irish Sea wells the top is taken by the writers at the lowest occurrence of salt of the Lower Saliferous Beds; in the Formby and Cheshire Basin, however, where the Keuper Marl intervenes, the location of the top is somewhat arbitrary (Fig 3).

Keuper Marl

The "Upper" Triassic of the Cheshire–Lancashire area consists of a sequence of red, brown and variegated mudstones and shales, sometimes partly calcareous or dolomitic. Sporadic inclusions or lenses of salt and anhydrite are often present. In addition there are two intervals of evaporitic deposits, consisting mainly of salt, comprising the Upper and Lower Saliferous Beds, as seen in the Prees and Wilkesley wells (Fig 4). These are considered to represent the deposits of extensive salt-lakes which came to occupy the low-lying plains of this region. It seems unlikely that these halites could have resulted from short-lived marine incursions as the carbonate–evaporite sequences characteristic of normal marine evaporitic series are not present.

The central Irish Sea wells contain only a clearly defined Lower Saliferous member which directly overlies the Waterstones equivalent in wells 110/2-1 and 110/8-1, although thin scattered salt-beds are present in the overlying Keuper Marl (Fig 4). It is therefore not clear whether the Lower Keuper Marl of the Cheshire Basin has been laterally replaced by Saliferous beds, or whether there is a local disconformity which has eliminated

the Lower Keuper marl in the Irish Sea. Given the transient nature of the saliferous–lacustrine facies the former appears more probable. When traced northwards from Lancashire around the Lake District uplift and into the Solway–Vale of Eden sub-basin, the entire Upper to Middle Triassic sequence is represented by the Stanwix Shales, a fairly thick sequence (up to 300m) with thin sandstones and thin salt or anhydrite beds and stringers in places. The Stanwix Shales appear to be lithologically very similar to the Keuper Marl, but lack extensive evaporites. A similar sequence of red beds occur in shallow boreholes in the northern Isle of Man (Eastwood, 1953).

Rhaetic

Thin shales, and siltstones with thin limestone stringers totalling some 13m crop out around the margin of the Prees–Wem (Jurassic) syncline at the southern end of the Cheshire Basin, and were penetrated in the Wilkesley borehole (Poole and Whiteman, 1966). A comparable thickness (21m) was also found in Prees No. 1 well. All of the other Irish Sea and Cheshire Basin wells under review started in stratigraphically older formations.

Although there is a small Lower Jurassic outlier around Carlisle in the Solway Basin, no Rhaetian has apparently been identified at the base of this sequence.

Jurassic

Two small outliers of Lower Liassic shales occur, one around Prees and Wem in the southern part of the Cheshire Basin, and the other around Carlisle in the Solway Basin. These are the youngest beds of the region. In Prees No. 1, 598m of Lower Jurassic shales were penetrated; the maximum thickness in the axis of the syncline is somewhat in excess of this figure. The presence of Lower Jurassic at the southern end of the Cheshire Basin, some distance from the deepest part of the Permian–early Triassic depositional basin suggests a relative displacement of the basin axis in late to post-Triassic times, as previously discussed under "Upper Mottled Sandstone".

Although no Jurassic has been found in any of the Irish Sea Wells—which were perforce drilled on structural uplifts—it is possible that some Rhaetian and Jurassic deposits may be present in the structurally deeper parts of the Irish Sea Basin.

AGES OF PERMO-TRIASSIC SEDIMENTS

Permian

The occurrence of the Manchester Marl and other fine clastic, carbonate and evaporite beds, above the largely coarse clastics of the Collyhurst and equivalent sandstones, is suggestive of a relationship similar to that of the Rotliegendes and the Zechstein of the southern North Sea Basin and north-west Germany.

Precise equivalence is difficult to demonstrate, in view of the virtual absence of reliable age indicators in the lower clastics, and the presence in the overlying beds of fossils of meagre stratigraphic significance. The finer clastics, carbonate and evaporites of the Manchester Marl and in other similar units appear on faunal grounds to have been laid down under marine conditions in the so-called "Bakevellia Sea", named after the bivalve *Bakevellia bicarinata*. If the transgression of this sea over the dune sands of the Collyhurst and other sandstones is taken to be a manifestation of a Zechstein transgression caused by a eustatic rise in sea-level, then the correlation suggested by the similarities of successions here and in the North Sea may be correct. Certainly, the general aspect of the *Bakevellia* fauna indicates Zechstein affinities (Pattison *et al.*, 1973).

Arthurton and Hemingway (1972) quote evidence from the St Bees evaporites that suggests an EZI to lower EZII age for these beds. These authors also state that similar floras to those from the St Bees evaporites have been recorded from the Hilton Plant Beds of the Vale of Eden sub-basin. The dolomite of this basin, however, is stated by Bennison and Wright (1969) to have affinities with the Upper Magnesian Limestone (EZIII) of the Zechstein Basin. Dr J. C. M. Taylor (personal communication) has noted in this dolomite the abundance of *Calcinema*, a form normally characteristic of the Upper Magnesian Limestone in the Zechstein basin.

It is impossible to discern in wells and boreholes in these two basins the four-fold subdivision of the Zechstein that is clearly seen over such a large area of the North Sea and the adjacent Dutch and German subsurface. This may mean that after the first transgression the subsequent eustatic sea-level changes producing the widespread Zechstein cycles did not affect the more isolated Bakevellia Sea, or that the Irish Sea evaporites, the Manchester Marl and their equivalents represent only a portion of the Upper Permian.

The Permo-Trias boundary is conventionally taken at the base of the Bunter Pebble Beds, where present, on account of the unconformity at the base of this formation in the Midlands (Wills, 1950; Pattison *et al.*, 1973). In the absence of pebble beds, however, even this imperfect subdivision is impossible.

Triassic

The subdivision of the Triassic sequence is beset by the same difficulties as that of the Permian, largely arising from the lack of diagnostic floras or faunas inherent in the prevailing continental facies.

Pattison *et al.*(1973) have reviewed the available evidence concerning the early Triassic, and Warrington (1970b) has discussed the Keuper stratigraphy of the British Isles and related it to that of the European sequences. The results of the recently drilled wells have not brought any new

evidence which would modify their conclusions, and the lithologic successions herein described conform tolerably well with the established stratigraphy. Warrington (1970a) suggested that the Waterstones Formation may be diachronous across the Irish Sea, Cheshire and Midlands basins, being older in the north than in the south. Unfortunately the well data so far examined are inconclusive on this point.

REFERENCES

Ager, D. V. 1970. The Triassic system in Britain and its stratigraphical nomenclature. *Q. Jl. geol. Soc. Lond.*, **126**, 3–13.

Arthurton, R. S. 1971. The Permian evaporites of the Langwathby Borehole, Cumberland. *Rep. Inst. Geol. Sci.*, **71/17**, 18 pp.

Arthurton, R. S. and Hemingway, J. E. 1972. The St Bees Evaporites—a carbonate–evaporite formation of Upper Permian age in west Cumberland, England. *Proc. Yorks. Geol. Soc.*, **38**, 565–92.

Audley-Charles, M. G. 1970a. Stratigraphical correlation of the Triassic rocks of the British Isles. *Q. Jl. geol. Soc. Lond.*, **126**, 19–49.

Audley-Charles, M. G. 1970b. Triassic palaeogeography of the British Isles. *Q. Jl. Geol. Soc. Lond.*, **126**, 50–81.

Balchin, D. A. and Ridd, M. F. 1970. Correlation of the younger Triassic rocks across eastern England. *Q. Jl. geol. Soc. Lond.*, **126**, 91–102.

Bennison, G. M. and Wright, A. E. 1969. *The geological history of the British Isles*. Edward Arnold (London), 406 pp.

Bott, M. H. P. 1964. Gravity measurements in the north-eastern part of the Irish Sea. *Q. Jl. geol. Soc. Lond.*, **120**, 369–96.

Bott, M. H. P. 1965. The deep structure of the northern Irish Sea, a problem of crustal dynamics. *Colston Pap.*, **17**, 179–201.

Bott, M. H. P. 1968. The geological structure of the Irish Sea Basin. *In: D. T. Donovan (Ed.), Geology of Shelf Seas*. Oliver & Boyd, Edinburgh, 93–113.

Bott, M. H. P. and Young, D. G. G. 1971. Gravity measurements in the north Irish sea. *Q. Jl. geol. Soc. Lond.*, **126**, 413–32.

Eastwood, T. 1953. Northern England. *British Regional Geology. Geol. Surv.* H.M. Stationery Office, London, 72 pp.

Edwards, W. and Trotter, F. M. 1954. The Pennines and adjacent areas. *British Regional Geology. Geol. Surv.* H.M. Stationery Office, London, 86 pp.

Falcon, N. L. and Kent, P. E. 1960. Geological results of petroleum exploration in Britain, 1945–57. *Geol. Soc. Lond., Mem. No.* **2**, 56 pp.

Fitch, F. J., Miller, J. A. and Thompson, D. B. 1966. The Palaeogeographic significance of isotopic age determinations on detrital micas from the Triassic of the Stockport–Macclesfield district, Cheshire. *Palaeogeogr. Palaeoclimatol. Palaeoecol.*, **2**, 281–313.

Geiger, M. E. and Hopping, C. A. 1968. Triassic stratigraphy of the Southern North Sea Basin. *Phil. Trans. R. Soc. Series B.*, **254**, 1–36.

Kent, P. E. 1948. A deep borehole at Formby, Lancashire. *Geol. Mag.*, **84**, 2–18.

Kent, P. E. 1970. Triassic rocks of the British Isles: Introductory remarks. *Q. Jl. geol. Soc. Lond.*, **126**, 1.

Magraw, D. 1961. Exploratory boreholes in the central part of the south Lancashire coalfield. *Min. Engr. (Lond.)*, **120**, 432–48.

Mykura, W. 1965. The age of the lower part of the New Red Sandstone of S.W. Scotland. *Scott. Jl. Geol.*, **1**, 9–17.

Pattison, J., Smith, D. B. and Warrington, G. 1973. A review of late Permian and early Triassic Biostratigraphy in the British Isles. In: A. Logan and L. V. Hills, The Permian and Triassic systems and their mutual boundary, *Can. Soc. Petrol. Geol. Mem.*, **2**, 220–60.

Poole, E. G. and Whiteman, A. J. 1966. Geology of the country around Nantwich and Whitchurch. *Mem. Geol. Surv. Gt. Br.* H.M. Stationery Office, London, 154 pp.

Poole, E. G. and Whiteman, A. J. 1955. Variations in thickness of the Collyhurst Sandstone in the Manchester Area. *Bull. geol. Surv. Gt. Br. No.* **9**, 33–41.

Smith, D. B. 1972. The Lower Permian in the British Isles. *In: G. H. Falke (Ed.), Rotliegend, Essays on European Lower Permian*, 1–33, E. J. Brill, Leiden, 229 pp.

Smith, D. B., Brunstrom, R. G. W., Manning, P. I., Simpson, S. and Shotton, F. W. 1974. Permian: a correlation of Permian rocks in the British Isles. *Geol. Soc. Lond., Special Report, No.* **6**, 45 pp.

Thompson, D. B. 1970a. Sedimentation in the Triassic Red Pebbly Sandstones in the Cheshire Basin and its margins. *Geol. Jl.*, **7**, 183–216.

Thompson, D. B. 1970b. The stratigraphy of the so-called Keuper Sandstone Formation. *Q. Jl. geol. Soc. Lond.*, **126**, 151–79.

Tonks, L. H., Jones, R. C. B., Lloyd, W., Sherlock, R. L. and Wright, W. B. 1931. Geology of the Manchester Area and the S.E. Lancashire Coalfield. *Mem. Geol. Surv., Gt. Br.* H.M. Stationery Office, London, 240 pp.

Warrington, G. 1970a. The "Keuper" Series of the British Trias in the northern Irish Sea and neighbouring areas. *Nature, Lond.*, **226**, 254–6.

Warrington, G. 1970b. The stratigraphy and palaeontology of the Keuper Series of the central Midlands of England. *Q. Jl. geol. Soc. Lond.*, **126**, 183–217.

Wills, L. J. 1970. The Triassic succession in the central Midlands and its regional setting. *Q. Jl. geol. Soc. Lond.*, **126**, 225–83.

Wright, J. E., Hull, J. H., McQuillin, R. and Arnold, S. E. 1971. Irish Sea Investigations 1969–70. *Rep. Inst. Geol. Sci.*, **71/19**, 55 pp.

DISCUSSION

J. G. Pratsch (Mobil Oil, Germany): What are the reasons for the term "Hardegsen" unconformity between Keuper and Buntsandstein in the Cheshire Basin, as the name "Hardegsen" already is in stratigraphic usage (for one of the Buntsandstein sand bodies) in onshore Germany? Continuation of the use of the name could lead to confusion.

V. S. Colter: The term "Hardegsen Disconformity" was used by Geiger and Hopping (1968) in their paper on the Triassic of the Southern North Sea Basin. The same name was used by Pattison *et al.* (1973) who refer also to Trusheim's use of the name (Trusheim, F. 1963, Zur Gliederung des Buntsandsteins; *Erdöl-Z. Bohr-u. Fordertech.*, **79**, pp. 277–92). Several other speakers at the Conference have also referred to the "Hardegsen Unconformity". It thus seems that the term is well established on both sides of the North Sea, despite Dr Pratsch's objections to its use. Dr Pratsch does not suggest a more correct or appropriate name for this stratigraphical break, whose existence seems well documented and which we assume that he does not question.

5

Tectonic and Stratigraphic Evolution of the Rockall Plateau and Trough

By DAVID G. ROBERTS

(Institute of Oceanographic Sciences)

SUMMARY

The Rockall Plateau is an extensive shallow-water area located between Iceland and the British Isles, from which it is separated by the 3000m deep Rockall Trough. Rockall Island is composed of 52 ± 9my aegirine-granite.

The Rockall Plateau is a continental fragment or micro-continent isolated during the sea-floor spreading evolution of the North Atlantic Ocean. Syntheses of the plate-tectonic evolution of the Atlantic show that three distinct phases of sea-floor spreading "hived off" and structurally isolated the Rockall Plateau. The earliest phase contemporaneously opened the Rockall Trough and Bay of Biscay in mid-early Cretaceous time and thereby split the Greenland–N. America–Rockall plate off Eurasia. By 76my, spreading had ceased in the Rockall Trough and spreading along the Labrador Sea split apart Greenland–Rockall and North America. At 60my, spreading about the incipient Reykjanes Ridge split Greenland and Rockall apart thereby completing the isolation of the Rockall Plateau.

Seismic reflection profiling has revealed three major sedimentary basins: The Hatton–Rockall Basin, the Rockall Trough and the Porcupine Seabight. All three basins are rifted and exhibit a history of progressive and/or intermittent subsidence that correlates closely with spreading rate. The structural and stratigraphic development of the basins is discussed in relation to these factors. Post Upper Eocene sedimentation throughout the area was characterized initially by widespread chert deposition and subsequently by differential deposition of Miocene-through-Recent oozes under the influence of ocean bottom currents.

INTRODUCTION

The continental margin west of the British Isles is a complex feature (Fig 1). The shelf area consists of an outer shelf and the partly enclosed Sea of the Hebrides, Celtic Sea, Bristol Channel and Irish Sea. West of Ireland, the Porcupine Ridge is separated from the shelf by the Porcupine Seabight. Farther west, the Rockall Plateau microcontinent and possibly contiguous Faeroes microcontinent (Roberts, 1971; Bott *et al.*, 1971, 1974) are separated from the continental margin of the British Isles by the Rockall Trough and Faeroe–Shetland Channel. The latter features are divided at 60° N by the Wyville–Thomson Ridge.

This review presents an outline of the tectonic and stratigraphic development of the Rockall Plateau, Rockall Trough and the continental margin west of the British Isles between 50° N and 60° N. Areas not discussed in detail include the Sea of the Hebrides, English Channel and the other shelf seas reviewed elsewhere in these proceedings. The review is divided into the following sections: pre-Permian structural framework; post-Carboniferous–pre-Rifting development; evidence of rifting from the shelf area; sea-floor spreading history; stratigraphic development of the Rockall Plateau and Trough. The data used in this review are presented in detail in Roberts (1974, *in press*). It should be borne in mind that, with the exception of two DSDP holes on Rockall Plateau, no deep stratigraphic data are yet available for the margin, Rockall Plateau and Trough. The tentative stratigraphy given herein is based largely on the DSDP data, the relationships of seismic reflectors to oceanic basement dated by magnetic anomaly identification and to the limited sampling referred to in the text.

PRE-PERMIAN STRUCTURAL FRAMEWORK

The major pre-Permian structural elements of onshore and offshore areas are shown in Fig 2. The north-east–south-west trending Caledonian mobile belt is the dominant feature and separates a north-western foreland composed of Precambrian granulites, overlain by Late Precambrian and Cambro–Ordovician sediments, from a south-east foreland shown by the Precambrian inliers of England and Wales. The fourth major structural division is the east-north-east–west-south-west trending Hercynian orogenic belt of southern England, Wales and Ireland.

Laxfordian granulites sampled from the Rockall Bank have established the continental nature of the Rockall Plateau and the westwards continuation of the north-west foreland (Roberts *et al.*, 1973a). Geophysical data and the occurrence of Grenvillian granulites on the Rockall Bank suggest the Grenville front may cross the Rockall Bank near 56° 20′ N ultimately to intersect the Caledonian front on the shelf west of Scotland (Miller *et al.*, 1973).

Important Upper Palaeozoic faulting (Fig 2)

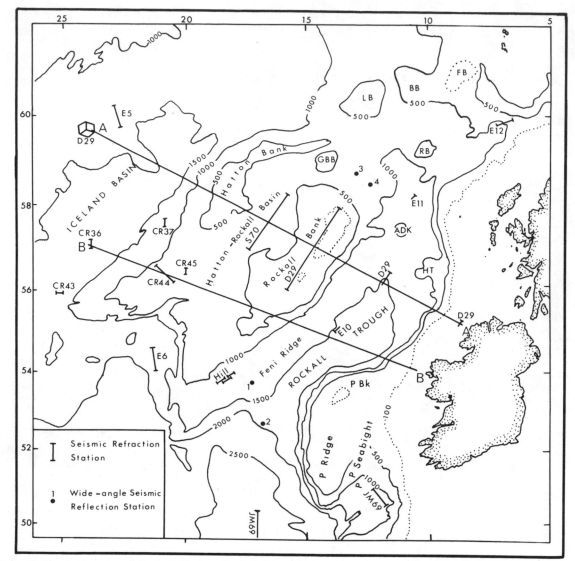

FIG 1.   Outline bathymetry of Rockall Plateau, Rockall Trough and continental margin west of the British Isles. Section AA is illustrated in Fig 4. FB: Faeroe Bank; LB: Lousy Bank; RB: Rosemary Bank; ADK: Anton-Dohrn Seamount; HT: Hebrides Terrace; PBk: Porcupine Bank; P. Ridge: Porcupine Ridge; P. Seabright: Porcupine Seabight.

includes north-east–south-west sinistral strike-slip faulting, important in subsequent Mesozoic basin development, exemplified by the Great Glen Fault, Minch Fault and Strathconnon Fault (Kennedy, 1946; Dearnley, 1962; Winchester, 1973; Garson and Plant, 1972; McQuillin and Binns, 1973). New geophysical data has clarified the offshore extension of these faults and their relationship to the margin. South-west of Mull, the Great Glen Fault splays into two branches which respectively continue west-south-west to the offset in the margin at 55° 30′ N and through Ireland as the Leannan–Leck Fault system (Riddihough, 1968; Roberts, 1969; Holgate, 1969; Pitcher, 1969; Bailey et al., 1974). West of Ireland, the important Highland Boundary Fault or Fold may be shown as a prominent west-south-west–east-north-east trending magnetic lineament (Vogt and Avery, 1974), whose southerly position may reflect

sinistral displacement by the Leannan–Leck Fault system (Fig 2).

In south-west Ireland, the Hercynian front is marked by overthrusting but to the north of the front the geology shows Lower Carboniferous limestone with many inliers of older rocks uplifted by Hercynian movements. The offshore extension of the Hercynian front is uncertain. Although Cherkis et al. (1973) have postulated a structural relationship with the Gibbs Fracture Zone, the eastwards truncation of the Gibbs Zone at anomaly-32 does not support such a connection (Roberts, 1974, in press). Day and Williams (1969) have suggested that free-air gravity minima in the western Celtic Sea show the Cornubian granites continue beyond the shelf to underlie the Goban spur. However, recent remapping of the free-air gravity by the Hydrographer of the Navy suggests that these minima may arise from small sub-basins

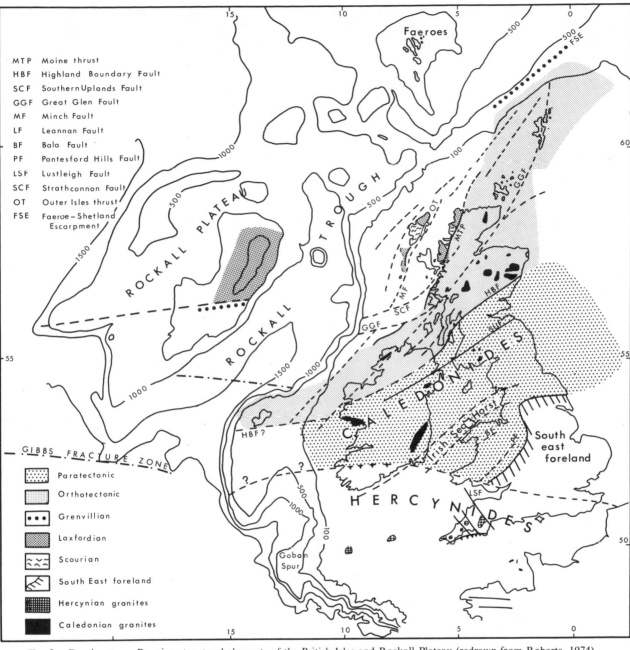

FIG 2.   Dominant pre-Permian structural elements of the British Isles and Rockall Plateau (redrawn from Roberts, 1974).

within the main Celtic Sea Basin. South of the English Channel, the geological and geophysical evidence suggests that the Hercynian (Variscan) arc extends to the shelf edge (*e.g.* Hill and Vine, 1965; Le Mouel *et al.*, 1971; Andreiff *et al.*, 1972).

POST-CARBONIFEROUS–PRE-RIFTING MOVEMENT

Kent (1969) has pointed out that the earliest basin-faulting in the North Sea occurred in Rotliegendes time and was followed by Zechstein evaporite deposition. However, the faulting and early basin development may be related to post-Hercynian orogenic collapse rather than extension on a continental scale. Audley-Charles (1970a, b) and Kent (1975) have reviewed the palaeogeography and stratigraphy of the Trias. The pattern of Triassic basin-development may have been largely controlled by Hercynian and Caledonian trends. In the Minch area for example, sedimentation studies by Steel (1971) are consistent with deposition in a subsiding trough controlled either by the Minch Fault or nearby Outer Isles thrust. Farther south, thick Permo-Triassic sediments are present in the Antrim Basin and may also be present in the Slyne Trough Basin and in basins developed adjacent to the Great Glen Fault (Eden *et al.*, 1970; Bailey *et al.*, 1974; McQuillin and Binns, 1973; Roberts, 1974).

The precise relationship of these basins to the margin and their extent beneath the slope is largely unknown. Gravity maps of the closure of gravity minima toward the outer shelf in the north-west approaches (Roberts, 1969, 1974) suggests that the basins are truncated in the shelf edge area. However, other basins lie sub-parallel to the slope, *e.g.* the Slyne Trough (Fig 6) and remnants may speculatively be present beneath the slope.

JURASSIC AND CRETACEOUS PRECURSORS OF MARGIN DEVELOPMENT

In post-Toarcian time, several stratigraphic and structural events that bear closely on the formation of the margin and the Rockall Trough took place.

In Middle Jurassic time, the main vertical movement on the Great Glen Fault (Bacon and Chesher, *in press*) was paralleled by rifting in the North Sea (Ziegler, *in press*). In the Minches, closures of isopachs drawn on the base of the Jurassic indicate Jurassic deformation. In the English Channel area, Donovan (1972) has noted a progressive easterly tilt that began in the Middle Jurassic and continued into the post-Wealden–pre-Albian folding. Important changes in sedimentation include the deposition of non-marine Bajocian and Bathonian. In north-west Scotland, estuarine deposition of the Middle Jurassic followed Caledonian lines (Hudson, 1964). Towards the end of the Jurassic, there was a period of widespread faulting that continued through the Wealden in the central English

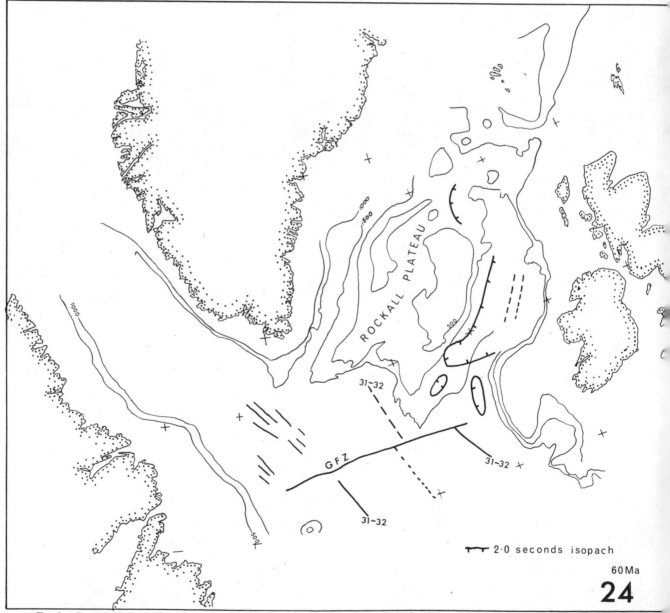

FIG 3.   Reconstruction of the Atlantic Ocean at 60my (anomaly-24). The fit has been made to the 2.0-sec isopach in the Rockall Trough. Heavy black lines indicate oceanic magnetic anomalies. GFZ: Gibbs fracture zone.

Channel (Curry *et al.*, 1970; Donovan, 1972; Boillot *et al.*, 1972), where it may be related to the post-Wealden–pre-Albian (pre-Aptian?) folding (Kent, 1975). In southern England and the English Channel, the restricted non-marine Wealden facies may have had a westerly provenance from a "high" oriented roughly parallel to the present continental margin. Allen (1969) has reported isotopic ages of detrital micas from the basal Cretaceous consistent with a North American or Iberian provenance. The period of late Jurassic to Lower Cretaceous folding and tilting was followed by a period of erosion succeeded by transgression from the east (and west) even onto the Precambrian.

The broad pattern of Jurassic and early Cretaceous stratigraphy and deformation, although observed distant from the margin, may weakly reflect the rifting and spreading that structured the continental margin (Roberts, 1974; Hallam, 1971). The faulting was evidently strongly influenced by Caledonian and Hercynian trends, and a rift-highland west of the British Isles may be evidenced by contemporaneous deposition of the Wealden facies. Evidence of associated volcanism is sparse but may include the 109 to 165my Jurassic volcanics of the North Sea (Howitt *et al.*, 1975), the 134 ± 2my Wolf Rock phonolite (Cowperthwaite *et al.*, 1972) and the possibly associated Bathonian bentonite tuff (Hallam and Sellwood, 1968). On the facing west side of the Atlantic, the Budgell Harbour Stock of Newfoundland has been dated at 137my (Helwig *et al.*, 1974). A late Jurassic to early Cretaceous age is also implied by the physical barrier perhaps shown by the striking differences between the Middle Jurassic ammonite faunas of East Greenland and the British Isles and by the absence of the Mediterranean faunas of the Valanginian of East Greenland in the North Sea (Callomon *et al.*, 1972). It is also worth noting that the better-documented Mesozoic stratigraphy of East Greenland contains a closely similar tectonic history to that reviewed above and may be in mirror image to the British Isles (*e.g.* see Birkelund *et al.*, 1974).

SEA-FLOOR SPREADING

Analyses of the plate-tectonics evolution of the North Atlantic Ocean based on the identification of oceanic magnetic anomalies show that three phases of spreading "hived off" and structurally isolated the Rockall Plateau continent (Vogt *et al.*, 1969, 1971, 1974; Le Pichon *et al.*, 1972; Laughton, 1971).

The spreading history has been deduced by successive closures of the Atlantic Ocean to isochrons defined by oceanic magnetic anomalies (Figs 3–5). The oldest magnetic anomaly abutting the west margin of the Rockall Plateau is 60my and trends north-eastwards to the west edge of the Faeroes Platform. These data and the closely comparable ages of Lower Tertiary igneous

activity in Greenland and the Faeroes (Beckinsale *et al.*, 1970; Tarling and Gale, 1968) suggest that the latest spreading between Greenland and the Rockall Plateau began at about 60my. Between 60my and the present, changes in spreading rate and direction occurred at about 35my and between 22 and 10my (Vogt and Avery, 1974).

The previous spreading phase began at about 76my (anomaly-32) and spread the Greenland–Rockall continent away from North America to form the embryo Labrador Sea (Fig 4). The rectilinear south-west margin of Rockall Plateau formed at this time and may be related to a fracture-zone localized perhaps by the Grenville front. Closure of the Labrador Sea to anomaly-32 effectively juxtaposes Greenland–Rockall and North America with the gaps between continents arising from the uncertain position of the transition between continental and oceanic crust in the Labrador Sea.

The first phase of spreading opened the Rockall Trough separating the European plate from the combined Rockall–Greenland–North America plate. There has been considerable difficulty in determining the age of this phase. However, the absence of post-76my oceanic magnetic anomalies in the Rockall Trough and Faeroe–Shetland Channel affirms that these areas and the margin west of the British Isles are older than 76my. Further the termination of the Gibbs Fracture Zone at anomaly-32 suggests a different pre-76my spreading pattern. There has also been difficulty in determining the evolution of the Rockall Trough during this phase because of the uncertain age and crustal structure.

Some of these difficulties can be resolved from gravity, seismic reflection, refraction and magnetic studies. In the central Rockall Trough, refraction stations give oceanic basement velocities and a Moho depth of 14km independently supported by gravity models (Jones *et al.*, 1970; Scrutton, 1972). Between 57° N and the Gibbs Fracture Zone, Jones and Roberts (in preparation) have defined a linear north to north-north-west trending magnetic anomaly pattern approximately contained within the 2.0-second isopach. On the west side of Rockall Trough, the 2.0-second isopach defines a prominent north-north-west trending sub-surface escarpment that separates areas of contrasting basement relief, magnetic character and crustal thickness and is believed to mark the approximate continent–ocean boundary (Figs 6 and 7). The width of oceanic crust so defined is approximately 200km. At 54° N changes in trend of basement isopachs and truncation of the magnetic anomaly pattern suggests a series of north-north-west–east-south-east trending fracture zones, perhaps marked by the westward offset in the margin north of Porcupine Bank. Recognition of these fracture zones permits an acceptable fit of the Rockall Plateau to the continental margin that does not require rotation of Porcupine Bank (Fig 5).

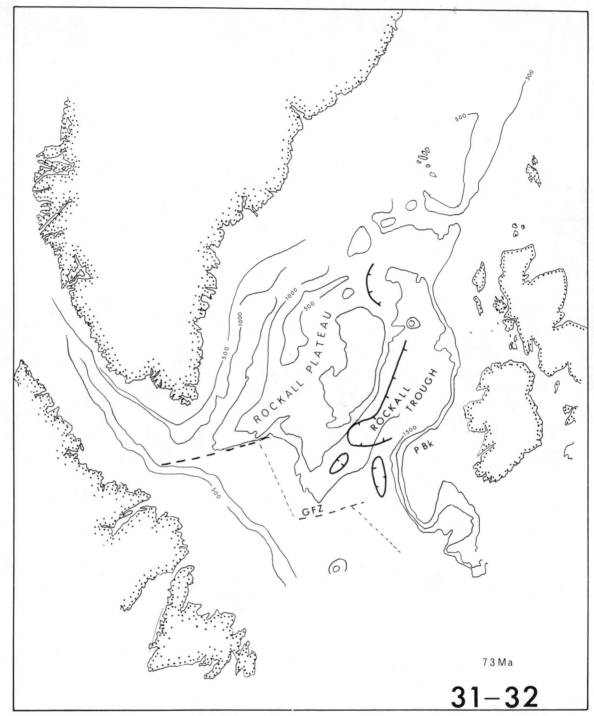

73 Ma

31–32

FIG 4.   Reconstruction of the Atlantic Ocean at 73my (anomaly-31-32). GFZ: Gibbs fracture zone; PBk: Porcupine Bank.

The age and duration of spreading are difficult to determine unequivocally. The vesicular basalt containing Maastrichtian foraminifera dredged from the Anton-Dohrn seamount and the microgabbros of the Rockall complex may be indicative of terminal volcanism in the late Cretaceous (Roberts, 1973; Roberts *et al.*, 1973a; Jones *et al.*, 1974). Roberts (1974) has noted that a reflector dated at 76my from its absence on ocean-crust younger than 76my pinches out against the basement of these seamounts. The sediments (c1.0 second thick) beneath this reflector thicken and contain more reflectors towards the basin margins, perhaps indicating deposition in a spreading ocean. The magnetic anomalies of the Rockall Trough can be followed southwards into the Bay of Biscay. Dewey *et al.* (1973) have argued the Biscay anomalies lie in the Keathley sequence (110–140my). However, the sequence is magnetically quiet and the seismic reflection data do not show evidence of any thicker sediments deposited in the 76–110my hiatus required by the latter argument.

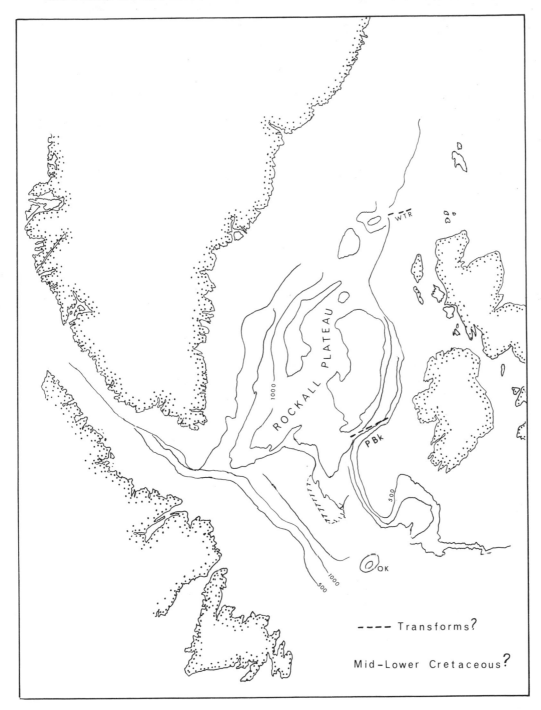

FIG 5.   Pre-spreading reconstruction of the Atlantic Ocean. Hachured area indicates possible subsided oceanic crust. OK: Orphan
Knoll; PBk: Porcupine Bank; WTR: Wyville—Thomson Ridge.

Tentatively, the magnetic anomalies may belong to the Cretaceous quiet magnetic zone (110–80my) (Larsen and Pitman, 1972). It should be noted that this date is compatible with the early Cretaceous deformation onshore and also with evidence of neritic Aptian in the Bay of Biscay and on Orphan Knoll (Montadert *et al.*, 1974; Laughton *et al.*, 1972). On the basis of the available evidence, the Rockall Trough may have been formed by sea-floor spreading between 110 and 80my although

the actual rifting and volcanism began perhaps in the Middle Jurassic.

SEDIMENTARY HISTORY OF THE ROCKALL PLATEAU AND TROUGH

West of the British Isles the Hatton–Rockall Basin, Rockall Trough and Porcupine Seabight comprise the major sedimentary basins. Although there are considerable differences in thickness and in the composition of the underlying basement, the

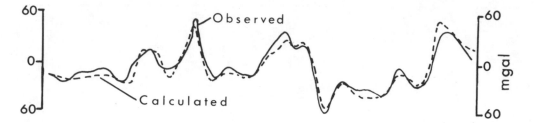

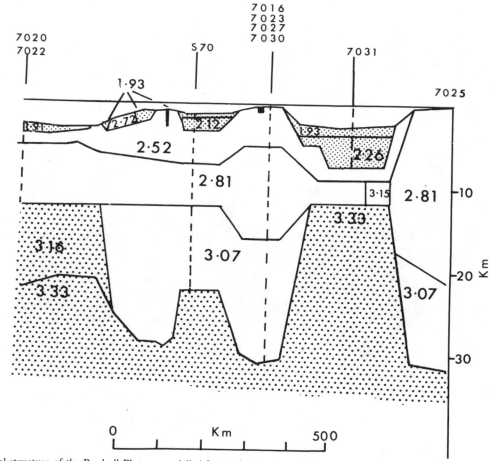

FIG 6.   Crustal structure of the Rockall Plateau modelled from seismic refraction and gravity data (Section AA of Fig 1). (Redrawn from Scrutton, 1972.)

structural framework, subsidence history and pattern of Tertiary sedimentation are all closely comparable. For example, early regional extension within each basin is indicated by down-faulting, and later progressive subsidence by unconformities that are greatest at the basin margins (Fig 7). A detailed study of the stratigraphy of these areas can be found in Roberts (1974).

In outline, the Rockall Plateau consists of a series of basement highs that partly enclose the Hatton–Rockall Basin. The basement geology is largely unknown although Precambrian granulites apparently form inliers in Cretaceous and Lower Tertiary igneous rocks. Mesozoic sediments have not been proved. Deep-sea drilling and seismic reflection studies of the Hatton–Rockall Basin have established the following units of variable thickness:

Acoustic	*Stratigraphy*
4. Post-R4 Series	Early Miocene to Recent oozes Oligocene to Early Miocene cherts
3. Reflector R4	Base of Oligocene unconformity
2. Pre-R4 Series	Late Cretaceous(?)–Upper Eocene
1. Basement	Precambrian granulites, or Tertiary and Cretaceous igneous

The basement structure can be considered as a down-flexure with superposed faulting or broad graben. The graben has a relief of c2km and is down-faulted by as much as 1.5 seconds at the

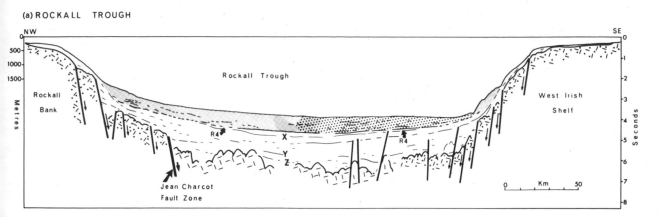

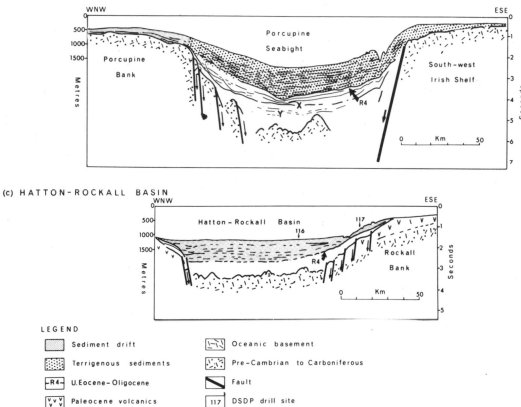

(a) ROCKALL TROUGH

(b) PORCUPINE BANK & SEABIGHT

(c) HATTON–ROCKALL BASIN

LEGEND

Sediment drift

Terrigenous sediments

R4 — U.Eocene–Oligocene

Paleocene volcanics

Oceanic basement

Pre-Cambrian to Carboniferous

Fault

117 DSDP drill site

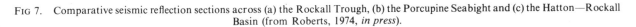

FIG 7.   Comparative seismic reflection sections across (a) the Rockall Trough, (b) the Porcupine Seabight and (c) the Hatton—Rockall Basin (from Roberts, 1974, *in press*).

basin margins. The basin-margin faults seem closely related to regional tectonic trends. The north-east–south-west trending faults of the central Hatton-Rockall Basin for example closely parallel the west margin of the Rockall Plateau and post-60my magnetic anomalies (Fig 8). South of 57° N the main basin-margin fault trends north–south and the basin comprises a half graben formed by tilting and rotation about a north–south axis parallel to the south-west margin of the Plateau. Beneath the basin, the basement is characteristically irregular indicating faulting that in some cases is antithetic to the main basin-margin faults.

The overlying pre-R4 series or Early Cretaceous(?) to Upper Eocene sequence is acoustically transparent. The top of the series is marked by a prominent reflector called R4. Within the basin, the post- and pre-Upper Eocene sequences form a conformable sequence but towards the basin margins, R4 comprises a major unconformity penetrated in DSDP hole 117 (Laughton *et al.*, 1972). The unconformity is between Oligocene cherts and Lower Eocene clays which pass downward into an Upper Palaeocene conglomerate overlying sub-aerial(?) basalt. Reflector R4 is a discrete reflector that is often masked by a thin (0.1 to 0.2 second) sequence of

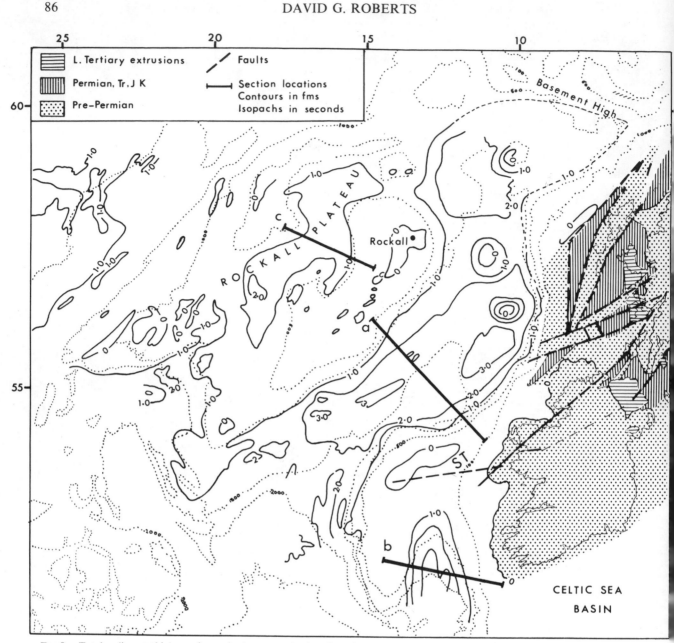

FIG 8.   Total sediment thickness isopachs (two-way time) and outline geology for the Rockall Plateau and Trough (after Roberts, 1974). ST: Slyne Trough.

very strong reflectors associated with numerous hyperbolae. These reflectors arise from the Oligocene to Early Miocene cherts penetrated in DSDP holes 116 and 117. The overlying sediments are predominantly oozes that thin progressively towards the basin margins, overstepping R4 to rest on the Precambrian or the Lower Tertiary. The post-chert sequence (early Miocene to Recent) is characterized by numerous non-sequences, unconformities and cuspate reflectors indicative of differential deposition by bottom currents.

In comparison to the basement and the pre-Upper Eocene sequence, R4 and the Post-Eocene sequence have been only affected by minor faults whose throws rarely exceed 0.1 second except along rejuvenated basin margin faults. The faulting however only appears to affect the lower parts of

the post-Upper Eocene–early Miocene sequence. The general character of the deformation is that of a broad downwarp controlled mainly by basement faulting.

Within the Hatton–Rockall Basin, R4 comprises the major unconformity, although the thinning of the post-Oligocene sequence towards the basin margins indicates substantial overstep, in part due to reduced or differential deposition. These unconformities are greatest at the basin margins but pass basinwards into an apparently conformable sequence. The stratigraphic data are thus consistent with deposition in an intermittently subsiding basin probably formed by extension in late Cretaceous(?)–Palaeocene time (cf. rifting between Greenland and the Rockall Plateau). Laughton et al. (1972) show that the subsidence took place in two phases during late Palaeocene–early Eocene time and during the Oligocene.

In the Rockall Trough (Fig 7a), the upper parts of the sequence are closely comparable to the Hatton–Rockall Basin although the section is thicker and older. In the preceding section, arguments have been reviewed in support of a Lower Cretaceous age for the oceanic basement. The deepest reflector "Z" is confined within the 2.0-second isopach and drapes the underlying oceanic basement relief. Towards the basin margins Z and the pre-Z sequence pass into prograded and tilted beds that may represent the earliest marginal clastics. Roberts (1974) has speculated that reflector Z may be Cenomanian in age. Three reflectors called X, Y and R4 occur above Z and are conformable except at the basin margins where faulting has locally affected them. Reflector Y as mentioned earlier pinches out on 76my oceanic basement and must therefore be about Santonian–Campanian in age. In the south west part of the Rockall Trough and adjacent to the Gibbs Fracture Zone this reflector is faulted and possibly associated with contemporaneous fan deposition. The faulting is apparently confined to the latter areas where it may be related to the major reorganization in plate motions responsible for the opening of the Labrador Sea. Roberts (1974) has noted that reflector X is absent above post-60my oceanic basement and has argued a 60my age on its stratigraphic position between reflector Y and R4. Reflector R4 of the Rockall Trough can be correlated with R4 of the Hatton–Rockall Basin. As in the latter area, it marks a major unconformity and change in depositional regime. On the west side of the Trough, the Oligocene cherts are overlain by the *c*1km-thick transparent sediments of the Feni Ridge deposited under the influence of southwards flowing Norwegian Sea Overflow Water (Jones *et al.*, 1970; Ellett and Roberts, 1973). These sediments have a remarkably uniform character indicative of deposition undisturbed by material derived from the Rockall Plateau. In conjunction with the DSDP data, this implies that Rockall Bank has probably been below sea level since at least Oligocene time. In contrast however, the Post-Oligocene sediments beneath the east side of the Trough contain more reflectors and apparently form two large fans. This influx of terrigenous material may reflect the late Tertiary uplift of the British Isles. On the other shelf and upper slope, west of the British Isles, reflector R4 is an unconformity dividing an older prograding series from an upper series that exhibits both progradation and onlap (Clarke *et al.*, 1970; Bailey *et al.*, 1974). Seismic profiles west of St Kilda show reflector R4 pinches out close to the Palaeocene St Kilda complex and is overlain by onlapping beds. Farther south, reflector R4 rests with marked unconformity on the Mesozoic(?) sediments of a small basin abutting against the Great Glen Fault

(Stride *et al.*, 1969; Roberts, 1974). The precise relationship of reflector R4 and X on the slope is not known and it is possible that there may have been a prolonged hiatus from 60my to the late Neogene. It is noteworthy that the pattern of down-warping of the outer shelf may reflect the late Tertiary history of uplift and faulting now becoming widely recognized as an important factor in the evolution of the British Isles.

In the Porcupine Seabight (Fig 7b) uplift of the Irish area is recorded by the influx of terrigenous material above R4. The reflector called Y has been tentatively identified as the 76my event of the Rockall Trough. The total post-76my sediment thickness in the Porcupine Seabight may exceed 2.0 seconds. However, it is remarkable that the sequence exhibits no major structure other than faulting at the basin margins. No diapirism has been observed. Until more deep refraction data are available, it would be unwise to comment on the origin of the Seabight. However, the reconstruction given in Fig 5 does not require rotation of Porcupine Ridge and the Seabight. Further, the orthogonal boundary faults support an origin by limited extension rather than spreading (*cf.* Bailey *et al.*, 1971).

CONCLUSIONS

In this brief review, an attempt has been made to summarize the important tectonic and sedimentary factors that have contributed to the development of the area. It would be inappropriate to conclude without mention of the petroleum potential. Although thick sediments are present in all three basins, comparatively little structure is present. Until deep stratigraphic data become available, areas with most potential may be downfaulted shelf basins, the early clastic rift margin facies and potential traps associated with regional unconformities.

ACKNOWLEDGEMENT

I thank Dr A. S. Laughton for discussion during the preparation of this review.

REFERENCES

Allen, P. 1969. Lower Cretaceous sourcelands and the North Atlantic. *Nature, Lond.*, **222**, 657–8.

Andreiff, P., Bouysse, Ph., Horn, R. and Monciardini, C. 1972. Contribution à l'étude géologique des approches occidentales de la Manche. *In* Colloque sur la géologie de la Manche. *Mem. Bur. Rech. géol. Min.*, **79**, 31–8.

Audley-Charles, M. G. 1970a. Stratigraphical correlation of the Triassic rocks of the British Isles. *Q. Jl. geol. Soc. Lond.*, **126**, 19–48.

Audley-Charles, M. G. 1970b. Triassic palaeogeography of the British Isles. *Q. Jl. geol. Soc. Lond.*, **126**, 49–90.

Bacon, M. and Chesher, J. 1974. Geophysical surveys in the

Moray Firth. *In* Geology of the North Sea and adjacent areas. *Norsk. geol. Surv. (In press).*

Bailey, R. J., Buckley, J. S. and Clarke, R. H. 1971. A model for the early evolution of the Irish continental margin. *Earth and planet Sci Lett.*, **13**, 79–84.

Bailey, R. J., Grzywacz, J. M. and Buckley, J. S. 1974. Seismic reflection profiles of the continental margin bordering the Rockall Trough. *Q. Jl. geol. Soc. Lond.*, **130**, 55–70.

Beckinsale, R. D., Brooks, C. K. and Rex, D. C. 1970. K-Ar ages for the Tertiary of East Greenland. *Meddr. dansk. geol. Foren.*, **20**, 27–37.

Birkelund, T., Bridgwater, D., Higgins, A. K. and Perch-Nielsen, K. 1974. An outline of the geology of the Atlantic coast of Greenland. *In "The Ocean Basins and Margins". 2. The North Atlantic*, Plenum Press, New York, 125–60.

Boillot, G., Horn, R. and Lefort, J. P. 1972. Evolution structurale de la Manche occidentale au Secondaire et au Tertiaire. *In "Colloque sur la géologie de la Manche". Mem. Bur. Rech. geol. min.*, **79**, 79–86.

Bott, M. P. H., Browitt, C. W. A. and Stacey, A. P. 1971. The deep structure of the Iceland–Faeroe Ridge. *Mar. geophys. Res.*, **1**, 328–51.

Bott, M. P. H., Sunderland, J., Smith, P. J., Casten, V. and Saxov, S. 1974. Evidence for continental crust beneath the Faeroe Islands. *Nature, Lond.*, **248**, 202–4.

Callomon, J. H., Donovan, D. T. and Trumpy, R. 1972. An annotated map of the Permian and Mesozoic formations in East Greenland. *Meddr. Grønland*, **168** (3), 35 pp.

Cherkis, N. Z., Fleming, H. S. and Massingell, J. V. 1973. Is the Gibbs Fracture Zone a westward continuation of the Hercynian front into North America? *Nature phys. Sci.*, **245**, 113–5.

Clarke, R. H., Bailey, R. J. and Taylor-Smith, D. 1970. Seismic reflection profiles of the continental margin west of Ireland. *In* F. M. Delany (Ed.), ICSU/SCOR Working Party 31 Symposium Cambridge 1970. "The Geology of the East Atlantic continental margin. 2. Europe." *Rep. Inst. geol. Sci.*, **70/14**, 69–76.

Cowperthwaite, I. A., Fitch, F. J., Miller, J. A., Mitchell, J. G. and Robertson, R. H. S. 1972. Sedimentation, petrogenesis and radioisotopic age of the Cretaceous Fuller's Earth of Southern England. *Clay. Min.*, **9**, 309–27.

Curry, D., Hamilton, D. and Smith, A. J. 1970. Geological evolution of the Western English Channel and its relationship to the nearby continental margin. *In* F. M. Delany (Ed.), ICSU/SCOR Working Party 31 Symposium Cambridge 1970. The geological evolution of the western English Channel and its relation to the nearby continental margin. *Rep. Inst. geol. Sci.*, **70/14**, 129–42.

Day, G. A. and Williams, C. A. 1969. Gravity compilation in the North East Atlantic and interpretation of gravity in the Celtic Sea. *Earth planet. Sci. Lett.*, **8**, 205–13.

Dearnley, R. 1962. An outline of the Lewisian complex of the Outer Hebrides in relation to that of the Scottish mainland. *Q. Jl. geol. Soc. Lond.*, **118**, 143–76.

Dewey, J. F., Pitman, W. C., Ryan, W. B. F. and Bonnin, J. 1973. Plate tectonics and the evolution of the Alpine system. *Bull. Geol. Soc. Am.*, **84**, 3137–80.

Donovan, D. T. 1972. Geology of the central English Channel. *In* "Colloque sur la géologie de la Manche". *Mem. Bur. Rech. geol. min.*, **79**, 215–20.

Eden, R. A., Wright, J. E. and Bullerwell, W. 1970. The solid geology of the east Atlantic continental margin adjacent to the British Isles. *In* F. M. Delany (Ed.), ICSU/SCOR Working Party 31 Symposium, Cambridge 1970. "Geology of the East Atlantic continental margin. 2. Europe". *Rep. Inst. geol. Sci.*, **70/14**, 114–28.

Ellett, D. J. and Roberts, D. G. 1973. The overflow of Norwegian Sea deep water across the Wyville–Thomson Ridge. *Deep-Sea Res.*, **20**, 819–35.

Garson, M. S. and Plant, J. 1972. Possible dextral movement on the Great Glen and Minch Faults in Scotland. *Nature phys. Sci.*, **240**, 31–5.

Hallam, A. 1971. Mesozoic geology and the opening of the North Atlantic Ocean. *Jl. geol.*, **79**, 129–57.

Hallam, A. and Sellwood, B. W. 1968. Origin of Fuller's Earth in the Mesozoic of Southern England. *Nature, Lond.*, **220**, 1193–5.

Helwig, J., Aronson, J. and Day, D. S. 1974. A late Jurassic mafic pluton in Newfoundland. *Canad. Jl. Earth Sci.*, **11**, 1314–9.

Hill, M. N. and Vine, F. J. 1965. A preliminary magnetic survey of the western approaches to the English Channel. *Q. Jl. geol. Soc. Lond.*, **121**, 463–75.

Holgate, N. 1969. Palaeozoic and Tertiary transcurrent movements along the Great Glen Fault. *Scott. Jl. geol.*, **5**, 97–139.

Howitt, F., Aston, E. R. and Jacque, M. 1975. Occurrence of Jurassic volcanics in the North Sea. *Paper 28, this volume.*

Hudson, J. D. 1964. The petrology of the sandstones of the Great Estuarine Series and the Jurassic palaeogeography of Scotland. *Proc. geol. Ass.*, **75**, 499–527.

Jones, E. J. W., Ewing, M., Ewing, J. I. and Eittreim, S. 1970. Influences of Norwegian Sea overflow water on sedimentation in the northern North Atlantic and Labrador Sea. *Jl. geophys. Res.*, **75**, 1655–80.

Jones, E. J. W., Ramsay, A. T. S., Preson, N. J. and Smith, A. 1974. A Cretaceous guyot in the Rockall Trough. *Nature, Lond.*, **251**, 129–131.

Jones, M. T. and Roberts, D. G. (in preparation). Marine magnetic anomalies in the eastern North Atlantic Ocean.

Kennedy, W. Q. 1946. The Great Glen Fault. *Q. Jl. geol. Soc. Lond.*, **102**, 41–72.

Kent, P. E. 1969. The geological framework of petroleum exploration in Europe and North Africa and the implications of continental drift hypotheses. *In* P. Hepple (Ed.), *"The Exploration for Petroleum in Europe and North Africa"*, Inst. Petrol., London, 3–17.

Kent, P. E. 1975. The tectonic development of Great Britain and the surrounding seas. *Paper 1, this volume.*

Larsen, R. L. and Pitman, W. C. 1972. World wide correlation of Mesozoic magnetic anomalies and its implications. *Bull. Geol. Soc. Am.*, **83**, 3645–62.

Laughton, A. S. 1971. South Labrador Sea and the evolution of the North Atlantic. *Nature, Lond.*, **232**, 612–6.

Laughton, A. S. and Berggren, W. A. *et al.* 1972. *Initial reports of the deep sea drilling project, XII.* US Government Printing Office, Washington, 1243 pp.

Le Mouel, J. L. and Le Borgne, E. 1971. Magnetic map of the Bay of Biscay. *In* E. Debyser, X. Le Pichon and L. Montadert (Eds.), *"Histoire structurale du Golfe de Gascogne", 2* (6), Editions Technip, Paris.

Le Pichon, X., Hyndman, R. and Pautot, G. 1972. Geophysical study of the opening of the Labrador Sea. *Jl. geophys. Res.*, **76**, 4724–43.

McQuillin, R. and Binns, P. E. 1973. Geological structure in the Sea of the Hebrides. *Nature phys. Sci.*, **241**, 2–4.

Miller, J. A., Roberts, D. G. and Matthews, D. H. 1973. Rocks of Grenville age from Rockall Bank. *Nature phys. Sci.*, **246**, 61.

Montadert, L., Winnock, E., Delteil, J. R. and Grau, G. 1974. Continental margins of Galicia–Portugal and the Bay of Biscay. *In* C. A. Burk and C. L. Drake (Eds.), *"Geology of continental margins"*, Springer-Verlag, New York.

Pitcher, W. S. 1969. North east trending faults of Scotland and Ireland and chronology of displacement. *In* North Atlantic—Geology and Continental Drift. *Am. Ass. Petrol. Geol., Mem.*, **12**, 724–33.

Riddihough, R. 1968. Magnetic surveys off the north coast of Ireland. *Proc. R. Ir. Acad.*, **66**, 27–41.

Roberts, D. G. 1969. Recent geophysical investigations on the Rockall Plateau and adjacent areas. *Proc. Geol. Soc. Lond.*, **1662**, 87–93.

Roberts, D. G. 1971. New geophysical evidence on the origins of the Rockall Plateau and Trough. *Deep-Sea Res.*, **18**, 353–9.

Roberts, D. G. 1973. The solid geology of the Rockall Plateau. *Rep. Inst. geol. Sci. (In press).*

Roberts, D. G. 1974. Structural development of the British Isles continental margin and Rockall Plateau. *In* C. A. Burk and C. L. Drake (Eds.), *"Geology of continental margins"*, Springer-Verlag, New York.

Roberts, D. G. *In press*. Marine geology of the Rockall Plateau and Trough. *Phil. Trans. R. Soc. A*.

Roberts, D. G., Ardus, D. A. and Dearnley, R. 1973a. Pre-Cambrian rocks drilled from the Rockall Bank. *Nature phys. Sci.*, **244**, 21–3.

Roberts, D. G., Flemming, N. C., Harrison, R. K. and Binns, P. E. 1973b. Helen's Reef: A Cretaceous microgabbroic intrusion in the Rockall intrusive centre. *Mar. geol.*, **16**, M21–30.

Scrutton, R. A. 1972. The crustal structure of Rockall Plateau microcontinent. *Geophys. Jl.,*, **27**, 259–75.

Steel, R. J. 1971. New Red Sandstone movement on the Minch Fault. *Nature phys. Sci.*, **234**, 158–9.

Stride, A. H., Curray, J. R., Moore, D. G. and Belderson, R. H. 1969. Marine geology of the Atlantic continental margin of Europe. *Phil Trans. R. Soc., Series A*, **264**, 31–75.

Tarling, D. H. and Gale, N. 1968. Isotopic dating and palaeomagnetic polarity in the Faeroe Islands. *Nature, Lond.*, **218**, 1043–44.

Vogt, P. R., Avery, O. E., Schneider, E. R., Anderson, C. N. and Bracey, D. R. 1969. Discontinuities in sea floor spreading. *Tectonophysics*, **8**, 285.

Vogt, P. R., Johnson, G. L., Holcombe, T. L., Gilg, J. G. and Avery, O. E. 1971. Episodes of sea floor spreading recorded by the North Atlantic basement. *Tectonophysics*, **12**, 211–34.

Vogt, P. R. and Avery, O. E. 1974. Detailed magnetic surveys in the north east Atlantic and Labrador Sea. *Jl. geophys. Res.*, **79**, 363–89.

Winchester, J. A. 1973. Pattern of regional metamorphism suggests a sinistral displacement of 160km along the Great Glen Fault. *Nature phys. Sci.*, **246**, 81–4.

Ziegler, P. A. 1974. The North Sea in a European Palaeogeographic framework. *Proc. Bergen North Sea Conference. Norg. geol. Unders. (In press)*.

DISCUSSION

Professor Arthur Whiteman (Aberdeen): Could Mr Roberts give the age sequence for the evolution of the Rockall Trough? Is not the sequence as follows: (1) crustal thinning and thermal uplift, (2) separation, (3) generation of new crust, (4) troughing and main period of sedimentation? Could not the process have started in the Triassic and have continued well up through the Mesozoic?

D. G. Roberts: I agree that the process described by Prof. Whiteman is that conceptually understood for passive rift-margin evolution. It has become clear during the meeting that the Triassic "rifting" phase was minor compared to the more important Middle Jurassic–Early Cretaceous phase that shaped the continental margin and indeed followed different structural trends. It may be that the Triassic phase should now be considered as a minor regional distension, possibly related to the distant opening of the Atlantic between North America and Africa rather than a true precursor of the North Atlantic rifting phase.

Professor M. P. H. Bott (Durham University): Although I used to favour an earlier age, present evidence seems to favour an early Cretaceous age, or thereabouts.

Dr R. A. Scrutton (Edinburgh University): The schematic cross-section of Rockall Trough shows only deposits of Cretaceous and younger age in the Trough and on the margins. Could Mr Roberts be sure that there were not older Mesozoic deposits undetected on the margins of the Trough where continental crust constitutes basement, these deposits having formed in the rift zone prior to the emplacement of oceanic crust?

D. G. Roberts: I think it is possible that down-faulted remnants of pre-rift sedimentary basins and sediments deposited contemporaneously with rifting may underlie the slope.

6

Structure and Stratigraphy of Sedimentary Basins in the Sea of the Hebrides and the Minches[1]

By P. E. BINNS, R. McQUILLIN, N. G. T. FANNIN, N. KENOLTY and D. A. ARDUS

(Institute of Geological Sciences, Edinburgh)

SUMMARY

Three principal north-easterly faults cross the Sea of the Hebrides and the Minches and control the western margins of asymmetric sedimentary basins.

The largest basin, the Sea of the Hebrides Basin, is bounded on the west by the Minch Fault; it is floored by late Precambrian and possibly Palaeozoic sediments and contains up to 3km of Mesozoic sediments. These have been intruded by Tertiary igneous rocks and covered by Tertiary lavas which locally have been downwarped to accommodate a basin of Tertiary sediments.

To the north, the North Minch Basin, also controlled by the Minch Fault, is separated from the Sea of the Hebrides Basin by a structural high; to the east the Inner Hebrides Basin is controlled by the Camasunary–Skerryvore Fault. The stratigraphy of both basins is comparable to that of the Sea of the Hebrides Basin. The Great Glen Fault is believed to affect both Mesozoic and Tertiary sediments but the precise structure and stratigraphy is still uncertain.

INTRODUCTION

The Sea of the Hebrides, the Minches and adjacent coastal waters lie between the islands of the Outer Hebrides and the Scottish mainland (Fig 1). The area includes the islands of the Inner Hebrides.

Following an aeromagnetic survey (Sheet 1 of the Aeromagnetic Map of Great Britain and Northern Ireland, scale 1:625 000), offshore work included deep seismic reflection and refraction surveys, shipborne gravity and magnetic surveys, and shallow coring.

Onshore geology has been mapped in detail by the Institute of Geological Sciences (IGS) and is well documented in publications by IGS and university workers. An interpretation of the offshore geology south of Skye, based on results to 1971, has already been published (Binns *et al.*, 1974). The present paper incorporates more recent results over the whole area and is mainly concerned with an interpretation of the structure and stratigraphy of the sedimentary basins. The study is based primarily on offshore geophysical data and onshore geology; for technical reasons shallow coring offshore has been restricted to the basin margins but future plans include operation of a deep-water drilling ship which will be able to occupy borehole sites over the whole area.

REGIONAL GEOLOGY

An important unconformity separating uncon-solidated Quaternary (glacial) sediments from consolidated pre-Quaternary rocks is recognized (Figs 5, 6 and 7). Quaternary sediment thickness varies greatly and locally exceeds 250m. This range in thickness reflects both the conditions of deposition and the glacial topography of the pre-Quaternary unconformity. Bathymetry is therefore complex and maximum depths of 323m have been recorded north-west of Coll (IGS survey 72/W2) and in the Inner Sound of Raasay (unpublished Hydrographic Department (M.O.D.) Original, K2657).

A pre-Quaternary subcrop map is presented in Fig 1 and a Bouguer anomaly gravity map in Fig 2. A generalized stratigraphic section (Fig 3) shows that all pre-Quaternary systems are represented in the area, commonly by a variety of lithologies. Geophysical contrasts are often small and in unsampled areas geological interpretation must be considered tentative.

Three principal faults cross the area, controlling the western margins of asymmetric sedimentary basins. The largest basin, the Sea of the Hebrides Basin, is bounded on the west by the Minch Fault. To the north the North Minch Basin, also controlled by the Minch Fault, is separated from the Sea of the Hebrides Basin by a structural high. To the east the Inner Hebrides Basin is controlled by the Camasunary–Skeeryvore Fault. Displacement on the Great Glen Fault affects sediments south of Mull.

The sedimentary basins are floored by late-Precambrian and possibly Palaeozoic rocks which are exposed on islands and on the mainland. An interpretation of seismic and gravity data suggests

---

[1] This account is published by permission of the Director of the Institute of Geological Sciences, London.

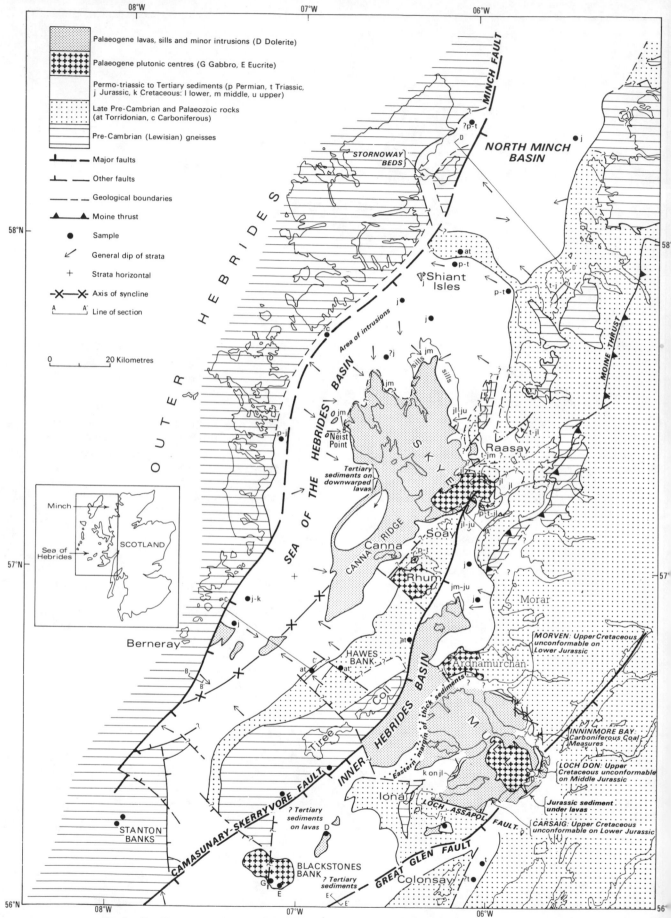

FIG 1. Pre-Quaternary geology of the area between the Scottish mainland and the Outer Hebrides.

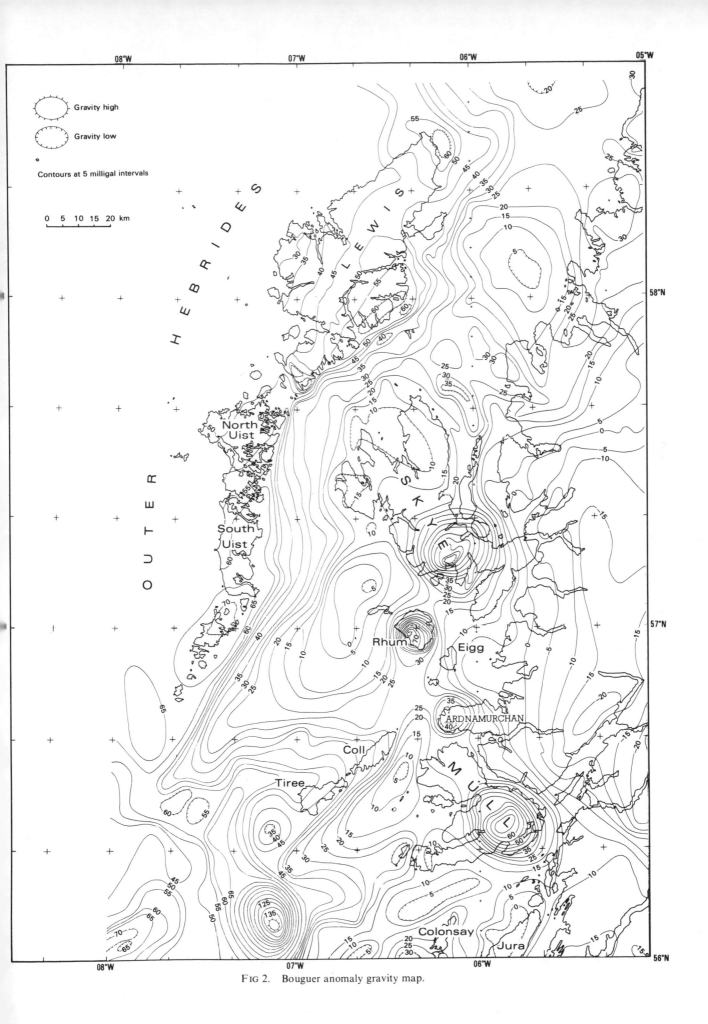

FIG 2. Bouguer anomaly gravity map.

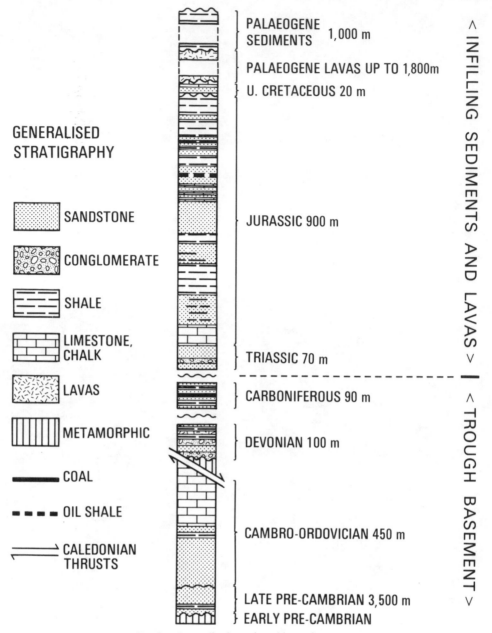

FIG 3.   Generalized stratigraphic section.

that the basins may contain up to 3km of infilling sedimentary rocks, of which substantial Mesozoic sections crop out onshore.

Palaeogene igneous rocks, including five plutonic centres, have intruded all older formations. There are also extensive remnants of a Palaeogene lava cover which locally has been downwarped to accommodate Palaeogene sediments (Smyth and Kenolty) and produce a secondary basin superimposed on the Sea of the Hebrides Basin.

THE INNER HEBRIDES BASIN

On Skye, Jurassic sediments are thrown down at least 650m to the east against the Precambrian

(Torridonian Sandstone) by the Camasunary Fault (Peach et al., 1910) (Fig 4). Pre-Upper Cretaceous movement on the fault is suggested by the fact that Upper Cretaceous sediments rest on the Lias 6.5km westwards of the fault whereas eastwards of the fault they lie directly on the Upper Jurassic. However, these exposures are separated by sufficient distance to allow the possibility of overstep prior to faulting. J. D. Hudson (personal communication) points out that more substantial evidence may be found through a study of the structural relationship between the Upper Jurassic and a remnant of Upper Cretaceous (Hudson and Morton, 1969) recently discovered close to the fault. Pre-lava movement is confirmed by the age of the lava subcrop on either side of the fault.

Northwards the fault is displaced by Palaeogene

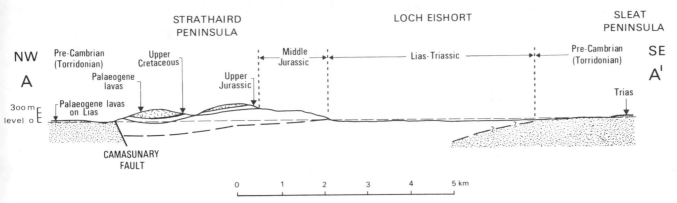

FIG 4.   Section A—A¹ across Loch Eishort and the Strathaird Peninsula. (Redrawn from Peach *et al.*, 1910, Fig 3.)

intrusions (Anderson and Dunham, 1966). It crosses the island of Raasay and probably controls the western scarp of the 323-m deep in the Inner Sound of Raasay (unpublished Hydrographic Department (M.O.D.) Original K2657). A final phase of post-lava movement on Raasay of at least 760m is indicated by the presence of lavas lying topographically below Precambrian rocks west of the fault.

The fault can be traced southwards from Skye as the Camasunary–Skerryvore Fault and a section similar to that across Strathaird (Fig 4) can be deduced from seismic profiles between Rhum, Eigg and Morar.

East of Coll the course of the fault is indicated by a steep gravity gradient (Fig 2). Gravity, magnetic and morphological evidence indicate that basaltic lavas are thrown down against the Lewisian and cover some 3km of sedimentary rocks. The gravity map also shows that these sediments thin eastwards towards the outcrop of a sub-Permo-Triassic unconformity on Mull and Ardnamurchan. A similar structure and stratigraphy to that on Skye is thus indicated, but here, as on Raasay, post-lava movement occurred on the fault.

North of Iona the basin narrows due apparently to cross-faulting of post-lava age. Aeromagnetic evidence shows that the metamorphic basement of Iona extends only a few kilometres to the west. Beyond this point resistant rocks of low magnetic susceptibility are present and are interpreted as Palaeozoic basement.

South of Tiree a prominent scarp marks the Camasunary–Skerryvore Fault and seismic profiles show bedded sediments dipping towards the fault. A possible Tertiary age for these sediments has been suggested by D. K. S. Smythe and N. Kenolty, who have interpreted underlying scarps of magnetized rock as the uppermost surface of a group of Palaeogene lavas. Further evidence for the presence of Tertiary sediments in this area is given by seismic reflection and refraction studies across the Blackstones Plutonic Centre (McQuillin *et al.*, in press). Seismic results also show that Tertiary sediments may extend eastwards to the Great Glen Fault.

Structure around the Blackstones Bank is complex and not fully understood. The main Inner Hebrides Basin probably terminates in this area but sediments of possible Mesozoic age are present to the south-west.

The stratigraphy of the onshore sections of the Inner Hebrides Basin has been established for many years (Richey *et al.*, 1961). A sequence of Permo-Triassic continental red-beds up to 90m thick passes conformably up into Jurassic sediments. Recent sedimentological studies (Bruck *et al.*, 1967; Steel, 1971, 1974) suggest alluvial fan and floodplain environments for the Permo-Triassic sediments. The direction of fan-building suggests contemporaneous movement on the Camasunary Fault as a cause. In the floodplain phase sediment transport paralleled the basin axis.

In the Jurassic, shallow-water marine sediments extend up to the Bajocian. An oil shale, 2–3m thick, forms the base of the Great Estuarine Series (Upper Bajocian to Upper Bathonian), interpreted by Hudson (1962) as a brackish-water lagoonal deposit.

A marine transgression in the Callovian initiates a sequence of shales with sandstones and siltstones which continues into the Kimmeridgian.

There is no evidence of contemporary faulting during the Lias in the studies of Hallam (1959) or Howarth (1956). Morton (1965) recognized a basin in north Skye during the Bajocian, separated from a basin in Raasay by "an area of relatively slow subsidence". This coincides with the present structural high between the Inner Hebrides and Sea of the Hebrides basins and suggests contemporary movement on the Camasunary Fault.

THE SEA OF THE HEBRIDES BASIN

The structure of the Sea of the Hebrides Basin has been investigated by IGS and Glasgow University seismic and gravity surveys (Binns *et al.*, 1974; Smythe *et al.*, 1972; McQuillin and Bacon, 1974). A deep reflection seismic survey offshore was commissioned by IGS and shot by Seismograph Service Ltd using a 4-gun Seisprobe to give 24-fold coverage. Operational details are described by McQuillin and Bacon (1974). Data quality was adversely affected by sea-floor and "base-Quaternary" multiples and by Palaeogene igneous intrusions and remnants of lava cover.

The basin is controlled on its western margin by the Minch Fault which throws down Mesozoic sediments to the east against Precambrian (Lewisian) gneisses and basal Mesozoic sediments. Its eastern margin is the Precambrian (Lewisian and Torridonian) structural high lying west of the Camasunary–Skerryvore Fault. This can be traced north-eastwards from Tiree and Coll through Rhum, southern Skye and Raasay. North of Raasay it is not clear if the Camasunary Fault dies out, resulting in the merging of the Inner Hebrides and Sea of the Hebrides basins. In the southern Minch however it is clear that the Torridonian of the mainland forms the eastern margin of the basin. The northerly margin of the basin is a structural high of Precambrian (Torridonian) rocks which crosses the Minch.

South of Skye the main structure of the basin is a complex syncline trending at an acute angle to the Minch Fault. In the extreme south therefore the north-western limb is truncated and sediments are thrown down against the Lewisian. This results in steep gravity (Fig 2) and topographic (Fig 5) gradients. Farther north the north-western limb is more fully developed and at the same time the throw on the Minch Fault apparently decreases.

The gravity gradient is therefore much lower and reflects the dip of the syncline as well as the fault. Here sediments lie on both sides of the fault which has no topographic expression. North-west of Skye the gravity gradient indicates a greater throw on the fault.

Two cores taken from the sediments lying unconformably on the Lewisian west of the Minch Fault have been identified as a red Permo-Triassic mudstone and a Rhaetian to Hettangian sandy limestone (Warrington, 1974). Reworked Carboniferous miospores in the former sample and in a sample taken farther north, together with the presence of Carboniferous glacial erratics on the Outer Hebrides (Jehu and Craig, 1925), suggest Carboniferous rocks also may occur west of the Minch Fault.

Sediments of a similar seismic character lie east of the fault and the succession, intruded by sills, can be traced eastwards, down-dip, into well-bedded sediments which pass up into Bathonian sediments on Neist Point, western Skye. The succession on the north-western limb has therefore been interpreted as Permo-Triassic to Jurassic.

A comparable sequence can be seen on seismic profiles across the south-eastern limb. Off Coll and Tiree (Fig 6) the synclinal structure of the basin is defined by a basal reflector, interpreted on the margin as the top of Precambrian (Torridonian)

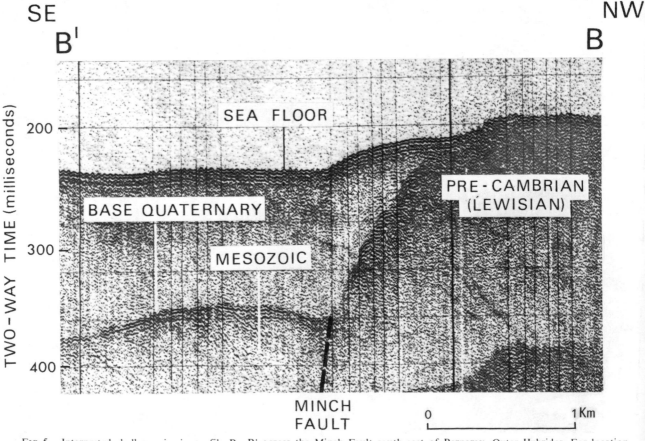

FIG 5.   Interpreted shallow seismic profile B—B¹ across the Minch Fault south-east of Berneray, Outer Hebrides. For location see Fig 1.

sandstone. There is no evidence that the reflector is of this age in the basin itself and it is possible that Palaeozoic sediments as young as the Carboniferous are present, beneath it. Though no cores of infilling sediments have been obtained from the south-eastern limb of the syncline, Permo-Triassic or Lower Jurassic sediments lie unconformably on the Torridonian on Skye and Rhum. Furthermore, off Coll and Tiree seismic records indicate that sediments of similar character to those west of the Minch Fault lie above the "top Torridonian" reflector and pass up into a well-bedded sequence which is capped by lavas (Fig 6). The whole succession reaches a maximum thickness of 2.5 to 3km. The possibility of a significant thickness of Tertiary sediments being present is eliminated by the capping of Eocene lavas and the evidence of Mesozoic sediments in cores. Onshore Jurassic and Cretaceous sediments vary little in thickness between localities. If this lack of variation is maintained across the basin it is implied that either there is a substantial thickness of Permo-Triassic sediment, as suggested for the Skye area by Smythe et al. (1972), or that Devonian to Carboniferous sediments lie above the basal reflector.

The structure at the southern margin of the basin is poorly understood. In part, the narrowing of Mesozoic outcrop evidence on the gravity map (Fig 2), can be ascribed to the convergence of the Minch and Camasunary–Skerryvore faults. However, the intersection of the Minch Fault with the south-eastern limb of the syncline together with north-westerly faulting or tilting are believed to be chiefly responsible for the wedging out of Mesozoic outcrop. Palaeozoic or Permo-Triassic sediments can be traced southwards into a down-faulted area east of Stanton Banks.

Immediately north and north-west of Skye, strata, shown by sampling to be of Jurassic age, dip radially in towards Skye, a structure which may be associated with the subsidence of the Palaeogene lava plateau on Skye.

The seismic and gravity results of Smythe and others (1972) show a substantial thinning of the Permo-Triassic under north-east Skye and offshore gravity north of Skye indicates that the basin remains shallow to the outcrop of the Torridonian structural high. As on the south-eastern margin of the basin, Permo-Triassic sediments lie unconformably on the Torridonian.

The Skye Eocene basaltic lavas extend southwards covering the sediments of the basin and forming a submarine plateau, the Canna Ridge. Remnants of the lava cover are found to the south and an outlier of lava lies close against the scarp of the Minch Fault, an indication of post-lava movement of the Fault.

North-west of Canna a small, steep-sided basin of "post-lava" sediment of probable Palaeogene age has recently been discovered (Smythe and Kenolty). About 1km of sediment lies on top of down-warped lava. The basin is comparable to the Lough Neagh Oligocene Basin in Northern Ireland (Wilson, 1972).

The Mesozoic stratigraphy of the Sea of the Hebrides Basin onshore is similar to that of the Inner Hebrides Basin. Remnants of Permo-Triassic red-beds on the structural high on Rhum and Skye have been interpreted as alluvial fan and floodplain deposits (Steel, 1971, 1974). The former suggest contemporaneous faulting. Evidence for contemporaneous faulting on the Minch Fault and on a fault under northern Skye is provided by the 2km of Permo-Triassic under north-west Skye (Smythe et al., 1972). This contrasts with the thin sequences (20m) on the structural high at Sligachan, Skye (Peach et al., 1910) and under north-east Skye (Smythe et al., 1972).

As in the Inner Hebrides Basin there is evidence of contemporaneous fault movement only in the middle of the Jurassic. Hudson (1964) shows that Bathonian sediments were derived from a Lewisian source west of Skye, a result which is consistent with contemporaneous movement on the Minch Fault.

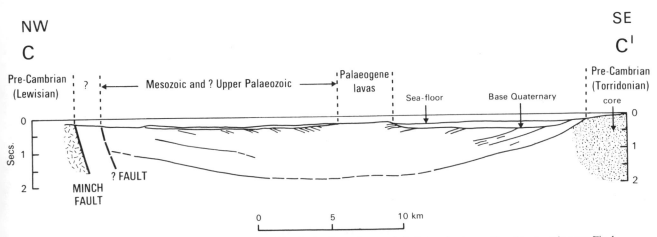

FIG 6. Section C—C¹ across the Sea of the Hebrides Basin; from deep and shallow seismic profiles. For location see Fig 1.

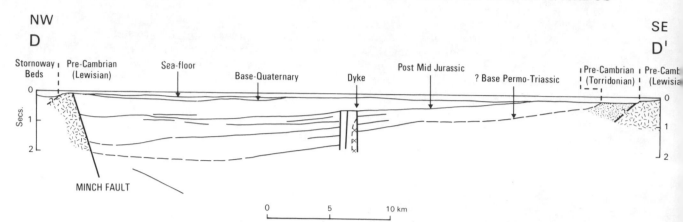

FIG 7.   D—D¹ across the North Minch Basin; from deep and shallow seismic profiles. For location see Fig 1.

THE NORTH MINCH BASIN

The North Minch Basin lies north of the Precambrian (Torridonian) structural high. It is bounded on the west by the Minch Fault and on the east by the outcrop of the Precambrian rocks of the mainland. Northwards it opens into Basin G of Bott and Watts (1970).

A thick sequence of red beds (the Stornoway Beds) lies unconformably on the Lewisian west of the fault. Steel's evidence (*in press*) that these sediments are of Permo-Triassic age is consistent with the palaeontological evidence of the sediments west of the Minch Fault farther south. The beds can be seen on shallow seismic profiles offshore and have been sampled in a borehole (Fig 1).

The only sample obtained within the main basin (Fig 1) is of Middle Jurassic age. Seismic profiles show that poorly-bedded, low-velocity sediments up to 0.7km thick underlie the Quaternary. Beneath this layer there are more prominent reflectors and a base to the infilling sediments has been interpreted on a deep reflection profile (Fig 7). The infilling sediments reach a maximum depth of 4km.

At present insufficient evidence is available to positively identify these layers. The upper layer is probably younger than Middle Jurassic and may be composed of Cretaceous or Tertiary sediments. The lower layer may be of Permo-Triassic to Lower Jurassic age.

In addition to showing this sedimentary succession and the asymmetric structure of the Basin, Fig 7 also shows the Loch Ewe Dyke, which penetrates to within about 2km of the sea-floor and is associated with faulting.

GREAT GLEN FAULT AND AREA TO SOUTH-EAST

The Great Glen Fault of the mainland is a major rectilinear fault transecting Precambrian and Palaeozoic rocks. It was believed by Kennedy (1946) to have a sinistral displacement of the order

of 100km. Offshore in the Moray Firth it loses its rectilinear character and has been shown (Bacon and Chesher, *in press*) to be a normal fault during the Mesozoic. A possible line of the fault southwards to Loch Buie on the south coast of Mull has been suggested (Lee and Bailey, 1925) but distortion of the area during Palaeogene igneous activity is such that the location of the fault with respect to the Permo-Triassic and Jurassic rocks of north-east Mull must be in doubt.

Rast *et al.* (1968) have suggested that a thickness of 2.1km of Permo-Triassic sediments lies west of the fault and Steel (1974) has interpreted these sediments as fault-controlled alluvial fan deposits. Their relationship to the Great Glen Fault however is not clear.

Cenomanian sediments lying unconformably on the Middle Lias at Carsaig, southern Mull, and on the Bathonian at Loch Don (Fig 1) suggest late Jurassic or early Cretaceous movement on the fault comparable to that on Skye. Offshore, a linear gradient on the aeromagnetic map coincides with a series of shoals which terminates in a scarp (Hydrographic Department (M.O.D.) Original K3228). Farther south-west the geology is poorly understood but a gravity gradient defines the fault which is also seen on a deep seismic profile (Fig 8). It is evident that the fault on the profile affects sediments but the age of these is uncertain, although their thickness and seismic character are consistent with a Tertiary age.

A possible splay from the Great Glen Fault is responsible for down-faulting Permo-Triassic sandstones east of Colonsay.

HISTORY OF THE SEDIMENTARY BASINS

The map of Bullard *et al.* (1965) proposing a pre-continental drift of the continents around the Atlantic has been used by Dewey (1969) to reconstruct the Caledonian–Appalachian orogenic belt of which the Hebridean region is a part. The development of the sedimentary basins west of

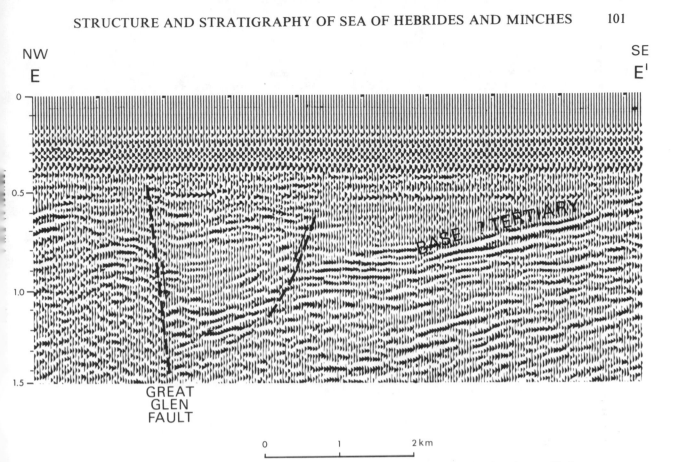

FIG 8.   Deep seismic profile E—E[1] across the Great Glen Fault, west of Colonsay. For location see Fig 1.

Scotland must therefore be considered in the context of continental drift in the North Atlantic.

The evidence outlined above suggests that the basins have developed as a result of normal movement on the three principal faults. A similar history of movement is suggested for each fault.

The presence of both Devonian and Carboniferous sediments in the lower levels of the basins or beneath the basal reflector must be considered a possibility. Onshore, along the Great Glen Fault, Devonian sediments were deposited in alluvial fans controlled by movement on, or parallel to, the fault (Stephenson, 1972). Carboniferous (Coal Measures) sediments occur at Inninmore Bay, north of Mull (Fig 1) (Lee and Bailey, 1925) and Carboniferous Limestone glacial erratics are present on the Outer Hebrides (Jehu and Craig, 1925).

Contemporaneous movement occurred on or parallel to the principal faults in Permo-Triassic times. Precise relationships are however not clear and it was not until the late Jurassic or early Cretaceous that the present configuration of the basins emerged.

During Palaeogene igneous activity the basins were covered with basaltic lavas up to 1.8km thick. Three tectonic episodes followed extrusion of the lavas. The normal movement on the principal faults may mark a final phase of north-west-south-east tension.

North-westerly faults with associated intrusion of dykes mark the change to a regime of north-east–south-west tension.

On Skye and north-west of Canna the lava plateau collapsed; on Skye by block-faulting and north-west of Canna by down-warping.

The history outlined above is consistent with current ideas on continental drift in the North Atlantic (Roberts, 1975). Regional extension commenced during Permo-Triassic times: a major phase of tension occurred during the late-Jurassic and early Cretaceous, possibly coinciding with the opening of the Rockall Trough.

In its Mesozoic history the area has structural, stratigraphical and lithological similarities to the North Sea Graben (Ziegler, *in press*) but during late Cretaceous/Tertiary times its history of igneous activity and uplift contrasts markedly with the history of subsidence and sedimentation which has made the North Sea so prospective. The area has therefore attracted little commercial activity and has been of interest chiefly to those making regional studies of Mesozoic rocks.

REFERENCES

Anderson, F. W. and Dunham, K. C. 1966. The geology of Northern Skye. *Mem. Geol. Surv. Gt. Brit.* HM Stationery Office, Edinburgh, 216 pp.

Bacon, M. and Chesher, J. A. C. *(In press.)* Evidence against post-Hercynian transcurrent movement on the Great Glen Fault. *Scott. Jl. Geol.*

Binns, P. E., McQuillin, R. and Kenolty, N. 1974. The geology of the Sea of the Hebrides. *Rep. Inst. Geol. Sci.*, **73/14**, 43 pp.

Bott, M. H. P. and Watts, A. B. 1970. Deep sedimentary basins proved in the Shetland–Hebridean Continental Shelf and Margin. *Nature, Lond.*, **255**, 265–68.

Bruck, P. M., Dedman, R. E. and Wilson, R. C. L. 1967. The New Red Sandstone of Raasay and Scalpay, Inner Hebrides. *Scott. Jl. Geol.*, **3**, 168–80.

Bullard, E. C., Everett, J. E. and Smith, A. G. 1965. The fit of the continents around the Atlantic. *Phil. Trans. R. Soc. Lond., Series A.*, **258**, 41–51.

Dewey, J. F. 1969. Evolution of the Appalachian/Caledonian Orogen. *Nature, Lond.*, **222**, 124–9.

Hallam, A. 1959. Stratigraphy of the Broadford Beds of Skye, Raasay and Applecross. *Proc. Yorks. geol. Soc.*, **32**, 165–84.

Howarth, M. K. 1956. The Scalpa Sandstone of the Isle of Raasay, Inner Hebrides. *Proc. Yorks. geol. Soc.*, **30**, 353–70.

Hudson, J. D. 1962. The stratigraphy of the Great Estuarine Series (Middle Jurassic) of the Inner Hebrides. *Trans. Edinb. geol. Soc.*, **19**, 139–65.

Hudson, J. D. 1964. The petrology of the sandstones of the Great Estuarine Series, and the Jurassic Palaeogeography of Scotland. *Proc. geol. Ass.*, **75**, 499–527.

Jehu, T. J. and Craig, R. M. 1925. Geology of the Outer Hebrides (Part II). *R. Soc. Edinb.*, **53**, 615–41.

Kennedy, W. Q. 1946. The Great Glen Fault. *Q. Jl. geol. Soc. Lond.*, **102**, 41–72.

Lee, G. W. and Bailey, E. B. 1925. Pre-Tertiary geology of Mull, Loch Aline and Oban. *Mem. Geol. Surv. Gt. Brit.*, HM Stationery Office, Edinburgh, 140 pp.

McQuillin, R. and Bacon, M. 1974. Preliminary report on seismic reflection surveys in the seas around Scotland, 1969–73. *Rep. Inst. geol. Sci.*, **74/12**, 7 pp.

McQuillin, R., Bacon, M. and Binns, P. E. *(In press.)* The Blackstones Tertiary igneous complex. *Scott. Jl. Geol.*

Morton, N. 1965. The Bearreraig Sandstone series (Middle Jurassic) of Skye and Raasay. *Scott. Jl. Geol.*, **1**, 189–219.

Peach, B. N. *et al.* 1910. The geology of Glenelg, Lochalsh and south-east part of Skye. *Mem. Geol. Surv. Gt. Brit.*, HM Stationery Office, Edinburgh, 206 pp.

Rast, N., Diggens, J. N. and Rast, D. E. 1968. Triassic rocks of the Isle of Mull, their sedimentation, facies, structure and relationship to the Great Glen Fault and the Mull calders. *Proc. geol. Soc. Lond.*, **1645**, 299–305.

Richey, M. C., MacGregor, A. G. and Anderson, F. W. 1961. *The Tertiary volcanic districts.* British Regional Geology. Geol. Surv.

Roberts, D. G. 1975. Tectonic and stratigraphic evolution of the Rockall Plateau and Trough. *Paper 5, this volume.*

Smythe, D. K., Sowerbutts, W. T. C., Bacon, M. and McQuillin, R. 1972. Deep sedimentary basins below Northern Skye and the Little Minch. *Nature phys. Sci.*, **236**, 87–89.

Smythe, D. K. and Kenolty, N. (In preparation). Tertiary sediments in the Sea of the Hebrides.

Steel, R. J. 1971. New Red Sandstone movement on the Minch Fault. *Nature phys. Sci.*, **234**, 158–9.

Steel, R. J. 1974. New Red Sandstone floodplain and alluvial fan sedimentation in the Hebridean province of Scotland. *Jl. sedim. Petrol.*, **44**, 336–57.

Steel, R. J. and Wilson, A. *(In press.)* Sedimentation and tectonism (?Permo-Triassic) on the western margin of the North Minch Basin at Stornoway, Isle of Lewis. *Jl. geol. Soc. Lond.*

Stephenson, D. 1972. Middle Old Red Sandstone alluvial fan and talus deposits at Foyers, Inverness-shire. *Scott. Jl. geol.*, **8**, 121–7.

Warrington, G. 1974. Mesozoic microfossil assemblages. Appendix 1 in Binns, P. E., McQuillin, R. and Kenolty, N. The geology of the Sea of the Hebrides. *Rep. Inst. Geol. Sci.*, **73/14**, 43 pp.

Wilson, H. E. 1972. *Regional geology of Northern Ireland.* British Regional Geology. Geol. Surv.

Ziegler, P. A. *(In press.)* The geological evolution of the North Sea area in the tectonic framework of northwestern Europe. *Bull. Norges. Geol. Unders.*

DISCUSSION

For relevant discussion see Paper 3 by D. R. Whitbread.

7

Structure and Evolution of the North Scottish Shelf, the Faeroe Block and the Intervening Region

By M. H. P. BOTT

(University of Durham)

SUMMARY

The three main structural units of the region are (1) the Shetland–Hebridean shelf, (2) the Faeroe Block and associated shallow banks to the south-west, and (3) the intervening Rockall and Faeroe–Shetland troughs which are separated from each other by the Wyville–Thomson Ridge.

Tertiary strata are generally absent from the Shetland–Hebridean shelf except on the slope and adjacent to it. Isolated and partially fault-bounded Mesozoic basins occur locally, the most spectacular of these being the West Shetland basin. The underlying crust is about 25km thick and is characterized by a 6.5km s[1] layer of generally shallow but variable depth, which almost reaches the surface beneath the prominent gravity high lying to the east of the West Shetland basin.

Recent crustal seismic investigations indicate that the early Tertiary lavas of the Faeroe Islands are underlain by continental crust, the Moho being about 30km or more in depth. The Faeroe Block thus forms the north-eastern part of the Rockall–Faeroe microcontinent, which probably includes Bill Bailey Bank, Lousy Bank, Faeroe Bank, and the subsided regions between them. In contrast, the Iceland–Faeroe Ridge is underlain by Icelandic type oceanic crust formed in the early Tertiary, and an anomalous type of continental margin formed about 65my separates the Ridge from the Faeroe Block.

Gravity and seismic investigations show that the crust beneath the Faeroe–Shetland Trough and the north Rockall Trough is thinner than continental but thicker than normal oceanic crust. A root of thickened crust occurs beneath the Wyville–Thomson Ridge. Thin sediments are draped over the crestal region of the Wyville–Thomson Ridge, and these are overlapped by later sediments which generally thicken along the troughs away from the Ridge. The troughs were probably formed during an early (120my or earlier) stage of opening of the North Atlantic, the split extending further north-eastwards beneath the North Sea slope and across the eastern part of the Vøring Plateau, a viewpoint which differs fundamentally from the recently published interpretation of Talwani and Eldholm.

INTRODUCTION

The region covered by this paper is divisible into three structural provinces, the west Shetland and Orkney shelf, the Faeroe Plateau, and the intervening Faeroe–Shetland Channel (Fig 1). This region forms the south-eastern extremity of a strip of anomalously shallow oceanic depths extending from Denmark Strait and Iceland to the north Scottish shelf.

Most of the results presented have been obtained during University of Durham marine geophysical cruises between 1967 and 1972, financed by the Natural Environment Research Council. Gravity, magnetic, seismic profiling and short refraction observations were made prior to 1972. In summer 1972 the sea-to-land North Atlantic Seismic Project (NASP) took place between north Scotland and Iceland, with shots fired at sea and seismic recording on land and at sea; results from this project provide the first firm evidence on the underlying crustal structure of the region.

THE WEST SHETLAND AND ORKNEY SHELF

General

A gravity and magnetic traverse made in 1965 by HMS Hecla and interpreted at Durham (Midford,

1966) indicated large gravity and magnetic anomalies on this shelf. This was followed up between 1967 and 1971 by Durham cruises on RRV John Murray and other ships. Over 10 000km of continuous gravity and magnetic data were obtained and a number of sparker profiles and refraction lines were observed (Bott and Watts, 1970a; Watts, 1971; Browitt, 1972; Bott and Browitt, 1975). The IGS aeromagnetic map also covers part of the shelf and has been qualitatively interpreted by Flinn (1969).

The Bouguer anomaly map (Bott and Watts, 1970a) shown in Fig 2 shows a conspicuous north-north-east trending gravity high marked A, west of the Shetlands and continuous for 250km. West of high A is gravity low E, with steep intervening gradients. Gravity low F occurs between the south-west coast of the Shetland Islands and Fair Isle. Two sharp local negative anomalies C and D separated by a north-north-east trending high occur about 100km west-north-west of the Orkney Islands. A broad gravity low G extends north from the Minch and low H lies between north Scotland and the Orkney Islands.

The results from NASP (Smith and Bott, 1975) show that the crust is about 26km thick beneath this shelf. Beneath the sediments and a 6.1km/s upper crustal layer, a 6.5km/s layer occurs at fairly shallow but variable depth.

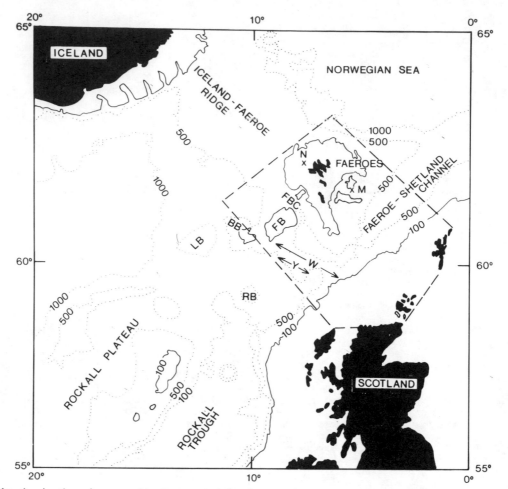

FIG 1.   Map showing the region covered by the paper and the adjacent region of the North Atlantic. FBC: Faeroe Bank Channel; FB: Faeroe Bank; BB: Bill Bailey's Bank; LB: Lousy Bank; RB: Rosemary Bank. N and M mark the positions of negative gravity anomalies on the Faeroes shelf (see text). W and Y indicate the extents of the Wyville—Thomson and Ymir Ridges respectively.

The Basement

The Caledonian front crosses the north Scottish coast onto the shelf. On land, it separates the Lewisian and Torridonian basement rocks of the foreland from the Moinian and Dalradian rocks of the Caledonian belt. This subdivision may be assumed to apply to basement rocks beneath the shelf. The Caledonian front probably passes between gravity high A and Foula, cutting the shelf edge north and slightly east of the Shetland Islands (Watts, 1971).

Occurrence of the metamorphic or igneous basement at the seabed or not far beneath it can be recognized variously by short wavelength magnetic anomalies, rough bathymetry, shallow seismic basement and high gravity. East of the inferred Caledonian front, shallow basement rocks occur in the vicinity of the Shetland Islands. West of it, an extensive belt of supposed Lewisian rocks has been recognized (Fig 3).

Basement crops out at the seabed beneath at least 110km of the length of gravity high A, which is associated with particularly large amplitude magnetic anomalies of short wavelength. A refraction line along the axis of the high indicates a shallow basement of apparent velocity 5.89km/s

(Browitt, 1972). Gravity high A is thus attributed to the shallow occurrence of strongly magnetic and highly dense metamorphic rocks, possibly Lewisian granulites of Scourian type (Bott and Watts, 1970a; Watts, 1971; Flinn, 1969). Results from NASP (Smith and Bott, 1975) support this interpretation by showing that the underlying crust is not thinned but that the 6.5km/s layer almost emerges at the surface here, deepening to about 9km beneath the adjacent low F to the east. Thus the spectacular drop of 110mgal in the Bouguer anomaly between high A and low F (Fig 2) is partly caused by deepening of the dense 6.5km/s granulite layer and partly by sediment thickening towards low F.

The belt of shallow basement rocks extends beyond the south end of high A, where it becomes displaced westwards. It passes between gravity lows C and D and encompasses the Islands of Skerries and North Rona (Fig 3). Himsworth (1973a) has shown that shallow or outcropping basement rocks of Lewisian type cover much of the shelf west of Lewis.

The north-east to north-north-east trend of the foreland basement rocks of the shelf is approximately parallel to the Caledonian front. It

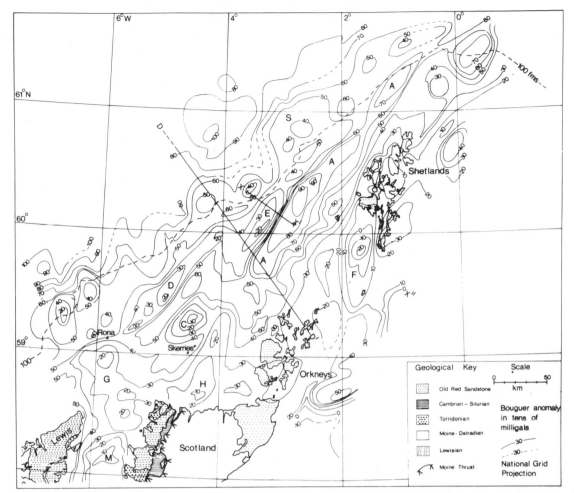

FIG 2.   Bouguer anomaly map of the west Orkney and Shetland shelf (after Bott and Watts (1970a) and Watts (1971)). Prominent anomalies are denoted by letters.

may have been implanted at the Caledonian orogeny or earlier. The slightly more easterly trend of the shelf edge gently truncates this basement trend and cuts the Caledonian front at an angle of about 15°.

Palaeozoic Sediments

Outcropping Palaeozoic sediments have not yet been recognized on the shelf west of the inferred Caledonian front, although they may occur. A region of smooth magnetic field about 25km north of Foula, lacking visible bedding on the sparker record or large gravity signature, may be caused by Torridonian or Palaeozoic (Old Red Sandstone) sediments. East of the Caledonian front, thick Old Red Sandstone is inferred to cover an extensive region between the Orkney and Shetland Islands where the magnetic basement is moderately deep (Watts, 1971; Bott and Browitt, 1975; Flinn, 1969). In this region, two main sediment types are visible on sparker records (Bott and Browitt, 1975): bedded strata with variable dips which are locally quite steep, and strata without visible bedding.

Mesozoic Basins

The series of local negative gravity anomalies C to H (Fig 2) have been interpreted as partially fault-bounded sedimentary basins on the basis of the gravity evidence supplemented by magnetic, sparker and refraction observations (Bott and Watts, 1970a; Watts, 1971; Bott and Browitt, 1975). Gravity lows C, D and E require a density contrast of at least $0.4g/cm^3$ to explain the steep gradients. Sparker records show that the thick sediments which cause the anomalies typically dip gently at about 5° to 10°, and that near the shelf edge (at the south end of low E) these are unconformably overlain by nearly flat-lying sediments which pass into the prograding sequence of the slope (Watts, 1971). The dipping sediments of low F have an estimated seismic velocity of 2.65km/s and unconformably overlie more strongly deformed bedded sediments of inferred Old Red Sandstone age (Bott and Browitt, 1975). Within the tectonic framework of Britain and the North Sea, this evidence indicates a Mesozoic age for the gently dipping sediments of the basins, and a Tertiary age for the overstepping slope sediments; confirmation by drilling is needed.

The postulated West Shetland Mesozoic basin causing gravity low E is separated from the shallow basement of high A by a normal fault or fault-belt.

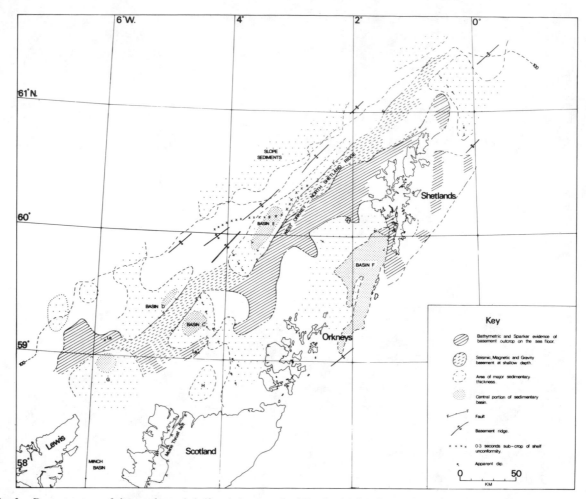

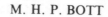

FIG 3.   Deep structure of the continental shelf and slope north of Scotland (after Bott and Watts (1971) and Watts (1971), with the
addition of the more recently discovered basin F (Bott and Browitt, 1975)).

The overlying slope sediments overstep the fault
onto the basement. Refraction lines just west of the
axis of the low indicate 700m of assumed 2.0km/s
Tertiary sediments, about 2.5km of velocity
3.1km/s (later Mesozoic) and 4km of velocity
4.7km/s (early Mesozoic or Palaeozoic) above
5.95km/s basement (Browitt, 1972). This type of
structure explains the gravity anomalies (Fig 4).
The basin is flanked on its western side by an
uplifted ridge of dense and highly magnetic
basement rocks giving rise to positive gravity and
magnetic features (Fig 4).

Gravity low F is partly caused by a sedimentary
basin containing up to 2km of Mesozoic sediments,
fault-bounded at its eastern side (Bott and Browitt,
1975). Gravity lows C and D can be explained by
probable Mesozoic infill of about 3 to 5km (Bott
and Watts, 1970a; Watts, 1971), and the sparker
profile across low C indicates an aggregate
sediment thickness of about 3km of north-westerly
dipping sediments across the north end of this
basin (Watts, 1971). The north-western margin of
basin C and the south-eastern margin of basin D
appear to be fault-bounded, with a ridge of shallow
magnetic basement rocks separating them and
extending round the southern margin of C to

include the Skerries. Gravity low H overlies the
immediate seaward extension of the Caledonian
front. Drilling by MV Whitehorn (Ann. Rep.
Inst. Geol. Sci. for 1972) reveals rocks of probable
Mesozoic age underlying a thin drift cover in this
region. Gravity low G forms a northward extension
of the Minch Basin.

The geophysical evidence indicates that basins
C, D, E and F are partially fault-bounded, three of
them on the eastern side and one (C) on the western
side. Sparker evidence where available indicates
strata dipping towards the faults and overlying flat-
lying Tertiary or Quaternary strata overstepping
the faults. Where gravity and sparker evidence is
combined, it indicates some stratigraphical
thinning towards the faults. This suggests that the
faults were active as hinge-lines during deposition,
with the possibility of some renewed post-
depositional movement prior to deposition of flat
overlying sediments.

Tertiary

Tertiary strata have not been recognized on this
shelf except on the slope and near to it. Sediments
of probable Tertiary and Quaternary age overstep
the older sediment of the west Shetland basin onto

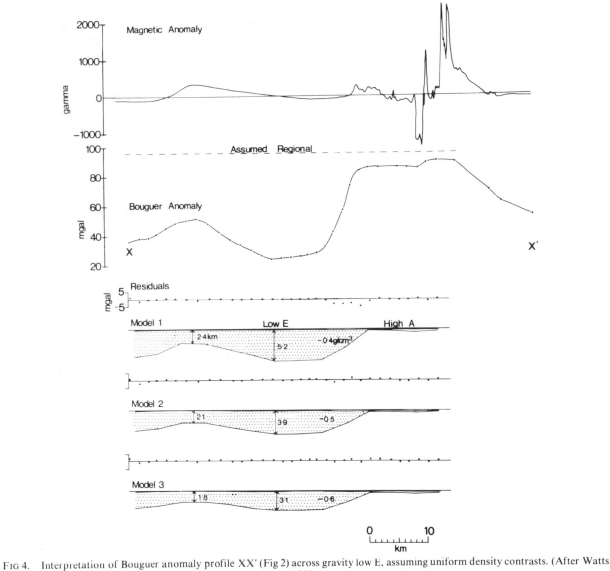

FIG 4. Interpretation of Bouguer anomaly profile XX' (Fig 2) across gravity low E, assuming uniform density contrasts. (After Watts (1971).)

the shallow basement to the east of it. These sediments thicken substantially seawards and pass into the prograding succession of the slope. Evidence is lacking for widespread early Tertiary volcanic activity on the shelf west of the Orkney and Shetland Islands, although the wide Minch dyke visible on the aeromagnetic map does cross this shelf.

Absence of marine Tertiary strata may be attributed to early Tertiary uplift which affected much of north-west Britain. The uplift was possibly caused by the hot underlying mantle associated with the early Tertiary igneous activity and accompanying split of Greenland from North Europe at about 60my. The unconformity seen near the shelf edge probably spans this period of uplift.

Transcurrent Faults

Dearnley (1962) postulated a major sinistral transcurrent fault of about 120km displacement along the Minch, and Avery et al. (1968) extended the fault-line northwards across the shelf. Gravity investigations, however, suggest that the North Minch sedimentary basin is separated from the Outer Hebrides by a normal fault (Allerton, 1968) and that this line may extend northwards to form the western boundary of basin C (Watts, 1971). The new evidence gives no support to Dearnley's hypothesis.

Flinn (1961) suggested that the Great Glen Fault passes west of Fair Isle and crosses the Shetland Islands as the Walls Boundary Fault. The gravity high A is not displaced along the northward extension of this line (Bott and Watts, 1970a) but this difficulty may be overcome by postulating an eastward deflection of the fault-line north of the Shetlands (Flinn, 1970). This would involve rather sharp bends for a transcurrent fault (Bott and Watts, 1970b). More recently, it has been shown that the Mesozoic basin corresponding to gravity low F has a linear tongue extending southwards to the latitude of the Orkney Islands along the expected fault line (Bott and Browitt, 1975) (Fig 3). This can be interpreted as a later subsidence feature along an underlying old line of weakness, and gives support to the hypothesis that the Great Glen Fault or a splay of it may lie along the line of the Walls Boundary Fault.

THE FAEROE PLATEAU

The Faeroe Block

The Faeroe Islands are formed by nearly flat-lying early Tertiary basalt lavas (Rasmussen and Noe-Nygaard, 1970). The lavas are 3000m thick in aggregate and their base is not seen. A thin coal-bearing sequence separates the lower and middle series and a minor unconformity occurs between the middle and upper series. The lava pile is cut by dykes and sills.

The Faeroe Block lies at the junction of the north-east trending Rockall–Faeroe Plateau and the south-east trending Iceland–Faeroe Ridge. Rockall Plateau is underlain by continental crust (Scrutton, 1972; Roberts et al., 1973) but the Iceland–Faeroe Ridge is underlain by anomalously thick oceanic crust of Icelandic type which probably formed at between 60 and 45my, during early Tertiary separation of Greenland from North Europe (Bott et al., 1971; Bott, 1975). The nature of the crust beneath the Faeroe Block has thus been problematical.

The first seismic refraction lines on the Faeroe Islands determined a 6.4km/s basement beneath 2.5 to 4.5km of lavas (Palmason, 1965), suggesting a similar upper crustal structure to that of Iceland. On the other hand the continental fit of Rockall to Greenland (Bott and Watts, 1971) and gravity anomalies between the Iceland–Faeroe Ridge and the Block (Bott et al., 1971) suggested an underlying continental crust. The more conclusive results from NASP (Bott et al., 1974) indicate a crustal thickness of 30 to 40km, a basement velocity of 6.0 to 6.2km/s or less, and lack of a 6.4 to 6.7km/s layer giving first arrivals. This suggests that continental crust underlies the early Tertiary lavas of the Faeroe Block. Results from NASP also indicate a lower velocity (5.5km/s) basement beneath the south-eastern Faeroe shelf.

Bouguer anomalies over the Faeroe Islands (Saxov, 1969) vary gently between +10 and +35mgal, and rise on the shelf towards the deeper water. A sharp local negative anomaly of 60mgal amplitude situated around point M (Fig 1) on the northern shelf (Fleischer, 1971) is probably caused by a deep Mesozoic basin or a granitic intrusion such as the "salic rock body" of Schrøder (1971). Another local negative gravity anomaly of about 20mgal amplitude occurs in the vicinity of position N (Fig 1) on the south-eastern Faeroe shelf and is attributable to thick sediments. It is possible that pre-Tertiary sediments are sandwiched between the base of the lavas and the basement elsewhere as a hidden layer, but there is no other indication of major sedimentary basins on the Faeroe Block.

The Faeroe Bank and Faeroe Bank Channel

The nearly flat-topped Faeroe Bank (Fig 1) occupies a roughly rectangular region of $80 \times 40km^2$ less than 100 fathoms (183m) deep and is separated from the west Faeroe shelf by the Faeroe Bank Channel. The bathymetry is smooth on a broad scale but may be rough on a small scale indicating crystalline rocks at the seabed (Dobinson, 1970). Seismic reflection profiling confirms absence of sediments on the bank (Lewis, J. E., private communication).

Highly magnetic rocks occur at or not far below the seabed (Bott and Stacey, 1967; Avery et al., 1968; Dobinson, 1970). A detailed marine magnetic survey made by Dobinson (1970) reveals conspicuous north to north-west trending dykes. It is possible that lavas may cover part or all of the top of the bank. A smoothed magnetic anomaly map indicates a more deep-seated west-south-west–east-north-east trend of longer wavelength magnetic anomalies which might be attributed to an underlying metamorphic basement (Dobinson, 1970).

The main problem of origin is to determine whether Faeroe Bank is a volcanic seamount like Rosemary Bank (Scrutton, 1971) and Anton Dohrn Bank (Jones et al., 1974), or an upstanding fragment of the Rockall–Faeroe microcontinent. The Faeroe Bank has a larger and less circular plan shape than these seamounts and it lacks their steep sides and large positive free air anomalies. The free air gravity anomalies over the Faeroe Bank are about +50mgal which is consistent with a similar underlying crustal thickness to that of the Faeroe Block. The continental fit between Greenland and Rockall (Bott and Watts, 1971) also suggests that the Faeroe Bank forms part of the Rockall–Faeroe microcontinent. These items of evidence, although not conclusive, suggest an underlying continental crust affected by vigorous volcanism of probable but unproved early Tertiary age. Himsworth (1973a, b) concluded that George Bligh Bank and Bill Bailey's Bank are also underlain by crust of continental thickness, but they lack evidence of volcanic activity.

The Faeroe Bank Channel joins the Faeroe–Shetland Channel to the main North Atlantic basin. At its narrowest it is 24km wide and has a sill depth of 410 fathoms (750m) (Stride et al., 1967). Sediments of the order of 1km in thickness form the floor of the Channel and are underlain by highly magnetic basement similar to that beneath the bank and the Faeroe shelf (Stride et al., 1967; Bott and Stacey, 1967). Traverses across the Channel and its extension towards the north-east indicate a conspicuous central positive magnetic

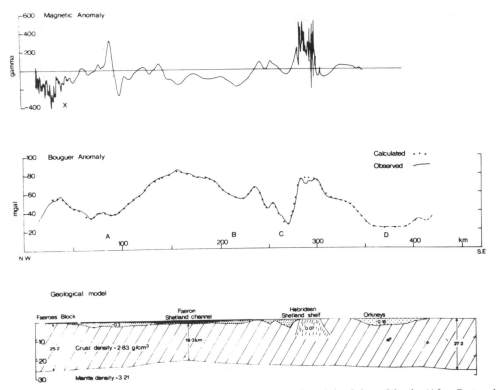

Fɪɢ 5.   Interpretation of a gravity profile between the Faeroe Islands and the Orkney Islands. (After Bott and Watts (1971).)

anomaly (Himsworth, 1973a; Bott and Stacey, 1967). The origin of the Channel is obscure, but possibilities include (1) subsidence on a pre-existing line of weakness, (2) erosion during early Tertiary uplift associated with the Faeroes volcanism and continental splitting, or (3) incipient sea-floor spreading with dyke injection.

FAEROE–SHETLAND CHANNEL AND WYVILLE–THOMSON RIDGE

Faeroe–Shetland Channel

The   smooth-bottomed   Faeroe–Shetland Channel is about 600 fathoms (1067m) deep directly between the Faeroe and Shetland Islands. It deepens slightly and widens towards its north-eastern end where it joins the Norwegian Sea. It is terminated at its south-western end by the Wyville–Thomson Ridge where it joins the Faeroe Bank Channel (Fig 1). The Channel is thus neither typically oceanic nor typically continental.

Flat lying sediments about 1.5km thick (Lewis, J. E., private communication) overlie a highly magnetic basement directly between the Faeroe and Shetland Islands. The magnetic basement generally shallows towards the south-west and deepens towards the north-east (Himsworth, 1973a; Talwani and Eldholm, 1972). Talwani and Eldholm (1972) have controversially identified the north-west margin of the Channel with their "Faeroe–Shetland escarpment" which occurs beneath the North Sea slope, separating shallow basement on the north-west from deep sediment-covered basement on the south-east side.

A gravity profile across the Channel shown in Fig 5 suggests that the crust is about 7km thinner than beneath the adjacent Orkney shelf (Bott and Watts, 1971), and preliminary results from NASP confirm the crustal thinning. The underlying crust is of anomalous type whatever its origin. It may have originated by thinning and subsidence of continental crust, or by sea-floor spreading of pre-Tertiary age with production of anomalously thick oceanic crust. Present evidence is not conclusive, but the highly magnetic basement beneath the sediments may favour the sea-floor spreading hypothesis. Furthermore, if the Rockall Trough formed by sea-floor spreading, then the plate geometry of its opening would imply that the Channel forms a north-eastern extension contemporaneous with the Rockall Trough. If this latter hypothesis is correct, then one would expect the Channel to extend considerably further north than its present junction with the Norwegian Sea. A speculative suggestion is that this extension may lie at the foot of the North Sea slope and continue further north to form the trough of thick sediments beneath the eastern part of the Vøring Plateau, which has been mapped by Talwani and Eldholm (1972).

Wyville–Thomson Ridge

This bathymetric ridge of west-north-west–east-south-east trend connects the south end of the Faeroe Bank to the North Scottish continental shelf and separates Rockall Trough from the Faeroe–Shetland and Faeroe Bank Channels (Fig 1). Detailed bathymetry (Himsworth, 1973a; Ellett and Roberts, 1973) shows that there are two ridges separated by a trough. The name Wyville-Thomson Ridge has been restricted by Ellett and Roberts (1973) to the northern ridge which is continuous between the Bank and the shelf, and the new name Ymir Ridge has been suggested for the southern ridge which extends about two-thirds way across, being separated from the shelf by an embayment at the north end of Rockall Trough.

The deep structure of the Wyville-Thomson Ridge *(sensu lato)* was studied specifically during the 1971 Durham cruise aboard MV Researcher, and the data obtained has been studied by Himsworth (1973a, b). Basement crops out along the crest of the northern ridge but sediments swamp the Ymir Ridge and the intervening trough. The two ridges are thus more pronounced as basement features than in the bathymetry. Two distinct sequences of sediments are seen. The underlying sequence consists of draped sediments in which the bedding is roughly conformable to the basement topography. The higher sequence consists of flat-lying sediment of the Rockall Trough and Faeroe-Shetland Channel which overlap the draped sequence against the flanks of the ridge. Based on ages of the prominent reflectors in Rockall Trough, the upper sequence is Tertiary and the lower sequence may be earliest Tertiary or Mesozoic.

The magnetic observations indicate that the basement is strongly magnetic and reversely magnetized, suggesting basic igneous composition. The gravity traverses indicate approximate isostatic equilibrium, with a deepening of the Moho of about 7km relative to the Rockall Trough and Faeroe-Shetland Channel.

The origin of the ridges is obscure on the basis of present evidence. One possibility is that intense igneous activity has penetrated and substantially thickened pre-existing crust, either continental or oceanic. An early Tertiary igneous origin might be suggested, but the unproved possibility of Mesozoic sediments draping the ridges and the untypical trend of the ridges may prove to be obstacles. Another possibility is that the ridge formed by unusually intense differentiation from the mantle during the evolution of the Rockall Trough and Faeroe-Shetland Channel by sea-floor spreading, analogous to the formation of the Iceland-Faeroe Ridge during the Tertiary. In support of this, the trend of the ridge is conformable to the expected direction of opening of Rockall Trough as indicated by the geometry of opposite margins of the Trough, but the lack of continuity of the Ymir Ridge to the shelf poses a problem. If this latter hypothesis is correct, a further interesting question is raised. Is it a coincidence that the Wyville-Thomson Ridge and the Iceland-Faeroe Ridge of supposed later age are nearly co-linear, or is there an underlying control of their development which is not yet understood?

STRUCTURAL SYNTHESIS

This paper covers a particularly complicated part of the North Atlantic where both aseismic ridges and microcontinents are found. It has been shown that the Faeroe Block and possibly the Faeroe Bank and related shoals form part of the Rockall-Faeroe microcontinent, and that a continental margin separates the Iceland-Faeroe Ridge from the Faeroe Block. There is less certainty over the origin of the Faeroe-Shetland Channel and the Wyville-Thomson Ridge, but a sea-floor spreading origin of Mesozoic age is favoured.

Thus three continental margins are believed to separate the main North Atlantic basin from the North Scottish continental shelf. Two of these on either side of the Faeroe-Shetland Channel formed contemporaneously and are probably of Mesozoic age. The most distant margin developed at about 60my as Greenland separated from the Rockall-Faeroe microcontinent, and it takes a particularly anomalous form where the Faeroe Block adjoins the Iceland-Faeroe Ridge.

The formation and development of these margins was probably an important influence on the structural history of the North Scottish shelf. The prominent, partially fault-bounded Mesozoic basins may be related either to the pre-existence of the adjacent Faeroe-Shetland Channel or to an early "rift-valley stage" in its development. The absence of marine Tertiary strata except near the slope may result from uplift associated with early Tertiary igneous activity roughly contemporaneous with the split-off of Greenland.

The main problems now centre on determining the origin and age of the Faeroe-Shetland Channel and the Wyville-Thomson Ridge, and the overlying sedimentary sequences. For further progress, one must now mainly look to a few well-placed drill holes through the sediments and into the basement, both on the above structures and on the shelf.

ACKNOWLEDGEMENTS

These investigations have been made jointly with several colleagues at Durham, and I am particularly grateful to Dr A. Dobinson, Dr E. M. Himsworth, Mr J. Lewis and Mr J. H. Peacock for allowing me to make use of data which are unpublished or in theses. I am grateful to the Natural Environment Research Council for providing research funds and ship time, and to Mr G. Dresser, Mrs Margaret Watson and Mrs Hilda Winn for technical assistance in preparation of the paper. Dr G. A. L. Johnson kindly read the manuscript.

REFERENCES

Allerton, H. A. 1968. An interpretation of the gravity of the North Minch, Scotland. *Univ. of Durham MSc dissertation.*

Avery, O. E., Burton, G. D. and Heirtzler, J. R. 1968. An aeromagnetic survey of the Norwegian Sea. *Jl. geophys. Res.,* **73,** 4583–600.

Bott, M. H. P. 1975. Deep structure, evolution and origin of the Icelandic transverse ridge. In: *L. Kristjansson (Ed.) Geodynamics of Iceland and the North Atlantic. (In press.)* Heidd Publishing Co.

Bott, M. H. P. and Browitt, C. W. A. 1975. Interpretation of geophysical observations between the Orkney and Shetland Islands. *Jl. geol. Soc. Lond. (In press.)*

Bott, M. H. P., Browitt, C. W. A. and Stacey, A. P. 1971. The deep structure of the Iceland–Faeroe ridge. *Marine Geophys. Res.,* **1,** 328–51.

Bott, M. H. P. and Stacey, A. P. 1967. Geophysical evidence on the origin of the Faeroe Bank Channel—II. A gravity and magnetic profile. *Deep-sea Res.,* **14,** 7–11.

Bott, M. H. P., Sunderland, J., Smith, P. J., Casten, U. and Saxov, S. 1974. Evidence for continental crust beneath the Faeroe Islands. *Nature, Lond.,* **248,** 202–4.

Bott, M. H. P. and Watts, A. B. 1970a. Deep sedimentary basins proved in the Shetland–Hebridean continental shelf and margin. *Nature, Lond.,* **225,** 265–8.

Bott, M. H. P. and Watts, A. B. 1970b. The Great Glen Fault in the Shetland area. *Nature, Lond.,* **227,** 268–9.

Bott, M. H. P. and Watts, A. B. 1971. Deep structure of the continental margin adjacent to the British Isles. *Rep. Inst. Geol. Sci.,* **70/14,** 89–109.

Browitt, C. W. A. 1972. Seismic refraction investigation of deep sedimentary basin in the continental shelf west of Shetlands. *Nature, Lond.,* **236,** 161–3.

Dearnley, R. 1962. An outline of the Lewisian complex of the Outer Hebrides in relation to that of the Scottish mainland. *Q. Jl. geol. Soc. Lond.,* **118,** 143–76.

Dobinson, A. A. 1970. A magnetic survey of the Faeroe Bank. *Univ. of Durham PhD thesis.*

Ellett, D. J. and Roberts, D. G. 1973. The overflow of Norwegian Sea deep water across the Wyville–Thomson Ridge. *Deep-sea Res.,* **20,** 819–35.

Fleischer, U. 1971. Gravity surveys over the Reykjanes Ridge and between Iceland and the Faeroe Islands. *Marine Geophys. Res.,* **1,** 314–27.

Flinn, D. 1961. Continuation of the Great Glen Fault beyond the Moray Firth. *Nature, Lond.,* **191,** 589–91.

Flinn, D. 1969. A geological interpretation of the aeromagnetic maps of the continental shelf around Orkney and Shetland. *Geol Jl.,* **6,** 279–92.

Flinn, D. 1970. The Great Glen Fault in the Shetland area. *Nature, Lond.,* **227,** 268–9.

Himsworth, E. M. 1973a. Marine geophysical studies between northwest Scotland and the Faeroe Plateau. *Univ. of Durham PhD thesis.*

Himsworth, E. M. 1973b. The Wyville–Thomson Ridge. *Jl. geol. Soc. Lond.,* **129,** 322–3.

Jones, E. J. W., Ramsay, A. T. S., Preston, N. J. and Smith, A. C. S. 1974. *Nature, Lond.,* **251,** 129–31.

Midford, R. L. 1966. A surface-vessel marine gravity and magnetic reconnaissance traverse from the Shetland Isles to Land's End. *Univ. of Durham MSc dissertation.*

Palmason, G. 1965. Seismic refraction measurements of the basalt lavas of the Faeroe Islands. *Tectonophysics,* **2,** 475–82.

Rasmussen, J. and Noe-Nygaard, A. 1970. Geology of the Faeroe Islands (pre-Quaternary). *Danm. geol. Unders.* raekke **25,** 1–142.

Roberts, D. G., Ardus, D. A. and Dearnley, R. 1973. Pre-Cambrian rocks drilled from the Rockall Bank. *Nature, phys. Sci.,* **244,** 21–3.

Saxov, S. 1969. Gravimetry in the Faeroe Islands. *Geodaet. Inst. Medd.,* **43,** 1–24.

Schrøder, N. F. 1971. Magnetic anomalies around the Faeroe Islands. *Annal. Societ. scient. Faeroensis,* **19,** 20–9.

Scrutton, R. A. 1971. Gravity and magnetic interpretation of Rosemary Bank, north-east Atlantic. *Geophys. Jl. R. astr. Soc.,* **24,** 51–8.

Scrutton, R. A. 1972. The crustal structure of Rockall Plateau microcontinent. *Geophys. Jl. R. astr. Soc.,* **27,** 259–75.

Smith, P. J. and Bott, M. H. P. 1975. Structure of the crust beneath the Caledonian foreland and Caledonian belt of the North Scottish shelf region. *Geophys. Jl. R. astr. Soc. (In press.)*

Stride, A. H., Belderson, R. H., Curray, J. R. and Moore, D. G. 1967. Geophysical evidence on the origin of the Faeroe Bank Channel—I. Continuous reflection profiles. *Deep-sea Res.,* **14,** 1–6.

Talwani, M. and Eldholm, O. 1972. Continental margin off Norway: a geophysical study. *Bull. geol. Soc. Am.,* **83,** 3575–606.

Watts, A. B. 1971. Geophysical investigations on the continental shelf and slope north of Scotland. *Scott. Jl. geol.,* **7,** 189 218.

DISCUSSION

W. E. Evans (S. and A. Geophysical): There seems to be a discrepancy between the Jurassic age ascribed by Professor Bott to the sedimentary fill of the West Shetlands (Scule Segir) Basin and the Devonian age given by Sir Peter Kent in his paper. Could we please have a clarification on this point? What are the bases for these Jurassic or Devonian designations?

Professor M. H. P. Bott: The gravity and shallow reflection evidence favours a Mesozoic age for the basins corresponding to gravity lows C and D, but one cannot be as specific as Jurassic on this basis alone. An extensive area of probable Old Red Sandstone rocks lies between these basins and the Orkney Islands.

W. H. Ziegler (Esso Europe Inc.): At what depth do you see the Conrad discontinuity under the high gravity anomaly?

Professor M. H. P. Bott: The shallow refraction line of Browitt (1972) indicates a depth of at least 2km, and the time-term analysis gives an estimate of about 2km, with an uncertainty of the order of 3km.

Dr R. A. Scrutton (Edinburgh University): Professor Bott's interpretation of the seismic reflection and magnetic data from the Wyville–Thomson Ridge appears to be consistent with a Cretaceous age for the oceanic basement in Rockall Trough and the Faeroe–Shetland Channel as suggested earlier by Mr Roberts. Would he agree with this?

Professor M. H. P. Bott: I agree with Dr Scrutton that present evidence is most consistent with an early Cretaceous age for the Rockall Trough and Faeroe–Shetland Channel.

J. K. Habicht (Consultant): Does Professor Bott have any suggestions regarding the sedimentary fill of the "F" basin?

Professor M. H. P. Bott: On the basis of the seismic velocity, the density contrast and structural relationships, basin "F" appears to be of Mesozoic age. It is underlain by more strongly folded sediments of probable Old Red Sandstone age. A detailed account is given by Bott and Browitt (1975).

R. McQuillin (Institute of Geological Sciences): Professor Bott suggests that his gravity high "A" is associated with a block-faulted zone which brings the Conrad discontinuity close to the seabed. Is it possible to establish any structural continuity between high "A" and the elongate gravity high mapped by IGS along the eastern coastal area of the Outer Hebridean island chain? McQuillin and Watson (1973. Large-scale basement structures of the Outer Hebrides in the light of geophysical evidence. *Nature Phys. Sci.*, **245,** pp. 1–3) have suggested that the Outer Hebridean gravity high is caused by a body of high-density granulite facies Lewisian Gneiss which is exposed mainly westwards of the Outer Isles Thrust. If a structural relationship between these two areas does exist it raises the interesting possibility that the granulites of the Outer Hebrides constitute a surface exposure of the sub-Conrad discontinuity rocks.

Is the sedimentary basin south of the Walls Peninsula in the Shetland Islands developed so that the Walls Boundary–Great Glen Fault line forms a margin to the basin, or does this fault line pass under the central axis of the basin?

Professor M. H. P. Bott: The belt of relatively shallow basement rocks including the gravity high "A" does extend more or less continuously to the Outer Hebrides, but exceptionally high Bouguer anomaly values in excess of 70mgal do not occur between high "A" and Lewis. However, the 6.5km/sec (Conrad) refractor, which has been interpreted by Smith and Bott (1975) as granulite facies basement, does appear to underlie the intervening region but at greater depth than beneath high "A". The interesting suggestion of Mr McQuillin that the 6.5km/sec refractor may again closely approach the surface beneath the high gravity region of the Outer Hebrides will, we hope, be tested during a further crustal seismic project planned for summer 1975 (Hebridean Margin Seismic Experiment).

The line of the southward extension of the Walls Boundary Fault which has been interpreted as the Great Glen Fault appears to underlie the central axis of the basin rather than its eastern margin. The eastern boundary of the basin is probably formed by a partly contemporaneous normal fault or fault zone (See Bott and Browitt, 1975).

8

The Geology of the Norwegian Continental Shelf

By H. RONNEVIK, E. I. BERGSAGER, A. MOE, O. ØVREBØ,
T. NAVRESTAD and J. STANGENES

(Norwegian Petroleum Directorate, Stavanger, Norway)

SUMMARY

The Norwegian continental shelf can be divided into three segments: The North Sea, the Møre–Lofoten area and the Barents Sea. The pre-Tertiary in the North Sea segment is divided into sub-basins: the Central Graben, the Viking Graben and the Danish Norwegian Basin. Two highs are dominant: the Vestland Ridge and the Ringkøbing–Fyn High. In the Møre–Lofoten area the following sub-basins are defined in the pre-Tertiary: the Møre Basin, the Vøring Basin and the Helgeland Basin. The ridge separating the last two basins is called the Nordland Ridge. In the Barents Sea the following sub-basins are identified south of 72° N: the Hammerfest Basin, the Tromsø Basin and the Harstad Basin. One highly disturbed ridge called the Senja Ridge was identified.

INTRODUCTION

The Norwegian shelf can be divided into three geological areas (Fig 1): the North Sea; the Møre–Lofoten System; the Barents Sea.

In this paper we give a summary of the geological framework of these three areas.

The areas are in different phases of exploration. In the Norwegian part of the North Sea 120 exploration wells have been drilled and 150 000km of reflection seismic profiles have been shot. North of 62° N there are no exploratory wells except the shallow Tertiary wells drilled by R/V "Glomar Challenger". The main contributions to our knowledge of the Møre–Lofoten and the Barents Sea areas are the geophysical investigations made on behalf of Norwegian authorities since 1969. The surveys during 1969–72 were organized by NTNFK,* but in 1973 the responsibility was transferred to the Norwegian Petroleum Directorate. With the data from 1974 included, 9700km of reflection seismic lines have been shot in the Møre–Lofoten area and 19 700km in the Barents Sea. The grid of lines is shown on Fig 2. The lines shot in 1974 are dotted, and have been used only partly in the interpretation thus far.

The grid of lines south of 72° N is fairly dense for regional work. For the northern part of the Barents Sea however, only a few reconnaissance lines are available. In addition to reflection seismic data, some refraction, gravimetric and magnetic data were obtained during the same programme.

A number of papers has been published for the North Sea, including a major contribution to the understanding of the geological evolution of the area by Ziegler (*in press*).

* The Royal Norwegian Council for Scientific and Industrial Research, The Continental Shelf Division.

Several Norwegian and foreign institutions have carried out geophysical research in the Møre–Lofoten area and the Barents Sea. Results from these investigations have been published, among others: Avery *et al.* (1968), Nysæther *et al.* (1969), Åm (1970, *in press*), Talwani and Eldholm (1972) and Sundvor (1973). In addition to the three areas mentioned, the shelf around the small Norwegian island of Jan Mayen can be included as a part of the Norwegian shelf area. This area is assumed to be partly a continental fragment in the oceanic crust of the Norwegian Sea (Laughton, 1973).

THE NORTH SEA

Structural Evolution

Figure 4 shows two schematic cross-sections in the northern part of the North Sea, locations of which are shown on Fig 3. These suggest a post Caledonian tectonic evolution of the northern North Sea basin as follows: (a) Permian–Jurassic, tensional tectonics culminating in the creation of a central rift zone; (b) Cretaceous, transitional period dominated by epeirogenic subsiding basin floors with some faulting; and (c) Tertiary, epeirogenic subsidence of the whole North Sea area with a contemporaneous uplift of bordering land areas, and some early faulting along boundary faults. This type of evolutionary sequence is known from many intracratonic basins in the world; *e.g.* the Sirte and Delaware basins.

During the Tertiary the North Sea area developed as a shallow oval basin (Heybroek *et al.*, 1967), with the axis of major subsidence running north–south in the central part. The structural framework of the beds underlying the Tertiary basin can in the Norwegian area be summarized as follows (Fig 5): Basins—the Central Graben, the

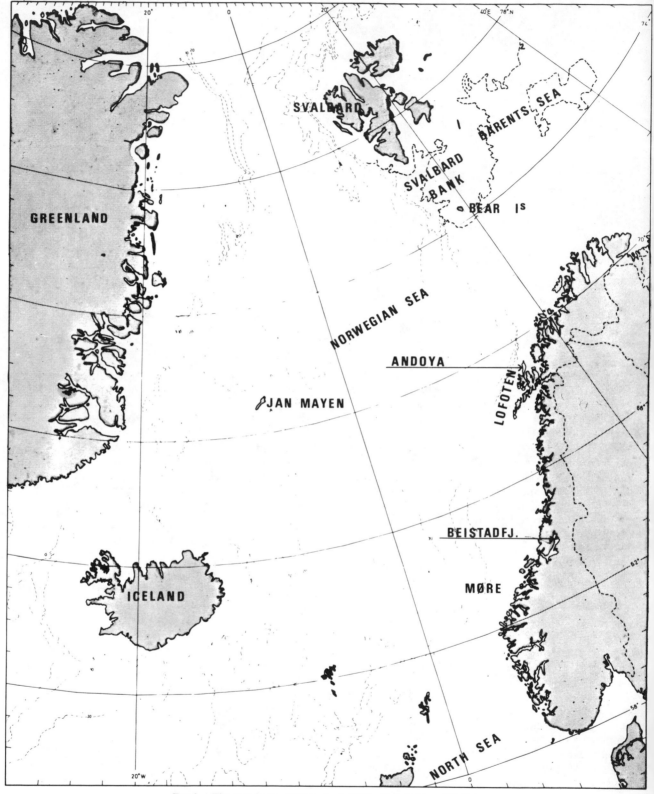

FIG 1.   The seas between Norway, Iceland, Greenland and Spitzbergen.

Viking Graben, and the Danish Norwegian Basin; Highs—the Vestland Ridge, and the Ringkøbing–Fyn High. The dominating Vestland Ridge is discontinuous and mainly a basement ridge, but it contains sedimentary rocks.

Sedimentation

All post Caledonian horizons, except the Carboniferous, have been penetrated in the Norwegian part of the North Sea. The sediments penetrated so far conform with the general

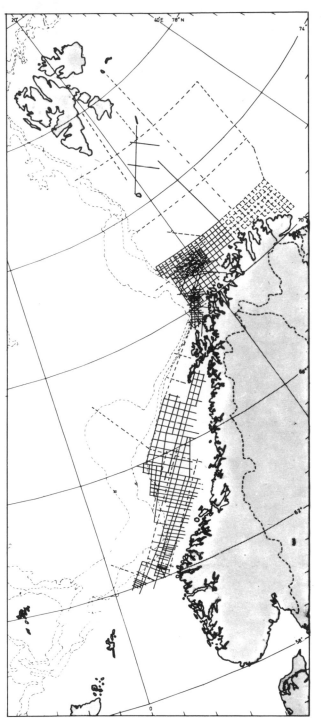

FIG 2.   Reflection seismic grid north of 62° N.

developed as a continental red-bed sequence without marked influence of the Middle Triassic transgression known in the southern North Sea.

At the onset of Jurassic a rapid transgression occurred, giving rise to a shaly facies in most of the Lower Jurassic. In the Middle Jurassic prograding deltas are present. In Upper Jurassic a new transgression occurred, and thick masses of dark shale were deposited in the central part of the basin. Marginal sand facies have not been penetrated so far.

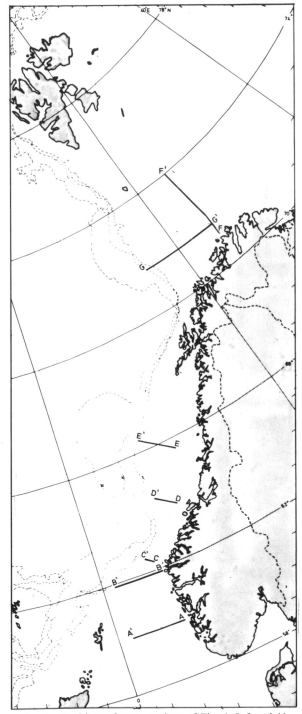

FIG 3.   Positions of cross-sections of Figs 4, 8, 9 and 11.

sedimentological sequence for north-western Europe. Figure 6 shows a generalized lithostratigraphy for the Norwegian North Sea basin. The sedimentological development is clearly related to major transgressive series in Upper Permian, Lower Jurassic, Upper Jurassic, Cretaceous and Holocene times.

Permian strata are known both in marginal facies as dolomite and red beds and in basinal evaporite facies. The whole Triassic seems to be

HORIZONTAL SCALE :

100 Km.

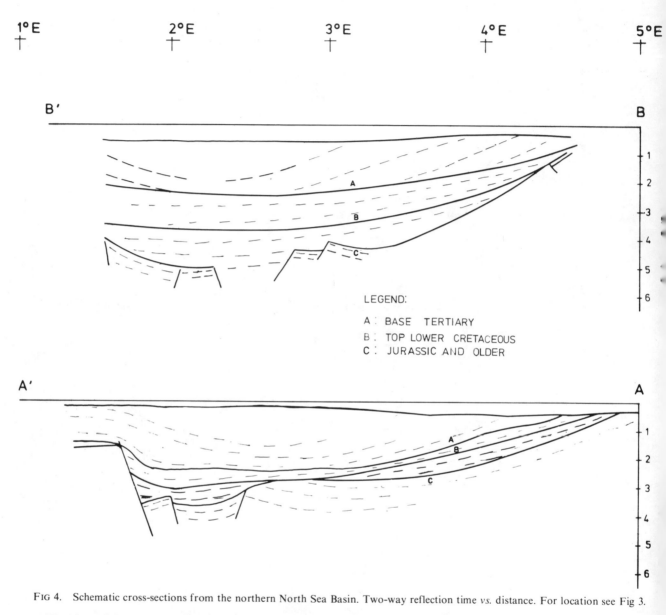

LEGEND:

A : BASE TERTIARY
B : TOP LOWER CRETACEOUS
C : JURASSIC AND OLDER

FIG 4. Schematic cross-sections from the northern North Sea Basin. Two-way reflection time *vs.* distance. For location see Fig 3.

The deposition of dark shale continued into the Lower Cretaceous. In the south red chalk and marl underlie the Upper Cretaceous chalk facies. Towards the north the chalk changes to a marly and clastic facies.

The Tertiary shows a prograding sequence basinwards where shale and turbiditic sand facies are developed. The shale is not compacted and in the deeper part of the basin shale diapirism is known.

During the Quaternary the North Sea basin was covered one or several times by glaciers, which left outwash and morainic material.

Exploration status

Several major oil- and gas-fields have been discovered in the Norwegian sector (Fig 7). So far the prospective horizons have been the following: (1) Palaeocene/Eocene sandstone, with the Frigg, Heimdal, and Cod fields, located along the axis of major subsidence in the Tertiary; (2) Danian/Maastrichtian Chalk, with fields of the Ekofisk complex concentrated in a part of the basin where extensive salt diapirism has occurred; (3) Middle and Lower Jurassic sandstone, with a number of large fields known in the Viking Graben. On the Norwegian side Statfjord is the major

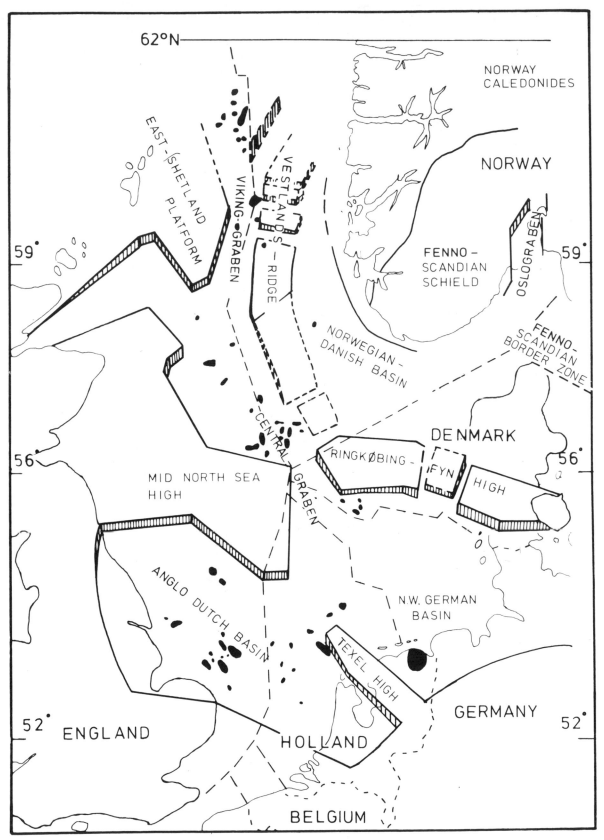

FIG 5. Major structural elements of the geology of the North Sea.

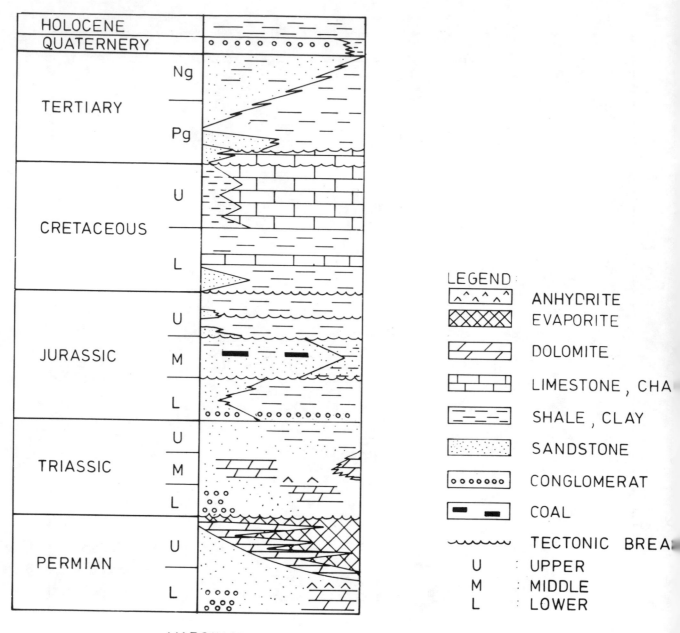

FIG 6.   Composite lithostratigraphy for the northern North Sea.

discovery in the Jurassic; oil from this period is also encountered in other fields like Bream and Brisling.

The recoverable reserves of the known fields amount to 1×10^9 tons of oil and oil equivalent. One early estimate of the total possible reserves south of 62° N on the Norwegian continental shelf is $3-4 \times 10^9$ tons oil and oil equivalent. This estimate is arrived at by an evaluation of the known structures in the three main producing horizons in the areas that have proved to be prospective in the light of similar types of basins elsewhere in the world.

Most of the activity and discoveries so far are concentrated along the median line with UK, that is in the area of the Central and Viking grabens. This region has turned out to be one of the major hydrocarbon provinces of the world. The discoveries of the Bream and Brisling fields show, however, that other areas of the Norwegian shelf are prospective.

THE MØRE–LOFOTEN AREA

The shelf outside Møre–Lofoten (Fig 1) is bordered oceanwards by the oceanic crust of the Norwegian Sea (Talwani and Eldholm, 1972). On the landward side the sediments are bordered by early Palaeozoic and Precambrian rocks. One exception is a smaller inlier of Upper Jurassic–Lower

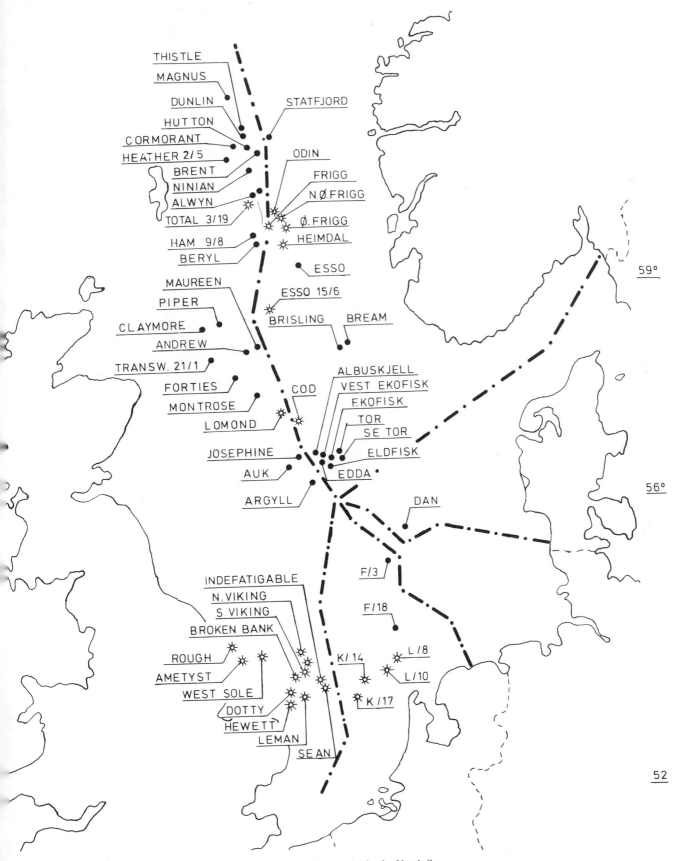

Fig 7. Hydrocarbon discoveries in the North Sea.

Cretaceous rocks on Andøya (Fig 1) (Dalland, 1973). A similar inlier is also indicated in Beistadfjorden, Trøndelag (Fig 1) (Oftedahl, 1972). Erratic boulders from this area have been correlated with the Middle Jurassic deltaic series of Yorkshire (Os Vigran, 1970).

The Møre–Lofoten shelf shows a development that is characteristic of passive (Atlantic) continental margins. Figures 8 and 9 show three profiles with locations indicated in Fig 3. These profiles suggest that the development following the Caledonian compressional regime can be summarized as follows: (a) sedimentation in block-faulted basins culminating with the creation of a major rift system; (b) sedimentation with clear onlap character in epeirogenic subsiding basins; (c) sedimentation of a prograding clastic wedge over an epeirogenic sinking basin floor contemporaneous with the uplift of the land. This seems to be the same development sequence as that in the North Sea.

The bottom of the clastic wedge (marker A, Figs 8, 9) can be correlated with a seismic reflector that in the North Sea is near base Tertiary. The pinchout of the wedge follows the shape of the coast (Fig 10).

The cross-sections indicate that the bottom of the wedge, on the average, has steeper dips in the south than in the north. This is also the case for the internal reflectors interpreted as top set, fore set and bottom set elements (Fig 9). The plate-tectonics model for the Norwegian Sea (Talwani and Eldholm, 1972; Laughton, 1973) suggests that the sedimentation of the wedge started simultaneously near the base of the Tertiary along the whole Møre–Lofoten shelf. The seismic sections indicate epeirogenic subsidence of the basin floor prior to any spreading activity.

The marked reflector B (Fig 9) underlying the base of the Tertiary reflector can be correlated with a reflector in the North Sea that is near the top of the Lower Cretaceous.

The following pre-Tertiary sub-basins have been identified (Fig 10): the Møre Basin, the Vøring Basin and the Helgeland Basin. This basin distribution is clearly indicated on the magnetic interpretation of Åm (1970). The ridge separating the Vøring and Helgeland basins is called the Nordland Ridge. With a possible exception of the southern part, the ridge contains sediments. The ridge is close to the shelf edge and near the gravimetric high described by Talwani and Eldholm (1972). They interpreted this, however, as an intrabasement density contrast.

The Møre Basin can be traced into the Viking Graben. One branch may also continue into the West Shetland Basin. The reflection seismic done so far defines only the eastern flank of the basin. The western border may be the Færøy–Shetland Escarpment as defined by Talwani and Eldholm (1972). In this basin we have not so far identified the rotated fault-blocks like those that characterize the flanks of the Viking Graben.

The fault trends in the Møre Basin are north-east–south-west. The faulting seems mainly to involve the pre-Upper Cretaceous layers.

The Vøring and Møre Basins are connected by a tectonically complex zone where some compressional structures are indicated. This tectonism can be thought of as a result of

HORIZONTAL SCALE :   10 Km.

LEGEND :   HORIZON A :   BASE TERTIARY
             HORIZON B :   TOP LOWER CRETACEOUS

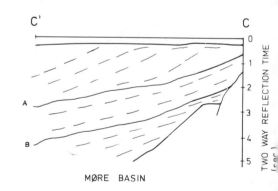

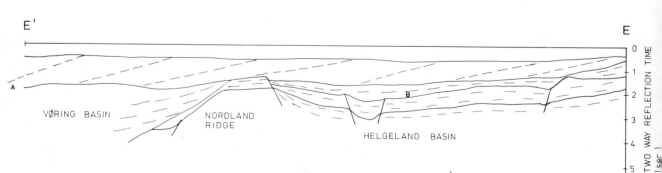

FIG 8.   Cross-sections in the Møre—Lofoten area. For locations see Fig 3.

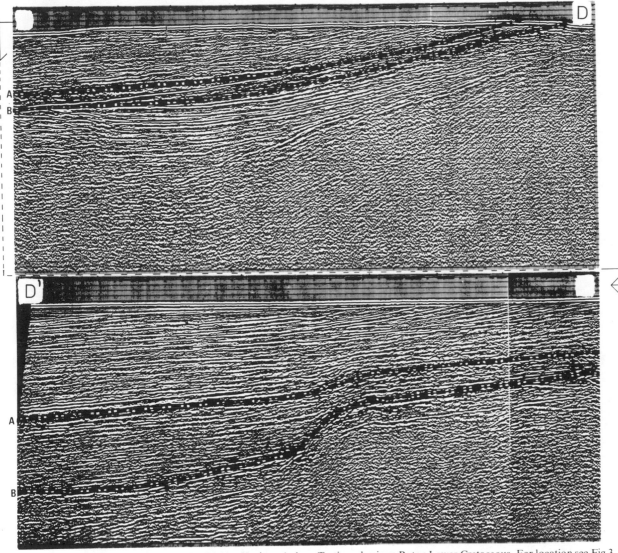

FIG 9.   Seismic section outside Møre/Trøndelag. Horizon A: base Tertiary; horizon B: top Lower Cretaceous. For location see Fig 3.

movement along shear-fractures combining rift-axes. The clastic Tertiary wedge and the supposed Upper Cretaceous interval are unaffected. This zone is landwards from the Jan Mayen Tertiary Fracture Zone that Talwani and Eldholm interpreted as a transform fault-zone. The reflection seismic profiles suggest that it follows an older zone of weakness.

As in the Møre Basin the reflection seismic data available define only the eastern faulted flank of the Vøring Basin. The western flank may be the Vøring Escarpment (Talwani and Eldholm, 1972).

The Helgeland Basin (Fig 10) is a shallow oval basin with a north-east–south-west axis. The faulting seems to stop at the reflector correlated with the near top Lower Cretaceous marker B (Fig 9). There is some uncertainty in correlating over the Nordland Ridge to the south.

We believe that the block-faulted basins are as old as those in the North Sea and East Greenland.

In East Greenland the basins are at least as old as Permian (Vischer, 1943). A radiometric age of 225my for diabases on the western coast of Norway (K. Storetvedt, personal communication) also indicates a pre-Triassic age.

The Møre and Vøring Basins are interpreted as resulting from the same rifting that created the Viking Graben and West Shetland Basin. This rifting is thought to have started in the first half of the Jurassic as in the North Sea (Ziegler, *in press*). There have been major rift pulses in the Lower Cretaceous.

The Mesozoic sediments of East Greenland (Haller, 1969) show great similarities with those in the northern part of the North Sea. In a pre-drift position, East Greenland would have had a central position opposite the Møre–Lofoten shelf. It is therefore reasonable to expect a sedimentological development not unlike that known from the North Sea (Fig 5).

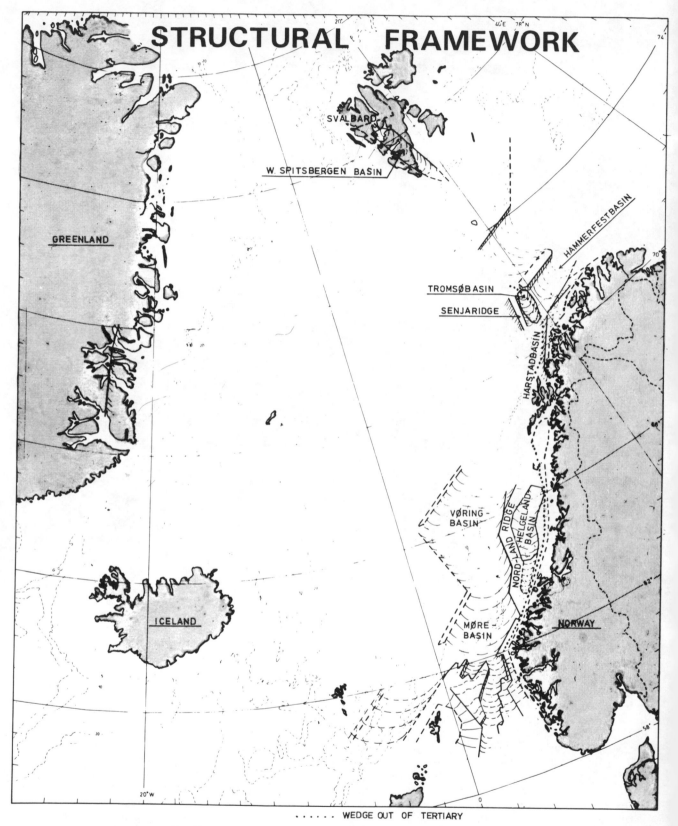

FIG 10.   Structural framework of the Norwegian Continental Shelf.

THE BARENTS SEA

The main structural elements so far identified in the Barents Sea are shown on Fig 10. The main geological basin is indicated in the area between the Svalbard Bank and the Norwegian mainland. Svalbard and the Svalbard Bank are looked upon as a sedimentary platform area. The West Spitsbergen basin is interpreted as a closed basin within this platform.

Two simplified structural cross-sections are presented in Fig 11; their positions are shown in Fig 3. The 6-second two-way reflection time indicates a sedimentary column of the order of at least 10 000m.

In the central part of the main basin there is a structural high with more than 6-second two-way time of sediments. This seismic high coincides with a magnetic high (Åm, *in press*). The reflection seismic data indicate that this high is inverted during a limited interval in the middle part of the basin's history. The older layers increase in thickness within the high, whereas the younger layers clearly thin over the high. The magnetic anomaly may indicate that the inversion was the result of deep-seated plutonic activity.

South of 72° N the following sub-basins are defined (Fig 10): the Hammerfest Basin, the Tromsø Basin and the Harstad Basin. The Hammerfest basin on the east, is bounded to the north by the above mentioned high. On the south it is separated from a platform area by a north-east-south-west fracture system. This is the least tectonically active of the three basins. The cross-section F-F' (Fig 11) in the eastern part of the basin indicates that the uppermost layers only were deformed by epeirogenic subsidence. The fault-system bordering the basin on the south dies out eastwards.

The Tromsø Basin lies west of the Hammerfest Basin and is separated from it by a marked north-south hinge line. The cross-section G-G' (Fig 11) shows that many of the seismic markers that can be followed in the Hammerfest Basin are deeper than 6-second two-way reflection time in the Tromsø Basin. The older layers are clearly down-faulted at the hinge line whereas the younger layers are deformed by epeirogenic subsidence. On the west, the Tromsø Basin is bordered by the Senja Ridge. The separation is related to a north-south fracture system where faulting has affected younger layers than farther east. Diapirs have been identified in the Tromsø Basin (Moe, 1974). The geophysical data and general geological setting indicate that they consist of evaporitic rocks. The source for the diapirs is deep and most probably below 6-second two-way time.

The Harstad Basin is a small but deep basin close to the continental margin (Fig 10). It is bordered on the east by a fracture-system with northerly trend.

The Senja Ridge is interpreted as a high containing thick sediments. Internal reflectors are seen, but are difficult to follow because of disturbances. The deeper sediments are highly disturbed. It seems reasonable to compare this disturbance with the Tertiary orogeny on West Spitsbergen. By using the concept of the transform de Geer fault-zone (Wilson, 1963; Harland, 1969) these disturbances can be related to shear between early spreading axes in the Norwegian Sea and the Arctic Ocean.

The plate-tectonics model implies that the shear lasted longer on West Spitsbergen than farther south. On West Spitsbergen rocks of Eocene-Oligocene age are deformed. However, the model allows rocks from rather early Tertiary to be undisturbed in the south-western part of the Barents Sea.

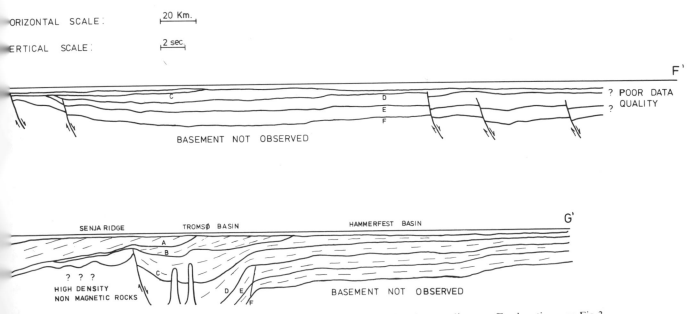

FIG 11. Structural cross-sections in the Barents Sea. Two-way reflection time *vs.* distance. For locations see Fig 3.

The data so far indicate that the Barents Sea south of 72° N can be divided into four different tectonic provinces. In the far west there is the highly disturbed Senja Ridge related to a north–south fault-system. East of this is the Tromsø Basin, also related to north–south faults, but with lesser and older disturbances. Farther east the Hammerfest Basin is related to the north-east–south-west fracture system. There seems to be a relationship between this fault-system and the high to the north. It is also indicated that the fault-system was active only for a short time, and that the Hammerfest Basin through most of its history has been an epeirogenic basin. East of the Hammerfest Basin the sedimentary basin seems to have been an epeirogenic basin throughout.

With the exception of the Hammerfest Basin the above tectonic setting is very similar to that found on Svalbard (Sokolov, 1965). This indicates that the model may be extended north of 72° N to include the whole Barents Sea. At present our conclusion is that the pre-drift rift-system between Norway and Greenland stops in the western part of the Barents Sea. It seems plausible that the old Devonian north–south fault system (Harland, 1969) became activated and relaxed the regional tension.

Dating of reflectors (Fig 11) must be speculative without well control, but on regional geological considerations it is possible to build up a hypothesis. The undisturbed wedge and the slightly disturbed wedge over the A and B reflectors are both considered as Tertiary. The great thicknesses of sediments and the plate-tectonics model for the Norwegian Sea are the main reasons for such dating. This dating also implies that most of the Barents Sea has been an area of erosion during deposition of the wedge. Reflector F is the oldest reflector that can be followed regionally in the Barents Sea. It correlates to the east with the top of a layer that is the source of salt pillows. The youngest evaporitic horizon from regional geological considerations is Permian.

The next marked regional reflector is horizon D. This is a characteristic double reflector that separates near the north-east–south-west faults. The general great thickness indicated between the D and F markers suggests a large time difference. The first marked hiatuses above the Permian on Spitsbergen are between Lower and Middle Jurassic and Middle and Upper Jurassic. An age near Middle Jurassic seems reasonable and is not in conflict with the occurrence of Upper Jurassic on Andøya.

Regional geological considerations suggest that the Permian and younger sediments have similarities with north-western Europe. The major difference is thought to be the existence of a humid climate and a marine environment during Triassic.

ACKNOWLEDGEMENTS

We would like to thank Dr B. J. Scull for his help in correcting the English text; R. Nonstad for drawing the figures and B. Meltveit for typewriting. Lastly we would like to thank the Norwegian Petroleum Directorate for permission to publish the paper.

REFERENCES

Åm, K. 1970. Aeromagnetic investigations on the continental shelf of Norway Stad-Lofoten (62°–69° N). *Norg. geol. Unders.*, **266**, 49–61.

Åm, K. (*In press*). Aeromagnetic basement complex mapping north of latitude 62° N, Norway. *Proc. Bergen North Sea Conference. Norg. geol. Unders.*

Avery, O. E., Burton, G. D. and Heitzler, J. R. 1968. An aeromagnetic survey of the Norwegian Sea. *Jl. geophys. Res.*, **73**, 4583–600.

Dalland, A. (*In press*). The Mesozoic rocks of Andøy, Northern Norway. *Proc. Bergen North Sea Conference. Norg. geol. Unders.*

Haller, J. 1969. Tectonics and neotectonics in East Greenland—review bearing on the drift concept. *Mem. Am. Ass. Petrol. Geol.*, **12**, 852–8.

Harland, W. B. 1969. Contribution of Spitsbergen to understanding of tectonic evolution of North Atlantic region. *Mem. Am. Ass. Petrol. Geol.*, **12**, 817–51.

Heybroek, P., Haanstra, U. and Erdmann, D. A. 1967. Observation of the geology of the North Sea area. *7th Wld. Petrol. Congr. Proc.*, **2**, 905–16.

Laughton, A. S. (*In press*). Tectonic evolution of the Northeast Atlantic. *Proc. Bergen North Sea Conference. Norg. geol. Unders.*

Moe, A. 1974. Continental shelf north of 62° latitude. *G-IV/S ONS 74.*

Nysæther, E., Eldholm, O. and Sundvor, E. 1969. Seismiske undersøkelser av den norske kontinentalsokkel, Sklinnabanken–Andøya. *Teknisk Rapport No. 3,* Jordskjelvstasjonen, Universitetet i Bergen, 15 pp.

Oftedahl, C. 1972. A sideritic ironstone of Jurassic age in Berstadfjorden, Trøndelag. *Norsk geol. Tidsskr.*, **52**, 123.

Os Vigran, J. 1970. Fragments of a middle Jurassic flora from northern Trøndelag, Norway. *Norsk geol. Tidsskr.*, **50**, 193–214.

Sokolov, V. N. (Ed.) 1965. *Geology of Spitsbergen.* 2 vols. Translated 1970, National Lending Library for Science and Technology, Boston Spa, 302 pp.

Sundvor, E. 1973. Seismic refraction and reflection measurement in Southern Barents Sea. *Teknisk rapport No. 7,* Seismological Observatory, University of Bergen.

Talwani, M. and Eldholm, O. 1972. Continental Margin off Norway: A geophysical study. *Bull. geol. Soc. Am.*, **83**, 3575–606.

Vischer, A. 1943. Die postdevonische Tektonik von Ostgronland zwischen 74°–75° N. *Meddr. Grønland*, **133**, 1–194.

Wilson, J. T. 1963. Continental drift. *Scient. Am.*, **208**, 86–100.

Ziegler, P. A. (*In press*). The geological evolution of the North Sea area in the framework of north-western Europe. *Proc. Bergen North Sea Conference. Norg. geol. Unders.*

DISCUSSION

Sir Kingsley Dunham (Institute of Geological Sciences): Would Mr Ronnevik care to comment on the origin and tectonic significance, if any, of the Norwegian Trench, which follows the east coast of Norway?

H. Ronnevik: The Norwegian Trench, as defined by bathymetric maps, curves around the whole Norwegian coast south of 62° N. At present, our knowledge of this geomorphological feature is limited to the west and south coast, where exploration has been concentrated. Here the Tertiary is unfaulted. Hence the only explanation we see for the trench feature is glacial erosion during the Quaternary. We would also suggest a similar explanation for the trench along the east coast.

M. F. Osmaston: The authors consider the West Spitsbergen Basin as closed to the south-east. Could Mr Ronnevik explain their reasons.

H. Ronnevik: The late Palaeozoic–Cenozoic West Spitsbergen basin as defined on geological maps is almost closed at the southern tip of Spitsbergen. The geophysical surveys carried out so far suggest that the Svalbard Bank area is a sedimentary platform north of the main basin in the Barents Sea. Hence we interpret the West Spitsbergen basin as a closed basin within this platform.

9

North Sea Basin History in the Tectonic Framework of North-Western Europe

By P. A. ZIEGLER

(Shell Internationale Petroleum Maatschappij B.V., The Hague)

SUMMARY

The tectonic history of the North Sea area can be divided into five stages:

1. *Caledonian geosynclinal stage (Cambrian–Devonian)*. Metamorphics and intrusives of Caledonian age form the basement complex for much of the North Sea area. The north-eastern boundary of the Caledonides fold belt can at this stage not be closer defined other than that it appears to trend from the central North Sea through northern Germany into Poland.

2. *Variscan geosynclinal stage*. Devonian and Carboniferous sediments transgress from the south over the eroded Caledonides and reach maximum thickness in the southern North Sea, an area which formed part of the Variscan foredeep.

3. *Permo-Triassic intracratonic stage*. Following the Variscan orogeny large parts of the North Sea were occupied by the rapidly subsiding intracratonic northern and southern Permian basins that contain thick sequences of clastics and evaporites.

4. *Rifting, taphrogenic stage*. Development of the North Sea rift system started during the Triassic and dominated the palaeographic setting of the area during the Jurassic and Cretaceous. The evolution of the North Sea rift is related to the development of the Arctic–North Atlantic rift zone. The latter reached the stage of crustal separation in the early Tertiary, at which time the North Sea rift became inactive.

5. *Tertiary, post-rifting stage of regional basin subsidence*. With the termination of rifting movements in the North Sea the area became subject to regional subsidence leading to the development of a symmetrical, saucer-shaped intracratonic basin. The late Tertiary Rhône–Rhine rift system does not extend into the North Sea and post-dates the North Sea rift.

INTRODUCTION

From Cambrian to Recent the North Sea area underwent a complex geological evolution during which it formed part of different tectonic provinces and sedimentary basins. From the viewpoint of basin development we distinguish the following stages:

1. Caledonian geosynclinal stage (Cambrian–Devonian);
2. Variscan geosynclinal stage (Devonian to Carboniferous);
3. Permo-Triassic intracratonic stage;
4. Rifting, taphrogenic stage (late Triassic to early Tertiary);
5. Post-rifting stage (Tertiary).

Superposition of these basins in various combinations in different parts of the North Sea controls the hydrocarbon potential of the respective provinces.

The geological history of the North Sea area can only be understood when viewed against the background of the tectonic evolution of north-western Europe. The aim of this paper is to retrace with a sequence of palaeogeographic maps in a kaleidoscopic fashion the main development stages of the North Sea within the framework of north-western Europe.

Geological information obtained during the recent exploration efforts in the North Sea is integrated with data available from the onshore areas. For the latter the author has drawn heavily from voluminous compilation literature dealing with the palaeogeographic evolution of Europe. The palaeogeographic maps presented cover a large area and span large time intervals, necessitating much generalization and simplification. These maps have not been palinspastically corrected; facies provinces are therefore distorted and crowded to various degrees in areas that have been subjected to compression during orogenic periods.

1. CALEDONIAN GEOSYNCLINAL STAGE

From Cambrian to early Devonian the North Sea area was occupied by the complex Caledonian geosynclinal sequence (Fig 1).

The Appalachian–Scottish–Norwegian geosyncline crossed the northern North Sea. Much of Sweden and the Baltic were occupied by a wide miogeosynclinal platform, the Fenno–Sarmatian shelf. The south-western boundary of this stable platform is paralleled by the Danish–Polish trough. This feature coincides roughly with the Tornquist Line, a major basement lineament that crosses eastern Europe from the Black Sea to the Skagerrak (Tectonic map of Europe, 1964; Shurawlew, 1965; Bush *et al.*, 1974).

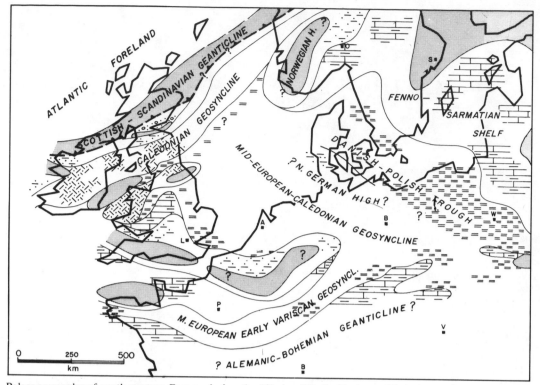

FIG 1.   Palaeogeography of north-western Europe during the Silurian (Wenlockian), modified after Walter (1972). For legend see
Fig 19.

The late Silurian palaeogeographic setting of Central Europe is known in a fragmentary way only. The occurrence of the North German high, postulated by Walter (1972) as separating the Danish–Polish trough from the largely hypothetical mid-European Caledonian geosyncline, remains to be proven by well data. Outlines of the mid-European "early Variscan" geosyncline are speculative.

The *Caledonian orogeny* (Gee and Wilson, 1974; Strand and Kulling, 1972; Bennison and Wright, 1969) resulted in the fusion of the North American–Greenland plate with the north-west European plate along the Appalachian and Scottish–Norwegian Caledonian fold belt (Wilson, 1966; Dewey, 1969). Isolated basement tests in the central North Sea give positive evidence for the existence of a mid-European Caledonian orogenic system (Fig 2) linking the Scottish–Norwegian Caledonides with those of the Ardennes, Germany (Odenwald, Harz), Czechoslovakia and Poland (Stille, 1950; Wills, 1950; von Gaertner, 1960; Znosko, 1964). However, to date insufficient control points are available to map the north-eastern boundary of the mid-European Caledonides (Busch and Kirjuchin, 1972; Bush *et al.*, 1974).

The pre-Permian Germano-type deformation (term used in the sense of foreland block-faulting and subordinate thrusting) of the Lower Palaeozoic sediments in the Skagerrak and in Denmark as well as in the western Baltic and Poland is interpreted to be of a Caledonian age (Bush *et al.*, 1974).

From a petroleum geologist's point of view metamorphics and intrusives of a Caledonian age form the economic basement for much of the North Sea. However, in the Skagerrak, in Denmark and the Baltic, Precambrian basement rocks are overlain by unmetamorphosed Cambrian to early Devonian sediments. These contain such potential hydrocarbon source rocks as *e.g.* the Cambrian Alum Shale (Schlatter, 1969).

2.   VARISCAN GEOSYNCLINAL STAGE

In Central Europe the Caledonides orogeny caused an accentuation of the Alemanic–Bohemian geanticline paralleled by a sharper definition of the Variscan geosyncline (Fig 3). North-western Europe obtained with this a new polarity. The Variscan geosyncline which crossed Europe from Ireland to beyond Poland developed, during the Devonian, into a dominating feature whereas the Caledonides mountain ranges were rapidly degraded.

Late to post-orogenic uplift, associated with a partial collapse of the Caledonian mountain system, led to the establishment of large intramontane basins such as the Orcadian and Caledonian cuvettes in which the thick, continental and in part lacustrine and bituminous series of the Devonian Old Red Sandstone were deposited

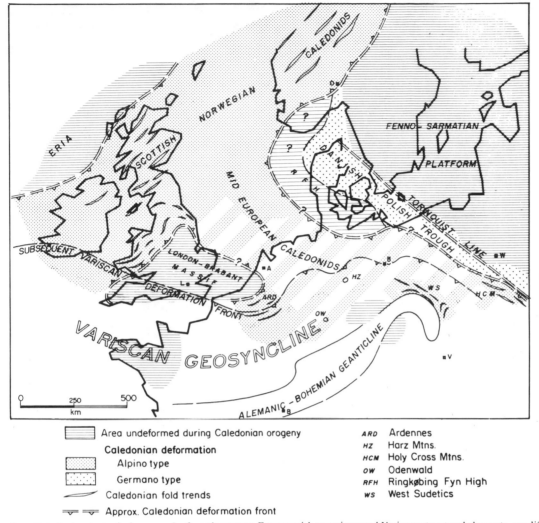

FIG 2.   Late Caledonian tectonic framework of north-western Europe with superimposed Variscan structural elements, modified after Znosko (1964).

(Bennison and Wright, 1969; Allen *et al.*, 1967). A connection between the Orcadian basin and the scattered Old Red Sandstone outcrops of the Norwegian coast (Nilsen, 1973) is likely. Old Red equivalent sandstones occur in the Baltic (Pajchlowa, 1970) and have also been encountered in some boreholes in the Central North Sea.

The Variscan geosyncline was flanked to the north by a wide, in part reef-bearing carbonate shelf, from which marine transgressions reached as far as the central North Sea. Control points are still too sparse to reconstruct details of the Devonian palaeogeographic framework of much of north-western Europe. However, the London–Brabant Massif appears as a major stable block, a role it had apparently already played during the Caledonian orogeny and which it continued to play during much of the subsequent development of the North Sea area.

The late Devonian–early Carboniferous *Bretonic orogeny* resulted in a consolidation of the Alemanic–Bohemian geanticline and in the emergence of the Armorican–Central German highs. During the Lower Carboniferous (Fig 4) these highs formed the source of the thick flysch-like Culm series that were deposited in the Variscan foredeep. During the early Carboniferous the distal northern parts of this foredeep were occupied by a wide carbonate shelf which extended from the Irish Waulsortian reef platform to Poland. Marine ingressions reached far to the north into the progressively degraded Caledonian chains. Viséan coal-bearing sequences were locally deposited in the central North Sea and northern England. Of special interest is the thick, essentially non-marine Oil Shale sequence in the Scottish Midland Valley (Macgregor and MacGregor, 1948; Bennison and Wright, 1969). Based on the sparse control available, the northern and eastern parts of the North Sea appear to have formed part of a large positive area during much of the Lower Carboniferous.

At the turn from the Lower to the Upper Carboniferous the *Sudetic orogeny* (Pfeiffer, 1971) led to a further consolidation of the Variscan mountain system with sedimentation continuing in

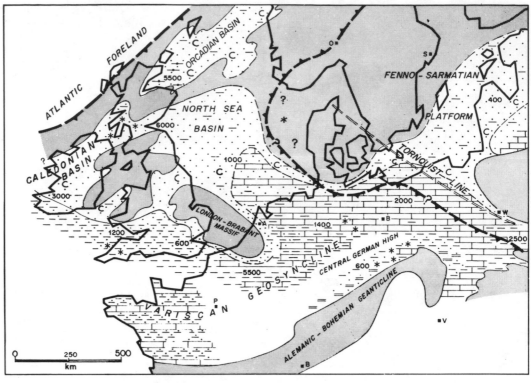

—▲—▲— Caledonian deformation front

FIG 3.   Palaeogeography of north-western Europe during the Upper Devonian (Upper Old Red). Thickness values refer to total Devonian. For legend see Fig 19.

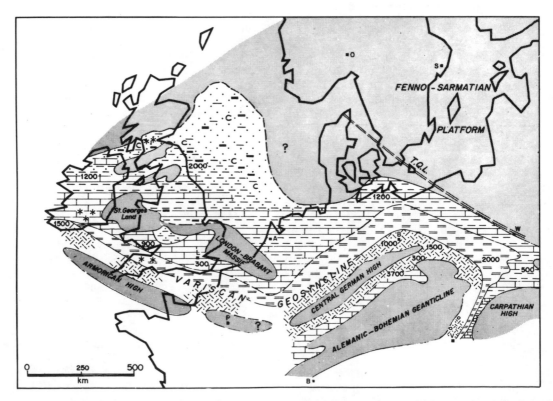

FIG 4.   Palaeogeography of north-western Europe during the Lower Carboniferous (Viséan). Thickness values refer to total Lower Carboniferous. For legend see Fig 19.

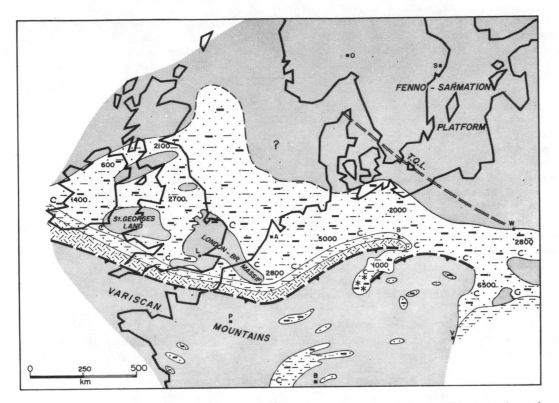

FIG 5.   Palaeogeography of north-western Europe during the Upper Carboniferous (Westphalian). Thickness values refer to total
Upper Carboniferous. For legend see Fig 19.

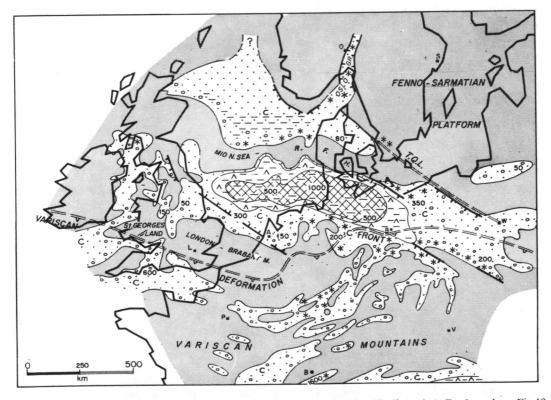

FIG 6.   Palaeogeography of north-western Europe during the Middle Permian (Rotliegendes). For legend see Fig 19.

intramontane successor basins. During the Upper Carboniferous deposition of the marine series was essentially restricted to a narrow trough flanking the rising Variscan mountain chains. Only during the Namurian did marine ingressions from the Variscan foredeep inundate much of the Baltic and north German shelf to reach also northern England and Scotland (Yoredale and Limestone Coal Group).

Figure 5 depicts the palaeogeographic set-up of north-western Europe during the Westphalian. Paralic conditions prevailed in much of the Variscan foredeep and the northward adjacent shelf areas, leading to the deposition of very thick coal-bearing sequences. These are of particular economic interest especially since they constitute the source rocks for the gas found in the southern North Sea and the adjacent Netherlands and German onshore area (Patijn, 1964; Lutz et al., in press). These Upper Carboniferous coal-bearing sequences do not extend into the central and northern North Sea and are also missing in the Danish onshore areas.

3.   PERMO-TRIASSIC INTRACRATONIC STAGE

The late Carboniferous Asturian and the early Permian Saalian orogenic phases (Variscan orogeny) resulted in a final consolidation of the Variscan chains with the north European craton. With this compressive stresses ceased to influence north-western Europe. The Permian tectonic framework (Figs 6 and 7) is characterized by an extensional setting resulting in the formation of two large, post-orogenic intracratonic basins in the Variscan foreland. These received a great thickness of red beds and evaporites.

During the Triassic a new megatectonic setting gradually came into being with the development of a set of narrow rifts and grabens.

The main features of the Permo-Triassic tectonic framework are the emplacement of the North Atlantic rift system and the subsidence of the Variscan mountain system resulting in the establishment of the Tethys Sea.

During the *Permian* the Variscan foreland was initially subjected to uplift, tilting and erosion, followed by differential subsidence along predominantly north-west–south-east trending normal faults (Hercynian strike, Knetsch, 1963) and widespread effusion of Lower Permian volcanics (Busch and Kirjuchin, 1972). The Tornquist Line appeared again as a major lineament. Progressive subsidence led to the establishment of the Permian Rotliegendes basins (Katzung, 1972) (Fig 6). The Mid North Sea–Ringkøbing–Fyn High appeared as the major positive element separating the less-well-known northern Permian basin from the southern Permian basin. In eastern Europe the southern

Permian basin encroached onto the Variscan fold belt.

In the southern Permian basin Rotliegendes dune sands (Glennie, 1972) are the primary gas reservoir in the southern North Sea and the adjacent Netherlands and German onshore areas. These sands grade northwards into sebka shales and evaporites; the latter were deposited in an inland sea. The Rotliegendes salts gave rise during the Mesozoic to salt diapirism in the German Bight and the adjacent onshore areas (Jaritz, 1973).

The configuration of the northern Permian basin is less known and for want of sufficient well data no reliable facies pattern can as yet be drawn up. Significant tectonic elements of the northern Permian basin are the volcanic Oslo graben (Wurm, 1973), the Danish trough and possibly the earliest traces of the Viking Graben in the northern North Sea.

Continued subsidence of the arid Rotliegendes basins possibly below sea level resulted finally in the catastrophic ingression of the Zechstein seas, probably through the northern North Sea (Fig 7).

The deposition of thin shelf carbonates and sulphate sequences was restricted to the margins of the Zechstein basins; these were offset by thick, basin-filling halite series that reached thicknesses in excess of 1000m, both in the southern and northern Permian basins.

Zechstein carbonates contain significant quantities of gas in the onshore areas of Germany and the Netherlands but in the North Sea play only a subordinate role as hydrocarbon reservoirs. Diapirism of the Zechstein salts both in the southern and northern Permian basins strongly influenced the post-Triassic sedimentation (Jaritz, 1973).

With the *Triassic* a continental regime of deposition returned to much of north-western Europe. The Permian structural pattern was modified by the emplacement of new graben systems (Fig 8), thus providing a transition to the Jurassic structural framework. Development of these new graben systems is thought to be related to early rifting movements along the Arctic–North Atlantic rift zone (Haller, 1970; Hallam, 1971).

The Viking–Central Graben system of the North Sea shows clear evidence of rapid differential subsidence during the Triassic. However, at this time it had not yet developed into the dominant structural feature it was to become during the Jurassic and Cretaceous. Similar rapidly subsiding Triassic grabens in the North Sea are the Horn Graben in the Danish offshore and the Glückstadt Graben in northern Germany.

The Danish–Polish furrow bounded to the north-east by the Tornquist Line subsided rapidly during the Triassic. Further grabens developed along the Atlantic sea-board of Scotland and Ireland as well as in the Celtic Sea and the Western Approaches. Normal to the northern margin of the Brabant–Rhenian massif the Emsland and Weser

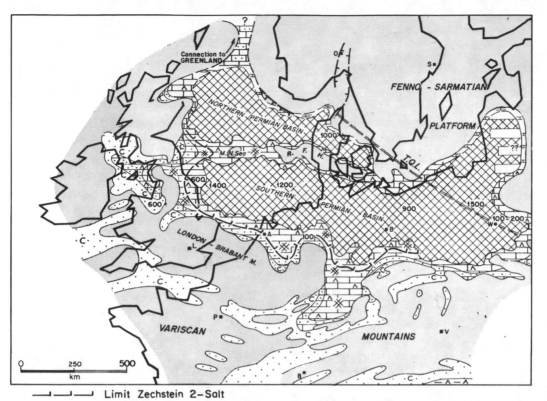

—⌐ ⌐—⌐— Limit Zechstein 2–Salt

FIG 7. Palaeogeography of north-western Europe during the Upper Permian Zechstein. For legend see Fig 19.

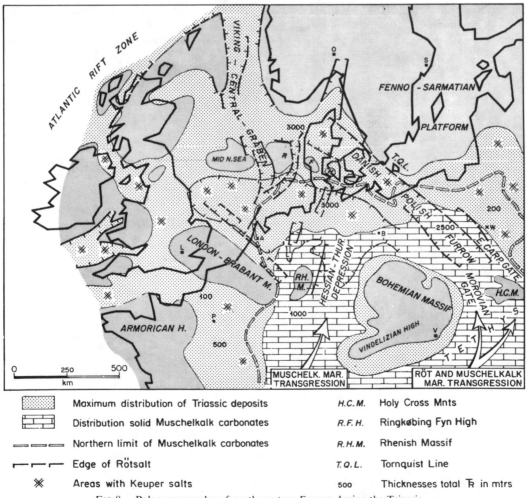

Maximum distribution of Triassic deposits	H.C.M. Holy Cross Mnts
Distribution solid Muschelkalk carbonates	R.F.H. Ringkøbing Fyn High
Northern limit of Muschelkalk carbonates	R.H.M. Rhenish Massif
Edge of Rötsalt	T.Q.L. Tornquist Line
Areas with Keuper salts	500 Thicknesses total Tr in mtrs

FIG 8. Palaeogeography of north-western Europe during the Triassic.

depressions formed. During the Lower Triassic Bunter the highs flanking these grabens were actively uplifted and subjected to erosion as illustrated by the Hardegsen unconformity (Wolburg, 1961, 1962).

Progressive down-warping and widening of the northern and southern Permo-Triassic basins resulted in the gradual burial of the Mid North Sea–Ringkøbing–Fyn High. Further degradation of the Variscan mountains and progressive subsidence of their intramontane basins brought about a link-up between the north European basin and the Tethian basin (Boselli and Hsü, 1973). From this marine incursions during Röt and Lower Muschelkalk times reached the southern North Sea *via* the Moravian and East Carpathian Gate (Polish furrow).

The Upper Muschelkalk marine transgression entered into the north European basin through the Hessian depression as well as through the Moravian and East Carpathian Gate (Tokarski, 1965; Wurster, 1968). However, to date no evidence is available for a Triassic marine connection between the North Sea and the marine areas of north-eastern Greenland (Birkelund, 1973) and Spitzbergen (Parker, 1967; Defrentin-Lefranc *et al.*, 1969; Harland, 1973; Cox and Smith, 1973).

In the north European basin where thick Zechstein salt deposits coincided with Triassic depocentres, halokinetic movements were triggered during late Triassic times. These movements continued to influence, at least on a local scale, the depositional pattern of the Jurassic, Cretaceous and Tertiary (Jaritz, 1973).

Several, and in places significant, Triassic Bunter gas accumulations occur in onshore Germany as well as in the southern North Sea. In the central and northern North Sea the hydrocarbon potential of the Triassic has as yet been evaluated by a limited number of tests only.

4. TAPHROGENIC, RIFTING STAGE (LATE TRIASSIC TO EARLY TERTIARY)

The late Triassic topography of north-western Europe was characterized by an extremely low relief (Trümpy, 1971). The early Cimmerian (Rhaetian) movements marked the transition from the Triassic depositional framework to the Jurassic sedimentary pattern (Rusitzka, 1967, 1968). With the Liassic transgression marine conditions returned to large parts of north-western Europe. Marine ingressions into the North Sea area originated chiefly from the newly established North Atlantic Seaway (Hallam, 1971) through the

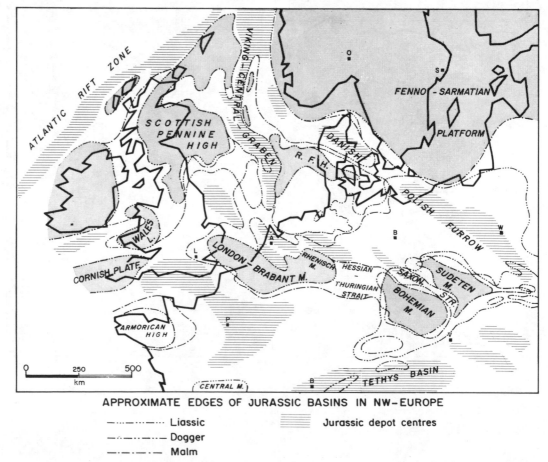

APPROXIMATE EDGES OF JURASSIC BASINS IN NW–EUROPE

—··—··— Liassic   ▨▨▨▨ Jurassic depot centres
—·—·—·— Dogger
—·—·—·— Malm

Fig 7. Palaeogeography of north-western Europe during the Upper Permian Zechstein. For legend see Fig 19.

northern North Sea, but also from Tethys *via* the Paris basin and southern England, *via* the Hessian–Thuringian depression, and in the Toarcian also through the Moravian Gate (Kölbel, 1968). Figure 9 presents an outline of the Jurassic framework of north-western Europe, indicating areas of maximum total Jurassic subsidence. A comparison of the approximate depositional edges of the Liassic, Dogger and Malm conveys an impression of the palaeogeographic changes that occurred in the general North Sea area during the Jurassic. Main features of the Jurassic palaeogeography are:

(a) Opening up of the Arctic North Atlantic Seaway during the Pliensbachian and its continued widening during the Middle and Upper Jurassic (Hallam, 1971); the Arctic North Atlantic remained, however, in a pre-drifting stage (Laughton, in press).
(b) The establishment of the North Sea Central Graben system as the dominant rift system. The Horn and Glückstadt grabens became largely inactive. The Danish–Polish furrow continued, however, to subside rapidly.
(c) Flanking the London–Brabant–Rhenish–Ardennes massif, as well as the Bohemian massif, marginal troughs (so-called "Randtröge") (Voigt, 1962) developed. Dur-

ing the Dogger and Malm the narrow Saxonian strait transected the Bohemian massif in a north-west–south-east direction.
(d) During the Jurassic the Alpine geosyncline reached a "Mediterranean paraoceanic", pre-orogenic stage, with eugeosynclinal (oceanic) troughs separated by more stable platform areas (Trümpy, 1971). Subsidence of large areas along flexures and normal faults is documented from the northern margin of the geosyncline and is indicative of extensional tectonics (Trümpy, 1958; Kölbel, 1968).

Viewed against this background one gains the impression that north-western Europe was subjected as a whole to extensional stresses during the Jurassic. This led to a progressive fragmentation of north-western Europe, a process that was already initiated during the Permian. During the Jurassic the Viking–Central North Sea Graben system and the Danish–Polish furrow developed into the dominant fracture zones in the Alpine foreland (Fig 10). The Variscan fold belt apparently prevented a southward extension of the North Sea graben system. Instead a series of *en echelon* tensional depressions or marginal troughs, "Randtröge" in the sense of Voigt (1962), displaying generally a Hercynian strike (north-west–south-east), developed mainly along the

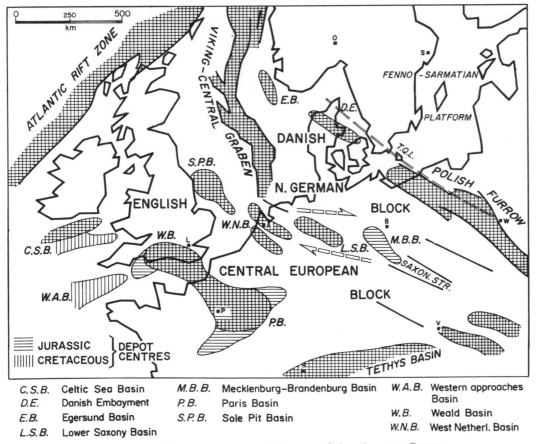

C.S.B.	Celtic Sea Basin	
D.E.	Danish Embayment	
E.B.	Egersund Basin	
L.S.B.	Lower Saxony Basin	
M.B.B.	Mecklenburg–Brandenburg Basin	
P.B.	Paris Basin	
S.P.B.	Sole Pit Basin	
W.A.B.	Western approaches Basin
W.B.	Weald Basin
W.N.B.	West Netherl. Basin

FIG 10. Jurassic—Cretaceous tectonic framework of north-western Europe.

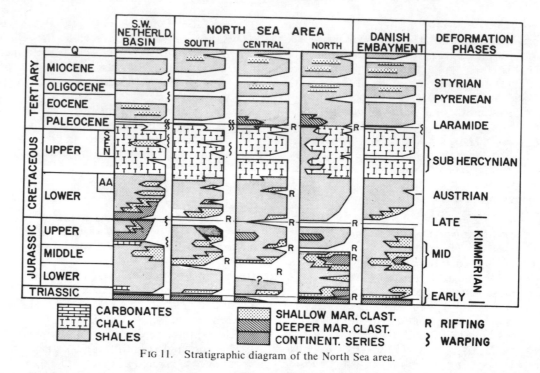

	S.W. NETHERLD. BASIN	NORTH SEA AREA			DANISH EMBAYMENT	DEFORMATION PHASES
		SOUTH	CENTRAL	NORTH		
TERTIARY	Q					
	Miocene					STYRIAN
	Oligocene					PYRENEAN
	Eocene					
	Paleocene					LARAMIDE
CRETACEOUS	Upper					SUB HERCYNIAN
	Lower					AUSTRIAN
JURASSIC	Upper					LATE
	Middle					MID
	Lower					EARLY
	TRIASSIC					

	CARBONATES		SHALLOW MAR. CLAST.	R RIFTING
	CHALK		DEEPER MAR. CLAST.	⟩ WARPING
	SHALES		CONTINENT. SERIES	

FIG 11. Stratigraphic diagram of the North Sea area.

northern margin of the already deeply eroded Variscan mountain system. This indicates a reactivation of a fault system that was already evident during the Permian. Similar north-west–south-east striking troughs developed in the Carpathians and their foreland (Roth, 1970), an area where Variscan and possibly Caledonian structural trends subparallel the north-west–south-east striking Tornquist Line. The latter formed again the boundary between the stable Fenno–Sarmatian platform to the east and the more mobile blocks of north-western Europe. Details of the fault pattern in the Tethys area, which formed the southern boundary of the metastable north-western European plate have yet to be un-ravelled (see also Dewey et al., 1973).

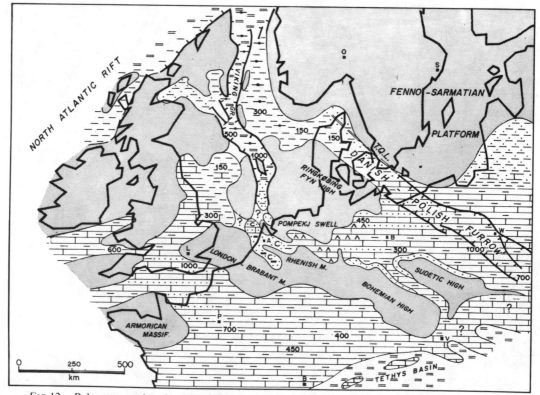

FIG 12. Palaeogeography of north-western Europe during the Upper Jurassic. For legend see Fig 19.

The Jurassic North Sea Viking–Central Graben

Despite the considerable number of well penetrations in the North Sea it is still difficult to develop a comprehensive story of the Jurassic evolution of the Viking–Central Graben system. Several stages of downfaulting and differential subsidence of its fragmented graben floor, coupled with uplifting of the rift margin as well as the effects of regional unconformities and halokinetic movements of the Zechstein salts make the deciphering of events rather difficult.

In the northern North Sea there is clear evidence for continuous differential subsidence of the graben floor since the Triassic and throughout the Jurassic. A major rifting phase locally corresponding to an angular unconformity within the Dogger is documented from the central North Sea (Fig 11). In the southern and northern North Sea this phase is expressed by regressive–transgressive clastic cycles often in concordance with the underlying marine Liassic sequences. A second phase of rifting, this time better documented in both the southern and northern North Sea, occurred during the transition from the Middle to the Upper Jurassic (Callovian–Oxfordian phase). In the Viking–Central Graben the Upper Jurassic is represented by deeper water shales that locally contain turbidite sands. This indicates that the rift system had developed into a submarine trough (Fig 12). Upper Jurassic shales are frequently rich in organic matter and constitute an important oil-source rock in the North Sea.

At the Jurassic–Cretaceous boundary a further major rifting phase occurred throughout the Viking–Central Graben system (late Cimmerian phase).

All these three main phases are recognizable through much of the North Sea area as either unconformities, disconformities or regressive–transgressive cycles (Fig 11). Only in the deepest parts of the Central Graben, a veritable taphrogeosyncline (term used in the sense of Trümpy, 1960), do more or less continuous Middle Jurassic–Lower Cretaceous sequences occur. During the successive period of down-faulting of the graben floor the graben margins were uplifted and subjected to erosion thus conforming to the rift model drawn up by Illies (1970). Erosion on the highs flanking the Central Graben cut as deep as the Zechstein (e.g. in the central North Sea) and locally even into the Devonian and basement. In the Moray Firth Basin and the Egersund–Danish Embayment basin, both located off the flanks of the North Sea central rift, more complete Mesozoic sequences are preserved.

Evidence for Jurassic rift volcanism is restricted to the general Central Graben area.

The Rhaetian to Lower Jurassic Statfjord sands and the Dogger Brent sands form the main reservoirs for the major oil accumulations in the Viking Graben. However, their equivalents are of less importance in the Central Graben, where (e.g. its north-western parts) Upper Jurassic shallow water sands constitute the major Mesozoic oil reservoir.

Jurassic Rifting Phases in North-west Europe and the North Atlantic

The major Jurassic rifting phases recognized in the Viking–Central Graben system are also evident in the sedimentary sequences of Spitzbergen, eastern Greenland, Andøya and the Hebrides. This correlation of tectonic events attests to the

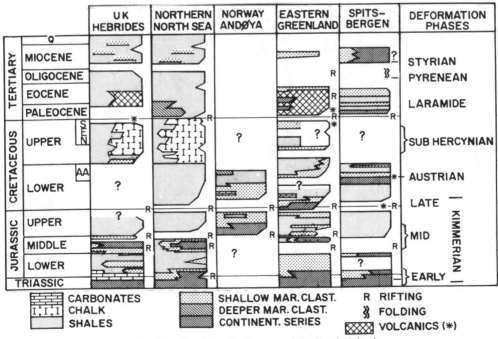

FIG 13. Stratigraphic diagram of the North Atlantic.

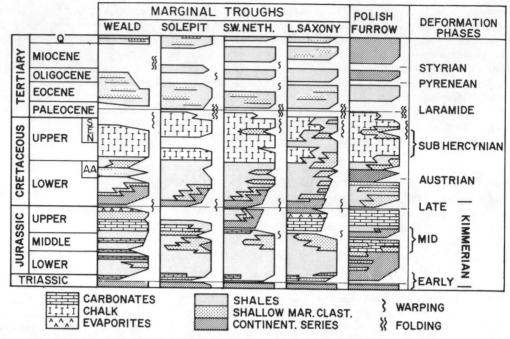

FIG 14. Stratigraphic diagram of the marginal troughs.

approximate contemporaneity of rifting movements in the Arctic–North Atlantic and the North Sea areas (Fig 13).

The tectonic phases recognized in the North Sea are also apparent in the "marginal troughs" flanking the Variscan massifs. In these basins they are evident as either regressive–transgressive cycles or as unconformities associated with minor warping of the basin floors followed by rapid subsidence (Fig 14). Warping and temporary uplifting of the basin floors may be due to slight jarring of these basins possibly in response to

minor transform movements between the Danish–north German block in the north and the central European block in the south (Fig 10).

In the Danish–Polish furrow similar extension phases are evident as in the North Sea. Here the Mid-Dogger phase corresponds to a basin-wide transgression, the Callovian–Oxfordian phase to a disconformity, and the late Cimmerian phase again to a basin-wide disconformity (Fig 14).

Thus the above regional tectonic model postulating that north-western Europe was subjected during the Jurassic to extensional

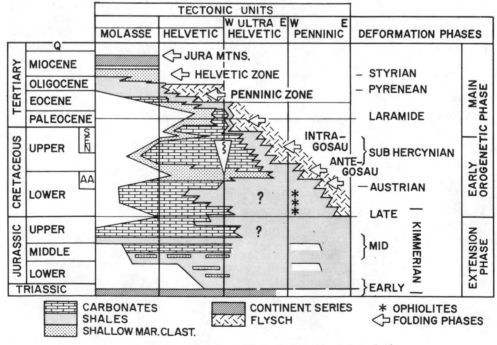

FIG 15. Stratigraphic diagram of the Eastern and Central Alps.

stresses, appears to be in good agreement with the occurrence of roughly correlative extensional phases in all of its major Jurassic basins. This applies not only to the extra-Alpine areas but is also valid for the Alpine geosynclinal realm, where during the Jurassic, extensional phases resulted in the accentuation of positive platforms and negative areas.

Cretaceous Rifting Phases in the North Sea and North Atlantic

During the *Lower Cretaceous* the overall stress pattern gradually changed in the Alpine geosyncline from an essentially extensional one to a compressional setting (Trümpy, 1965) (Fig 15).

However, this gradual change in the regional stress pattern during the Lower Cretaceous remained restricted to the Alpine geosyncline. In the northern Alpine foreland the Jurassic setting which was accentuated by the late Cimmerian phase persisted through much of the Cretaceous (Fig 16). Rifting movements continued during the Cretaceous in the Arctic–North Atlantic (Fig 13) causing the progressive widening of the Proto-Iceland basin and of the Rockall trough (Laughton, in press). In the North Sea graben system several minor rifting phases are recognized during the Cretaceous (Fig 11). The Viking and Central grabens continued, however, to subside rapidly and received thick Lower Cretaceous shales (up to 500m) followed by up to 1200m of Upper Cretaceous chalks and marls. The last phase of rifting, accompanied by uplifting of the graben

margins and rapid subsidence of the graben itself occurred during the Palaeocene (Laramide phase).

Overall, rifting movements gradually abated in the North Sea during the Cretaceous. This was accompanied by the gradual subsidence of the uplifted rift flanks resulting in their inundation during the Upper Cretaceous. This process was only interrupted by the sub-Hercynian and Laramide rifting phases, both of which are widely recognized in the North Sea, but are of minor importance only when compared to the Jurassic phases.

With the Laramide phase the deposition of chalk came to an end in the North Sea. In the Viking and Central grabens the Upper Palaeocene is represented by deeper water shales and turbidite sands. The latter, as well as Maastrichtian and Danian chalks, contain major oil and gas accumulations in both Viking and Central grabens.

Cretaceous Development of Onshore North-Western Europe

As a result of the late Cimmerian phase the London–Brabant, Rhenish and Bohemian massifs were consolidated into a continuous positive area stretching from England to Poland. This barrier separated the Alpine geosynclinal domain from the North Sea area. In the marginal troughs the late Cimmerian movements are readily recognizable as basin-wide unconformities preceded by minor warping (Fig 14). Also in the Polish–Danish basin the Jurassic–Cretaceous boundary is marked by a regional unconformity. During the early Lower

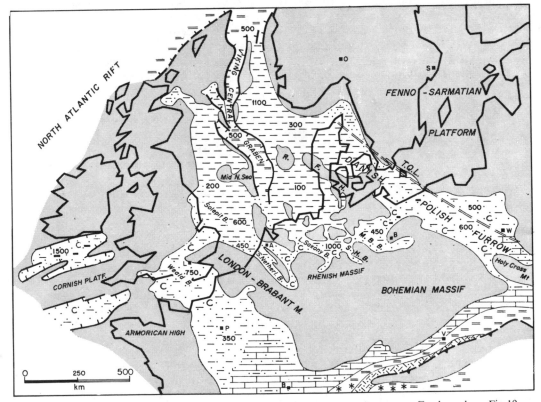

FIG 16. Palaeogeography of north-western Europe during the Lower Cretaceous. For legend see Fig 19.

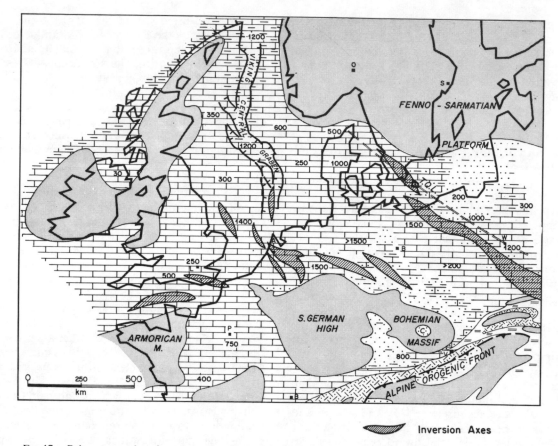

FIG 17.   Palaeogeography of north-western Europe during the Upper Cretaceous. For legend see Fig 19.

Cretaceous, subsidence resumed in the marginal troughs, as well as in the Polish–Danish furrow, leading to the deposition of thick clastic sediments (Wealden facies) in initially narrow basins. These basins gradually overstepped their margins during the Lower Cretaceous.

The Aptian–Albian (Austrian) phase, which is only vaguely reflected in the North Sea, interrupted this process, resulting in renewed erosion of the basin margins and the deposition of regressive–transgressive sand sequences (Kemper, 1972; Marek and Raczynska, 1972) (Figs 11 and 12).

As in the North Sea area the Upper Cretaceous transgression is very widespread in central Europe (Fig 17). During the Senonian (sub-Hercynian phase) first inversion movements (term used in the sense of Voigt, 1962) occurred in the marginal troughs as well as in the Polish Lowland, leading to the uplifting of the axial zones of the basins and the development of secondary depocentres flanking the uplifted basin centres (Voigt, 1962; Heybroek, 1974; Pozarski, 1960) (Fig 14). Large-scale inversion involving also the southern part of the Central North Sea Graben took place during the Maastrichtian to Upper Palaeocene (Laramide phase), resulting in uplifting of the basin-fill either along steep reversed, or normal faults and/or regional warping above the erosional level. These inversion movements are time-correlative with the

last phase of downfaulting in the central and northern parts of the North Sea Graben (Fig 11) and are thus not compatible with the above developed tectonic model (p. 142 and Fig 10). However, in the Alpine geosynclinal realm compressive movements became dominant during the Upper Cretaceous, with flysch deposition spreading during the Maastrichtian from the Penninic into the Ultra-Helvetic facies domain (Fig 15). It is conceivable that the compressive stresses which led during the late Cretaceous and early Tertiary (sub-Hercynian and Laramide phases) to crustal shortening in the Alps may also have affected the Alpine foreland to distances up to some 800km, causing the inversion of the marginal troughs and the tilting and uplifting of the Variscan massifs. A model for this type of foreland deformation is provided by the mountain flank thrusting in the Rocky Mountain foreland of Wyoming and Colorado (USA).

This interpretation of the inversion tectonics in the marginal troughs is supported by the occurrence of steep reversed faults and thrusts of a late Cretaceous age in, for example, the sub-Hercynian and Lower Saxony basin (Wolburg, 1953; Boigk, 1968; Keller, 1974). Steep reversed faults of a similar age have been mapped in the subsurface of the West Netherlands Basin and the Sole Pit Trough. The southern parts of the Central North Sea Graben display less intensive inversion tectonics than the marginal troughs. Inversion in the North Sea Graben dies out towards the north near the Dutch–German border. Similarly the inversion tectonics die out northwards in the Polish–Danish trough in the area of Scania. In view of the oblique strike of many of the marginal troughs relative to the Alpine strike a transcurrent component in their deformation pattern is likely and may even be dominant as, for example, in the Sole Pit Trough and the West Netherlands Basin.

5. POST-RIFTING, INTRACRATONIC STAGE (TERTIARY)

The taphrogenic stage of the North Sea that had lasted since the Triassic came to a close in the late Palaeocene. Crustal separation between the European and the North American–Greenland plate was effected during the late Palaeocene or Eocene (Pitman and Talwani, 1972; Laughton, 1974), thus initiating the drifting stage in the

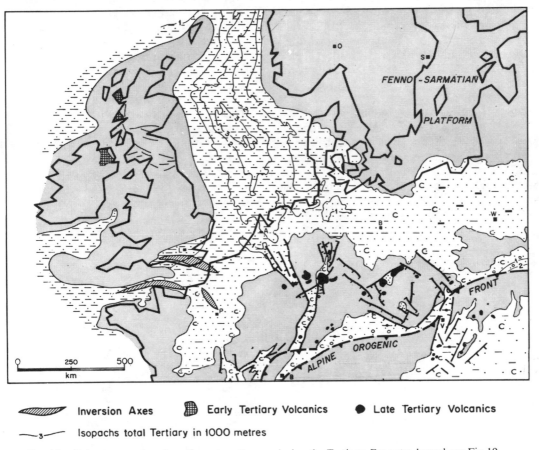

| ▨ | Inversion Axes | ▧ Early Tertiary Volcanics | ● Late Tertiary Volcanics |

—₃— Isopachs total Tertiary in 1000 metres

FIG 18. Palaeogeography of north-western Europe during the Tertiary. For extra legend see Fig 19.

Once inverted, these basins became largely inactive. Only minor inversion movements are recognized in the marginal troughs flanking the Variscan massif during the post-Laramide main Alpine orogenic phases with the exception of, for example, the Weald basin (Gallois and Edmunds, 1965), which was deformed mainly during the Miocene and in which only mild warping occurred during the Laramide phase (Fig 14).

An explanation for the contemporaneity of rifting movements in the North Sea and in the North Atlantic and compressive, orogenic movements in the Alpine geosyncline has to be sought on a scale beyond the scope of this paper (see *e.g.* Dewey *et al.*, 1973; Laughton, 1974).

Arctic–North Atlantic. Late Palaeocene to Eocene flood basalts extruded during the final rifting and early drifting phase are known from Scotland (Rayner, 1967; Richey, 1961; Mitchell and Reen, 1973), the Faeroe Islands (Tarling and Gale, 1968), as well as from eastern Greenland (Brooks, 1973).

With the Arctic–North Atlantic entering the drifting stage during the Eocene (Laughton, 1974) extensional stresses apparently ceased to influence north-western Europe.

During the Eocene to Recent the North Sea was dominated by regional subsidence resulting in the formation of a symmetrical saucer-shaped intracratonic basin the axis of which coincides with the now inactive Central Graben. Maximum

Tertiary thicknesses of up to 3.5km (Dunn *et al*, 1973; Heybroek *et al.*, 1967) occur in the central parts of the North Sea (Fig 18). A possible geodynamic model explaining this Tertiary subsidence pattern of the North Sea has been discussed by Ziegler (1975).

During the Late Cretaceous and Tertiary Central Europe was dominated by the Alpine orogeny, the main phases of which are summarized by Fig 15. In the North Sea regional disconformities correlate roughly to these main orogenic phases, which are reflected in the marginal troughs by minor inversion movements. A notable exception is the Weald basin which underwent major inversion during the Miocene.

In the northern Alpine foreland the Rhône–Rhine graben system started to subside rapidly during the Eocene (Illies, 1970; Horn *et al.*, 1972). Its further development during the Oligocene and Upper Tertiary is largely concomitant with the late Alpine folding phases. Volcanic activity in the Rhône–Rhine rift started in the early Tertiary and persisted until sub-Recent times. This graben system is presently in the active rifting stage.

Other graben systems developed during the

LEGEND TO PALEOGEOGRAPHIC MAPS

Positive areas

C Continental series

520 Thicknesses in metres

DOMINANT LITHOLOGIES

Sandstones and conglomerates

Sandstones

Deeper marine sandstones, Flysch

Shallow marine shales

Organic shales

Deeper marine shales

Carbonates

Halites (※)

Gypsum, Anhydrite

Coal

Volcanics

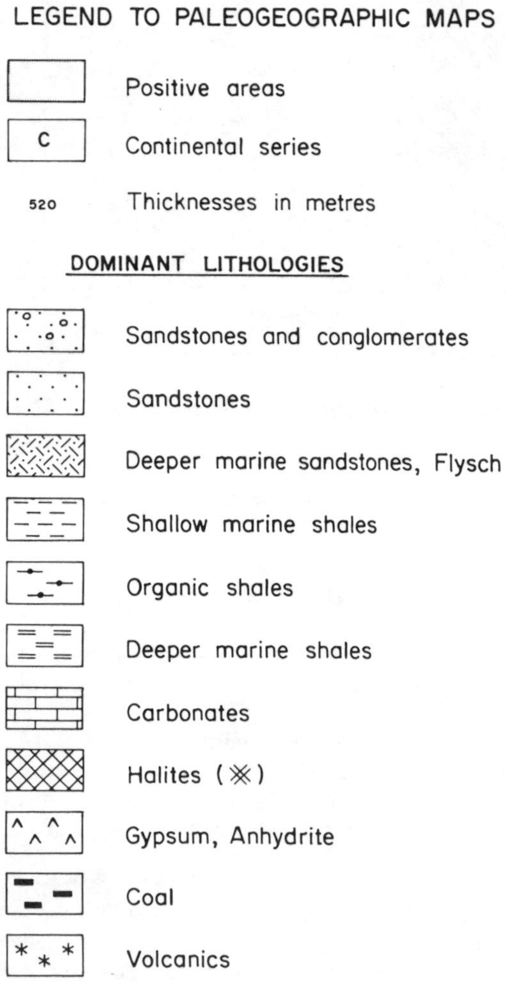

FIG 19. Legend to palaeogeographic maps.

Tertiary in, for example, the Panonic basin with the Vienna basin practically intersecting the Alpine orogenic front (Fig 18).

Uplifting and fracturing of the Variscan massif led to widespread volcanism during the Mio-Pliocene (Knetsch, 1963). The northernmost branch of the Rhône–Rhine graben system is the Ruhr Graben, which reaches the southern Netherlands (Heybroek, 1974). There is no evidence for a late Tertiary reactivation of the North Sea central rift system. In view of the age disparity between the Rhône–Rhine graben system which is in an active rifting stage and the North Sea central rift system which is in a post-rifting stage, the two should not be considered as part of one megafracture system dissecting western Europe. However tempting such a speculation may be, their apparent continuity is apparently largely fortuitous.

The Tertiary taphrogenic system of which the Rhône–Rhine graben and the Vienna basin form part is a large collapse system which affects the entire Alpine fold belt as well as the Mediterranean and the African plate. This system was apparently intermittently active during the late phases of the Alpine orogeny. An analogue to this late to post-orogenic extension phase in the Alpine domain may be seen in the early Permian collapse of the Variscan fold belt and its foreland (Fig 6).

REFERENCES

Allen, J. R. L., Dineley, D. L. and Friend, P. F. 1967. Old Red Sandstone Basins of North America and North-West Europe. *In:* D. H. Oswald (Ed.), International Symposium on the Devonian System, **1**, 69–98. *Alberta Soc. Petrol. Geol., Calgary.*

Bennison, G. M. and Wright, A. E. 1969. *The geological history of the British Isles.* Edward Arnold Ltd, London.

Birkelund, T. 1973. Mesozoic geology of east Greenland (summary). *In:* M. G. Pitcher (Ed.), Arctic Geology. *Mem. Am. Ass. Petrol. Geol.*, **19**, 149.

Boigk, H. 1968. Gedanken zur Entwicklung des Niedersachsischen Tektogens, *Geol. Jb.*, **85**, 861–900.

Boselli, A. and Hsü, K. J. 1973. Mediterranean plate tectonics and Triassic palaeogeography. *Nature, Lond.*, **244**, 144–46.

Brooks, C. K. 1973. Tertiary of Greenland—A volcanic and plutonic record of continental break-up. *In:* M. G. Pitcher (Ed.) Arctic Geology. *Mem. Am. Ass. Petrol Geol.*, **19**, 150–60.

Busch, V. A. and Kirjuchin, L. G. 1972. Über die verbreitung subsequenter Effusiva des Jungpalaeozoikums in Mitteleuropa. *Z. angew. Geol.*, **18**, 323–8.

Bush, V. A., Garetskiy, R. G. and Kirjuchin, L. G. 1974. Buried zone of Caledonian folding along the southwestern margin of the East European platform, Doklady. *Earth Science Sect.*, **208**, 65–9.

Cox, C. B. and Smith, D. G. 1973. A review of the Triassic vertebrate faunas of Svalbard. *Geol. Mag.*, **110**, 405–18.

Defrentin-Lefranc, S., Grasmück, K. and Trümpy, R. 1969. Notes on Triassic stratigraphy and palaeontology of North-Eastern Jamesonland (East Greenland). *Meddr. Grønland*, **168**, 136 pp.

Dewey, J. F. 1969. Evolution of the Appalachian/Caledonian Orogen, *Nature, Lond.*, **222**, 124–29.

Dewey, J. F., Pitman, W. C., Ryan, B. F. and Bonnin, J. 1973. Plate tectonics and the evolution of the Alpine System. *Bull. geol. Soc. Am.*, **84**, 3137–80.

Dunn, W. W., Eha, S. and Heikkila, H. H. 1973. North Sea is a tough theater for the oil-hungry industry to explore. *Oil and Gas Jl.*, **71** (2), 122–28, (3), 90–3.

Falke, H. (Ed.) 1972. Rotliegend, essays on European Lower Permian. *International Sedimentary Petrographical Series V.* **15**, Leiden: Bull.

Gaertner, H. R. von, 1960. Über die Verbindung der Bruchstücke des Kaledonischen Gebirges im nordlichen Mitteleuropa. *Int. Geol. Congr. Rep. 21, Sess. Norden, Proc. Sect. 19*, Copenhagen 1960, 96–101.

Gallois, R. W. and Edmunds, F. H. 1965. The Wealden District. *British regional geology. Inst. Geol. Sci.* HM Stationery Office, London, 101 pp.

Gee, D. G. and Wilson, M. R. 1974. The age of orogenic deformation in the Swedish Caledonids; *Am. Jl. Sci.*, **274**, 1–9.

Glennie, K. W. 1972. Permian Rotliegendes of northwest Europe interpreted in light of modern desert sedimentation studies. *Bull. Am. Ass. Petrol. Geol.*, **56**, 1048–71.

Hallam, A. 1971. Mesozoic geology and opening of the North Atlantic. *Jl. Geol.*, **79**, 129–57.

Haller, J. 1969. Tectonics and neotectonics in East Greenland. *In:* North Atlantic—Geology and Continental Drift. *Mem. Am. Ass. Petrol. Geol.*, **12**, 852–58.

Haller, J. 1970. Tectonic map of East Greenland (1:500 000), an account of tectonism, plutonism and volcanism in East Greenland. *Meddr. Grønland*, **171**, 286 pp.

Harland, W. B. 1961. An outline structural history of Spitzbergen. *In:* G. O. Raasch (Ed.), *Geology of the Arctic. Proceedings of the first International Symposium on Arctic Geology.* Toronto Press. 68–132.

Harland, W. B. 1973. Mesozoic geology of Svalbard. *In:* M. G. Pitcher (Ed.), Arctic Geology. *Mem. Am. Ass. Petrol. geol.*, **19**, 135–48.

Hesse, R. 1974. Long distance continuity of turbidites: possible evidence for an early Cretaceous trench-abyssal plain in the Eastern Alps. *Bull. Geol. Soc. Am.*, **85**, 859–70.

Heybroek, P. 1974. Explanations to tectonic maps of the Netherlands. *Geologie. Mijnb.*, **53**, 43–50.

Heybroek, P., Haanstra, U. and Erdman, D. A. 1967. Observations on the geology of the North Sea area. *Wld. Petrol. Congr. 7, Mexico. Proc.*, **2**, 905–16.

Horn, P., Lippolt, H. J. and Todt, W. 1972. Kalium–Argon Altersbestimmungen an Tertiären Vulkaniten des Oberrheingrabens—I. Gesamtgesteinsalter. *Eclogae geol. Helv.*, **65**, 131–56.

Illies, J. H. 1970. Graben tectonics as related to crust–mantle interaction. *In: J. H. Illies and S. Mueller (Eds.), Graben Problems, Internat. Upper Mantle Project Scientific report No. 27.* Schwizerbart'sche Verlagsbuchhandlung, Stuttgart, 4–27.

Jaritz, W. 1973. Zur Entstehung der Salztrukturen Nordwest Deutschlands. *Geol. Jb. Reihe A.* (10) 77.

Katzung, G. 1972. Stratigraphie und Palaeogeographie des Unterperms in Mitteleuropa. *Geologie*, **21**, 570–84.

Keller, G. 1974. Die Fortsetzung der Osningzone auf dem Nordwestabschnitt des Teutoburger Waldes. *Neues Jb. Geol. Palaont. Mh.*, **74**, 72–95.

Kemper, E. 1972. The Valanginian and Hauterivian stages in North-west Germany. *In:* R. Casey and P. F. Rawson (Eds.), The Boreal Lower Cretaceous. *Geol. Jl. Special Issue No. 5*, 327–44.

Kemper, E. 1972. The Albian and Aptian stage in Northwest Germany. *In:* R. Casey and P. F. Rawson (Eds.), The Boreal Lower Cretaceous. *Geol. Jl. Special Issue No. 5*, 345–60.

Knetsch, G. 1963. *Geologie von Deutschland.* Ferdinand Enke Verlag, Stuttgart.

Kölbel, H. 1968. Regionale Stellung der DDR im Rahman Mitteleuropas. *Grundriss der Geologie der DDR. Band 1 Geologische Entwicklung des Cresamtgebietes.* Akademie-Verlag. Berlin, 18–66.

Laughton, A. S. *(In press).* A review of the tectonic evolution of the northeast Atlantic Ocean. *Proc. Bergen North Sea Conference. Norg. geol. Unders.*

Lutz, M., Kaasschieter, J. and Wijhe, J. *(In press).* Geological factors controlling gas accumulations in the Mid-European Rotliegend basin. *Wld. Petrol. Congr. 9 Tokyo, Proc.*

Macgregor, M. and MacGregor, A. G. 1948. The Midland Valley of Scotland. *British regional geology. Geol. Surv.* HM Stationery Office, Edinburgh, 95 pp.

Marek, S. and Raczynska, A. 1972. The stratigraphy and paleography of the Lower Cretaceous deposits of the Polish Lowland area. *In:* R. Casey and P. R. Rawson (Eds.), The Boreal Lower Cretaceous, *Geol. Jl. Special Issue No. 5*, 369–86.

Mitchell, J. G. and Reen, K. P. 1973. Potassium–argon ages from the Tertiary ring complexes of the Ardnamurchan Peninsula, Western Scotland. *Geol. Mag.*, **110**, 331–40.

Nilsen, T. H. 1973. Devonian (Old Red Sandstone) sedimentation and tectonics of Norway. *In: M. G. Pitcher* (Ed.) Arctic Geology. *Mem. Am. Ass. Petrol. Geol.*, **19**, 471–81.

Pajchlowa, M. 1970. The Devonian. *In: Geology of Poland, Vol. 1. Stratigraphy, Part I: Pre Cambrian and Palaeozoic.* Publishing House Wydawnictwa Geologiozne, Warsaw, 321–70.

Parker, J. R. 1967. The Jurassic and Cretaceous sequence in Spitzbergen. *Geol. Mag.*, **104**, 487–505.

Patijn, R. J. H. 1964. Die Entstehung von Erdgas. *Erdöl Kohle Erdgas Petrochem.*, **17**, 2–9.

Pfeiffer, H. 1971. Die variscische Hauptbewegung (sogenannte sudetische Phase) im Umkreis der äusseren Kristallinzone des variszischen Bogens, *Geologie*, **8**, 945–58.

Pitman, W. C. and Talwani, M. 1972. Seafloor spreading in the North Atlantic. *Bull. Geol. Soc. Am.*, **83**, 619–46.

Pozarski, W. 1960. An outline of the stratigraphy and palaeogeography of the Cretaceous in Polish Lowland. *Pr. Inst. geol. Warsaw*, **30**, 377–440.

Rayner, D. H. 1967. *The stratigraphy of the British Isles.* Cambridge University Press.

Richey, J. E. 1961. Scotland: the Tertiary volcanic districts. *British regional geology. Geol. Surv.* HM Stationery Office, Edinburgh.

Roth, Z. 1970. Alpine remobilization in the West Carpathians of Czechoslovakia. *Vest. ustred. Ust. geol. Prague*, **45**, 331–38.

Rusitzka, D. 1967. Paläogeographie der Trias im Nordteil der DDR. *Ber. dt. Ges. geol. Wiss. R.A.*, **12**, 268–89.

Rusitzka, D. 1968. Trias. *In: Grundriss der Geologie der DDR. Band 1. Geologische Entwicklung des Gesamtgebietes.* Akademie-Verlag, Berlin, 268–89.

Schlatter, L. E. 1969. Oil shale occurrences in Western Europe. *In:* P. Hepple (Ed.), *The exploration for petroleum in Europe and North Africa.* Institute of Petroleum, London; Elsevier, Amsterdam, 251–8.

Shurawlew, W. S. 1965. Vergleichende Tektonik der Petschora-, Kaspi- und Polnish-Norddeutschen Senke. *Geologie*, **14**, 11–25.

Stille, H. 1950. Die Kaledonische Faltung Mitteleuropas im Bilde des gesamteuropäischen. *Z. dt. geol. Ges.*, **100**, 221–66.

Strand, T. and Kulling, O. 1972. *Scandinavian Caledonoids*, Wiley-Interscience, J. Wiley & Sons Ltd, London.

Tarling, D. H. and Gale, N. H. 1968. Isotopic dating and palaeomagnetic polarity in the Faroe Islands. *Nature, Lond.*, **218**, 1043–44.

Tektonika Jewropy, 1964. Legende zur Internationalen Tektonischen Karte von Europa, Masstab 1:2 500 000. Moskou.

Tokarski, A. 1965. Stratigraphy of the salinary Röt of the fore-Sudetic monocline. *Acta geol. pol.*, **15**, 105–29.

Trümpy, R. 1958. Remarks on the pre-orogenic history of the Alps. *Geologie Mijnb. (nw. serie)*, **20**, 340–52.

Trümpy, R. 1965. Zur geosynklinalen Vorgeschichte der Schweizer Alpen. *Umschau 65/g*, **18**, 573–77.

Trümpy, R. 1971. Stratigraphy in mountain belts. *Q. Jl. geol. Soc. Lond.*, **126**, 293–318.

Voigt, E. 1962. Über Randtroge vor Schollenrändern und ihre Bedeutung im Gebiet der Mitteleuropäischen Senke und angrenzender Gebiete. *Z. dt. geol. Ges.*, **114**, 378–418.

Walter, R. 1972. Palaogeographie des Siluriums in Nord-, Mittel- und Westeuropa. *Geotekt. Forsch.*, **41**, 1–180.

Wills, L. J. 1950. *A paleogeographic Atlas of the British Isles and adjacent parts of Europe*. Blackie & Son Ltd, London and Glasgow.

Wilson, J. T. 1966. Did the Atlantic close and then re-open? *Nature, Lond.*, **211**, 676–81.

Wolburg, J. 1953. Der Nordrand der Rheinischen Masse. *Geol. Jb.*, **67**, 83–114.

Wolburg, J. 1961. Sedimentationszyklen und Stratigraphie des Buntsandstein in NW Deutschland. *Geotekt. Forsch.*, **14**, 7–74.

Wolburg, J. 1962. Über Schwellenbildung im mittleren Buntsandstein des Weser-Ems Gebietes. *Erdöl Z.*, **78**, 183–90.

Wurm, F. 1973. Neue Ergebnisse radiometrischer Altersbestimmungen an präquartären Gesteinen in Scandinavien (Finnland, Schweden, Norwegen). *Zentlb. Geol. Paläont.*, **1**, 257–73.

Wurster, P. 1968. Palaeogeographie der deutschen Trias und die palaeogeographische Orientierung der Lettenkohle in Südwest-deutschland. *Eclogae geol. Helv.*, **61**, 157–66.

Ziegler, P. A. 1975. The North Sea in a European Paleogeographic Framework. *Proc. Bergen North Sea Conference. Norg. geol Unders.* (In press.)

Znosko, J. 1964. Opinions sur l'etendue des Caledonides en Europe. *Kwart. geol. Warsaw*, **8**, 697–720.

DISCUSSION

Dr M. E. Badley (Burmah Oil (North Sea) Ltd): Several contributors to the Conference have assigned the term Cimmerian, in its various forms of early, mid and late, to a number of unconformities in the North Sea Mesozoic succession. In particular, "Mid Cimmerian" has been assigned to unconformities ranging from the Toarcian to Callovian and "Late Cimmerian" to unconformities in the Upper Jurassic and Lower Cretaceous. This practice has resulted in considerable confusion and can only lead to lack of definition in the geological analysis of North Sea stratigraphic relationships.

It is suggested that the procedures for the naming of unconformities suggested by W. B. Harland *et al.* (1972, A concise guide to stratigraphical procedure; *Jl. geol Soc. Lond.*, **128**, pp. 295–305) should be adopted. They propose that a tectonic event should never be derived from a locality or region where the resulting deformational structure is most evident. Instead, wherever possible, the age of such an event should be translated into stratigraphic terms. An unconformity would thus be dated by the oldest beds overlying it.

P. A. Ziegler: The terms "early" and "late Cimmerian phases" for tectonic movements around the Triassic–Jurassic boundary and around the Jurassic–Cretaceous boundary respectively have been adopted from the German literature where they are referred to as "Alt-kimmerische" and "Jung-kimmerische Phases" (Knetsch, 1963).

The term "Middle Cimmerian" is new and has been applied to the tectonic pulses in the Dogger.

Inasmuch as the early and late Cimmerian phases are readily recognized in the German part of the north-west European basin (Stille, 1924; Diener *et al.*, 1968), there should be no major objection to extend this terminology to the North Sea.

Names are generally used for basinwide recognized tectonic phases or, say, periods of increased tectonic activity for the sake of ease of reference. If such a tectonic phase is expressed as a basinwide unconformity or disconformity, it seems appropriate to name this unconformity after the respective tectonic phase.

Thus the name "late Cimmerian unconformity" has been systematically used for the unconformity between the Upper Jurassic and Lower Cretaceous sediments in the North Sea in all papers presented at this conference by Shell representatives.

In the North Sea area the early Cimmerian pulse can be mainly recognized as a basinwide transgression and only locally as an unconformity.

For the North-Sea-wide Dogger tectonic pulses the term "Middle Cimmerian phase" has been proposed. This is not only for the reason that this important phase is bracketed by the early and late Cimmerian phases but also for the fact that the tectonic deformations associated with these three phases seem to be in response to the same stress pattern that resulted in the reactivation of the same megatectonic features. In the North Sea these three phases appear as periods of increased tectonic activity in the overall development history of the Viking–Central Graben system.

As regards the naming of unconformities associated with the Middle Cimmerian phase I feel that the stratigraphical inventory of the North Sea has not yet far enough progressed to distinguish the local from the regional events.

I am gladly leaving the task of naming these unconformities to the stratigraphical nomenclature committees of the United Kingdom and of Norway.

REFERENCES

Diener, I., Kölbel, H. and Rusitzka, H. 1968. *Grundriss der Geologie der Deutschen Demokratischen Republik, Band 1*, Akademie Verlag Berlin, 316–9.

Knetsch, G. 1963. *Geologie von Deutschland.* Ferdinand Enke Verlag, Stuttgart.

Stille, H. 1924. *Grundfragen der vergleichended Tektonik.* Verlag von Gebruder Borntraeger, Berlin, 443 pp.

10

A Proposed Standard Lithostratigraphic Nomenclature for the Southern North Sea

By G. H. RHYS

(Institute of Geological Sciences, Leeds)

SUMMARY

A standard lithostratigraphic nomenclature is proposed for the Permian, Triassic, Jurassic and Cretaceous rocks of the southern North Sea. The Permian Rotliegendes is considered to contain two groups which pass laterally into each other; the Zechstein is subdivided into four groups based on the four evaporite cycles recognized in northern Europe and traceable across the North Sea into NE England. The Triassic is divided into two groups, the lower comprised of Bunter sandstones and shales and the upper of mudstones with associated halite, anhydrite and dolomite; the Rhaetic is included as a separate formation. The Jurassic is divided into three groups broadly equivalent to the Lower, Middle and Upper Jurassic and the Cretaceous into two groups broadly equivalent to the Lower and Upper Cretaceous. New names drawn from the southern North Sea are proposed for many of the groups and formations and some existing names from Britain and the continent are also used.

INTRODUCTION

A report (Rhys, 1974) published by the Institute of Geological Sciences (IGS) contained the results of deliberations by a committee that met at intervals during 1973 to discuss the lithostratigraphic nomenclature of the southern North Sea. This committee included representatives of six companies with interests in the North Sea and chosen by the Petroleum Exploration Society of Great Britain together with representatives from the IGS. The report includes summaries of the rock sequences encountered below the Permian, for which there was insufficient information to form a basis for a classification, and summaries of the Tertiary rocks, which will be more appropriately considered at a later date when the nomenclature of the northern North Sea is considered. It also includes an outline of the structural geology of the UK portion of the North Sea. The major part of the report is taken up with a lithostratigraphic nomenclature for the Permian, Triassic, Jurassic and Cretaceous rocks for which type-well sections are set up and are illustrated on a scale of 1:1000, showing the lithology and the gamma ray and sonic log responses.

The present paper outlines the proposed nomenclature for the Permian, Triassic, Jurassic, and Cretaceous rocks and discusses some of the decisions taken in its formulation.

PERMIAN

The two major divisions of the Permian are the Rotliegendes, comprising sandstones, siltstones, mudstones and evaporites, and the Zechstein, comprising carbonate-evaporite sequences and only minor detritus (Fig 1, Table I). The Rotliegendes is accorded group status and is considered to include two main formations. The Zechstein is divided into four groups each corresponding to a major evaporite sequence and each containing a number of formations.

The two formations of the Rotliegendes are the Leman Sandstone Formation, characteristic of the southern and western parts of the basin and named after the major southern North Sea gas-field where it is the reservoir rock (type-well section 49/26-4), and the Silverpit Formation, characteristic of the central part of the basin and named after the Outer Silver Pit near well 44/21-1 which includes the type-well section. The Leman Sandstone Formation passes laterally, by interdigitation, into the mudstones and siltstones of the Silverpit Formation and wells drilled through thick sequences of the latter encounter developments of the Leman Sandstone Formation at the top and base. It is possible that the sandstone at the top may prove to have been deposited under water, unlike the bulk of the Leman Sandstone Formation which is considered to be largely of aeolian origin. In this eventuality a separate name, of Member status would be required for this development although difficulties might arise in recognising it in areas where the water-lain sand lies directly on aeolian sand without the intervention of any representative of the Silverpit Formation.

The Zechstein of the North Sea Basin lies between, and is continuous with, similar extensive deposits in Germany and more localized deposits in north-east England. It exhibits the same major cycles of sedimentation that have been described in those two countries and each cycle has been accorded the status of a Group, each of which is subdivided into formations. The names given to the

TYPE SECTIONS- ROTLIEGENDES
(a) SANDSTONE FACIES

SHELL ESSO 49 26-4 WELL

Co-ordinates 53 07' 52''N
02 04' 57''E

R.T.E. 80ft. (24m.) A.M.S.L.
Drilled 10th August-13th October 1966

Scale 1:1000

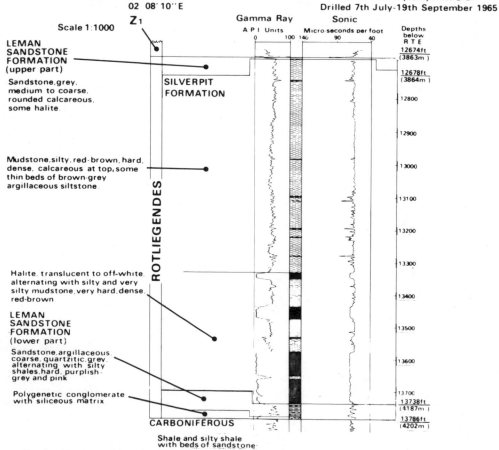

(b) MUDSTONE/ SILTSTONE FACIES

TOTAL 44 21-1 WELL

Co-ordinates 54 10' 55''N
02 08' 10''E

R.T.E. 92ft. (28m) A.M.S.L.
Drilled 7th July-19th September 1965

Scale 1:1000

FIG 1. Photographic reductions of the type-well sections of the Permian. Scale approximately 1:3500.

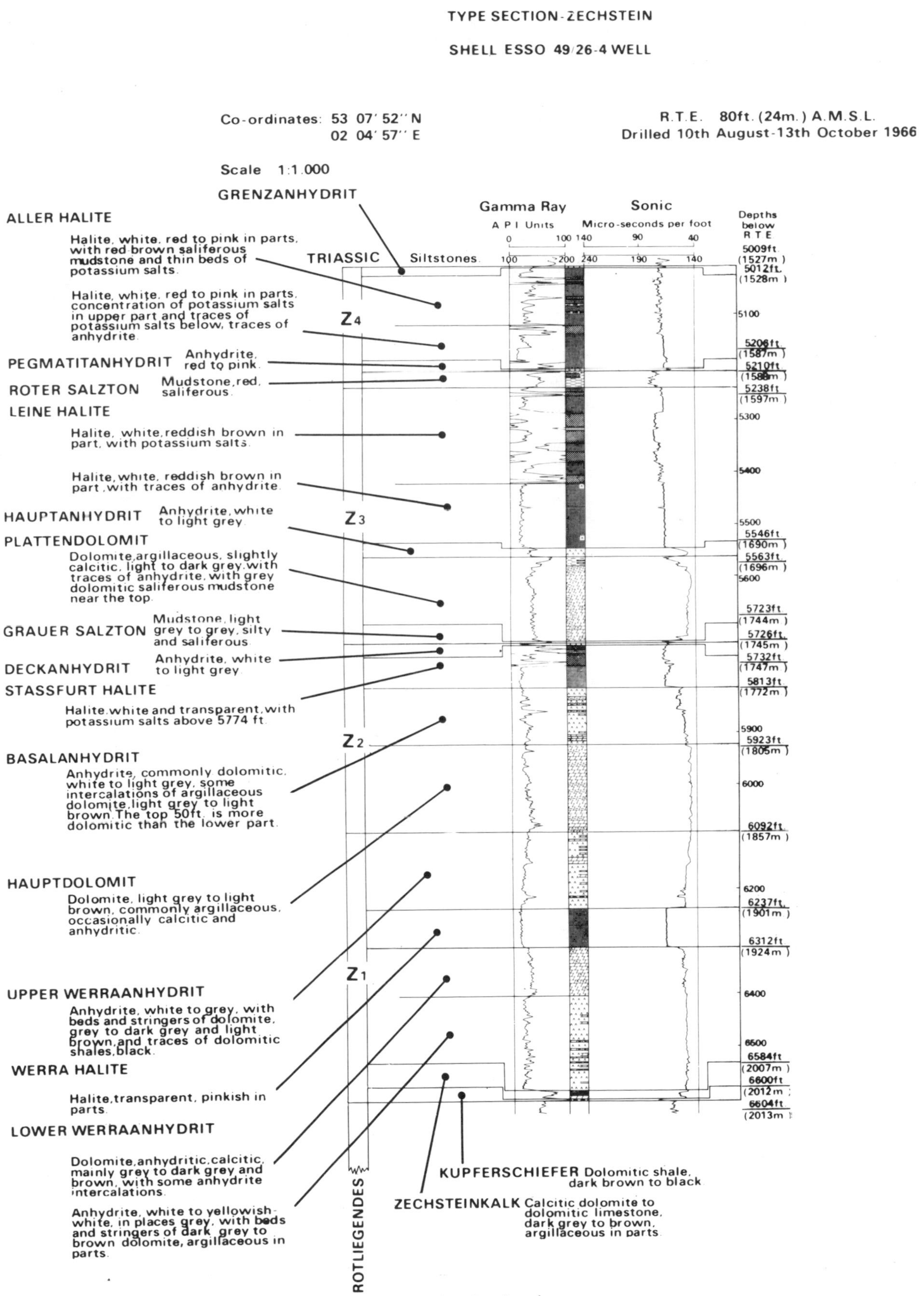

TYPE SECTION-ZECHSTEIN

SHELL ESSO 49/26-4 WELL

Co-ordinates: 53 07' 52" N
02 04' 57" E

R.T.E.   80ft. (24m.) A.M.S.L.
Drilled 10th August-13th October 1966

Scale   1:1.000

GRENZANHYDRIT

ALLER HALITE

Halite, white, red to pink in parts,
with red brown saliferous
mudstone and thin beds of
potassium salts.

Halite, white, red to pink in parts,
concentration of potassium salts
in upper part and traces of
potassium salts below, traces of
anhydrite.

PEGMATITANHYDRIT Anhydrite,
red to pink.

ROTER SALZTON Mudstone, red,
saliferous.

LEINE HALITE

Halite, white, reddish brown in
part, with potassium salts.

Halite, white, reddish brown in
part, with traces of anhydrite.

HAUPTANHYDRIT Anhydrite, white
to light grey

PLATTENDOLOMIT

Dolomite, argillaceous, slightly
calcitic, light to dark grey, with
traces of anhydrite, with grey
dolomitic saliferous mudstone
near the top.

GRAUER SALZTON Mudstone, light
grey to grey, silty
and saliferous.

DECKANHYDRIT Anhydrite, white
to light grey

STASSFURT HALITE

Halite, white and transparent, with
potassium salts above 5774 ft.

BASALANHYDRIT

Anhydrite, commonly dolomitic,
white to light grey, some
intercalations of argillaceous
dolomite, light grey to light
brown. The top 50ft. is more
dolomitic than the lower part.

HAUPTDOLOMIT

Dolomite, light grey to light
brown, commonly argillaceous,
occasionally calcitic and
anhydritic.

UPPER WERRAANHYDRIT

Anhydrite, white to grey, with
beds and stringers of dolomite,
grey to dark grey and light
brown, and traces of dolomitic
shales, black.

WERRA HALITE

Halite, transparent, pinkish in
parts.

LOWER WERRAANHYDRIT

Dolomite, anhydritic, calcitic,
mainly grey to dark grey and
brown, with some anhydrite
intercalations.

Anhydrite, white to yellowish-
white, in places grey, with beds
and stringers of dark grey to
brown dolomite, argillaceous in
parts.

TRIASSIC   Siltstones

Z_4

Z_3

Z_2

Z_1

ROTLIEGENDES

KUPFERSCHIEFER Dolomitic shale,
dark brown to black

ZECHSTEINKALK Calcitic dolomite to
dolomitic limestone,
dark grey to brown,
argillaceous in parts

Gamma Ray
A P I Units
0   100
100   200

Sonic
Micro-seconds per foot
140   90   40
240   190   140

Depths
below
R T E

5009ft
(1527m)
5012ft.
(1528m)

5100

5206ft
(1587m)
5210ft
(1588m)
5238ft
(1597m)

5300

5400

5500
5546ft
(1690m)
5563ft
(1696m)
5600

5723ft
(1744m)
5726ft.
(1745m)
5732ft
(1747m)
5813ft
(1772m)

5900
5923ft
(1805m)

6000

6092ft
(1857m)

6200
6237ft.
(1901m)

6312ft
(1924m)

6400

6500
6584ft
(2007m)
6600ft
(2012m)
6604ft
(2013m)

FIG 1.—*Continued*

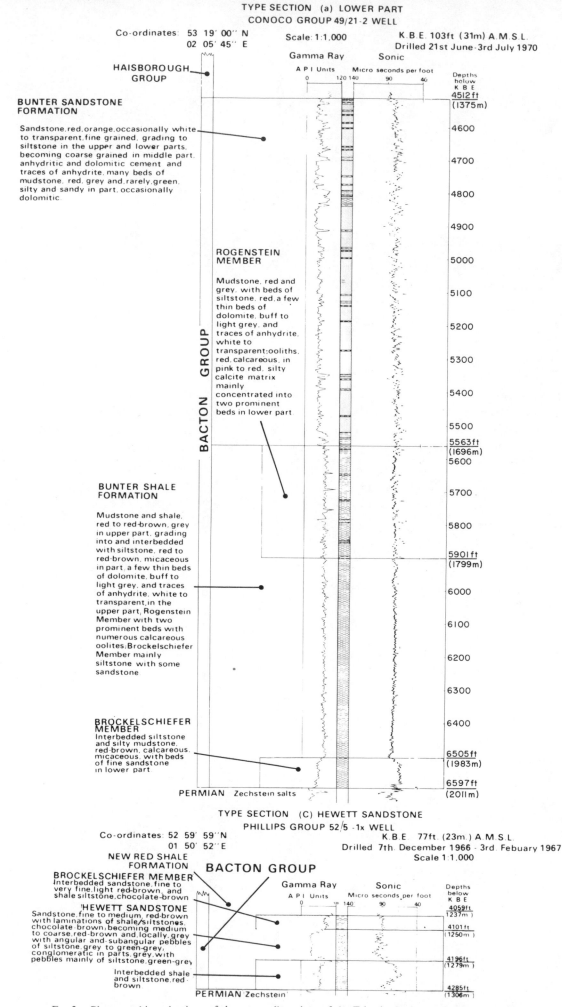

FIG 2. Photographic reductions of the type-well sections of the Triassic. Scale approximately 1:3500.

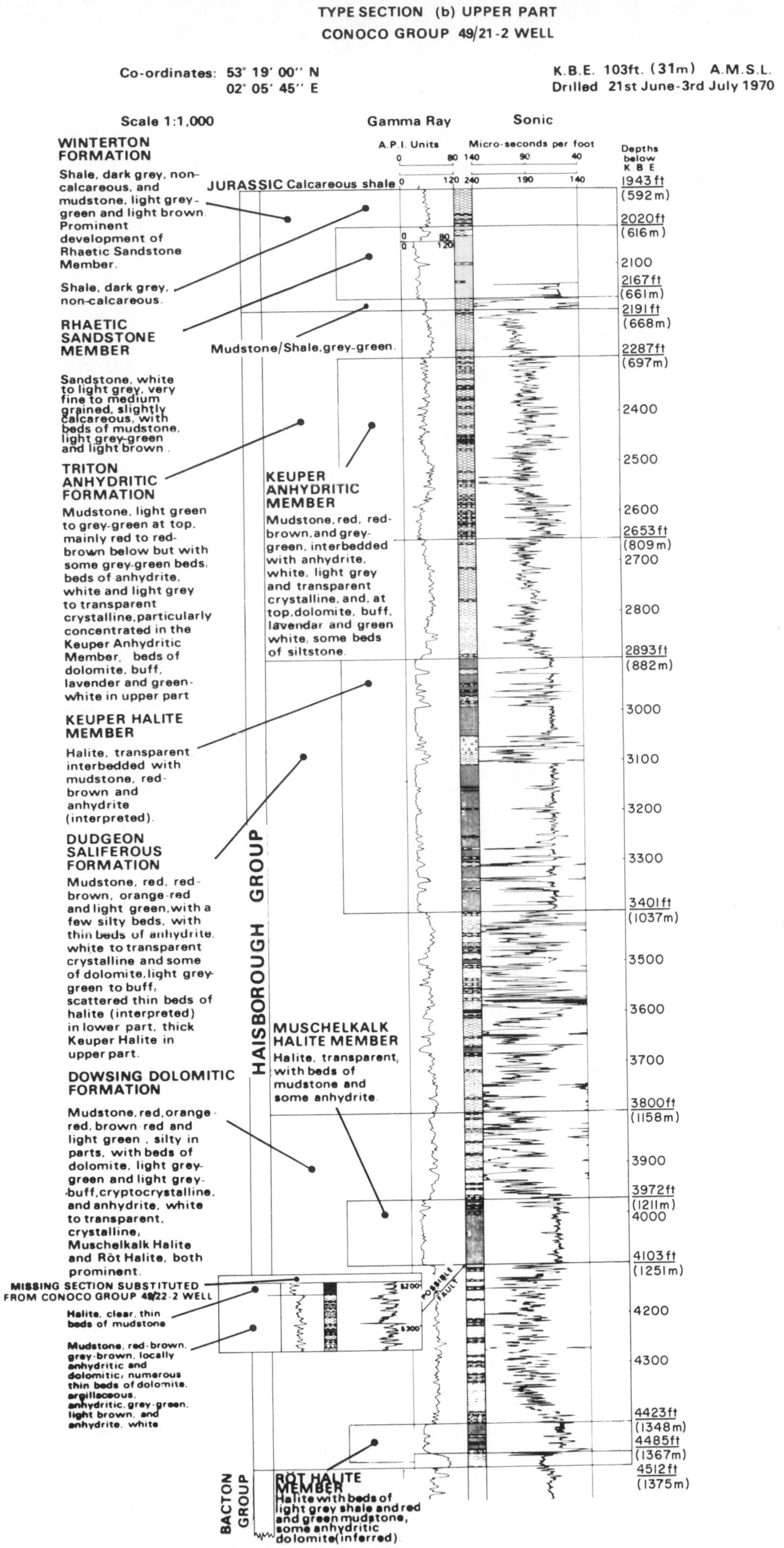

TYPE SECTION (b) UPPER PART
CONOCO GROUP 49/21-2 WELL

Co-ordinates: 53° 19′ 00″ N
02° 05′ 45″ E

K.B.E. 103ft. (31m) A.M.S.L.
Drilled 21st June-3rd July 1970

Scale 1:1,000

Gamma Ray   Sonic

A.P.I. Units   Micro-seconds per foot

WINTERTON FORMATION

Shale, dark grey, non-calcareous, and mudstone, light grey-green and light brown. Prominent development of Rhaetic Sandstone Member.

JURASSIC Calcareous shale

Shale, dark grey, non-calcareous.

RHAETIC SANDSTONE MEMBER

Mudstone/Shale, grey-green.

Sandstone, white to light grey, very fine to medium grained, slightly calcareous, with beds of mudstone, light grey-green and light brown.

TRITON ANHYDRITIC FORMATION

Mudstone, light green to grey-green at top, mainly red to red-brown below but with some grey-green beds, beds of anhydrite, white and light grey to transparent crystalline, particularly concentrated in the Keuper Anhydritic Member, beds of dolomite, buff, lavender and green-white in upper part

KEUPER ANHYDRITIC MEMBER

Mudstone, red, red-brown, and grey-green, interbedded with anhydrite, white, light grey and transparent crystalline, and, at top, dolomite, buff, lavendar and green white, some beds of siltstone.

KEUPER HALITE MEMBER

Halite, transparent, interbedded with mudstone, red-brown and anhydrite (interpreted).

DUDGEON SALIFEROUS FORMATION

Mudstone, red, red-brown, orange-red and light green, with a few silty beds, with thin beds of anhydrite, white to transparent crystalline and some of dolomite, light grey-green to buff, scattered thin beds of halite (interpreted) in lower part, thick Keuper Halite in upper part.

MUSCHELKALK HALITE MEMBER

Halite, transparent, with beds of mudstone and some anhydrite.

DOWSING DOLOMITIC FORMATION

Mudstone, red, orange-red, brown-red and light green, silty in parts, with beds of dolomite, light grey-green and light grey-buff, cryptocrystalline, and anhydrite, white to transparent, crystalline, Muschelkalk Halite and Röt Halite, both prominent.

MISSING SECTION SUBSTITUTED FROM CONOCO GROUP 49/22-2 WELL

Halite, clear, thin beds of mudstone.

Mudstone, red-brown, grey-brown, locally anhydritic and dolomitic; numerous thin beds of dolomite, argillaceous, anhydritic, grey-green, light brown, and anhydrite, white

HAISBOROUGH GROUP

POSSIBLE FAULT

RÖT HALITE MEMBER

Halite with beds of light grey shale and red and green mudstone, some anhydritic dolomite (inferred).

BACTON GROUP

Depths below K B E
1943ft (592m)
2020ft (616m)
2100
2167ft (661m)
2191ft (668m)
2287ft (697m)
2400
2500
2600
2653ft (809m)
2700
2800
2893ft (882m)
3000
3100
3200
3300
3401ft (1037m)
3500
3600
3700
3800ft (1158m)
3900
3972ft (1211m)
4000
4103ft (1251m)
4200
4300
4423ft (1348m)
4485ft (1367m)
4512ft (1375m)

FIG 2. *Continued*

TABLE I

		Proposed nomenclature for North Sea	Geological Society nomenclature for Eastern England	
	Z4 GROUP	Grenzanhydrit Aller Halite Pegmatitanhydrit Roter Salzton	Top Anhydrite Upper Halite Upper Anhydrite Carnallitic Marl	Eskdale Group Staintondale Group
ZECHSTEIN SUPER-GROUP	Z3 GROUP	Leine Halite Hauptanhydrit Plattendolomit Grauer Salzton	Boulby Halite Billingham Main Anhydrite Upper Magnesian Limestone	Teeside Group
	Z2 GROUP	Deckanhydrit ⎫ Stassfurt Halite ⎬ Basalanhydrit ⎭ Hauptdolomit	Fordon Evaporites Kirkham Abbey Formation	Aislaby Group
	Z1 GROUP	Upper Werraanhydrit ⎫ Werra Halite ⎬ Lower Werraanhydrit ⎭ Zechsteinkalk Kupferschiefer	Hayton Anhydrite Lower Magnesian Limestone Marl Slate	Don Group
ROTLIEGENDES GROUP		Leman Sandstone Formation Silverpit Formation		

groups and formations are those of the German nomenclature, the decision having been taken that these names, having the stability resulting from use over the years by workers on the Permian, are to be preferred to English names which are as yet unfamiliar. These English names, recently recommended by the Permian Working Group of the Geological Society of London (Smith *et al.*, 1974), are included in the report to facilitate correlation with that sequence. It may be that as the English names gain acceptance and become familiar they will be preferred by some workers to the proposed German names. Before arriving at the decision to recommend the existing German names due consideration was given to nomenclatures already being employed by the operating companies in the North Sea. It was found that there was no common usage and in many instances individual operators were not consistent in the nomenclatures they employed. Variations on a Lower, Middle and Upper Magnesian Limestone theme were common as was a scheme involving carbonates I, II and III together with their corresponding halites. Some operators used the strict German classification and others used anglicizations of it, in some cases mixed in a hybrid scheme with English names. The nomenclature now proposed introduces some minor anglicizations, such as "Upper", "Lower" and "Halite" in the hope that they will make it easier to use by English speaking geologists.

The section chosen as the type-well section is Shell 49/26–4, situated on the shelf where, in contrast to the basin proper, there is little disturbance resulting from halokinesis. In most respects the section is typical of many so situated but it is atypical in that it shows the Werraanhydrit divided into Upper and Lower parts separated by a Werra Halite. This halite occurs only locally and its presence in 49/26–4 was a major reason for choosing it as the type-section capable of illustrating all the Zechstein formations. Though the section is complete, the relative thicknesses of some of the formations are not truly typical and in this respect mention should be made of the relatively thin Stassfurt Halite (less than 100ft thick) and the relatively thick Werraanhydrit, features which are common near the outer edge of the shelf but not over the shelf as a whole.

The boundary between the Permian and Triassic has been taken at the top of the Permian evaporites. This is an ideal lithostratigraphic boundary though the continental and hypersaline nature of the deposits results in uncertainty as to the degree of correspondence with the chronostratigraphic boundary.

TRIASSIC

The Triassic deposits of the southern North Sea (Fig 2, Table II) are readily divisible into three: a lower unit largely comprised of clastics—sandstones, siltstones, and mudstones—with only minor anhydrite and carbonate; a middle unit of shales, mudstones and siltstones with well-developed carbonates and evaporites; and an upper unit of shales and mudstones with local developments of sandstone. In the proposed classification, illustrated by type-well section 49/21–2 (Conoco Group), the two lower units have been accorded group status. They are named the Bacton and Haisborough groups and include the major deposits of red sediments commonly associated with the Trias. They are subdivided into formations some of which include named members. The upper unit which contains the deposits formerly known as Rhaetic has been

TABLE II

Proposed nomenclature for North Sea	Names in current use in North Sea	
Winterton Formation Rhaetic Sandstone Member		RHAETIC
HAISBOROUGH GROUP Triton Anhydritic Formation Keuper Anhydritic Member		KEUPER
Dudgeon Saliferous Formation Keuper Halite Member		
Dowsing Dolomitic Formation Muschelkalk Halite Member		MUSCHELKALK
Rot Halite Member	Rot	BUNTER
BACTON GROUP Bunter Sandstone Formation	Middle Bunter Sandstone	
Bunter Shale Formation Rogenstein Member Brockelschiefer and Hewett Sandstone Members	Lower Bunter Shale Rogensteins Brockelschiefer and Lower Bunter Sandstone	

accorded formation status and named the Winterton Formation.

This division of the Triassic differs fundamentally from those of existing classifications in the position of a major boundary. Whereas the proposed scheme separates the largely clastic lower part of the Triassic from the carbonate- and evaporite-bearing middle part, existing classifications place the major boundary—that between Bunter and Muschelkalk—above the Rot and thus include the lower part of the evaporitic sequence in the lower division with the clastic deposits. It is felt that a more logical lithostratigraphic grouping is achieved by the new proposals.

Within the Bacton Group the division into two formations accords with current practice in the North Sea. The Bunter Shale Formation and the Bunter Sandstone Formation perpetuate names that have long been associated with units which are traceable from the North Sea into northern Europe. Within the Haisborough Group new names have been introduced for the three new formations that have been recognized. The subdivision of this group is based on the recognition of sequences rich in dolomite, halite and anhydrite which are named, in ascending order, the Dowsing Dolomitic Formation, the Dudgeon Saliferous Formation and the Triton Anhydritic Formation. The Dowsing Dolomitic Formation includes both the Rot and Muschelkalk of existing classifications.

These units have proved incapable of precise separation, for while it can be demonstrated that locally there is a greater concentration of dolomite in the upper part of the formation dolomite persists in varying proportions into the lower part and any division that might be attempted would be artificial and incapable of sustaining over any distance. The

Dudgeon Saliferous Formation and the Triton Anhydritic Formation make up the greater part of the Keuper of some existing classifications (those that include the Rhaetic) and equate with the entire Keuper of others. Because of the ambiguity the name "Keuper" has not been retained in the proposals.

Established Triassic names have, therefore, been dropped at formation level on the grounds either of the inability to recognize the units to which they refer, as in the case of the Rot and Muschelkalk, or on the grounds of ambiguity as to their precise meaning, as in the case of the Keuper. The introduction of new names, not only overcomes these objections but obviates any confusion that arises over the use of existing Triassic names in both lithostratigraphic and chronostratigraphic senses, which has led to recommendations to adopt new names for Triassic lithostratigraphic divisions in Britain (Warrington in Rayner and Hemingway, 1974 p. 158). At the member level, however, where names are used for smaller units and in an obviously lithostratigraphic sense, use of established names has been continued for the Rot Halite Member, the Muschelkalk Halite Member, the Keuper Halite Member and the Keuper Anhydritic Member. It is felt that the use of the names in this restricted sense will help with familiarization of the new nomenclature.

JURASSIC

The proposed nomenclature is shown below:

HUMBER GROUP:
    Kimmeridge Clay Formation
    Corallian Formation
    Oxford Clay Formation
WEST SOLE GROUP
LIAS GROUP

TYPE SECTION

PHILLIPS 47/15 – 1x WELL

Co-ordinates: 53°36'57" N
00°54'52" E

K.B.E.　84ft (26m.) A.M.S.L.
Drilled 3rd January- 4th March 1968

Scale 1:1000

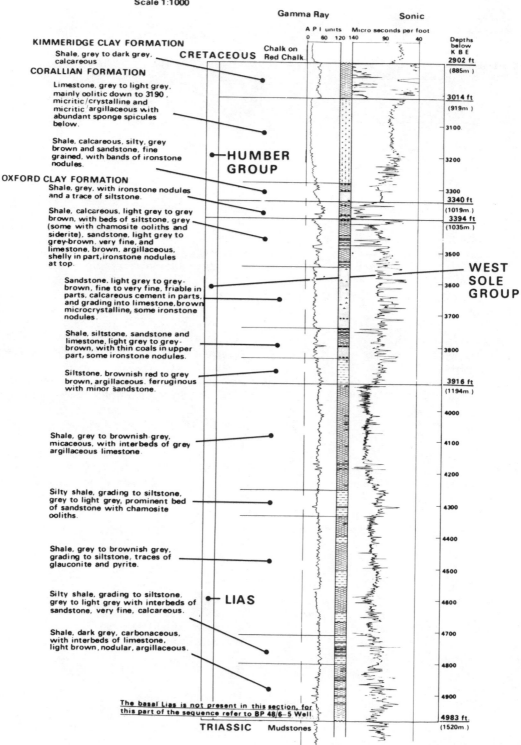

FIG 3.　Photographic reduction of the type-well section of the Jurassic. Scale approximately 1:3500.

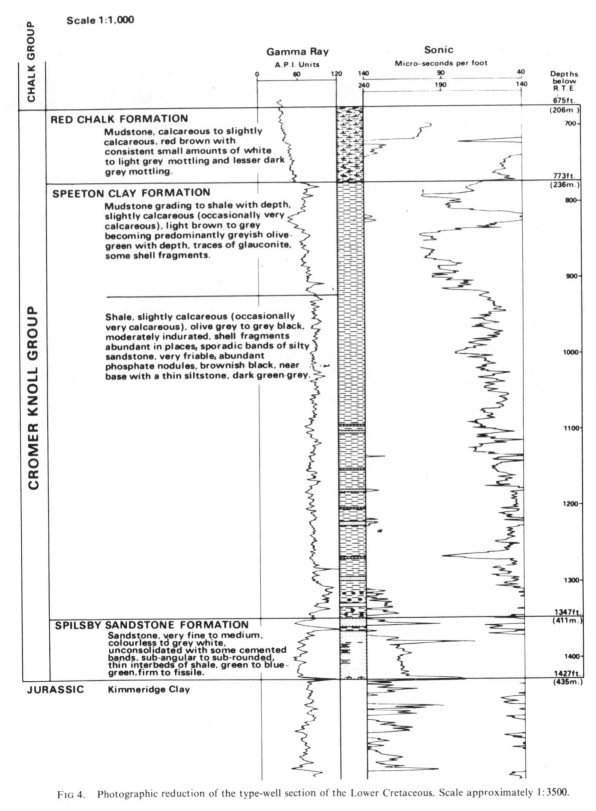

TYPE SECTION —   TOP OF RED CHALK TO BASE OF CRETACEOUS

BURMAH 48/22–2 WELL

Co–ordinates: 53 15′ 34″ N
              01 22′ 36″ E

R.T.E.   93 ft (28m.) A.M.S.L.
Drilled 11th November 1967 – 28th January 1968

Scale 1:1,000

CHALK GROUP

CROMER KNOLL GROUP

Gamma Ray
A.P.I. Units
0   60   120   140

Sonic
Micro–seconds per foot
90   40
240   190   140

Depths
below
R.T.E.
675ft.
(206m)
700
773ft.
(236m.)
800
900
1000
1100
1200
1300
1347ft.
(411m.)
1400
1427ft.
(435m.)

RED CHALK FORMATION
Mudstone, calcareous to slightly
calcareous, red brown with
consistent small amounts of white
to light grey mottling and lesser dark
grey mottling.

SPEETON CLAY FORMATION
Mudstone grading to shale with depth,
slightly calcareous (occasionally very
calcareous), light brown to grey
becoming predominantly greyish olive-
green with depth, traces of glauconite,
some shell fragments.

Shale, slightly calcareous (occasionally
very calcareous), olive grey to grey black,
moderately indurated, shell fragments
abundant in places, sporadic bands of silty
sandstone, very friable, abundant
phosphate nodules, brownish black, near
base with a thin siltstone, dark green-grey.

SPILSBY SANDSTONE FORMATION
Sandstone, very fine to medium,
colourless to grey white,
unconsolidated with some cemented
bands, sub-angular to sub-rounded,
thin interbeds of shale, green to blue-
green, firm to fissile.

JURASSIC   Kimmeridge Clay

FIG 4.   Photographic reduction of the type-well section of the Lower Cretaceous. Scale approximately 1:3500.

SHELL ESSO 49/24–1 WELL

Co-ordinates: 53 16' 50" N
02 41' 30" E

R.T.E. 91 ft. (28m.) A.M.S.L.
Drilled 4th. November 1967 - 27th. January 1968

Scale 1:1,000

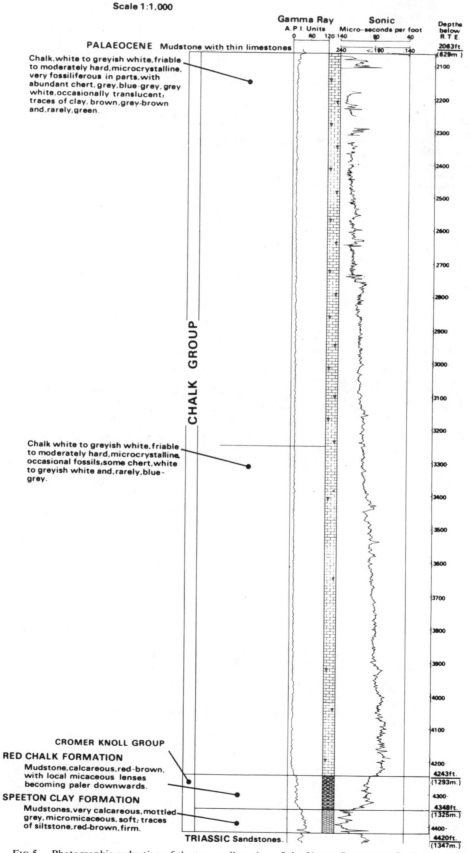

PALAEOCENE Mudstone with thin limestones

Chalk, white to greyish white, friable to moderately hard, microcrystalline, very fossiliferous in parts, with abundant chert, grey, blue-grey, grey white, occasionally translucent, traces of clay, brown, grey-brown and, rarely, green.

Chalk white to greyish white, friable to moderately hard, microcrystalline, occasional fossils, some chert, white to greyish white and, rarely, blue-grey.

CHALK GROUP

Gamma Ray
A.P.I. Units
0 40 120 140

Sonic
Micro-seconds per foot
240 <190 140

Depths below R.T.E.

2063ft (629m)
2100
2200
2300
2400
2500
2600
2700
2800
2900
3000
3100
3200
3300
3400
3500
3600
3700
3800
3900
4000
4100
4200

CROMER KNOLL GROUP

RED CHALK FORMATION
Mudstone, calcareous, red-brown, with local micaceous lenses becoming paler downwards.

4243ft.
(1293m.)
4300

SPEETON CLAY FORMATION
Mudstones, very calcareous, mottled grey, micromicaceous, soft; traces of siltstone, red-brown, firm.

4348ft.
(1325m.)
4400

TRIASSIC Sandstones.

4420ft.
(1347m.)

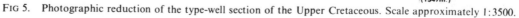

Fig 5. Photographic reduction of the type-well section of the Upper Cretaceous. Scale approximately 1:3500.

The major subdivisions correspond in general terms with the Lower, Middle and Upper Jurassic of Britain and, with the Lias, Dogger and Malm of continental classifications. They reflect the simple three-fold division of which the Jurassic of the southern North Sea is capable on the limited evidence available from wells lacking cored material and in consequence, macro-fossils. That all three of these divisions have been accorded group status while only one of them includes named formations is the result of a belief that ultimately further subdivision will be possible.

The Lias of the type-well sections 47/15–1X (Fig 3) and 48/6–5, and of the majority of southern North Sea wells consists of argillaceous sediments with thin beds of limestone and sandstone. At first glance it appears as a somewhat monotonous sequence but the variations in characters such as colour and silt content, the presence or absence of pyrite and local concentrations of thin beds of sandstone and limestone suggest that lithological subdivisions will eventually emerge. Whether the subdivision can ever be as detailed as that encountered in the nearby land sequences of Yorkshire and Lincolnshire however, must be doubted.

The West Sole Group comprises sediments which, for the most part, are coarser than those of the other two groups; they are also extremely variable, beds of siltstone and sandstone being difficult to trace between adjacent wells. In the type-well section, moreover, limestone is a minor constituent, appearing only near the base and top of the group, whereas in certain other wells it becomes an important and even a major feature. Thus, while it is clear that there is sufficient variation in lithology to warrant the erection of formations, there is as yet insufficient faunal evidence on which to trace the changes of lithology and to make reliable correlations.

The Humber Group, like the West Sole Group, is variable in composition though not to the same degree. Locally it is entirely mudstone or shale but in other localities as in the type-well section, Phillips 47/15–1x, it includes a thick development of limestone and some sandstone. The Humber Group of the type-well section is divided, on the basis of the three distinct lithologies exhibited, into the Oxford Clay Formation and the Kimmeridge Clay Formation, respectively at the base and top, separated by the mainly limestone-bearing Corallian Formation. In areas where the limestone is not developed the argillaceous sequence that then makes up the Humber Group is incapable of division on a lithostratigraphic basis.

CRETACEOUS

The proposed nomenclature is shown below.

CHALK GROUP
CROMER KNOLL GROUP:

Red Chalk Formation
Speeton Clay Formation
Spilsby Sandstone Formation

The two groups into which the Cretaceous deposits have been divided correspond roughly to the Lower and Upper Cretaceous and comprise argillaceous and arenaceous sediments with minor calcareous matter in the lower group and calcareous sediments, mainly in the form of chalk, with minor argillaceous matter and variable amounts of flint or chert in the upper group.

The Cromer Knoll Group of the type-well section, Burmah Well 48/22–2, Fig 4, contains three formations. The lowest, the Spilsby Sandstone Formation, may be partially Upper Jurassic in age and is a marginal deposit of the western part of the depositional basin. The Speeton Clay Formation is widespread and is represented in all wells that penetrate the Lower Cretaceous sequence. The Red Chalk Formation comprises calcareous mudstones commonly interbedded with and passing up into chalky limestone; the sediments have a distinctive red coloration.

The Chalk Group (Fig 5) is mainly comprised of chalk varying from soft to hard and with a variable amount of flint (or chert). Future work is likely to result in subdivision into formations, one of which might well have its top boundary at a bed of dark grey to grey-green mudstone that is commonly found (though not in the type-well section Shell/Esso 49/24–1) about 100ft above the base of the group and may equate with the "Plenus Marls" of eastern England.

ACKNOWLEDGEMENTS

The author wishes to express his appreciation to Burmah Oil (North Sea) Limited, Conoco Europe Limited, Phillips Petroleum Company, Shell UK Exploration and Production Limited and Total Oil Marine Limited for giving permission to publish the photographic reductions of the well sections which originally appeared in the IGS Report No. 74/8. The paper is published by permission of the Director, Institute of Geological Sciences.

REFERENCES

Arkell, W. J. 1933. *The Jurassic System in Great Britain.* Clarendon Press, Oxford, 681 pp.
Casey, R. 1971. Facies, faunas and tectonics in late Jurassic–early Cretaceous Britain. *In* Faunal provinces in space and time. *Geol. Jl. Spec. Issue, No. 4,* 153–68.
Colloque sur le Trias de la France et des régions limitrophes. 1963. Comptes rendus du Congrès des sociétés savantes de Paris et des départements, Montpellier 1961. *Mem. Bur. Rech. Geol. Minieres,* **15.**
Geiger, M. E. and Hopping, C. A. 1968. Triassic stratigraphy of the southern North Sea Basin. *Phil. Trans. Roy. Soc. Lond., Series B,* **254,** 1–36.

Glennie, K. W. 1972. Permian Rotliegendes of N.W. Europe interpreted in the light of modern desert sedimentation studies. *Am. Ass. Petrol. Geol.*, **56**, 1048–71.

Lamplugh, G. W. 1889. On the subdivisions of the Speeton Clay. *Q. Jl. geol. Soc. Lond.*, **45**, 575–618.

Rayner, D. H. and Hemingway, J. E. (Eds.) 1974. *The Geology and Mineral Resources of Yorkshire*. Yorkshire Geological Society, 405 pp.

Rhys, G. H. (Compiler) 1974. A proposed standard lithostratigraphic nomenclature for the southern North Sea and an outline structural nomenclature for the whole of the (UK) North Sea. A report of the joint Oil Industry–Institute of Geological Sciences Committee on North Sea Nomenclature. *Rep. Inst. Geol. Sci.*, **74/8**, 14 pp.

Richter-Bernburg, G. 1972. Saline Deposits in Germany; a review and general introduction to the excursions. In *Geology of Saline Deposits. Proc. Hanover Symp. 1968*. (Earth Sciences 7) UNESCO, Paris, 316 pp.

Smith, D. B., Brunstrom, R. G. W., Manning, P. T., Simpson, S. and Shotton, F. W. 1974. A correlation of the Permian rocks of the British Isles. Report of the Permian Working Group of the Geological Society's Stratigraphy Committee. *Spec. Rep. geol. Soc. Lond.*, **No. 5**, 45 pp.

Swinnerton, H. H., 1935. The rocks below the Red Chalk of Lincolnshire and their cephalopods faunas. *Q. Jl. geol. Soc. Lond.*, **91**, 1–46.

DISCUSSION

F. Plumhoff (Deutsche Texaco AG): Are you sure that the four Zechstein evaporate cycles can be correlated time-stratigraphically from Germany to England across the North Sea?

G. H. Rhys: The oil industry representatives on the sub-committee dealing with the Permian have access to the details of the Zechstein sequences as proved in the intervening areas (both seawards and landwards) between the UK North Sea and Germany. These representatives had no doubt that the sequence could be traced in detail between these localities. The chronostratigraphic equivalence of the various sub-divisions is more difficult to establish and Taylor and Colter (Paper 18, this volume) postulate time-equivalence for certain anhydrites in the basin with part of the carbonate sequence on the shelf. Nevertheless in comparing the shelf succession in the UK North Sea with the shelf succession in Germany we are comparing similar environments within the same basin. The basin is known to contain four major evaporite cycles and the formations within these compare closely on lithological and faunal evidence so that there can be little room for doubt as to their time-equivalence.

Another member of the Permian sub-committee, Dr D. B. Smith, has investigated the problem in some detail and I refer the questioner to his correlation of the English and German Zechstein sequences (Smith *et al.*, 1974, pp. 32–3 and Table vii).

11

Outline of the Geological History of the North Sea
By WALTER H. ZIEGLER

(Esso Europe Inc.)

SUMMARY

The North Sea area lies on cratonic crust, which has been affected by a series of diastrophic events from Cambrian to present.

The *Caledonide Diastrophism* built a mountain chain which trended from Scotland to Norway and Greenland. Most of the North Sea area lay in the southern foreland of these mountains. Late orogenic sediments are present in the Devonian Old Red Sandstone and in part of the Carboniferous in the North Sea area.

The *Hereynian Diastrophism* resulted in a mountain chain which stretched from southern Ireland through England to Germany and Poland. The Upper Carboniferous, Permian and Triassic basins under the North Sea occupied the northern foreland of this mountain belt. The time from Rotliegendes, through Zechstein and Bunter was essentially a quiet tectonic period. The sediments of this interval form a coherent stratigraphic wedge with a clastic base and top and a carbonate–evaporite wedge in the middle.

The *Hardegsen movements* at the end of Bunter time caused widespread faulting and mark the start of halokinetic deformation in the southern North Sea. The upper Triassic is a mainly fine clastic sequence.

The *Cimmerian movements*, which were intermittent, lasted from Rhaetian–Liassic time through a main phase in the upper Dogger and ended in late Malm to early Cretaceous time. They produced complex fault and rift structures in the North Sea and were accompanied by localized volcanic intrusions and strong salt movements.

The Lias and Dogger sediments are developed in a deltaic sandstone facies in the northern part of the area. Further to the south they grade into shales. Carbonates rim the London–Brabant Platform. The Upper Jurassic and Lower Cretaceous series are composed of deltaic, turbiditic and pelagic sediments which cover a complex fault morphology.

The Upper Cretaceous was a period of tectonic quiet during which widespread chalks were deposited. The *Laramide phase* at the Cretaceous–Tertiary boundary was marked by wrench compressions, especially along the southern and south-western rim of the North Sea. These led to the inversion of the marginal "Weald" troughs. The subsidence of the deep north–south trending Tertiary Basin that established itself along the axis of the present North Sea began in the early Tertiary but accelerated in Eocene and Oligocene time. Danian chalks and Palaeogene turbidites accumulated in the deepest parts of this basin, which was later filled by a predominantly argillaceous upper Tertiary sequence.

The principal hydrocarbon occurrences in the North Sea are found in the Rotliegendes, the Jurassic and the lower Tertiary, which relate to the Hercynian, Cimmerian and Laramide diastrophic episodes respectively.

INTRODUCTION

In the short time of under ten years the North Sea area has moved from a tough wildcat frontier to the biggest and most successful oil exploration play in recent times. Without doubt this venture is the largest offshore operation ever undertaken by the oil industry and takes place in one of the most severe marine environments in the world. On top of this comes the more than usual complexity of the geological history of the North Sea, which is the subject of this paper.

To begin with, there are some points, which are fundamental to the understanding of the geology of the North Sea and indeed to that of Europe.

1. The great North Sea exploration adventure takes place in a marine environment—but geologically speaking, the entire North Sea area lies on an ancient craton. Only that part which is underlain by the Caledonides was ever close to the continental margin, and the North Sea basin proper has remained an intracratonic basin from the Devonian to the present. Hence, the history of the North Sea is one of cratonic deformation.

2. The earth's crust under most of Europe and the North Sea appears to be formed by a complex checkerboard of cratonic fragments which lies along the south-west margin of the Fenno-Scandia–Russia–Asia craton, and extends eastwards to Iran. The Tornquist Line and its various *en echelon* successors mark the boundary between this "Fragmented Europe" and the more cohesive craton to the north-east.

3. The cratonic elements which make up this mosaic are limited by crustal alignments or weakness zones, some of which appear to be so ancient as to be features of the primordial crust. The primary directions of these zones are north-west–south-east as shown by the example of the Tornquist Line and north-east–south-west, as in the main Caledonoid trend. These directions limit the majority of the craton fragment boundaries. They have been rejuvenated time and time again in the geological history of the area. The geology of these individual tectonic elements records a common sequence of events but each element responded to the various stress patterns in its own way. Therefore the elements show great differences in facies as well as in structure, internally and along their boundaries. The latter particularly records the relative movements between the block boundaries.

4. The geological events on the cratonic plates can be related to worldwide tectonic events, from Cambrian to present. The manner in which a craton was deformed can be related to various exterior dynamic settings and to changes in the interaction of the plates. Hence, the geological evolution of the North Sea area can be told as it relates to global plate dynamics. Before discussing the area in more detail, an attempt is made to summarize what is known about the formation of the craton, after which the further development of the area will be traced.

THE FORMATION OF THE CRATON

Caledonide Diastrophism (Fig 1)

This lasted from Cambrian to Devonian and was caused by the collision of the Asian and North American plates. This process probably involved the subduction of a hypothetical Proto North Atlantic Oceanic Plate, which is now obliterated, and may have been followed by some sialic subduction. It resulted in the accretion of a cratonized belt of mainly continental material between the Russian and American platforms. The

strike of the main Caledonian orogenic zone may record the attitude of the original Cambro-Silurian plate margins, whereas relative shear movements between the America and Asia plates during this orogeny may have caused the Great Glen wrench-fault system. A possible conjugate to this Caledonoid trend is the Tornquist Line. Too little is yet known about its exact nature but it appears to have been an important structural and stratigraphic boundary between the stable Russian platform and the complex puzzle of cratonic fragments and deep sedimentary basins under what is now Western Europe. The extensive granitic activity in this area suggests processes of cratonization which lasted from the Precambrian through Caledonian and even Hercynian time into the Upper Carboniferous.

Hercynian Diastrophism (Fig 2)

This lasted from Devonian to Carboniferous and resulted in the collision of America and Africa and in parts also Eurasia. It has been suggested that the driving mechanism was the subduction of an early Central Atlantic Oceanic plate under America and Africa—thus building the Appalachians in America and the Mauretanides and the Atlas in

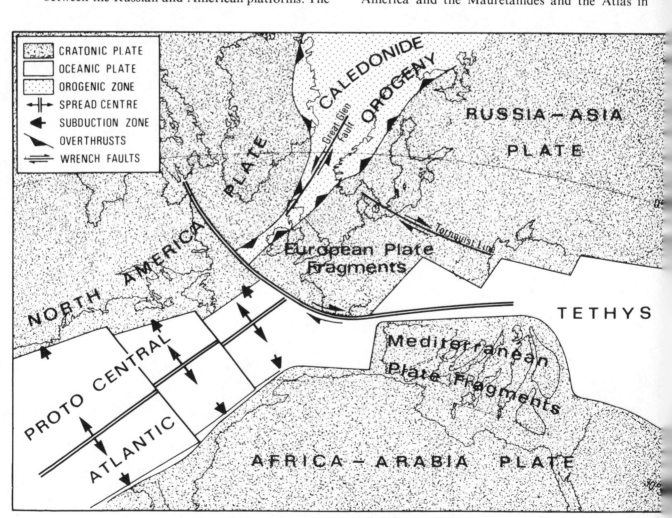

FIG 1.   Caledonide palaeo-tectonic framework.

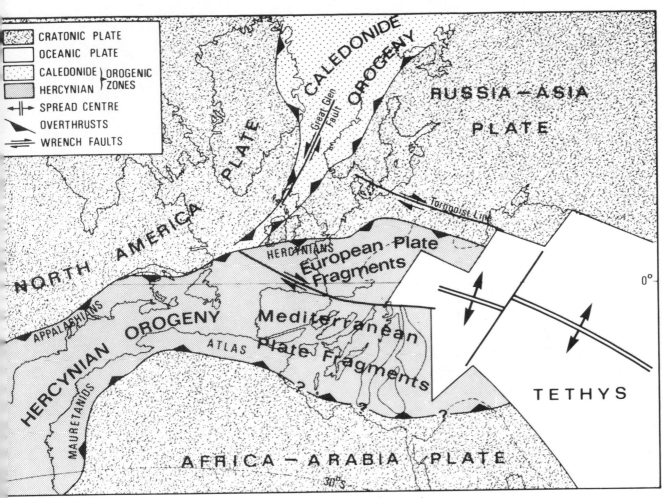

FIG 2.   Hercynian palaeo-tectonic framework.

Africa. These processes also affected the crustal checkerboard of Europe, and resulted in further consolidation of the craton in Central Europe, west of the Tornquist Line. In northern and western Europe there developed the curiously discontinuous and sinusoidal band of the Hercynian mountains, which wrap themselves around the old cratonic blocks in Europe. To the south-east the Tethys was now opening up between Africa–Arabia and the Asia plate. In this way the Caledonian and Hercynian phases are responsible for the formation and consolidation of the European craton on which the subsequent evolution of the North Sea basin took place.

DEFORMATION OF THE CRATON

In Permian and Triassic times, late Hercynian movements affected most of the Atlantic realm. During this time the main Hercynian–Appalachian orogenic zone between the Africa and America plates rifted and collapsed. These were the first manifestations of movements which led to the opening up of a new Central Atlantic in the Jurassic.

Cimmerian Diastrophic Cycle (Fig 3)

This lasted from early Jurassic to Lower Cretaceous. These movements were spasmodic but reached their climax during the Dogger. The start of the spreading and opening of the central Atlantic falls into this time. Concurrent with these movements was a phase of north–south rifting in the Atlantic area, to the north of the Newfoundland/Azores fracture zone.

Simultaneously and possibly as a consequence of this Atlantic spreading, Africa rotated counter-clockwise. Subduction of the oceanic Tethys plate below the Asiatic Plate Complex started to the east as evidenced by the fold belt which stretches from the Carpathians and Dinarids through Turkey to Iran. The bay of Biscay began to open up and complex spreading centres formed in the western Mediterranean. All this caused some rearrangement of the fragmented European craton: it reactivated old fault trends and led to wrenching and, above all, to rifting in the extensional quadrants of the shear ellipse. The north–south trending Viking Graben and the marginal troughs along the northern flank of the Hercynides are explained as tension features of this nature in this hypothesis (see Fig 4).

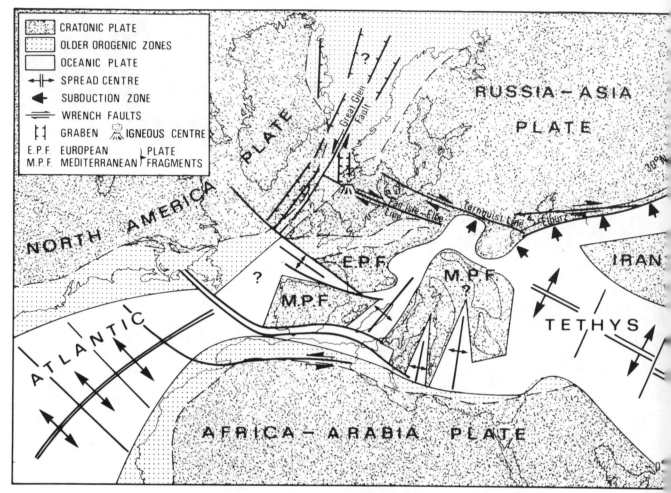

FIG 3.   Cimmerian palaeo-tectonic framework.

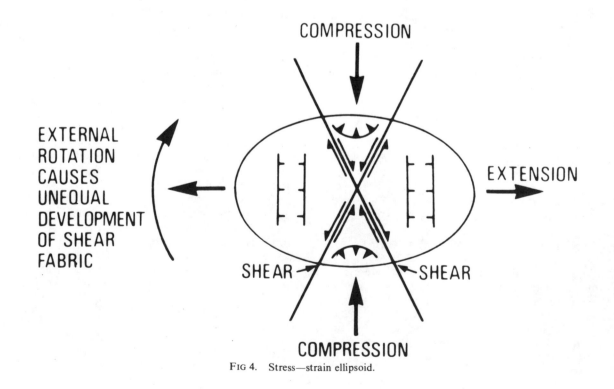

FIG 4.   Stress—strain ellipsoid.

Laramide and Alpine Movements (Fig 5)

These movements, which lasted from late Cretaceous to mid-Tertiary are related to two separate major tectonic developments. In the north-west, the northern North Atlantic began to open by the early Tertiary, thus removing one of the lateral constraints to the European plate mosaic. Contemporaneously, but farther to the south, the Mediterranean Tethys became more and more restricted between the European cratonic mass and the northwards moving African craton. North-dipping subduction zones under the Alps carried a complex array of plate fragments, some of which were of oceanic origin, against the central European craton. On the north flank of the old Hercynides these stresses resulted in wrench movements and compression on the old shear conjugates, especially on those that are parallel to the Tornquist Line. The result of these shears was the inversion or "extrusion" of the marginal "Weald" basins of Germany, Holland and Britain.

Later followed widespread rifting and volcanicity on the continent, and the formation of the deep Tertiary North Sea basin. They can be attributed to foreland spread-tensions in the craton, caused by the arcuate nature of the Alpine subduction zone.

PRE-PERMIAN SURFACE

The pre-Permian still forms the economic basement in the North Sea. Therefore, the detailed discussion of the area starts with an analysis of the land surface as it must have existed at this time (Fig 6). By late Carboniferous time the compressive forces of the Hercynian tectonic phase had passed their peak and were followed by a period of rifting and wrenching. The southern rim of the North Sea was formed by the highlands of the Hercynian mountains and their foreland basins which stretched from south Ireland to Germany. The main North Sea area was rifted into high and low blocks which were limited by reactivated old fault trends. Among these predominates the old Caledonoid north-east–south-west trend represented by the Midland Valley and the Great Glen Fault complex. Conjugate to it are trend-faults which are parallel to the Tornquist Line, i.e. north-west–south-east. A particularly persistent

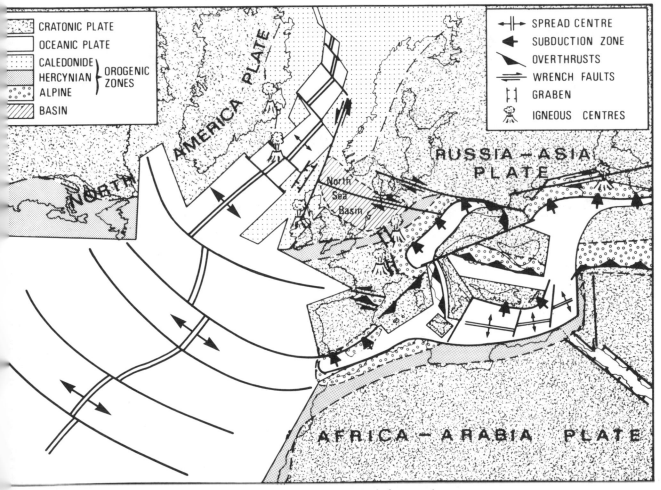

FIG 5.   Palaeogene palaeo-tectonic framework.

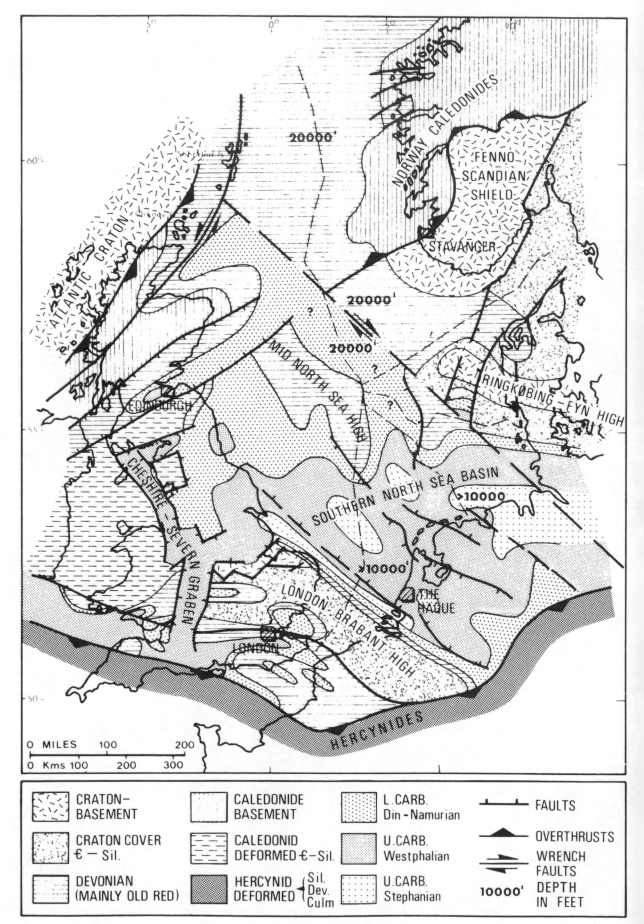

FIG 6.   Pre-Permian subcrop.

line trends from Fair Isle south-eastwards past the Fladen Ground Spur and may line up with the Elbe Line in north-west Germany. The same direction occurs also in the Ringkøbing–Fyn High, the Mid North Sea High, as well as in the London–Brabant Platform and the Pennine High. Some indications of wrenching can be found along both of these conjugate directions but dip-slip movements are more common. Nevertheless the assumption of a left-lateral displacement of the Great Glen Fault in the Upper Palaeozoic is accepted as substantially correct.

The approximate north–south direction, which is equally important in the tectonic fabric of the North Sea lies in the extensional quadrant of the shear-ellipse (Fig 4) and pure grabens such as the Oslo–Horn Graben and the Severn Graben system follow this direction. In the north the Caledonides mountains were much reduced, but still formed a regional arch linking Scotland and Norway.

Subcropping on this complex pre-Permian surface was a great variety of rocks. Basement and unmetamorphosed or only slightly altered Cambro-Silurian and Devonian sediments occur on high cratonic blocks, as for instance on the Ringkøbing–Fyn and Mid North Sea Highs, on the Fenno-Scandian Shield and on the London Brabant Platform. They are also preserved in the Oslo Graben. Over the Old Caledonides arch between Scotland and Norway the Cambro-

Silurian occurs in a metamorphic facies, but is largely covered by Devonian Old Red Sandstone sediments. The Lower Carboniferous subcrops mainly in the northern part of the area, whereas the main Westphalian to Stephanian Basin lies beneath the southern North Sea.

The main elements of the tectonic fabric that was established at this time will reappear time and time again in the geological history of the North Sea as features controlling facies and sediment distribution. Perforce, much is still very hypothetical in this interpretation, particularly in the Central Graben region.

PERMIAN TO BUNTER

The Permian and Bunter sediments (Fig 7) form a coherent stratigraphic wedge, with a distinct base, middle and top. This wedge overlies the complex late Hercynian landscape. It reflects the preceding episode of craton formation and consolidation. The period is marked by post-orogenic rifting and long episodes of quiescence.

The Rotliegendes

The Rotliegendes sands at the wedge-base lie unconformably on rocks which can range from Precambrian igneous to Upper Carboniferous. Within the Rotliegendes one can distinguish a variety of facies types (Fig 8). The most important

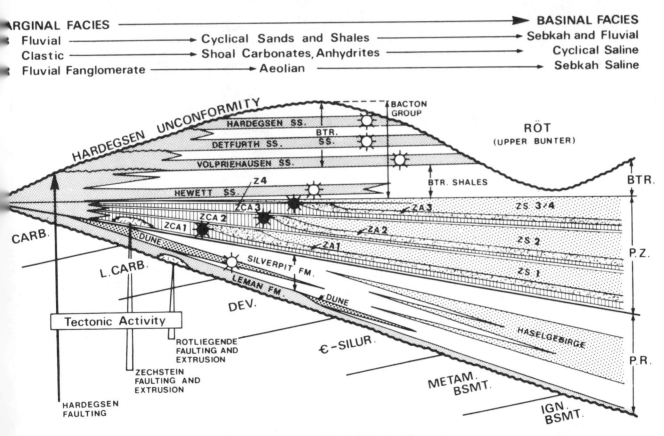

FIG 7. Permian—Bunter facies wedge-diagram.

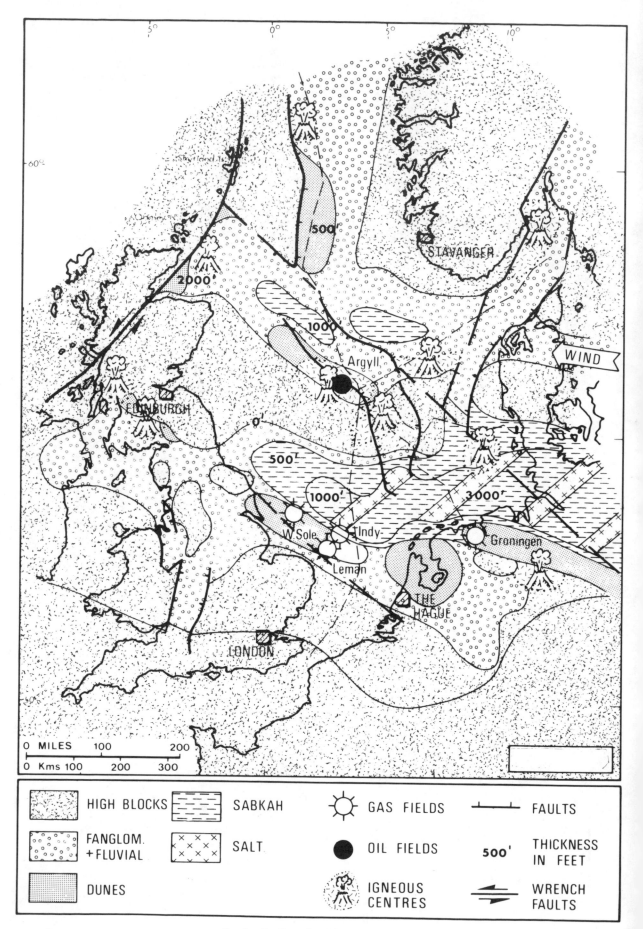

Fig 8.   Rotliegendes palaeogeography.

of these is a fanglomeratic–fluvial rim facies which grades basinwards, through a fluvial and eolian facies, into a claystone and salt–sabkah facies in the basin centre.

During the Rotliegendes there were two main sedimentary basins: one to the south of the Mid North Sea—Ringkøbing–Fyn High, the other to the north of it.

The southern basin received most of its sediments from the Hercynian Mountains to the south, but the London–Brabant Platform and elements of the Pennine High were also emergent and shed detritus. There is a clear facies differentiation into basin centre and basin margin sediments. The gas-fields in Holland and the UK sector of the North Sea occur in the best-sorted parts of the marginal facies.

The facies distribution in the northern basin is much less well known. The sediments are probably thinner and without much of a saline facies. The dune sands which occur in big fields in both basins were blown against the western basin edges by the prevailing east winds.

Igneous effusives are relatively widespread in the lower Rotliegendes but occur also higher in the section. They are found preferentially along the boundaries of major cratonic elements. This igneous activity is indicative of continuous rifting and fracturing in the area.

Active faults occur in the north-east and north-west directions—but especially important were approximately north–south trending faults. The Oslo–Horn Graben complex subsided slowly during this time and was the focus of major igneous activity. Igneous rocks indicate that the Central Graben broke through the Mid North Sea High by late Rotliegendes times. Far to the north there is evidence of some faulting and igneous activity in the Viking Graben.

The margins of the Moray Firth Basin, the north-east flank of the Mid North Sea High, as well as the outlines of an incipient Texel–IJsselmeer High in Holland and the London–Brabant–Pennine block seem to be controlled by faults following the two trends of the conjugate couple. Again most fault movements were dip-slip although some wrench displacements may have occurred.

Zechstein

Zechstein sediments occupy a middle wedge position (Fig 7). At the basin margin there is a variably developed but usually thin clastic rim facies which is followed basinwards by tidal carbonate and evaporite shelf-banks and finally by thick salts in the basin centre (Fig 9). Within this section a four-fold vertical cyclicity can be recognized, which is best developed in the slope and basin facies of the Southern North Sea Basin. Each of these cycles consists of an upwards sequence of shale, carbonate, anhydrite, salt and potash. Depending on basin position and proximity to the clastic source and to the shelf, the cycles and facies belts are variably developed.

The fourth cycle is usually rudimentary only. The carbonates of the basin margins and of the basin centre shoal form gas and oil reservoirs.

Igneous manifestations are few in Zechstein times and indicate very stable tectonic conditions. One basalt flow occurs on the Mid North Sea High in the Auk area where downfaulting of the Central Graben continued.

The facies distribution map (Fig 9) shows two salt basins, one to the north, the other to the south of the Mid North Sea High. A north–south segment of the Central Graben is evident by now and connects the two basins. Carbonate and anhydrite shoals ring the basin on all sides but are best developed on the flanks of the Pennine High and the London–Brabant Platform and in Holland and onshore Germany. The Mid North Sea High, together with the Ringkøbing–Fyn high formed an extensive basin centre shoal. Zechstein shoal-carbonates and evaporites also occur in the Moray Firth Basin and have been discovered over the Caledonides arch far to the north in the Viking Graben. The German and Dutch stratigraphic breakdown and nomenclature can be applied with little difficulty in the Southern North Sea Basin, but not enough detail is known as yet in the northern basin to confirm the same cycles in more than their outlines.

Overall the tectonic plan remained the same as inherited from the preceding epoch. The marine invasion probably came from the north, but an opening to the Tethys may have existed far to the east in Moravia. At the end of Zechstein time nearly the whole North Sea area, northern Germany and parts of Poland must have been one large saltpan with few elevations and low-lying borderlands.

Bunter or Bacton Group

The Bunter clastics (Fig 10) are the wedge-top sediments. Before discussing their facies development and distribution it is necessary to define what is meant by Bunter in the context of this paper. The term as used here refers only to the lower and middle Bunter of the classical German nomenclature of the Triassic. This is that part of the section that lies between the Zechstein top and the Hardegsen unconformity or in other words the "Bacton Group" as recently defined by the joint Oil Industry–Institute of Geological Sciences Committee on North Sea nomenclature (see G. H. Rhys, 1975). There is some reason to believe that the "Bunter" as here defined may form the uppermost part of the Permian and that the true Mesozoic base lies at the Hardegsen unconformity but this point is only mentioned in passing.

During Bunter time, thick clastic sediments spread widely over the Zechstein peneplain. The main source for the sediments in the southern North Sea area was the Massif Central in France. The principal sediment transport routes reached the North Sea trough through the Severn Valley system on England and the Hessian depression in

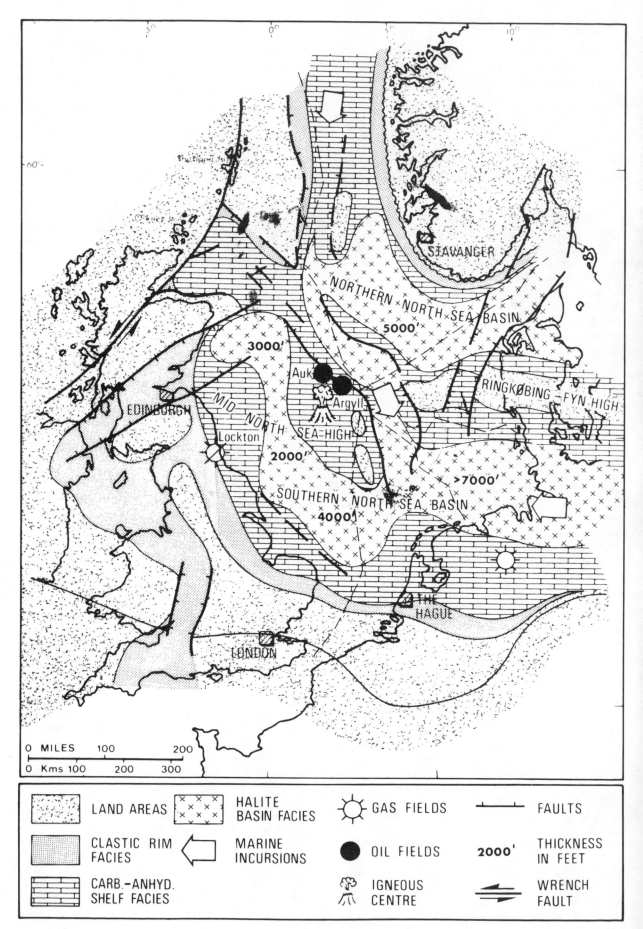

FIG 9. Zechstein palaeogeography.

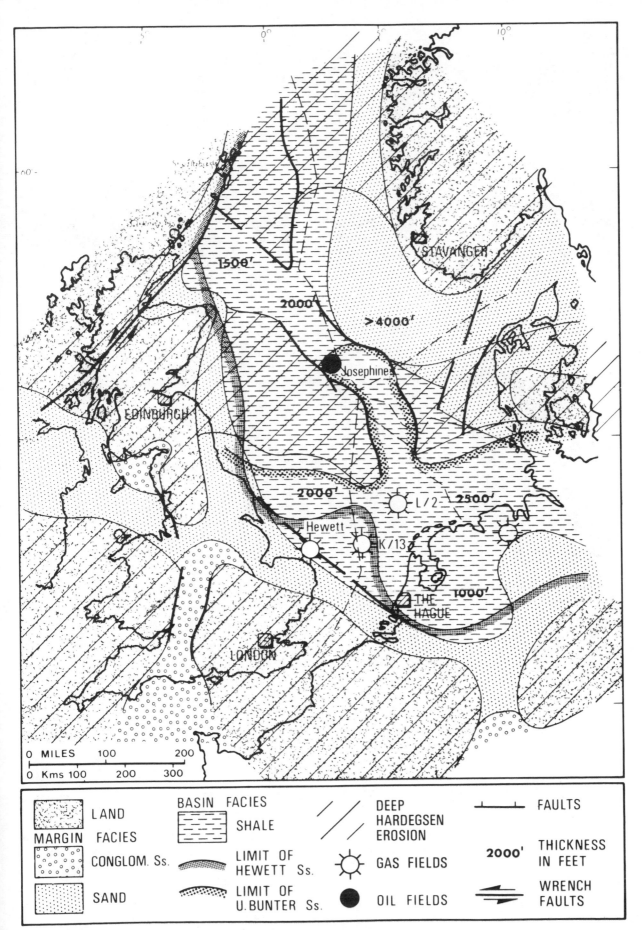

FIG 10.   Bunter (Bacton Group).

Germany. Other active sediment sources were the Fenno-Scandian Shield, the Caledonides, and the Pennines. The London–Brabant platform remained a high area, but the Mid North Sea and Ringkøbing–Fyn Highs were probably covered during Bunter time.

The Bunter consists of a sandy and coarse clastic rim-facies and a more shaly basin-centre facies. Away from the basin margins, the lower part of the Bacton Group is predominantly a red-brown shale with thin oolitic limestone stringers. However in earliest Bunter time the Hewitt Sand spread as a thin sheet far into the Southern North Sea Basin from a source on the London–Brabant Platform. It was caused by a widespread tectonic pulse which can be correlated to equivalent sand-pulses in such distant places as in the Moray Firth area and in southern and central Germany where the Hewitt Sand corresponds more or less to the Eck Conglomerate of the Palatinate and the Black Forest.

In the upper part of the Bacton Group there are a number of cyclical sand-pulses which reached far out into the basin. The three principal sands are the Volpriehausen, the Detfurth, and the Hardegsen sands. This lithologic breakdown can be traced on well-logs from southern to northern Germany, through the Dutch offshore to the UK and into the area of the Mid North Sea High, and possibly as far north as the southern Viking Graben.

Shortly after the deposition of the Hardegsen sand series occurred a tectonic pulse of regional significance. This pulse rejuvenated the Triassic morphology by uplifting some of the cratonic blocks and caused widespread erosion of the just-deposited Bunter on the highs. The resulting hiatus is the Hardegsen Unconformity. The Hardegsen movements are well known and documented in Poland, Germany and Holland and can be seen equally well in the North Sea area. Major tectonic elements that were uplifted at this time are the Mid North Sea and the Ringkøbing–Fyn Highs, the Caledonides arch between Scotland and Norway, the Pennine High and the London–Brabant platform. Farther north–south rifting and subsidence also marked the Severn Valley, the Central Graben and the Horn Graben. There is also some evidence of early faulting in the area of the Viking Graben.

The Hardegsen pulse had a long-lasting effect on the future structure of the southern part of the North Sea area, because these movements triggered the first salt deformations which later gave rise to the halokinetic phenomena which are so widespread in that part of the North Sea which is underlain by Zechstein salt. It is also probable that the Hardegsen faulting and erosion exposed Zechstein salt at the surface, where it was eroded, leached and later reincorporated in the Röt and other younger salts of the upper Triassic.

Economically the Bunter sands have some importance as gas-reservoirs in the Hewitt field as well as in offshore Holland and in Germany. Some

of the oil in the Josephine field occurs in Triassic sands which may be equivalent to the Bunter Sands.

MIDDLE AND UPPER TRIASSIC

The term Middle and Upper Triassic is here applied to what remains of the Triassic, after excluding the Bunter (Fig 11). This sedimentary section can be subdivided in ascending order into the Röt, Muschelkalk and Keuper of the classic German order, or, following the joint Oil Industry–IGS Committee on North Sea nomenclature, the Haisborough Group and the Winterton Formation. The gamma log correlations within this Triassic complex are astonishingly good throughout Germany, Holland and the southern North Sea. Excellent chronostratigraphic markers allow very detailed correlations with the German and Dutch sections and these correlations are supported by palynology. However, a major lithologic change occurs in the middle Triassic, because of the "shale out" of the Muschelkalk facies from the German type areas into the UK offshore sector. Nevertheless, Röt, Muschelkalk and Keuper log-markers can quite clearly be carried right up to the UK shore and the continental nomenclature is therefore used in this paper.

The Hardegsen pulse resulted in considerable morphology and produced a number of subsidiary basins and highs. It was followed by a period of regional subsidence during which the large cratonic depression in the North Sea area was filled with a great thickness of predominantly continental variegated muds. Thickness and facies varied according to proximity to sediment source areas and to high or low position of the various tectonic blocks. They are also affected by halokinesis. The Mid North Sea and Ringkøbing–Fyn Highs were probably covered by sediments but, as evident on palaeogeographic maps, formed the northern limit to the various evaporatic pan-systems in the Röt, Muschelkalk and Keuper. The main sediment source throughout Triassic time was in all probability the Scandinavian Shield, although the London–Brabant–Pennine High and the Scottish Caledonides Highlands shed some sands as well. Strongly subsiding were the Horn Graben in the Danish–German offshore and others parallel to it, onshore in Germany. There is also some evidence of movement along the Severn Graben System and in the Moray Firth areas.

In the southern North Sea area the thickness of the Röt section is variable. Increases are mainly caused by the presence of thick salt-lenses in the shale section. These lenses occur in the depressions in the morphology that were caused by the Hardegsen movements. Probably, much of this salt is leached and redeposited Zechstein salt.

In *Muschelkalk* time a shallow sea entered the

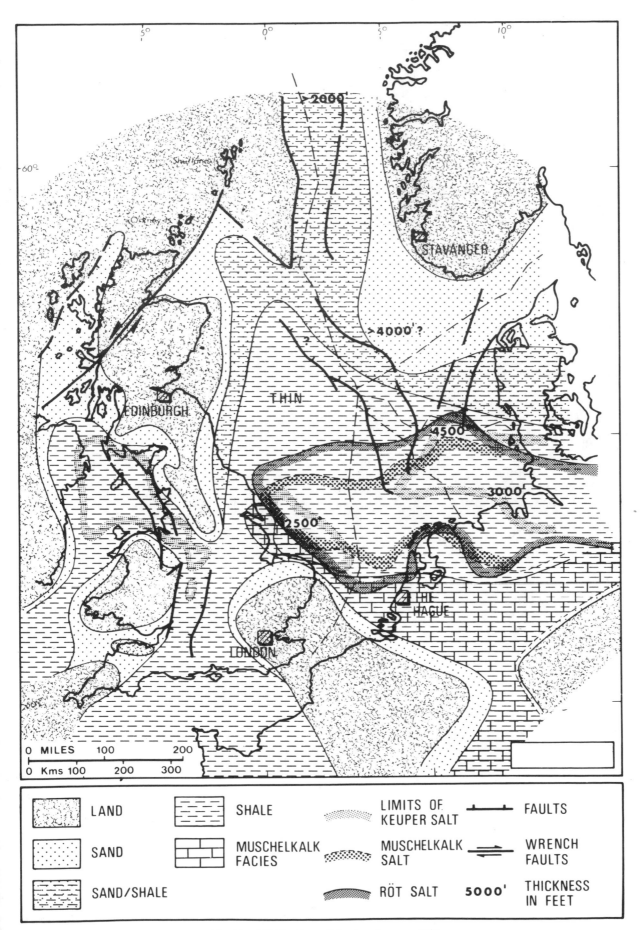

FIG 11. Middle/Upper Triassic palaeogeography.

southern North Sea basin from the south and east. Only the south rim of the North Sea basin is developed in the classic carbonate facies of Germany. Evaporites and halites are widespread in the middle Muschelkalk of the southern North Sea, but overall the clastic muds prevailed.

By *Keuper* time the Muschelkalk Sea again retreated and the Scandinavian Shield began to shed more detritus in deltaic fans. Localized but widespread evaporites and salt pans developed again in the southern North Sea.

North of the Mid North Sea High direct log-correlations become more difficult but this may be largely due to the sparseness of information at this time. On palaeo-evidence the occurrence of Norian-age sediments equivalent to the Keuper has been confirmed in the very northern part of the North Sea.

The Middle and Upper Triassic reflects a period of tectonic quiet. So far no oil or gas of importance has been found in these series in the North Sea area and finds are unlikely in much of the area due to the lack of reservoir or source facies in this part of the geological section.

RHAETIC, LIAS AND DOGGER

The first minor tectonic pulse of the Cimmerian diastrophic cycle (Fig 12) that affected the European checkerboard occurred at the Rhaetic–Keuper boundary and indicates the start of a new phase of crustal instability. These Cimmerian

movements recur throughout the Jurassic and last into the Lower Cretaceous. The Rhaetic–Liassic pulse resulted in the reactivation of fault-trends in the direction of the Old Caledonides and Tornquist Line conjugate couple (north-east by north-west) and led to rifting in the north–south extensional direction. These early rift movements are noticeable over the old Caledonides Arch and in the Central Graben but wrenches are more difficult to identify, mainly because of the usually small lateral displacement that resulted. As a result of this pulse, the cratonic elements which are marginal to the North Sea basin such as the London–Brabant Platform, the Scandinavian Shield, the Ringkøbing–Fyn High and others became uplifted.

In the basin centre there is little hiatus between Keuper, Rhaetic and Lias, but a noticeable wedge-base unconformity is developed over the high elements, with Rhaetic and Liassic sands on their flanks (Fig 13). In mid-Lias time a widespread marine transgression followed which reached its maximum extent at the beginning of Dogger time. After that, deltaic sand-complexes started to spread into the North Sea from the marginal highlands. These deltas developed especially in the north of the area. Further to the south, the London–Brabant Platform was rimmed by a carbonate shoal.

This situation lasted through Bajocian and Bathonian time but by Callovian time, another shear-pulse of much greater intensity set the European plate mosaic in motion once more. This

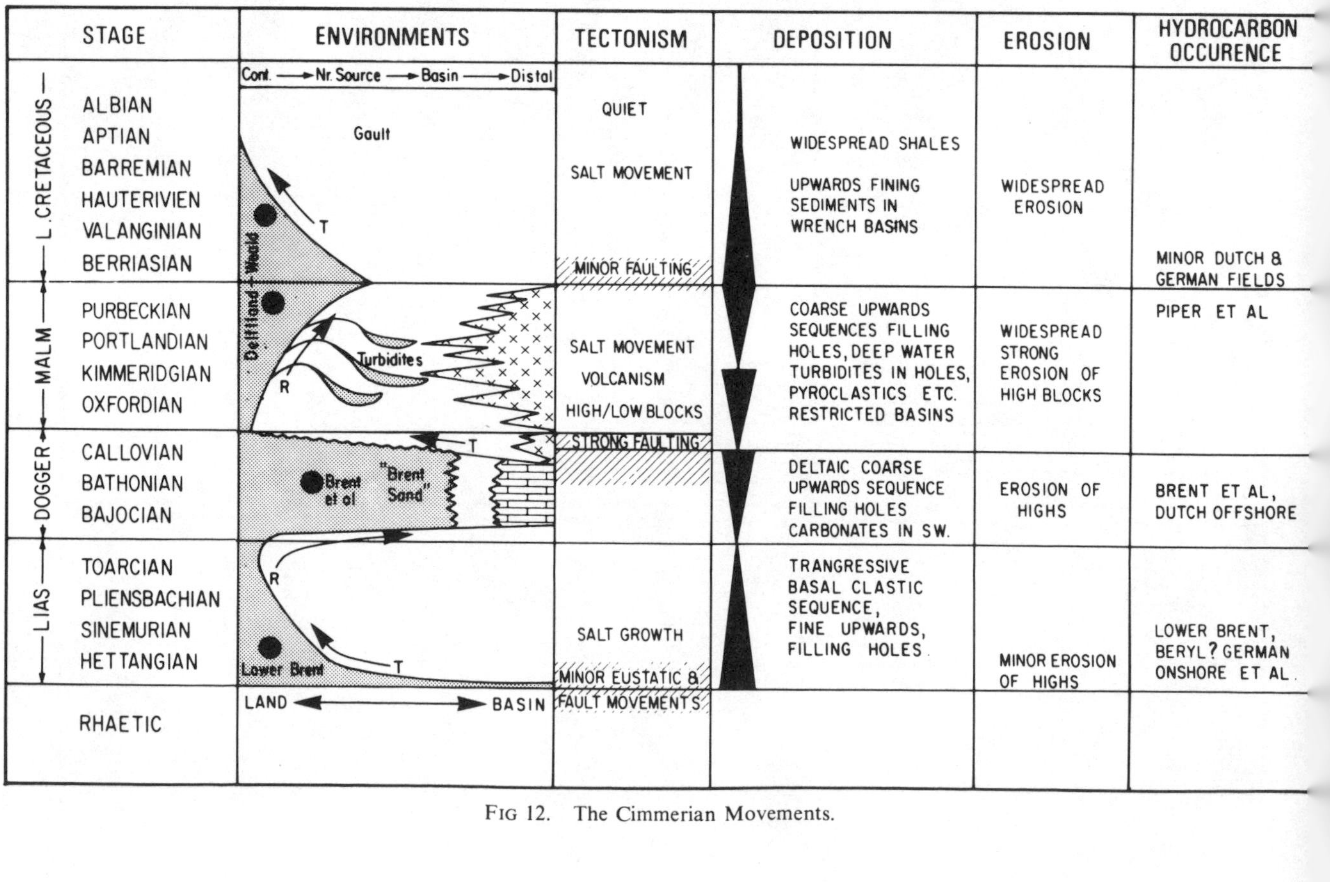

	STAGE	ENVIRONMENTS	TECTONISM	DEPOSITION	EROSION	HYDROCARBON OCCURENCE
L. CRETACEOUS	ALBIAN APTIAN BARREMIAN HAUTERIVIEN VALANGINIAN BERRIASIAN	Cont. → Nr. Source → Basin → Distal Gault Delfland Wield T	QUIET SALT MOVEMENT MINOR FAULTING	WIDESPREAD SHALES UPWARDS FINING SEDIMENTS IN WRENCH BASINS	WIDESPREAD EROSION	MINOR DUTCH & GERMAN FIELDS
MALM	PURBECKIAN PORTLANDIAN KIMMERIDGIAN OXFORDIAN	Turbidites R	SALT MOVEMENT VOLCANISM HIGH/LOW BLOCKS STRONG FAULTING	COARSE UPWARDS SEQUENCES FILLING HOLES, DEEP WATER TURBIDITES IN HOLES, PYROCLASTICS ETC. RESTRICTED BASINS	WIDESPREAD STRONG EROSION OF HIGH BLOCKS	PIPER ET AL
DOGGER	CALLOVIAN BATHONIAN BAJOCIAN	Brent et al "Brent Sand" T		DELTAIC COARSE UPWARDS SEQUENCE FILLING HOLES CARBONATES IN S.W.	EROSION OF HIGHS	BRENT ET AL, DUTCH OFFSHORE
LIAS	TOARCIAN PLIENSBACHIAN SINEMURIAN HETTANGIAN	R Lower Brent T	SALT GROWTH MINOR EUSTATIC & FAULT MOVEMENTS	TRANGRESSIVE BASAL CLASTIC SEQUENCE, FINE UPWARDS, FILLING HOLES	MINOR EROSION OF HIGHS	LOWER BRENT, BERYL? GERMAN ONSHORE ET AL
	RHAETIC	LAND ↔ BASIN				

FIG 12.   The Cimmerian Movements.

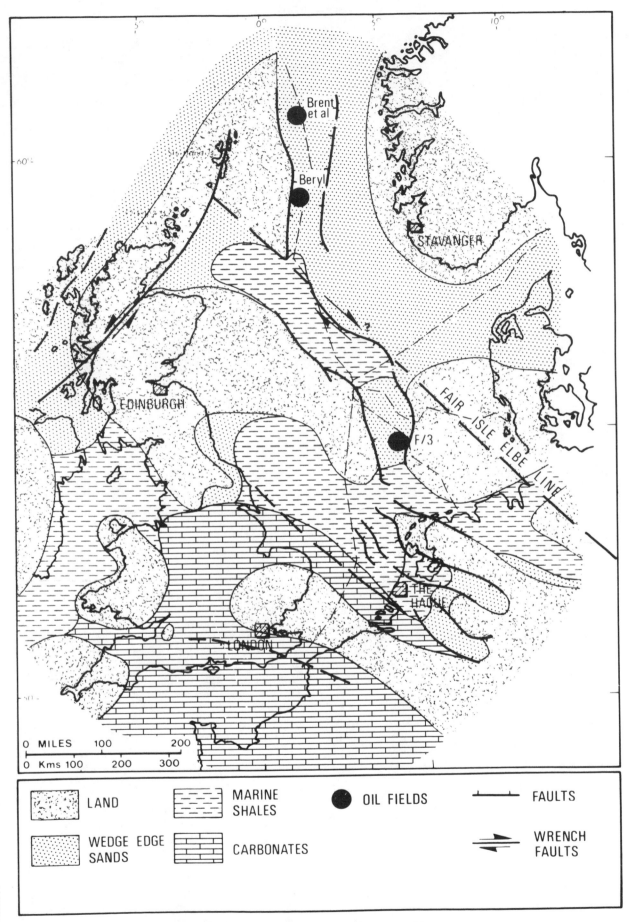

FIG 13.   Rhaetic/Lias/Dogger palaeogeography.

pulse probably originated somewhere far to the east where the Tethys Plate was subducted below the Fenno-Scandian–Russian Cratonic Plate. These disturbances were transmitted along the south-west margin of the Russian master-plate and caused some wrenching and considerable rifting in the North Sea area.

As mentioned before, the south-west flank of the Russian Craton is marked by the Tornquist Line which can be traced from the Black Sea to the Kattegat, where it stops. Farther to the west, but parallel to it, the Elbe–Fair Isle Line takes over as one of the main structural trends in the North Sea area. Both of these lines are suspected wrench-fault zones, but are still not well documented as such. North-east of the Elbe–Fair Isle line the structural style is dominated by the extensional north-south relay faults which delineate the Viking Graben, whereas in the area to the south-west the dominant structural features follow the old north-west–south-east trend. In the Moray Firth area there is considerable evidence of left-lateral wrenching of mid and upper Jurassic age in the direction of the Great Glen Fault.

Alternative explanations, interpreting the Viking–Central Graben complex as a complicated failed arm of a triple junction are possible. The North Sea Graben then could be viewed as part of the Jurassic rift-system that developed in the suture between the Fenno-Scandian and North American plates where north-south faults reactivated old Devonian and Triassic fault-systems.

All these Callovian movements again caused uplift of some blocks and subsidence of others. Very deep holes developed in the northern parts of the Central Graben and in the southern Viking Graben. This is the area of Jurassic volcanism in the North Sea. Here the main extensional north-south fault trend intersects the north-west trending shear fault system and fissures in the crust allowed the extrusion of subaquatic volcanic rocks into the post-Bathonian section.

The widespread extensional normal and wrench faulting, the high heat flow and the localized volcanism may indicate a general attenuation of the earth's crust in the area of the Jurassic Graben systems of the North Sea.

As a further complicating factor, deformation by buoyant salt was very strong in Lias and Dogger time in those parts of the basin that are underlain by Zechstein salt. Many of the saltwall structures of the southern North Sea trend roughly in a north–south direction. Perhaps they, too, utilize the weakness zones caused by extensional relay faults which follow this direction. To a considerable extent, these salt structures control Jurassic facies and sediment distribution in the Southern North Sea Basin.

The sands at the base, edge and top of this Rhaetic–Liassic–Dogger wedge are excellent reservoir rocks and contain the giant oil-fields in the Viking Graben. Minor oil-fields in the same section are also known from offshore Netherlands and are well known in the onshore of Germany. In the south the traps are essentially caused by salt-structuring whereas in the Viking Graben the oil-fields occur in large fault-blocks which were caused by the Callovian fault movements and which are sealed by shales of Upper Jurassic and Cretaceous age.

UPPER JURASSIC–LOWER CRETACEOUS

During and after the mid-Jurassic faulting episode the high blocks were attacked by erosion which modified their morphology in parts. The detritus of this erosion started to fill the lows and grabens. Other sediments were brought into the area from the bordering land, but only a few discreet depocentres developed close to shore, such as in the Moray Firth, along the east flank of the London–Brabant Platform and in the south-western Netherlands area. Basinwards from the source areas, thick sequences of upwards coarsening sediments which are a multi-phase mixture of turbidites and pelagic clays accumulated in the deeper parts of the Viking Graben and in other depressions. In yet more distal basin parts and over high blocks only a thin Malm section in a starved facies of pelagic shales is present (Fig 14). The maximum regression of the sea occurred in Purbeckian to Berriasian times.

The volcanic rocks which occur offshore southern Norway, in the Forties and in the Piper areas of the Central Graben are often of great thickness and are interspersed with marine sediments.

The oil in Piper and related fields is found in Upper Jurassic sands, which lie at the top of an upward coarsening shale–sand sequence that onlaps pre-existing high blocks.

At the end of the Malm renewed tectonic pulsation affected the European plate mosaic and led to further warping, rifting and wrenching. All of this structuring is more in evidence south-west of the Fair Isle–Elbe Line than to the north-east of it. In addition to causing the opening and rapid subsidence of the areally restricted Wealden basins along the southern rim of the North Sea, these late Jurassic to Lower Cretaceous wrench movements are also suspected for a complex shear zone which lies in the offshore of southern Norway and trends about east–west, between the Tornquist and Fair Isle–Elbe Lines.

During the lower Cretaceous the sea became more widespread again and depocentres of sediments established themselves along the margins of the North Sea, in the offshore of southern Norway and especially to the south, in the marginal wrench-induced troughs along the north rim of the Hercynian mountains. A lower Cretaceous basin also occupied the Moray Firth area. Like the marginal basins to the south it was filled with a considerable thickness of coarse continental clastics in a "Weald" facies. In the basin

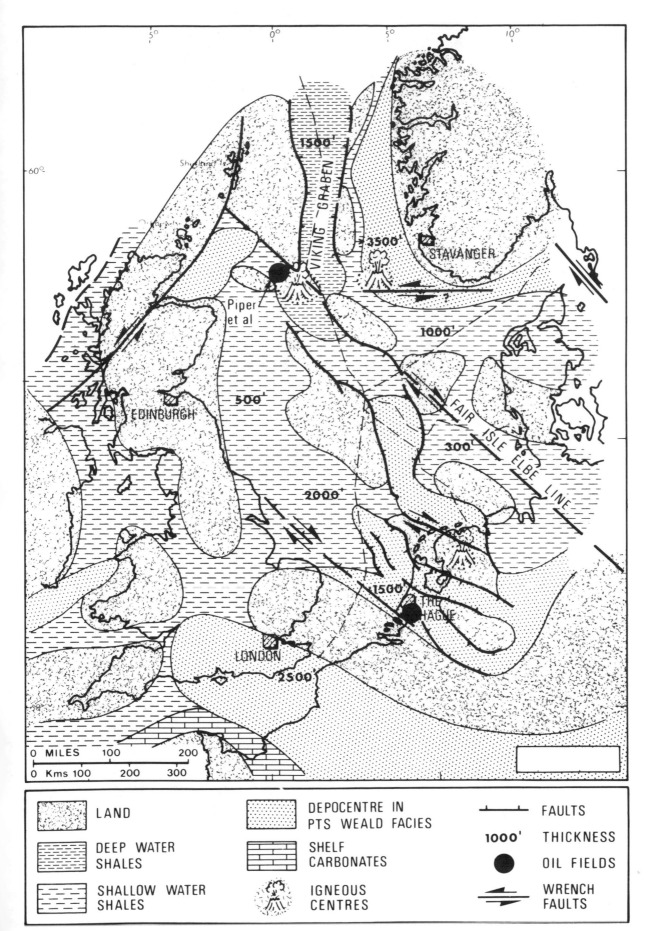

FIG 14.   Malm/Lower Cretaceous palaeogeography.

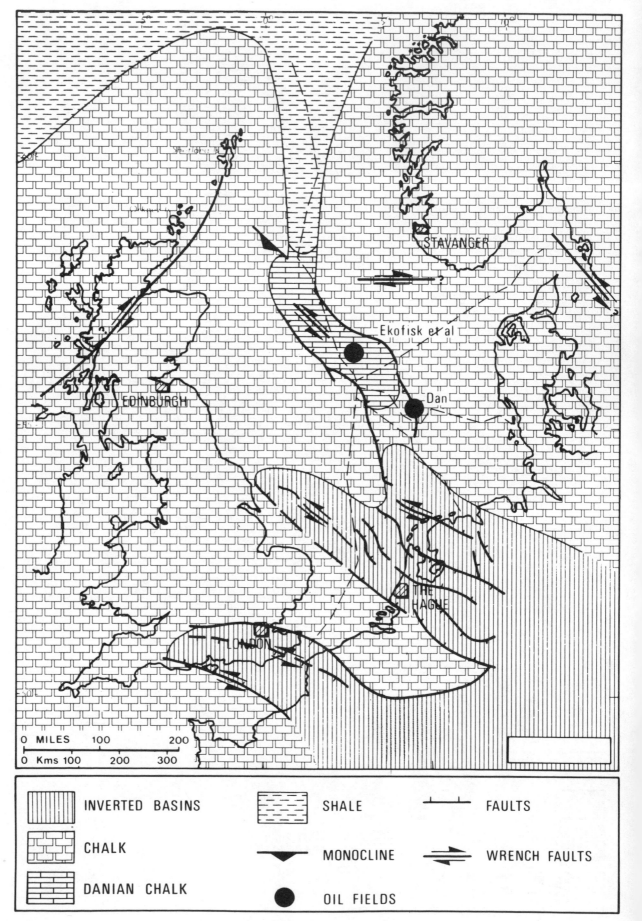

▦ INVERTED BASINS	▦ SHALE	┼ FAULTS
▦ CHALK	▼ MONOCLINE	⇒ WRENCH FAULTS
▦ DANIAN CHALK	● OIL FIELDS	

FIG 15.   Upper Cretaceous/Laramide palaeogeography.

centre parts of the North Sea the Lower Cretaceous is developed as relatively thin shallow-water shales and some carbonates, which gradually thicken and grade to a deeper water facies in the Viking Graben.

UPPER CRETACEOUS–LARAMIDE PHASE

With the transgression of the Albian–Aptian Sea the recurring episodes of the Cimmerian diastrophism gave way to a period of tectonic quiet. Much of Europe was covered with a shallow sea in which chalk was laid down. In the Viking Graben this chalk grades into shales. This graben system probably opened into an incipient North Atlantic rift in the very far north (Fig 15).

Except for salt-structuring there is little evidence for tectonic movements until Senonian time, when the precursors of the Laramide disturbances show up. These movements reached their climax in late Maastrichtian time and rapidly faded away after the Palaeocene.

The Laramide structures are essentially confined to the south-west of the Fair Isle–Elbe Line. They are particularly strongly developed on the flanks of the Hercynian complex along the south and south-west rim of the North Sea Basin. The structural style is dominated by wrench associated compression. Up to Laramide time the Hercynian sense of right-lateral movement had persisted on the north-west trending shear-zones. It apparently was temporarily reversed at this time. The Weald basins which had opened up as wrench-troughs got squeezed by this reversal of shear-direction and adjusted to it by extruding the contents of their basin centres. These are the inverted Weald Basins. Such inversions of relative movements along the Caledonides shear couple in Laramide time seem also possible along the Great Glen Fault system.

Very strongly affected by these movements was the Anglo-Dutch basin which now contains the major gas-fields of the southern UK sector of the North Sea. The complex anticlinal features on which these fields are located were caused by these compressional forces.

In the central parts of the North Sea, the effects of the Laramide movements were relatively minor. There was some renewed down-faulting and warping along the margins of the main Central Graben and the southern Viking Graben. Erosion removed considerable amounts of Chalk over high blocks near those margins. These detrital products were washed into the deepest graben parts and there formed the reservoirs of the Danian Chalk fields.

TERTIARY

The Laramide time also marks the beginning of a new episode of subsidence of the whole North Sea Basin, which lasted throughout Tertiary times until the present (Figs 5, 16). The cause for the postulated temporary shear-reversal during the Laramide episode and the subsequent strong basin-subsidence in the North Sea is to be sought in the fundamental changes which occurred in the North Atlantic plate-pattern during this time and in the Alpine orogenic phase which is slightly younger. The main new element in the geological framework at this time is provided by the oceanic spreading centres that established themselves in about Palaeocene time in parts of the northern North Atlantic rift systems. These centres lay between southern Greenland and north-western Europe and provided the motor to move the European plate complex south-eastwards towards the Alpine subduction zones in the western Tethys. This shifting resulted in left-lateral shear movements along the boundaries of the European plate checkerboard. The movements were finally overcome in the middle Tertiary by the increasing strength of the Alpine subduction zones which dipped under the central European plate complex.

The early Tertiary was accompanied by great volcanic activity and igneous extrusions throughout western Scotland, the Atlantic region and Greenland. The widespread "Tuff" at the Eocene–Palaeocene boundary is a record of the opening of the Atlantic whereas the rifting and slightly younger volcanicity of central Europe may well be more closely related to the Alpine orogeny. Together with the North Sea subsidence we attribute the latter to foreland-spread tensions in the Central European craton in front of the arcuate Alpine subduction zones.

During the Tertiary the North Sea Basin continued to subside and great thicknesses of sediments accumulated in it. The deep basin axis trends roughly north–south but includes a number of doglegs which reflect the underlying graben systems. During Palaeocene and Eocene times sands and shales poured into this trough from a mainly north-western source. The sands formed a deltaic shelf along the Scottish coast but developed into deep-water turbidite tongues and lobes in the deep basin parts. These sands now form the reservoirs for the important oil and gas accumulations of the Forties, Frigg and related fields. By late Eocene time the source areas of these sediments, which lay in the Atlantic realm, finally foundered or drifted away; and from then on the North Sea Basin was mainly filled with fine clastics which sealed the trough.

CONCLUSIONS

The history of the North Sea area records a whole sequence of events in the development of the earth's crust (Fig 17). It is a story of purely cratonic deformation but each event can be related to worldwide tectonic processes in the ever-shifting and partly regenerating patchwork of oceanic and cratonic plates. It is recorded on the essentially permanent cratonic plates as a complex sequence

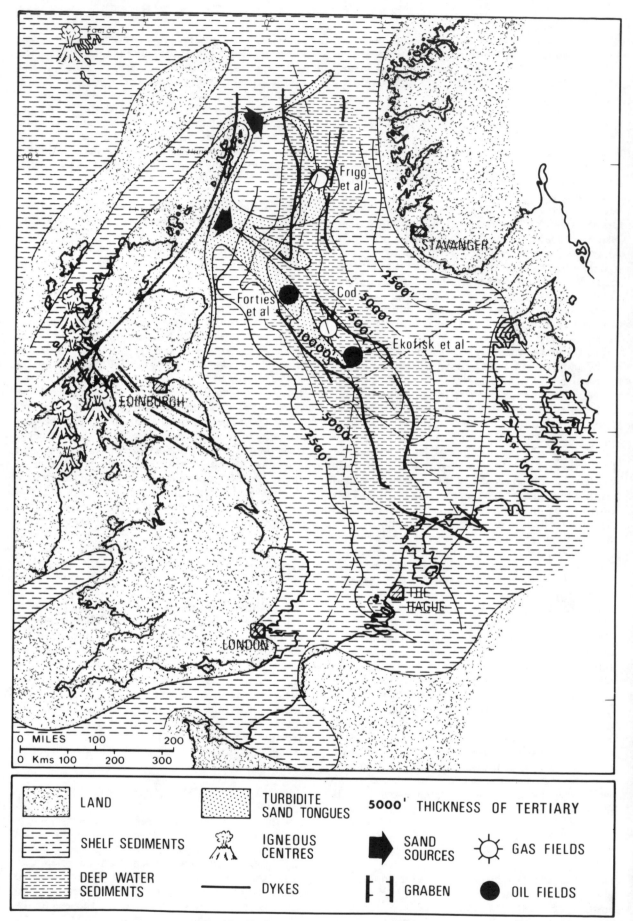

FIG 16. Tertiary, Palaeogene palaeogeography.

LAND

SHELF SEDIMENTS

DEEP WATER
SEDIMENTS

TURBIDITE
SAND TONGUES

IGNEOUS
CENTRES

DYKES

5000' THICKNESS OF TERTIARY

SAND
SOURCES

GRABEN

GAS FIELDS

OIL FIELDS

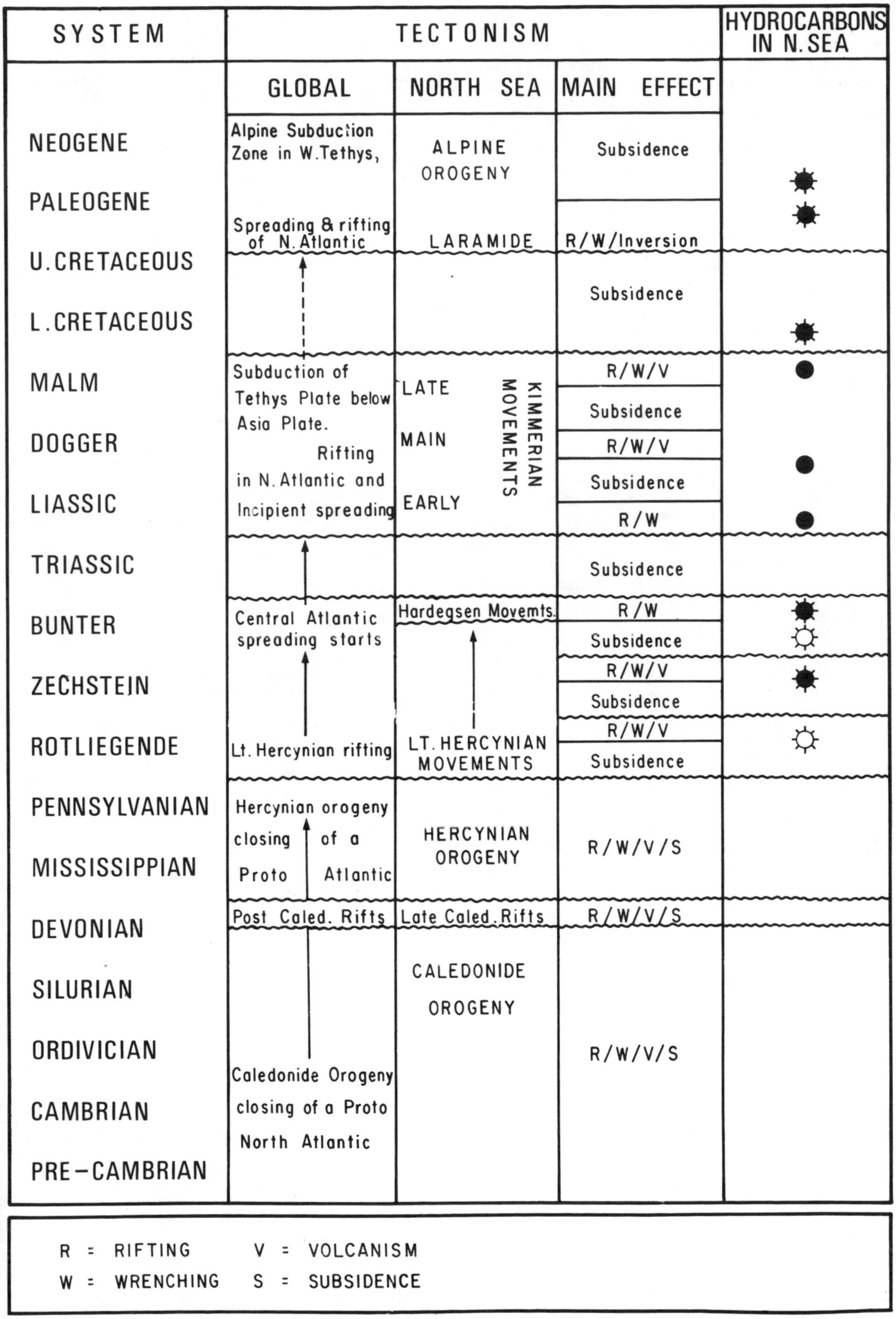

SYSTEM	TECTONISM			HYDROCARBONS IN N.SEA
	GLOBAL	NORTH SEA	MAIN EFFECT	
NEOGENE	Alpine Subduction Zone in W.Tethys,	ALPINE OROGENY	Subsidence	✴
PALEOGENE	Spreading & rifting of N.Atlantic	LARAMIDE	R/W/Inversion	✴
U.CRETACEOUS			Subsidence	
L.CRETACEOUS				✴
MALM	Subduction of Tethys Plate below Asia Plate.	LATE	R/W/V	●
			Subsidence	
DOGGER	Rifting in N.Atlantic and Incipient spreading	MAIN	R/W/V	●
			Subsidence	
LIASSIC		EARLY	R/W	●
TRIASSIC			Subsidence	
BUNTER	Central Atlantic spreading starts	Hardegsen Movemts.	R/W	✴
			Subsidence	☼
ZECHSTEIN			R/W/V	✴
			Subsidence	
ROTLIEGENDE	Lt. Hercynian rifting	LT. HERCYNIAN MOVEMENTS	R/W/V	☼
			Subsidence	
PENNSYLVANIAN	Hercynian orogeny closing of a Proto Atlantic	HERCYNIAN OROGENY	R/W/V/S	
MISSISSIPPIAN				
DEVONIAN	Post Caled. Rifts	Late Caled. Rifts	R/W/V/S	
SILURIAN		CALEDONIDE OROGENY	R/W/V/S	
ORDIVICIAN				
CAMBRIAN	Caledonide Orogeny closing of a Proto North Atlantic			
PRE-CAMBRIAN				

(North Sea column, spanning Malm–Liassic: KIMMERIAN MOVEMENTS)

R = RIFTING   V = VOLCANISM
W = WRENCHING   S = SUBSIDENCE

FIG 17.   Summary chart of geological events in the North Sea.

of alternating periods of great tectonic stress and quiet. It resulted in a variety of structural phases, each with its own style of deformation, and in the superposition of multiple sedimentary basins. It is therefore no wonder that the geology of Western Europe and the North Sea area is so complex. But in this very complexity also lies the reason for the unusual number of major exploration plays in one area. A succession of basins formed and filled rapidly, were structured and sealed before the hydrocarbons escaped. The Rotliegendes and Zechstein gas reserves of the southern North Sea are trapped in late Hercynian sediments, the big Jurassic oil play owes its existence to the Cimmerian events and the Danian and Palaeogene plays relate to the Laramide movements.

ACKNOWLEDGEMENTS

Thanks are expressed to my colleagues in the exploration team in the Walton-on-Thames office of Esso Europe, in EPRCo. and elsewhere in the Company. I have freely drawn on their experience and much of the data presented has been compiled from their contributions. Particular thanks are expressed to Jim Brown, John Parmenter, Ian Cook and Al Caan. Many thanks also to Bill Clarke and Martin Watts for the preparation of the illustrations.

BIBLIOGRAPHY

Ager, D. V. 1970. The Triassic System in Britain and its stratigraphical nomenclature. *Q. Jl. geol. Soc. Lond.*, **126**, 3–17.

Allen, P., Sutton, J. and Watson, Janet V. 1974. Torridonian tourmaline-quartz pebbles and the Precambrian crust north-west of Britain. *Jl. geol. Soc. Lond.*, **130**, 49–89.

Audley-Charles, M. G. 1970. Triassic palaeogeography of the British Isles. *Q. Jl. geol. Soc. Lond.*, **126**, 49–89.

Audley-Charles, M. G. 1970. Stratigraphical correlation of the Triassic rocks of the British Isles. *Q. Jl. geol. Soc. Lond.*, **126**, 19–47.

Bailey, R. J., Grzywacz, J. M. and Buckley, J. S. 1974. Seismic reflection profiles on the continental margin bordering the Rockall Trough. *Jl. geol. Soc. Lond.*, **130**, 55–69.

Balchin, D. A. and Ridd, M. F. 1970. Correlation of the younger Triassic rocks across eastern England. *Q. Jl. geol. Soc. Lond.*, **126**, 91–101.

Bartenstein, H. 1968. Present status of the Palaeozoic paleogeography of north Germany and adjacent parts of north-west Europe. *In* D. T. Donovan (Ed.) *Geology of Shelf Seas.* Oliver and Boyd, Edinburgh, 31–54.

Bartenstein, H. 1968. Palaeogeographische Probleme beim Aufsuchen von Kohlen-Wasserstoff-Lagerstatten im Palaozoikum und in der Untertrias von Mittel- und Nordwest-Europa Einschliesslich des Nord-See-Raumes. *Erdoel und Kohle-Erdgas-Petrochem*, **21**, 2–7,

Belloussov, V. V. 1971. On possible forms of relationship between magmatism and tectogenesis. *Jl. geol. Soc. Lond.*, **127**, 57–68.

Birks, J. 1973. The oil potential of the North Sea. *Pet. Times*, **1428**, 33–7.

Boigk, H. 1968. Gedanken zur Entwicklung des Nieder-sachsischen Tektogens. *Geol. Jb.*, **85**, 861–90.

Brand, E. and Hoffman, K. 1963. Stratigraphy and facies of the north-west German Jurassic, and genesis of its oil deposits. *Wld. Petrol. Cong. 6 Frankfurt*, **1**, 1–23.

Clarke, R. H. 1973. Cainozoic subsidence in the North Sea. *Earth and Planet. Sci Lett.*, **18**, 329–32.

Collette, B. J. 1968. On the subsidence of the North Sea area. *In* D. T. Donovan (Ed.) *Geology of Shelf Seas.* Oliver and Boyd, Edinburgh, 15–30.

Dewey, J. F. 1969. Continental Margins: a model for conversion of Atlantic type to Andean type. *Earth and Planet. Sci. Lett.*, **6**, 189–97.

Dewey, J. F. 1969. Evolution of the Appalachian/Caledonian orogen. *Nature, Lond.*, **222**, 124–9.

Dewey, J. F. 1971. A model for the Lower Paleozoic evolution of the southern margin of the early Caledonides of Scotland and Ireland. *Scott. Jl. Geol.*, **7**, 219–40.

Dewey, J. F. and Horsfield, Brenda. 1970. Plate tectonics, orogeny and continental growth. *Nature, Lond.*, **225**, 521–5.

Dewey, J. F., Pitman, W. C., Ryan, W. B. F. and Bonnin, J. 1973. Plate tectonics and the evolution of the Alpine System. *Bull. geol. Soc. Am.*, **84**, 3137–80.

Donovan, D. T. 1968. Geology of the Continental Shelf around Britain: Survey of progress. *In* D. T. Donovan (Ed.) *Geology of the Shelf Seas.* Oliver and Boyd, Edinburgh, 1–14.

Dunn, W. F., Eha, S. and Heikkila, H. H. 1973. North Sea is tough theater for the oil-hungry industry to explore. *The Oil and Gas Jl.*, **71** (2 and 3), 90–3, 126–8.

Emmons, R. C. 1968. Strike-slip rupture patterns in sand models. *Tectonophys.*, **7**, 71–87.

Evans, T. R. and Coleman, N. C. 1974. North Sea geothermal gradients. *Nature, Lond.*, **247**, 28–30.

Evans, W. B. 1970. The Triassic salt deposits of north-western England. *Q. Jl. geol. Soc. Lond.*, **126**, 103–23.

Fedynskiy, V. V. and Levin, L. E. *The tectonics and oil and gas potential of the marginal and inland seas of the USSR.* Lamont-Doherty Geological Observatory of Columbia University. Translation.

Flinn, D. 1973. The topography of the seafloor around Orkney and Shetland and in the northern North Sea. *Jl. geol. Soc. Lond.*, **129**, 35–59.

Fuglewicz, R. 1973. Megaspores of Polish Buntersandstein and their stratigraphical significance. *Acta. Palaeont. Pol.*, **18**, 401–53.

Garson, M. S. and Plant, Jane, 1972. Possible dextral movement on the Great Glen and Minch Faults in Scotland. *Nature phys. Sci.*, **240**, 31–5.

Haanstra, U. 1963. A review of Mesozoic geological history in the Netherlands. *Verh. K. ned. geol. mijnb. Genoot.*, **21**, 35–57.

Heybroek, P. 1974. Explanation to tectonic maps of the Netherlands. *Geologie Mijnb.*, **53**, 43–50.

Heybroek, P., Haanstra, U. and Erdman, D. A. 1967. Observations on the geology of the North Sea area. *Wld. Petrol. Congr. 7 Mexico Proc.*, **2**, 905–16.

Hinz, K. 1968. A contribution to the geology of the North Sea according to geophysical investigations by the Geological Survey of German Federal Republic. *In* D. T. Donovan (Ed.) *Geology of Shelf Seas.* Oliver and Boyd, Edinburgh, 55–71.

Kent, P. E. 1968. Geological problems in North Sea exploration. *In* D. T. Donovan (Ed.) *Geology of Shelf Areas.* Oliver and Boyd, Edinburgh, 73–86.

Kent, P. E. and Walmsley, P. J. 1970. North Sea Progress. *Bull. Am. Ass. Petrol. Geol.*, **54**, 168–81.

Krebs, W. and Wachendorf, H. 1973. Proterozoic-Palaeozoic geosynclinal and orogenic evolution of Central Europe. *Bull. geol. Soc. Am.*, **84**, 2611–30.

Laubscher, H. P. 1973. Alpen und Plattentektonik: Das Problem der Bewegungsdiffusion und Kompressiven Plattengrenzen. *Z. dt. geol. Ges.*, **124**, 295–308.

Lefort, J. P. 1973. La Zonale. Biscaye-Labrador: Mise en evidence de cisaillements dextres anterieurs a l'ouverture de l'Atlantique nord. *Marine Geology*, **14**, 33–8.

Marston, R. J. 1970. The Foyers granitic complex, Inverness-shire, Scotland, *Q. Jl. geol. Soc. Lond.*, **126**, 331–68.

Naylor, D., Pegrum, D., Rees, C. and Whiteman, A. 1974. Nordsjøbassengene: The North Sea trough system. *Noroil*, **2**, 17–22.

Pearson, D. A. B. 1970. Problems of Rhaetian stratigraphy with special reference to the lower boundary of the stage. *Q. Jl. geol. Soc. Lond.*, **126**, 225–85.

Rhys, G. H. (Compiler) 1974. A proposed standard lithostratigraphic nomenclature for the southern North Sea and an outline structural nomenclature for the whole of the (UK) North Sea. A report of the joint Oil Industry Institute of Geological Sciences Committee on North Sea Nomenclature. *Rep. Inst. geol. Sci.,* **78/8,** 14.

Roberts, D. G., Matthews, D. A. and Eden, R. A. 1972. Metamorphic rocks from the southern end of the Rockall Bank. *Jl. geol. Soc. Lond.,* **128,** 501–6.

Sanderson, D. J. and Dearman, W. R. 1973. Structural zones of the Variscan fold belt in SW England, their location and development. *Jl. geol. Soc. Lond.,* **129,** 527–36.

Simpson, S. 1962. Variscan orogenic phases. *In K. Coe (Ed.) Some aspects of the Variscan fold belt. 9th Inter-university geological congress, Exeter 1961.* Manchester University Press, (4), 65–73.

Smirnov, V. I. and Kazanski, V. I. 1973. Ore-bearing tectonic structures on geosynclines and activized platforms in the territory of the USSR. *Z. dt. geol. Ges.,* **124,** 1–17.

Smith, G. A. and Briden, J. C. 1973. Phanerozoic world maps in 5 organisms and continents through time. Special papers in palaeontology, *Pal. Ass. Lond.,* **12,** 1–42.

Sorgenfrei, T. 1969. Geological perspectives in the North Sea area. *Meddr. dansk. geol. Foren.,* **19,** 160–96.

Sorgenfrei, T. and Buch, A. 1964. Deep test in Denmark 1935–59. *Danm. geol. Unders. Raeke III,* **36,** 1–146.

Talwani, M. and Eldholm, O. 1972. Continental margin off Norway: A geophysical study. *Bull. geol. Soc. Am.,* **83,** 3575–606.

Talwani, M. and Eldholm, O. 1973. Boundary between continental and oceanic crust at the margin of rifted continents. *Nature, Lond.,* **241,** 325–30.

Trumpy, R. 1970. Stratigraphy in mountain belts. *Q. Jl. geol. Soc. Lond.,* **126,** 293–318.

Trusheim, F. 1971. Zur Bildung der Salzlager im Rotliegenden und Mesozoikum Mitteleuropas. *Beih. geol. Jb.,* **112,** 1–51.

Voigt, E. 1962. Über Randtroge vor Schollenrandern und ihre Bedeutung im Gebeit der Mitteleuropaischen Senke und angrenzender Gebiete. *Z. dt. geol. Ges.,* **114,** 378–418.

Warrington, G. 1970. The stratigraphy and palaeontology of the 'Keuper' series of the central midlands of England. *Q. Jl. geol. Soc. Lond.,* **126,** 183–223.

Wills, L. J. 1970. The Triassic succession in the central midlands in its regional setting. *Q. Jl. geol. Soc. Lond.,* **126,** 225–85.

Woodrow, D. L., Fletcher, F. W. and Ahrnsbrak, W. F. 1973. Paleogeography and paleoclimate at the deposition sites of the Devonian Catskill and Old Red facies. *Bull. geol. Soc. Am.,* **84,** 3051–64.

Wolburg, J. 1969. Die epirogenetischen Phasen der Muschelkalk-und Keuper-Entwicklung Nordwest-Deutschlands, mit einem Rueckblick auf den Buntsandstein. *Geotekt. Forsch.,* **32,** 1–65.

Ziegler, P. A. (*In press*). The North Sea in a European palaeo-geographic framework. *Proc. Bergen North Sea Conference. Norg. geol. Unders.*

DISCUSSION

John Ancock (Amoco Europe): What direct evidence do you have for wrench-movements on the Great Glen Fault in the Jurassic of the Moray Firth and how do you reconcile your proposed basin-outline with on-shore exposures of Jurassic controlled by the Helmsdale Fault? What effect would this wrench-movement have on your proposed tectonic lineament through Fair Isle?

W. H. Ziegler: Interpretation of some Jurassic wrench-fault movement along the Great Glen Fault is based on seismic correlations, well-data and surface geological information. Since some of these data belong to other companies and were obtained by us by confidential data-trade I cannot furnish you with detailed and concrete evidence. I regret this as much as you do, as one cannot conduct a scientific discussion by saying 'I told you so!'. The faults along the Great Glen lineament and in the Moray Firth are *en echelon* and show vertical and horizontal splays which we interpret as being wrench-induced. The fault-dating is done by use of well-data and seismic correlation. I regret the oversimplification of the geological sketch-maps which omit the Helmsdale Fault and the well-known Jurassic section along the coast near Brora.

Since I interpret the Fair Isle and Great Glen lineaments as conjugate shear-directions, movements along one would be accompanied by movement along the other, as long as the stress pattern is unchanged. External rotation of the pattern however will result in unequal amounts of displacement along the conjugate pair. The problem of tracing the north-east continuation of the Great Glen Fault needs a lot more work—but it may well be that later movement did not revive the fault trend over its whole Caledonide length. This is a fascinating subject for further study.

G. H. Brown (Brown's Geological Information Service): On Dr Ziegler's Zechstein map we saw an "Auk igneous centre". On the Auk sections such volcanics are not to be seen. Do these exist, and if so what is their geological significance?

W. H. Ziegler: A few wells in the Auk area show igneous intrusives. These basalts occur in the Rotliegendes and Zechstein. I did not have them age-dated. Presumably they are related to post-Hercynian rifting, like so many others in the Permian of north-west Europe. If they are not Permian in age I would guess that they are of upper Jurassic age.

R. P. F. Hardman (Amoco Europe): Could Dr Ziegler give a specific example of the Hardegsen Pulse in the North Sea?

W. H. Ziegler: In the well Conoco 49/21-2, which is published by G. H. Rhys (1974) as a type section for the North Sea, the contact between the Bacton Group and the Haisborough Group lies at 4512ft (1375m). This is the Hardegsen Unconformity. Start your detailed correlations from there—you can carry the Triassic markers right over to Denmark, the UK, Holland and the Viking Graben. Further information is given by T. P. Brennand (Paper 22, this volume), and if you can obtain them study the German papers by Trusheim and Wolburg who settled this point ten years ago.

N. Morton (Birbeck College): I would like to ask Dr Ziegler about movement along the Great Glen Fault in the Palaeocene. In his lecture and in one diagram he referred to right-lateral movements, but in the general tectonic map of the North Atlantic he showed left-lateral displacement. Given the angular relationships of the direction of the northern end of the Great

Glen Fault with the North Atlantic spreading direction, a right-lateral movement might be difficult to reconcile.

W. H. Ziegler: Before the whole story is properly understood a lot more work needs to be done about the motion history of the Great Glen Fault. In general its movements were left-lateral during most of its history whereas the Tornquist Line shows right-lateral tendencies. I propose that some "standard" wrenching was involved in the formation of the Weald Troughs along the southern margin of the North Sea–German basin. This happened from Upper Jurassic to Lower Cretaceous. Later, mainly in early Tertiary time a renewal of this "wrench system" squeezed and dissected these basins. We find evidence of such reversed movements along the structures in the Anglo-Dutch basins. As the Great Glen Fault forms part of the conjugate shear-system, these movements must also have had some expression on it. That is why I propose a short reversal of relative movement in about Laramide time. Overall the first opening of the North Atlantic and the inversion of the Weald Basin are contemporaneous—and for that reason I would like to attribute both to the same geotectonic regimes.

J. C. Pratsch (Mobil Oil, Germany): What is your definition of the Mauretania belt? It was originally defined as the Precambrian trend from the Reguibat to the Bové syncline.

W. H. Ziegler: Since this paper is about the North Sea the Mauretanids are not really germane to the argument.

On my paleotectonic reconstructions of the Hercynian diastrophism I show an orogenic belt along the west side of north-west Africa from the Anti-Atlas to Liberia. The rocks in this belt range from Precambrian to Carboniferous and were deformed in a series of events during the Taconic/Caledonian and Hercynian phases. The Mauretanids form the southern part of this belt. The excellent papers by Professor Sougy are a valuable source for any further information on the area.

Dr B. W. Sellwood (Reading University): I am anxious to lay the ghost of the Welsh landmass which has so consistently appeared in Jurassic reconstructions during this Conference. In the higher and middle parts of the Lower Jurassic, British onshore sediments thicken from east to west with the thickest published section occurring at Mochras on Cardigan Bay (100m+). In general, the sediments become less sandy and more basinal in this direction. On these grounds, it seems inconceivable that a substantial clastic source-area should have existed over Wales. One could make similar comments—though less certainly—concerning the supposed Irish and Cornish landmasses. A positive region may have supplied late Lias sediments from the South-west Approaches of the English Channel. These considerations have important repercussions from the point of view of Celtic Sea prospects as coarse sands are not likely to have been supplied from non-existent land areas! Thick Jurassic sands are unlikely in the Celtic Sea before late Jurassic (or even Early Cretaceous).

On the other hand, the important and well-known structurally controlled slump-beds and turbidites of the Moray Firth (Kimmeridgian) have been omitted from the Ziegler models. Thus, these and other inaccuracies (*e.g.* the extensive, but non-existent carbonate spreads of the Lias) lead me to be extremely suspicious of Ziegler's broad-brush approach.

W. H. Ziegler: I acknowledge the comments about the Celtic fringe and assure the questioner that I am aware of the Jurassic section near Brora.

Dr R. A. Scrutton (Edinburgh University): It seems to me that for a graben or rift-system of the size of the North Sea system, the occurrence of igneous material is, in fact, scarce. In comparison to the grabens of the Rhine, Oslo and the East African Rift System, volcanic eruptions are fairly isolated both chronologically and geographically.

Dr W. Ziegler agreed with this observation and this led to further discussion on the "r-r-r theory" for the origin of the North Sea graben.

W. H. Ziegler: The understanding of the structural evolution of an area presupposes considerable knowledge of the stratigraphy of the area. Rift-systems of the East African type usually are preceded by an axial uplift and erosion, prior to rifting and igneous activity. There is no such evidence in the North Sea. The ideas that I presented here evolved precisely for this reason. In addition they are also an alternative to all the esoteric and outlandish speculations involving a "*deus ex machina*" in the deeper mantle and I feel that they adequately explain a complex tectonic evolution.

I would very much agree with your observation about the relative scarcity of igneous loci in the North Sea area.

12

Coal Rank and Gas Source Relationships—Rotliegendes Reservoirs

By T. D. EAMES

(Conoco Europe Ltd, London)

SUMMARY

There is little doubt that the gas found in the Rotliegendes reservoirs of the southern North Sea area has its source in the coal and carbonaceous material in the underlying Carboniferous. The geological conditions under which devolatilization of the coals has occurred is therefore an important factor in the understanding of the petroleum geology of the area.

The view is advanced that the high rank coals with less than about 30 percent volatiles (d.a.f.) in the Carboniferous have been formed by high temperature effects of igneous intrusives, whether locally or over larger regions where large masses at depth have significantly raised the palaeo-geothermal gradient. Alternatively strong orogenic folding and thrusting could have produced comparable high temperatures.

A pre-Permian subcrop map in the southern North Sea area is presented. A map showing the distribution of high and low rank Carboniferous coals in the area is also shown, and its significance discussed.

INTRODUCTION

There is little doubt that the gas found in the Rotliegendes reservoirs of the Southern North Sea area has its source in the coal and carbonaceous material in the underlying Carboniferous. The geological conditions under which devolatilization of the coals has occurred is, therefore, an important factor in the understanding of the petroleum geology of the area. The objective of this paper is to summarize briefly the evidence for the source of the gas and to advance some geological explanations to account for what has been found.

RANK OF COAL

First we must understand what is meant by the term "rank". This word is applied to coals to indicate the overall maturity or grade, but it gives no indication of what geological processes may have caused increases in the rank of the coal. There are various methods of measuring the rank. These include chemical analysis (known as Ultimate Analysis) of the important elements Carbon, Hydrogen, Oxygen, Nitrogen and Sulphur, measurement of calorific value, determination of coking properties, reflectance of vitrinite and spore coloration. These methods are useful in various fields, but they all have the disadvantage that none of them gives a direct measurement of the volatile content of coal, which is what the exploration geologist is really interested in. The only method which does this is the Proximate Analysis, which determines directly inherent moisture, volatiles, ash and fixed carbon, by heating the coal under standardized laboratory conditions described in the Fuel Research Publication No. 44 (Anon, 1940).

MOISTURE CONTENT

There is plenty of evidence to show that the moisture content in the low rank coal series is directly related to compaction with depth of burial. Trotter (1952) gives the following figures:

Rank	Moisture content	Estimated maximum depth of burial (approx.)
Peat	45–60 percent	
Brown lignite	30–45 percent	2000–3000ft (610–914m)
Sub-bituminous coal	10–30 percent	6000–7000ft (1829–2134m)
Bituminous coal	10 percent	

These estimates are supported by borehole evidence, and data have been published on the Notown borehole near Greymouth in New Zealand (Gage, 1952), on several boreholes in the Taranaki area, New Zealand (Elphick and Suggate, 1964) and on the Prestwich boreholes near Manchester (Trotter, 1954). Analyses of Carboniferous coals in the Tetney Lock No. 1 Well, on the Lincolnshire east coast, show a progressive loss of moisture with depth of burial.

Burial to extreme depths can produce high volatile bituminous coals with very low moisture contents. For example, Eocene coals from boreholes in the Taranaki area, New Zealand, between 10 500ft and 14 750ft (3200–4496m) have a moisture content decreasing with depth and volatiles 39–51 percent (m.m.f.) (Elphick and Suggate, 1964), and a Jurassic coal in a core at 16 193ft (4936m) in Richardson-1 borehole, Mississippi (USA), has 0.44 percent moisture and 39.9 percent volatiles (d.a.f.) (Kuyl and Patijn, 1961).

It is concluded that in the low rank coal series burial and compaction has caused the loss of inherent moisture, and this is associated with changes in colour, density, reflectivity, and an increase in the Total Carbon percentage.

VOLATILE MATTER

Volatile Matter is usually calculated on a dry-ash free (d.a.f.) or dry-mineral matter free (m.m.f.) basis, so that the effects of moisture and ash can be eliminated when comparing different coals. The most important fact is that there is no direct relationship between volatiles and moisture. South Wales anthracites, for example, contain about 1 to 3 percent moisture, and similar percentages are common in bituminous coals, cannel coals and torbanites with a range from less than 5 percent to more than 80 percent volatiles (d.a.f.). Since the moisture percentage is essentially related to depth of burial, there must be other geological processes to account for this variation in volatile percentage.

Low Rank Series

Published analyses show that the range in volatile content of Pleistocene peats is from 38 to 94 percent (d.a.f.). Analyses of the immature Top Peat from the top 3ft (0.9m) of bogs in the northern hemisphere show a range from 66 to 94 percent (d.a.f.), so the variation must be partly due to original differences in the composition of the vegetation. The lower values from 38 to 66 percent are from levels up to 30ft (9m) below the surface of the peat, so it is clear that there has been substantial loss of volatiles within the thickness of peat bogs. This is the result of chemical changes caused by reducing bacteria during maturation of the peat.

Analyses of all low rank coals show a considerable variation in volatile content (see Trotter, 1952). Much of, if not all, of the variation can be attributed, first, to difference in composition of the original vegetation and, second, to the initial rate of accumulation and burial, which controlled the period of time that bacteria were active in reducing the original peat. This initial variation is referred to as the peat or coal-type band.

The lower limits of the peat and coal-type bands are roughly constant for all the low rank series from peats to bituminous coals at approximately 35 percent (d.a.f.)—between 30 and 40 percent—and this has been termed the "critical limit of coal substance volatiles". It represents the approximate minimum volatile content due to

devolatilization of the original peat by bacterial activity, and initial variations in the vegetation. This shows that the progressive decrease of moisture with depth of burial in the low rank series is not accompanied by a significant decrease in the volatiles. There may be a slight loss of volatiles but this is apparent only at the upper limit of the volatile range for each peat or coal type. This can be explained by the continued presence of a few bacteria, which are known to survive up to temperatures as high as 65°C below the surface.

Analyses of bituminous coals (with less than 10 percent moisture) give evidence to show that depth of burial has no effect on the volatile matter at all. Figure 1 illustrates this with numerous analyses of Carboniferous coals from the Prestwich Nos. 1 and 2 boreholes, and from the much deeper Eocene coals in four boreholes in the Taranaki area of New Zealand.

There are other examples. In the Sidney coalfield, Australia, 12 coal seams all have a volatile content of 35 to 38 percent (mineral matter free) over a stratigraphic interval of 3500ft (1067m) (Hacquebard, 1961). In the Northumbrian Trough, there is a stratigraphic thickness of about 7000ft (2134m) in the Lower and Middle Carboniferous with a range of 37 to 48 percent volatiles (d.a.f.) which shows no decrease through the section. Similar results were obtained in analyses by the Coal Survey Laboratory of five Carboniferous Coal seams in the Tetney Lock No. 1 Well covering a vertical interval of 1820ft (555m).

The fact that high volatile bituminous coals occur in very deep boreholes is also evidence that depth of burial by itself has no effect on the volatile content. Apart from the New Zealand coals shown in Fig 1, the following have been published by Kuyl and Patijn (1961):

Borehole	Depth	Volatile matter (d.a.f.)
Richardson-1 (Mississippi)	16 193ft (4936m) (core)	39.9 percent
S.P. No. 48 (Texas)	12 804ft (3903m) (core)	51.3 percent
Zulia 36-E-2 (Venezuela)	11 200ft (3414m) (core)	47.8 percent

The coal in the Richardson-1 is of particular interest since it consists of nearly pure vitrinite, and ultimate analysis gave 5.5 percent Hydrogen, which is the average figure for a low rank bituminous coal, and 89.0 percent Total Carbon.

Finally, Carboniferous torbanites with volatiles around 80 percent (d.a.f.) and low moisture are found in close association with true bituminous coals, and there are earlier records with up to 91 percent volatiles (d.a.f.) from Fifeshire. These are comparable with volatiles of the most recent Top Peats.

Figure 2 shows the ranks of Carboniferous coals in the Southern North Sea area at their maximum

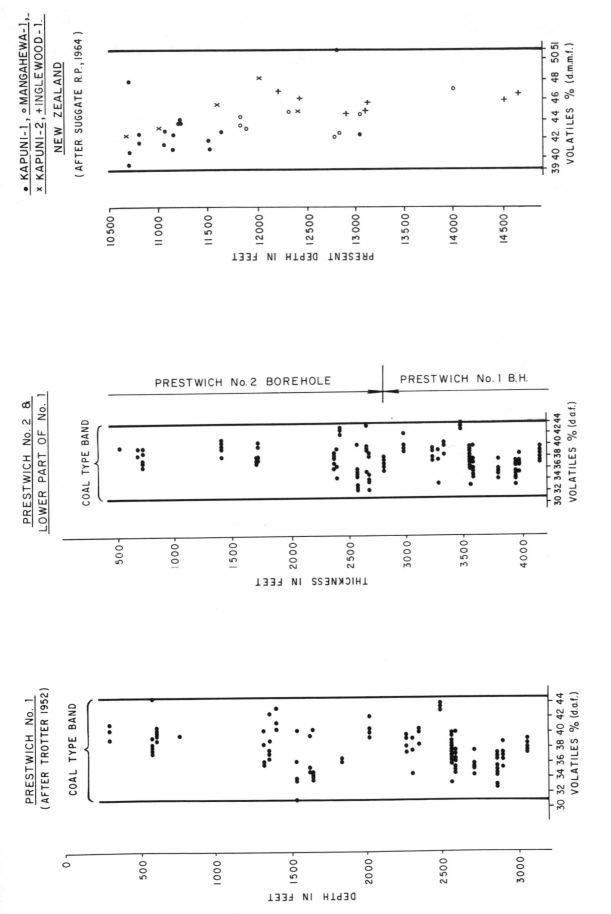

Fig 1. Graph showing relationship between depth and volatile matter in coal.

depths of burial based on the detailed reconstruction of the section. The estimates of volatiles have to be regarded with some reservations because they are all based on vitrinite reflectance and not direct measurements of volatiles. However, it is suggested that there is reasonable evidence for a coal type band from 30 to 38 percent volatiles (d.a.f.), which is comparable to that seen onshore in the Nottinghamshire coal fields. Several wells have low rank coals in the 10 000 to 13 000ft (3048–3962m) maximum depth of burial range. In contrast, many high rank low volatile coals have never been buried so deep, and an explanation for this phenomenon must now be given.

High Rank Series

The high rank solid fuel series is subdivided on the basis of volatile content approximately as follows:

Rank	Volatiles (d.a.f.)
Coking coals	24–30 percent
Semi-bituminous coals	15–24 percent
Carbonaceous coals (or semi-anthracites)	9–15 percent
Low rank anthracites	7–9 percent
High rank anthracites	4.5 (or less)–7 percent

Most workers on rank in coal now agree that heat is the principal factor, which has caused devolatilization of the high rank coals. Before discussing the geological conditions for devolatilization, I shall briefly summarize the results of laboratory tests.

When bituminous coals are heated, there are no visible changes below about 300°C. Hoffman (in Gropp and Bode, 1933) found that bituminous coals undergo a change at temperatures above 325°C, and that the temperature at which the change occurs is higher for higher ranks of coal tested. He also converted a low rank coal into a substance petrographically comparable to anthracite at about 550°C. In experiments where both pressure and temperature have been applied, the liberation of volatiles is inhibited. Lewis (1936) showed that the greater the pressure the higher temperature required to bring about a given increase in rank, and this conforms to Le Chatelier's Principle when applied to chemical reactions which give an increase in volume.

Determination of the temperatures at which different ranks of coal have been formed in geological conditions has been attempted by McFarlane (1929) in the Yampa coalfield, Colorado, where the heat from basalt sill intrusions has caused devolatilization of the coals. The crystallization temperature of the basalt was estimated at about 1200°C. Samples from each rank of coal were heated to different temperatures and then specific gravity measurements and fixed carbon analyses were made for each recorded maximum temperature. The results of these

numerous tests showed that there had been no loss of volatiles from low rank bituminous coals below 260°C, and between 260°C and 330°C there was a marked increase in the fixed carbon percentage corresponding to the loss of volatiles. Tests on successively higher ranks of coals give similar results except that the sharp increase in fixed carbon occurred at progressively higher temperatures, and it was concluded that semi-anthracites were formed at temperatures over 400°C.

The fact that coal does not give off volatiles until a certain temperature is reached, and that this temperature increases with the rank of coal, strongly suggests that high rank coal was metamorphosed at about the temperature at which volatiles are liberated on heating. It is of interest that there is some chemical confirmation of this: Illingworth (in Lewis, 1936) found that pyridine-soluble constituents of coal are destroyed when heated to 450°C, and that the semi-anthracites and anthracites do not contain these pyridine-soluble constituents.

Figure 3 is a graph which shows the rank, maximum depth of burial and maximum temperatures of coals in selected Southern North Sea boreholes. The maximum temperatures have been estimated by taking the present borehole temperature, calculating the maximum depth of burial from geological reconstructions of the area, and extrapolating the average geothermal gradient of the well. Although there is some margin of error possible, the overall relationships between the temperature and rank of different coals are reasonably comparable. There are two important observations:

First   Many low rank coals have been subjected to much higher temperatures compared to many high rank coals than can be accounted for by depth of burial alone.

Second There is evidence for a decrease in volatiles with increasing temperature in high rank coals with less than 30 percent volatiles (d.a.f.) in individual boreholes, but there is no evidence for this in the low rank coals with more than 30 percent volatiles (d.a.f.).

When these observations and the results of laboratory tests are taken together, there is strong evidence to conclude that devolatilization of the high rank coals is essentially a high temperature phenomenon, probably starting around 300°C. We must now examine what natural sources of heat have produced these high temperatures, with particular reference to the North Sea area.

Magmatic Heat Source. The effects of igneous alteration of coal seams by contact metamorphism are well known. The type and extent of alteration depends on the temperature of the magma, which is determined mainly by its composition, and the size and nature of the igneous body. Extrusives

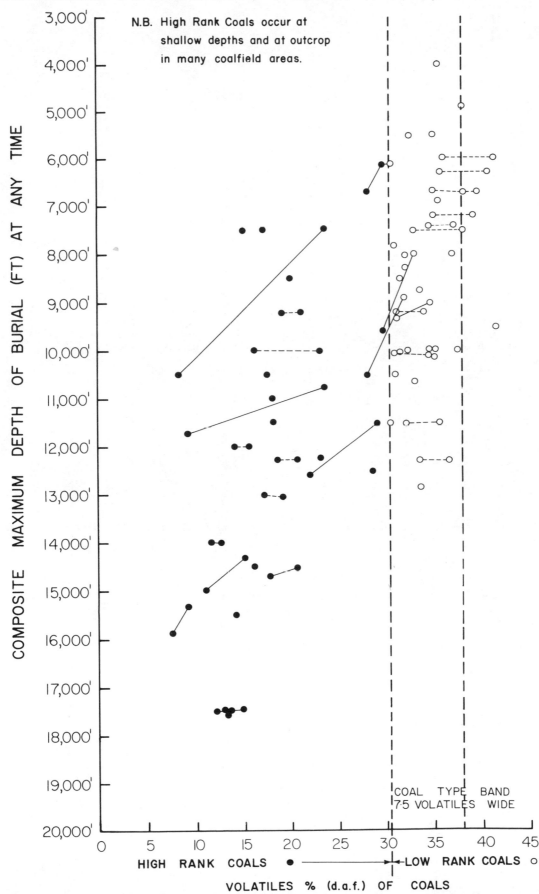

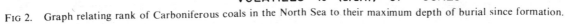

FIG 2.   Graph relating rank of Carboniferous coals in the North Sea to their maximum depth of burial since formation.

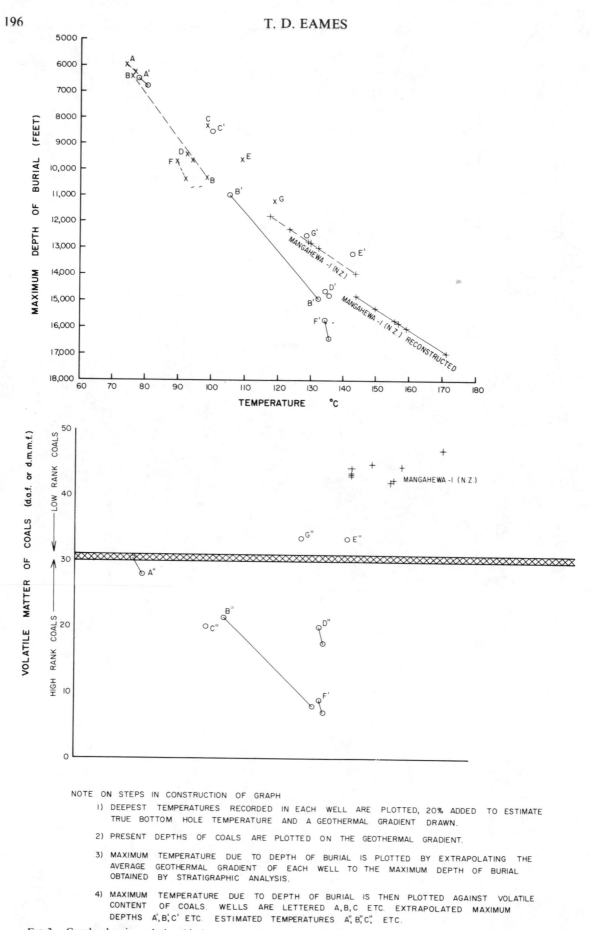

NOTE ON STEPS IN CONSTRUCTION OF GRAPH

1) DEEPEST TEMPERATURES RECORDED IN EACH WELL ARE PLOTTED, 20% ADDED TO ESTIMATE TRUE BOTTOM HOLE TEMPERATURE AND A GEOTHERMAL GRADIENT DRAWN.

2) PRESENT DEPTHS OF COALS ARE PLOTTED ON THE GEOTHERMAL GRADIENT.

3) MAXIMUM TEMPERATURE DUE TO DEPTH OF BURIAL IS PLOTTED BY EXTRAPOLATING THE AVERAGE GEOTHERMAL GRADIENT OF EACH WELL TO THE MAXIMUM DEPTH OF BURIAL OBTAINED BY STRATIGRAPHIC ANALYSIS.

4) MAXIMUM TEMPERATURE DUE TO DEPTH OF BURIAL IS THEN PLOTTED AGAINST VOLATILE CONTENT OF COALS. WELLS ARE LETTERED A,B,C ETC. EXTRAPOLATED MAXIMUM DEPTHS A', B', C' ETC. ESTIMATED TEMPERATURES A", B", C", ETC.

FIG 3.   Graphs showing relationship between rank, maximum depth of burial and temperatures of selected coals.

generally have a very small zone of alteration since the heat is rapidly dissipated at the surface. Where there are dykes and sills, devolatilization is usually local and restricted to a narrow zone parallel to the intrusion, beyond which the coal is unaltered. For example, in the Northumberland coal field, the Brockwell seam is only altered in a zone up to about 10ft (3m) from the Westerhope dyke (Raistrick and Marshall, 1939). In the Whitley Bay No. 1 and Harton No. 1 boreholes in Northumberland, a series of sills was penetrated in the Carboniferous, and the rank of the coal seams shows a sharp increase in a narrow zone both above and below each sill.

The width of the zone of alteration is essentially proportional to the thickness of the intrusive. In the Natal coalfield, South Africa, Heslop (in Briggs, 1935) found the coal to be unaffected by a dyke at a distance from it roughly equal to the thickness of the dyke, and provided data for a 1200ft (366m) thick dyke.

Devolatilization of coal seams by large plutonic intrusions can be compared with the effects of very large sills or dykes, except that the alteration zone extends a greater distance from the contact. Since plutonic intrusions are large and usually deep seated, the period of cooling was longer than for dykes or sills with the general result that the regional geothermal gradient was probably raised substantially.

One example is the devolatilization of Carboniferous coals by the Weardale granite on the Alston block in Durham, described by Trotter (1954). The isovols decrease from 40 percent to 10–12 percent (d.a.f.) towards the Alston block, and these can be correlated with five roughly concentric zones of hydrothermal mineralization, and also to the gravity contours. The Rookhope borehole was drilled into the Weardale granite at 1281ft (390m) below the surface and, although the Carboniferous is unconformable on the granite, there is reasonable evidence from the mineralization that the granite has been rejuvenated later. The composition of pyrrhotite in veins between 1083ft (330m) and 1562ft (476m) has been used to calculate temperatures of formation of 450°C to 485°C (Sawkins et al., 1964), and this is in remarkable agreement with the laboratory estimates of temperature for semi-anthracites previously discussed.

A second example is the devolatilization of both Carboniferous and Lower Cretaceous coals on the Bramsche Massif in the Osnabrück area of Germany, described by Teichmüller and Teichmüller (1958). In this case, the isovols in both the Wealden and the Carboniferous coals are virtually parallel to the magnetic contours, which show a positive anomaly over the Bramsche Massif, and there is no correlation with the stratigraphic or sedimentary position of the coals. The existence of this intrusive has not been proved by the drill, since it is probably about 10 000ft

(3048m) below the surface anthracite area of Osnabrück. The time of this intrusion can be fixed during the early Upper Cretaceous, because Campanian lignites overlie Lower Cretaceous anthracites in the area.

It is concluded that large plutonic intrusions have probably raised the temperature sufficiently high to exceed the critical limit of about 300°C at which devolatilization of coals begins. Since, in effect, the regional geothermal gradient has been raised during the period of intrusion of a large pluton, the decrease of volatile content of coals with depth commonly observed in the high rank coals can be correlated with the palaeotemperature.

Frictional Heat Source. Many workers have noted the progressive rise of rank of coals towards zones of intense tectonic deformation. White (1913) was the first to demonstrate the progressive decrease in volatiles from bituminous coals to anthracites in the Pennsylvanian of the Appalachian coal belt towards the highly folded and thrusted Allegheny Front. In Canada, Cretaceous sub-bituminous coals become anthracites as the Rocky Mountains are approached from the east, and in Bavaria Oligocene and Miocene lignites show an increase in rank towards the Alpine fold belt, with the carbon content varying with the degree of folding.

According to the White theory, the volatile content of a high rank coal is determined by the intensity of the orogenic forces and by the natural heat treatment to which the coal has been subjected. Roberts (1950) considered that the loss of volatiles was caused by frictional heat generated by differential movement along bedding planes during folding, and along thrust planes.

Firstly, I will discuss the evidence for devolatilization of coals due to thrusts. Progressive decrease in volatile content of coals locally towards thrusts has been recognized by various workers. For example, Teichmüller and Teichmüller (1958) have shown that this decrease in volatiles occurs towards the Sutan Thrust in the Ruhr coalfield, the loss of volatiles occurring both above and below the thrust plane. Also in Liège coalfield in Belgium, where there is extensive low angle thrusting, the horizontal decrease in volatiles is much more pronounced than the vertical decrease, and geothermal gradients with depth of burial cannot alone account for the devolatilization (Geny, 1911).

In the South Wales coalfield isovols show a general north-westerly decrease in volatile content from bituminous coals to high rank anthracites. Trotter (1948; 1954) has suggested that the volatile content is determined by proximity of the various coal seams to a major low-angle thrust beneath the coal field, outcropping along the line of the Cennen disturbance to the north of the coalfield. This hypothesis appears to be reasonable for the southern margin of the Wales–London–Brabant

Massif, since major thrusting is recognized in other coal fields in this geological setting. An alternative theory is that a huge thickness of Mesozoic sediments of up to 18 000ft (5486m) may have been originally deposited over the anthracite area and subsequently completely eroded. This alternative looks very implausible in the light of the present knowledge of Mesozoic basins in the Bristol Channel–Celtic Sea area, and the evidence that much higher temperatures than usually found in sedimentary basins would be required for anthracite formation.

I will now discuss the possibility that moderately strong folding may have also caused a limited amount of devolatilization of coals, although evidence for this is not so conclusive. An example of this is apparently shown by the Potteries syncline in the North Staffordshire coalfield, where dips range from 20° to 60°. Millot (1941/42) plotted the variation in volatile content of one seam, the Cockshead, and demonstrated the close relationship between volatiles and the geological structure, the seam contours being parallel to the isovols. He concluded that the forces, which determined the principal tectonic features of the coalfield, probably also determined, directly or indirectly, the properties of the seam.

In the Pie Rough Borehole (Keele No. 1) drilled near the centre of the syncline, the coal volatiles decrease progressively with depth from about 40 to 28 percent (d.a.f.) at 4200ft (1280m) near the base of the Coal Measures (Millot et al., 1945/46). However, there is no apparent decrease in the volatiles of coals penetrated by boreholes on the complementary Western and Eastern anticlines (Trotter, 1954). In this particular case, it appears that high rank coals are found only in the syncline where the strata have been compressed, but not in the complementary anticlines, where there has been relative relief from tectonic pressure due to radial tension over the crests. It must also be emphasized that the coals in the Potteries syncline are only at the lowest end of the high rank coal series, so that the amount of devolatilization which may have occurred is at best only small.

In the Ruhr coalfield, folding is much stronger and associated with some thrusting, and the coal rank is correspondingly higher with volatiles ranging from 30 percent to about 9 percent (d.a.f.). Isovols roughly parallel the structure contours on individual seams, but with lower volatiles in the synclines than in the anticlines.

SUMMARY OF CONCLUSIONS

(1) In the Low Rank series (peat to bituminous coal), increase in rank is due to compaction and loss of moisture.
(2) Variation in the volatile content of Low Rank series is due to initial variation in the vegetation and bacterial decomposition within the original peat bog.

(3) A slight reduction of volatile content in the Low Rank series is evident only at the upper limit of the volatile range, and this is attributed to a limited amount of continued bacterial reduction up to temperatures of about 65°C below the surface. It is suggested that this is the probable source of the immature gas commonly seen in the rapidly buried shallow Tertiary–Quaternary sections in the North Sea and elsewhere..

(4) Increase in rank in the High Rank series (coking coals to anthracites) is due to progressive devolatilization at high temperatures above about 300°C, and the amount of devolatilization is dependent on the maximum temperature it has reached at any time.

(5) Devolatilization of the High Rank coals has been caused by heat from igneous intrusions and by frictional heat generated during strong tectonic folding and thrusting.

(6) Depth of burial by itself has not subjected coals to sufficiently high temperatures to have caused devolatilization, and the occurrence of high volatile bituminous coals in deep boreholes at high temperature substantiates this.

APPLICATION TO EXPLORATION FOR GAS IN THE SOUTHERN NORTH SEA AREA

Figure 4 is a subcrop map of the pre-Permian floor in the southern North Sea area. I would like to express my thanks to all the companies operating in this area—too numerous to mention individually—for making the basic data available to me, so that the map constructed is probably as accurate as can be made at this time. Published data in the Netherlands by Thiadens (1963), a map of the Netherlands by Heybroek of NAM (1974), and the map of the UK onshore area by Wills (1971) have also been used.

The most significant observation for the explorer is that all the major gas-fields in the Rotliegendes overlie or are very close to the subcrop of the Westphalian B Coal Measures, and to a lesser extent the Westphalian A. This is perhaps not surprising since the greatest development of coals is in this part of the Coal Measures and consequently there is a large volume of source material for large gas-fields.

Figure 5 shows the approximate regional distribution of high and low rank coals in the Upper Carboniferous, and an interpretation is proposed. Five major provinces are distinguished, which will be discussed from south to north:

(1) Zone of Strong Hercynian Folding and Thrusting
This zone is characterized by mainly high rank coals, formed by devolatilization during the strong tectonic folding and thrusting of the Hercynian

orogeny. The Coal Measures have been compressed and thrust northwards onto the Wales–London–Brabant Massif. The volatiles were therefore essentially given off before the deposition of Permo-Trias reservoirs and consequently mainly lost to the atmosphere. There is, therefore, not much chance of finding major gas accumulations in this area from the Carboniferous source.

(2) Wales–London–Brabant Massif

This stable massif was mainly a land area during the Carboniferous, and therefore there are essentially no source beds in this province.

(3) Zone of Gentle Hercynian Folding and Faulting

The Coal Measures are well developed in this area, but are almost entirely of low rank, and have not therefore been devolatilized. This area is north of the main Hercynian orogenic front, and has mainly been shielded from the strong Hercynian tectonics by the Brabant Massif. Coals in a very few folds, such as the Potteries Syncline, may have had a limited amount of devolatilization, but since this was mainly in the synclines and occurred before the deposition of Permo-Trias reservoirs, there is little prospect of finding major gas fields from the Carboniferous source.

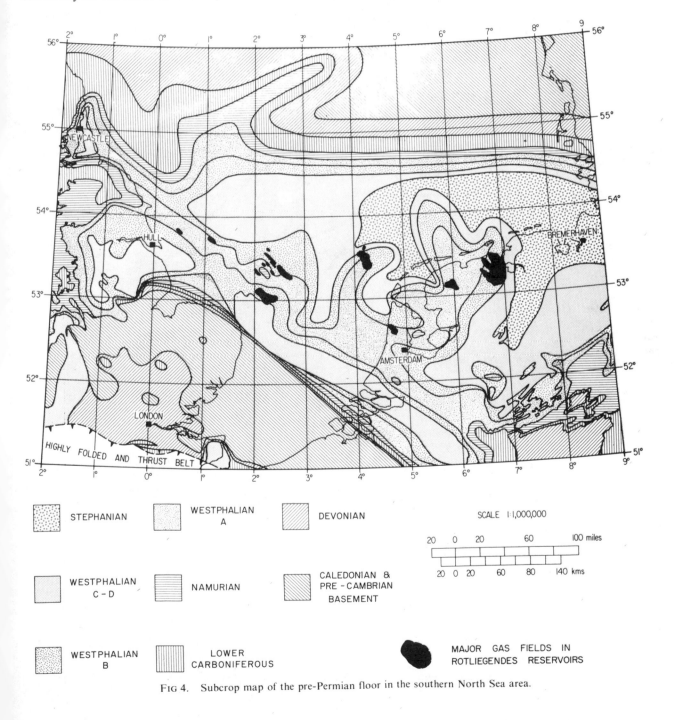

FIG 4. Subcrop map of the pre-Permian floor in the southern North Sea area.

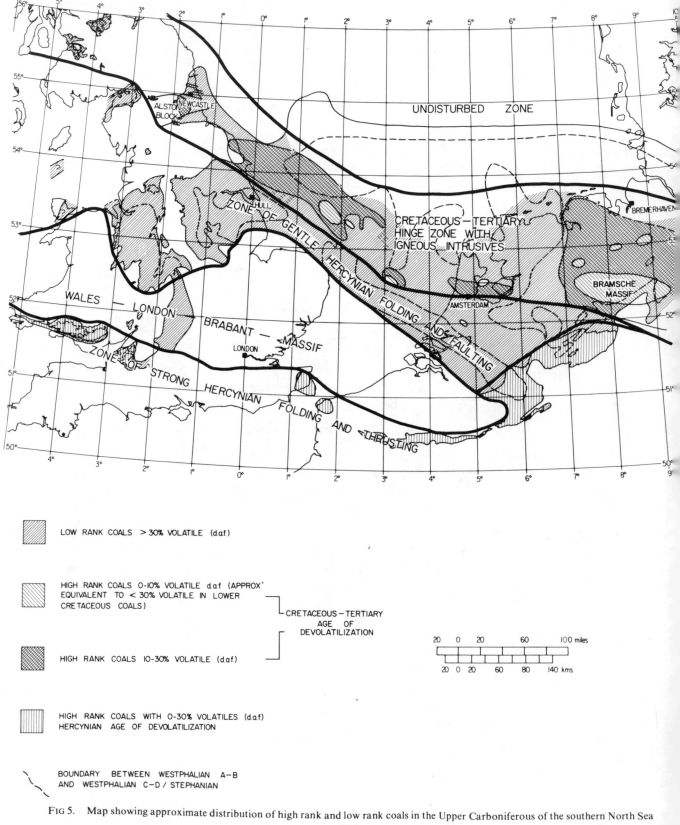

LOW RANK COALS > 30% VOLATILE (d.a.f.)

HIGH RANK COALS 0-10% VOLATILE d.a.f (APPROX'
EQUIVALENT TO < 30% VOLATILE IN LOWER
CRETACEOUS COALS)

⎱
⎰ CRETACEOUS — TERTIARY
         AGE   OF
    DEVOLATILIZATION

HIGH RANK COALS 10-30% VOLATILE (d.a.f.)

HIGH RANK COALS WITH 0-30% VOLATILES (d.a.f.)
HERCYNIAN AGE OF DEVOLATILIZATION

BOUNDARY   BETWEEN WESTPHALIAN A—B
AND WESTPHALIAN C—D / STEPHANIAN

FIG 5.   Map showing approximate distribution of high rank and low rank coals in the Upper Carboniferous of the southern North Sea
area.

(4) Cretaceous–Tertiary Hinge Zone with Igneous Intrusives

This province is a hinge zone in the foredeep between the Alpine orogenic belt to the south and the centre of the Upper Cretaceous–Tertiary basin to the north. Igneous intrusions due to crustal weakness have caused devolatilization of the Carboniferous coals, and the gas has migrated at a relatively late stage into Permo-Trias reservoirs in structural traps of pre-Cretaceous age.

The configuration of the high rank zones is very approximate due to the sparse data, but a trend of major igneous intrusives can be postulated based on the Weardale granite intrusion, the Bramsche Massif, and some major magnetic and gravity anomalies in the southern North Sea. In addition, there is probably a certain amount of dyke and sill intrusion since several boreholes in the area have a mixture of high and low rank coals in the Carboniferous. In these boreholes the coals are in two groups, one low rank with 31–35 percent volatiles (d.a.f.) and one high rank with 17–23 percent volatiles (d.a.f.). These can be compared to the coals seen in the Harton No. 1 and Whitley Bay No. 1 boreholes onshore. In one borehole, a high rank coal sample was found at a shallower depth than low rank coals.

Further evidence for igneous intrusions being the cause of devolatilization of coals in this area is provided by detailed study in the Viking–Indefatigable gas field area. Here the distribution of high and low rank coals clearly has no relation to geological structure or geological history. Evidence of an Upper Cretaceous–Tertiary age of intrusion and devolatilization is provided by the occurrence of high rank Lower Cretaceous coals in the Bramsche Massif and Tertiary intrusives in the Scottish coal fields. In addition Lower Cretaceous shales form a seal for Rotliegendes/Zechstein gas reservoirs in some structures in the Southern North Sea, indicating that gas migration and accumulation must have been later.

All the important gas fields are found in this province, and the usual requirements of structural trap, reservoir and seal are conditional for the location of gas fields.

(5) Undisturbed Zone

North of the hinge zone with igneous intrusions there is very little data available. However, the coals so far analysed in the Carboniferous are all of low rank, and this is probably true for Mesozoic coals as well. The structure is essentially horst and graben block-faulting in the Carboniferous, indicating a tensional regime, and the coals are therefore likely to be unaltered. It is suggested, therefore, that the Carboniferous coals in this province may not be particularly attractive as a source for late stage gas generation and migration.

ACKNOWLEDGEMENTS

I wish to thank BP, Conoco and the National Coal Board for encouragement in this field of study and for permission to publish this paper.

REFERENCES

Anon. 1940. Methods of analysis of coal and coke. *Fuel Research Publication No. 44.*

Briggs, H. 1935. Alteration of coal seams in the vicinity of igneous intrusions and associated problems. *Trans. Inst. Min. Eng.*, **89**, 187–219.

Elphick, J. O. and Suggate, R. P. 1964. Depth/rank relations of high volatile bituminous coals. *N.Z. Jl. Geol. Geophys.*, 7, 594–601.

Gage, M. (Ed.) 1952. The Greymouth coalfield. *Bull. geol. Surv. N.Z.*, **45**, 232 pp.

Geny, P. 1911. Etude sur le distribution des teneurs en matières volatiles dans les veines de la Concession de Courrières. *Ann. Soc. Geol. Nord.*, **40**, 147–57.

Gropp, W. and Bode, H. 1933. The metamorphosis of coals and the problem of artificial coalification. *Fuel*, **12**, 341–55.

Hacquebard, P. A. 1961. Discussion p. 364. *In:* Kuyl and Patijn. 1961. Coalification in relation to depth of burial and geothermic gradient. *C.R. 4ieme Congr. Strat. Geol. Carbonifère, Heerlen.*

Heybroek, P. 1974. Explanation to tectonic maps of the Netherlands. *Geologie Mijnb.*, **53**, 43–50.

Kuyl, O. S. and Patijn, R. J. H. 1961. Coalification in relation to depth of burial and geothermic gradient. *C.R. 4ieme. Congr. Strat. Geol. Carbonifère, Heerlen.* 357–65.

Lewis, E. 1936. The formation of coal with particular reference to its behaviour under heat and pressure. *Jl. Inst. Fuel*, **9**, 233–50.

McFarlane, G. C. 1929. Igneous metamorphism of coal beds. *Econ. Geol.*, **24**, 1–14.

Millot, J. O'N. 1941/2. Regional variations in the properties of the Eight-Foot Banbury or Cockshead seam in the North Staffordshire coalfield. *Trans. Inst. Min. Eng.*, **101**, 2–24.

Millot, J. O'N., Cope, F. W. and Berry, H. 1945/6. The seams encountered in a deep boring at Pie Rough, near Keele, North Staffordshire—Part II. The relationship between seam properties and depth. *Trans. Inst. Min. Eng.*, **105**, 534–70.

Raistrick, A. and Marshall, C. E. 1939. *The nature and origin of coal and coal seams.* English Universities Press: London, 282 pp.

Roberts, J. 1950. The thermodynamics of Hilt's "Law". *Colliery Guardian*, **180**, 325–29.

Sawkins, F. J., Dunham, A. C. and Hirst, D. M. 1964. Iron-deficient low-temperature pyrrhotite. *Nature, Lond.*, **204**, 175–6.

Teichmüller, M. and Teichmüller, R. 1958. Inkohlungsuntersuchungen und ihre Nutzanwendung. *Geol. Mijnb., N.S.*, **20**, 41–66.

Thiadens, A. A. 1963. The Palaeozoic of the Netherlands. *Verh. Ned. Geol. Mijnb. Genoot., Trans. Jubilee Convention*, 9–28.

Trotter, F. M. 1948. The devolatilization of coal seams in South Wales. *Q. Jl. geol. Soc. Lond.*, **104**, 387–437.

Trotter, F. M. 1952. The genesis of a fuel series of rising rank; top peat to fat bituminous coal. *Proc. Yorks. Geol. Soc.*, **28**, 125–63.

Trotter, F. M. 1954. The genesis of high rank coals. *Proc. Yorks Geol. Soc.*, **29**, 267–303.

White, D. 1913. The origin of coal. *Bull. U.S. Dept. Int. Bur. Mines*, **38**, 105–30.

Wills, L. J. 1971. England and Wales, Palaeogeographical map of the Palaeozoic floor, *Geol. Soc. Lond. Memoir No. 7,* 1973.

DISCUSSION

Sir Kingsley Dunham (Institute of Geological Sciences) said that he was very glad that Dr Eames had taken up the question of coal devolatilization, which we all believed to be fundamental to the North Sea gas-fields. However, it was necessary to disagree with the temperature of 450°C cited for the mineralization in the Weardale granite on the basis of the occurrence of pyrrhotite. The pyrrhotite in the presumed Hercynian veins cutting the proved Caledonian granite is of the monoclinic low-temperature type, formed below 200°C (Dunham, Ansel, C. 1964, *Nature Lond.*, **204**, pp. 175–6). This is consistent with the temperatures derived from homogenizing large numbers of fluid inclusions in minerals from the veins which range from a maximum of 220°C down to below 100°C (Sawkins, S. J. 1960, *Econ. Geol.*, **61**, pp. 385–401). Moreover, at the time of mineralization, believed to be mainly during the Westphalian–Zechstein interval, though continuing into the Zechstein, the very well-developed mineral temperature-zoning indicates that the mineralizing fluids were at temperatures appreciably above those of the host rocks. Nevertheless, it is true that there is a general coincidence between high-rank coal occurrence in West Durham and the area underlain by the Weardale granite, with the rank falling to the east. The laboratory results on coal devolatilization appear suspect and it is suggested that this proceeds at much lower temperatures given adequate time. It is worth noting that almost the first production of usable methane in Britain was from Point of Air Colliery, Flintshire, from a sandstone reservoir directly overlying a low-rank bituminous coal, adjacent to the low-temperature lead–zinc mineral field of Halkyn-Minera. Nevertheless, a modest extra contribution of mantle heat, in places conducted through old granite offers the best explanation for devolatilization in the present unsatisfactory state of the art.

T. D. Eames: The temperature evidence given by the pyrrhotite was quoted from an article in *Nature*, and I would not dispute any contrary evidence there may have been given at a later date, as this is outside my field of study. However, I still believe that the high rank of the coals is related to rejuvenation of the Weardale granite at a later stage, probably in Cretaceous–Tertiary times.

Professor Arthur Whiteman (Aberdeen University): I should like to ask Mr Eames if Conoco have considered that there could be a relationship between devolatilization of coals and the sub-Upper Cretaceous trough system which is now being further unveiled at this conference and which has been described by others elsewhere, *e.g.* P. A. Ziegler (Bergen Conference, 1973), Exploration Consultants Ltd and myself (*Noroil*, 1974 and *Tectonophysics* (In press)). At the northern end of the North Sea, between Shetland and Sognefjord, the Viking Graben appears to be underlain by thin crust (Bergen University data); the Oslo Graben is underlain by thin crust (gravity and refraction seismic evidence); and I think that everyone will agree that the Rhine Graben is underlain by thinned crust. It may follow that beneath much of the sub-Upper Cretaceous trough system that there is thinned crust. This and the associated volcanicity (also in the process of being unveiled at this conference) will have an effect on the geothermal gradient.

Have Conoco done any work, that Dr Eames is in a position to tell us about, on the relationship of the trough system, geothermal gradient, devolatilization of coals and the maturation of kerogen (another source for methane)?

T. D. Eames: There appears to be no relationship between coal rank and the Cretaceous trough-system in the northern North Sea. All the coals examined so far in the northern area are of low rank indicating a relatively low temperature regime. This has to be qualified since the various techniques for measuring palaeotemperature show wide differences. In particular, vitrinite determination of fixed carbon shows low maturity in over-pressured Jurassic coals, which can be explained by the low rank coal maturation being related essentially to compaction and not to temperature. Present day temperature gradients are generally slightly higher in the basin areas compared to the shelf areas, but this is not exclusive.

P. Braithwaite (Sun Oil Co.): Is there any relation between the nitrogen content of the gas in the Rotliegendes of the German North Sea and the rank of the coal or igneous activity?

T. D. Eames: There is no apparent relationship between the rank of coal and composition of the gas in the North Sea area. The large methane gas-fields in the Rotliegendes lie over or close to the subcrop of the Westphalian B, the main coal-productive formation. The nitrogen-rich area overlies the subcrop of the Stephanian–Upper Westphalian red-measures, and it is suggested that the composition of the gas is related to the facies of the subcropping formations.

13

Rotliegendes Stratigraphy and Diagenesis

By J. P. P. MARIE

(Shell UK Exploration and Production Limited)

SUMMARY

The quality of the Rotliegendes gas-reservoir in the southern North Sea is the result of the interaction of facies, early cementation, depth of burial and tectonics.

The bulk of the Rotliegendes consists of unfossiliferous red beds in which aeolian, fluvial, sabkha and desert-lake environments are recognized.

Sandstone-quality trends correlate acceptably with the interpreted depositional environments outside the main areas of diagenetic pore destruction. Nevertheless in those areas in which the history of burial and later inversion is known to have been severe, diagenesis seems to have more strongly affected those sandstones with poorer primary qualities than the more porous aeolian sands.

INTRODUCTION

The Permian Rotliegendes of the southern North Sea, which is the main gas-producing horizon of the area, is a westwards extension of that found in the German/Dutch Rotliegendes basin. In the UK area the southern limit is formed by the northern edge of the London–Brabant Massif, marginal to the Variscan orogenic belt which was uplifted in the late Carboniferous/early Permian times (the Hercynian phase) (Fig 1A). The northern edge of these highlands must have formed an escarpment against which Rotliegendes sediments abutted. The northern boundary was the Mid North Sea High over which the Rotliegendes was essentially not deposited. To the west, the present boundary is largely erosional but it is likely that the depositional basin was limited by the Pennine High.

The Variscan Orogeny was followed by a prolonged period of denudation of the highlands during which the adjacent lowland basins were infilled with Rotliegendes sediments. These sediments consist of unfossiliferous red beds that were deposited under arid continental conditions. The deposits in the Southern North Sea Basin are developed as two main facies:

(i) Sands, largely of aeolian origin (Glennie, 1972) with minor water-laid sediments. These are restricted to a north-west to south-east trending belt in the southern part of the basin (Fig 1C).

(ii) Siltstone and claystones with evaporites, interpreted as having been deposited in a lacustrine area (Glennie, 1972). They characterize the deposits of the northern and central parts of the basin.

In the UK area these two facies were recently designated the Leman Sandstone Formation and the Silverpit Formation of the Rotliegendes by the joint Oil Industry-Institute of Geological Sciences Committee on North Sea stratigraphic nomenclature (Rhys, 1974). In this paper, Rotliegendes stratigraphy and diagenesis are discussed in the light of additional data provided by well-cuttings and core and especially from the interpretation of electric logs.

DISTRIBUTION AND PALAEO-HISTORY

The Top Rotliegendes structural contours (Fig 1A) reflect roughly the same configuration as those of the Top Carboniferous with the latter deepening rather more rapidly from the margin of the basin towards the centre. Indeed the Rotliegendes isopach map (Fig 1B) shows a maximum thickness of sediments of around 1000ft (300m) in the centre of the basin. In the main sandstone belt of the Rotliegendes (Fig 1C) thick sandstones, about 900ft (270m), are found on the site of the present Sole Pit High along the north-west–south-east trending Dowsing Fault Zone and to the east in Dutch waters on the site of the present Broad Fourteen's High. These greater thicknesses of sandstone imply that already in Rotliegendes time, deposition over the Sole Pit and Broad Fourteen's areas was influenced by a locally faster rate of subsidence, which was probably controlled by contemporaneous faulting.

During the Mesozoic, greater thicknesses of sediments were deposited in these faster subsiding areas of the Sole Pit and Broad Fourteen's than elsewhere in the area. The Late Mesozoic tectonic phases resulted in the inversion of these more deeply buried local basins and resulted in erosion and truncation of part or all of the Jurassic and Triassic (Fig 1A).

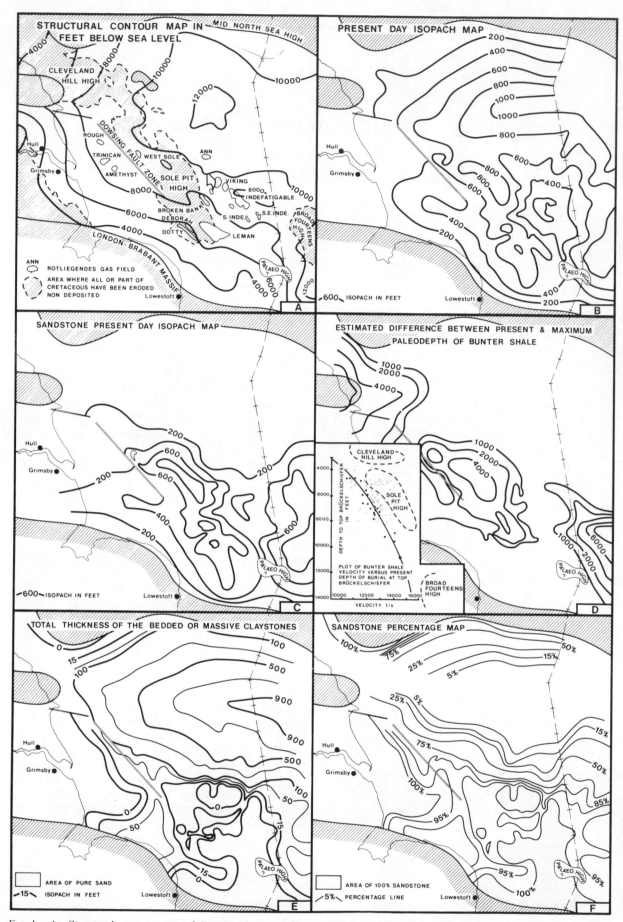

Fig 1. A—Structural contour-map of Top Rotliegendes in Southern North Sea Basin. B—Present day isopach map of the Rotliegendes. C—Cumulative thickness-map of the Rotliegendes sandstones. D—Estimated difference between present and maximum palaeo-depth of Bunter Shale, together with a plot of Bunter Shale velocity *versus* present depth of burial at top Bröckelschiefer. E—Cumulative thickness of the bedded and massive claystones of the Rotliegendes. F—Rotliegendes sandstone-percentage map.

The minimum uplift suffered by these former, deeply buried, basins was calculated by comparing the maximum palaeo-depth derived from sonic velocities and the present depth of burial of the Triassic Bunter Shale (Fig 1D). The method is calibrated using a plot of the Bunter Shale acoustic velocities for wells with a normal burial history against the present depth of burial at Top Bröckelschiefer (Basal Triassic level) (Faust, 1951; Jankowsky, 1970; Marr, 1971). The difference between the palaeo-depth and present depth indicates the minimum thickness of sediments eroded during and after the late Mesozoic inversion and so represents the minimum uplift of the Rotliegendes at that period. It is clearly seen (Fig 1D) that the Sole Pit, Broad Fourteen's and Cleveland Hills areas were more deeply buried as a result of the Cimmerian tensional tectonic phases (respectively of 4000 to 6000ft (1200 to 1800m), 4000 to 10 000ft (1200 to 3050m), 4000 to 6000ft (1200 to 1800m)) than at present. This supports the idea that the Permian synsedimentary faults continued to influence sedimentation during the later Permian, Triassic and Jurassic and probably acted as hinges during the subsequent major tectonic phases and were reactivated during inversion movements.

DEPTH RELATED DIAGENESIS OF SANDSTONES

In the areas where the Rotliegendes Sandstones were buried under at least 12 000ft (3650m) of sediments, e.g. the Sole Pit and Broad Fourteen's basins, and in the area known as the Cleveland Hills high (Fig 1A), the temperature and degree of compaction were sufficiently high for the formation of illite and chlorite crystals (Stadler, 1973) and later quartz cement which together seriously damaged the primary sandstone porosity and permeability. There is also evidence for a later phase of dolomite and anhydrite cementation that post-dates the illite formation. This further reduced the porosity and permeability.

Porosity and permeability destruction in good primary reservoirs by the above secondary diagenetic processes is generally confined to those areas in which the history of burial and later inversion was severe.

SAND AND SHALE/CLAYSTONE DISTRIBUTION AND INTERPRETATION

The northern limit of the main Rotliegendes Sandstone facies is defined by the shaling-out of the sands to the north and corresponds to the shore-line of a desert lake. A cumulative thickness map of the thin-bedded and massive claystones (Fig 1E) outlines three areas in which the Rotliegendes

sands contain a significant proportion of thin claystone interbeds.

Two of these claystone trends are directed toward the north and north-east from the southern basin rim, while the third, in the Dutch area, trends to the north. A detailed Rotliegendes Sandstone percentage map (Fig 1F) shows a similar pattern, thereby indicating that the increase in clays is not thickness dependent.

The thin claystone beds found within thick sections of mainly aeolian sandstones were most probably deposited in areas of low relief from turbid water associated with intermittent desert flooding. Their distribution can be considered as consistent with a fluvial wadi system which drained flood-water away from the highlands and which was probably also partially inundated at times with salt waters from the central lacustrine basin.

DISTRIBUTION AND EVOLUTION OF LITHOFACIES AND ENVIRONMENT OF DEPOSITION

The lithology and facies of the Rotliegendes sediments vary from pure aeolian sandstones to claystone/halite deposits.

Since each facies has a limited areal extent, there is no basin-wide vertical sequence or marker horizon that permits correlation from well to well by use of electric logs. Moreover, there is no stratigraphically determined age-subdivision. Because of these limitations, a study of the evolution of sedimentation, together with the associated reservoir qualities, was attempted by arbitrarily subdividing the Rotliegendes sequence in each well into three equal parts ("Thirds"). Each "Third" approximates to the sequence found in much of the sand belt—Lower Leman Sandstone, Silverpit Sandstone, Upper Leman Sandstone. They give rise to "Lower", "Middle" and "Upper" Rotliegendes layer maps (Fig 2, A, B, C respectively).

Sands were separated quantitatively from the non-sandy claystones, siltstones and evaporites by using the gamma-ray log; this permitted the construction of the detailed cumulative claystone thickness map (Fig 1E) and sandstone percentage map (Fig 1F) discussed earlier. These, combined with data from cutting and core descriptions and a gamma-ray differentiation between "clean" and "dirty" sands, permitted a qualitative description of the sandstones and an interpretation of their depositional environments.

Lithofacies and Depositional Environments

By these means it was possible to confirm the presence of four primary facies which correspond to the sediments of the main depositional environments: aeolian, fluvial/wadi, sabkha, and desert-lake (Haselgebirge facies of north Germany).

Aeolian. 100 percent sandstone; clean, with minor anhydrite and dolomite cement. The aeolian clean sandstones were derived mainly from unconsolidated alluvial-fan/wadi flood-plain deposits fringing the southern highlands, by a prevailing easterly wind (Glennie, 1972). Fluvial flood-waters reworked the marginal aeolian sands and resulted locally in some interbedded argillaceous and conglomeratic sediments. There exists a transition zone where both fresh water from the wadis and salt water from the desert lake (both surface and ground-waters) invaded the marginal aeolian deposits and added argillaceous, silty, anhydritic, dolomitic and conglomeratic contaminant material.

Fluvial/wadi. 100 percent sandstone; slightly argillaceous and conglomeratic, with calcite cement. Alluvial-fan and flood-plain deposits (sand, conglomerate and thin clay beds) derived their sediments from the southern Variscan Highlands, probably through canyons and gorges. They were transported towards the deepest part of the basin to the north (desert lake) during the wet season when sporadic violent and heavy rainfall over the highlands gave rise to floods. The flow of water followed the lines of low relief and when the supply of water diminished, much of the water soaked into the underlying wadi and aeolian sediments. Evaporation of ground water at the air/water interface, resulted in carbonate cementation of the sediments. During the succeeding dry period, wind transport was resumed and sand dunes again migrated across these areas of low relief. This sequence of events was repeated throughout Rotliegendes time thereby giving an alternation of water-laid and aeolian sediments.

Sabkha. 0–80 percent sandstone; argillaceous nodular, anhydritic, dolomite; interbedded with massive claystone. The sabkha sediments (clays, silts and argillaceous sands, commonly evaporitic and dolomitic) were deposited over a fluctuating lake shoreline of low relief (Glennie, 1970). They derive their evaporite content partly from ground-water evaporation and partly from evaporation of water left behind by high lake-levels. Continued wind-action resulted in the adhesion of silt and sand grains to the damp surface of the sabkha with the consequent formation of adhesion ripples (Glennie, 1972). Sand dunes may have formed during prolonged periods of lake contraction. This alternation of sabkha and aeolian sediments comprises a common association.

Desert Lake. 95 percent claystone and halite; occasional thin sandstone intercalations; anhydritic clays and cement. The water that flowed down the wadis from the southern highlands accumulated in the north central part of the basin and formed a desert lake of fluctuating areal extent. Mostly clays and silts were deposited together with occasional sand intercalations (fluvio-lacustrine). A high rate of evaporation was only roughly balanced by inflow of water so that during extremes of lake contraction, bedded halite was

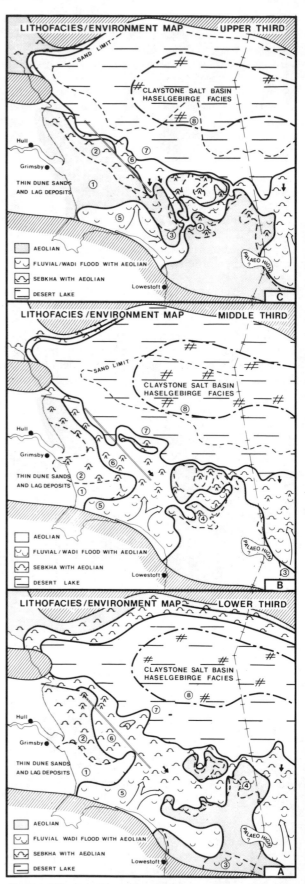

FIG 2. Lithofacies/deposition/environment maps of the three arbitrary subdivisions of the Rotliegendes—Lower (A), Middle (B) and Upper (C) "Thirds".

precipitated. Lake contraction implies a reduced fluvial input and coincident deflation of the exposed surrounding sabkha surface; fluvio-lacustrine sands (point bars, deltas, etc.) are therefore probably of limited extent.

Associations of the deposits of these main environments can also be recognized by the interdigitation of two or more of the main facies. In the vertical sequence they indicate long-term fluctuations in rainfall and lake-level or changes in the distribution of aeolian sands. In all, eight facies or facies-associations have been tentatively defined and their distributions are shown on Fig 2A–C. They are:

(1) *Aeolian:* characteristics defined above.

(2) *Shore-line dunes* and *interdune sabkhas:* the former consist of shore-line sabkhas (7, below) with significant thicknesses (>5ft) of intercalated clean foreset sandstones; the latter consist of aeolian deposits with significant thicknesses containing anhydrite nodules, anhydrite or dolomite cement, and slightly argillaceous adhesion ripples—gradational to or within sequences of facies 1.

(3) *Aeolian with wadi intercalations:* aeolian deposits as defined above, with intercalations of argillaceous, calcite-cemented sandstone and minor conglomerates (clay-flakes and pebbles common) and thin clay beds.

(4) *Transition wadi/shore-line sabkha/aeolian:* varying proportions of wadi facies (see above) with aeolian deposits (1, above) and shore-line sabkhas (7, below).

(5) *Wadi with aeolian intercalations:* dominantly wadi facies with intercalations of mainly horizontally laminated beds of aeolian facies.

(6) *Shore-line sabkha with aeolian intercalations:* commonly anhydrite cemented; gradational to shore-line dunes facies.

(7) *Shore-line sabkha:* characteristics defined above; locally thick sandstone intervals are confined to the lower part of the section and may represent a fluvio-lacustrine facies.

(8) *Desert lake:* characteristics defined above.

Facies Distribution

The region covered mainly with fluvial and aeolian sediments is bordered to the north by a desert lake with a shoreline that probably fluctuated considerably depending on the water input, and this resulted in broad sabkhas. In the early stages (Lower and Middle Thirds, Fig 2A, B) fairly extensive patches of aeolian sands (100 percent sandstone) were deposited over the sabkha and possibly represent the remnants of an important shoreline dune-complex. During deposition of the Upper Third (Fig 2C), the desert lake must have received considerable volumes of water from sources other than the UK wadis, which had almost disappeared, resulting in a slight

recession of the shoreline and so to a narrower shoreline sabkha. This phase was succeeded by the Zechstein marine transgression.

CONCLUSIONS

1. In the desert environment of the southern North Sea Rotliegendes Basin, there was a normal primary depositional sequence from the Variscan Highlands northwards, of alluvial fans and wadi flood-plain grading into a sabkha that was marginal to a permanent desert lake of fluctuating extent. Across and interbedded with this sequence of sediments are aeolian sands which locally reach almost 100 percent of the total thickness, that were transported by prevailing winds from the east.

2. Several different Rotliegendes environments and associations of environments have been recognized from a study of cores and electric logs.

3. Apart from aeolian sands, which form the best reservoirs, the porosity and permeability of other sand-facies is reduced by their primary argillaceous content and by the early formation of calcite, dolomite and anhydrite cements.

4. In three areas, the Cleveland Hills, Sole Pit and Broad Fourteen's Highs, primary porosity and permeability have been reduced by the combined diagenetic effects of initial deep burial followed by considerable uplift. Even so, local preservation of fair porosity in these areas suggests that the damaging effects of diagenesis on porosity and permeability is also controlled by the pattern of depositional environments.

ACKNOWLEDGEMENTS

This paper is published by permission of Shell UK Exploration and Production Co Ltd, and Esso Exploration and Production (UK) Inc. Valuable discussions with colleagues in Shell UK Exploration are acknowledged.

REFERENCES

Faust, L. Y. 1951. Seismic velocity as a function of depth and geologic time. *Geophysics*, **16**, 192–206.

Glennie, K. W. 1970. *Desert sediment environments. Developments in Sedimentology 14.* Elsevier, Amsterdam, 222 pp.

Glennie, K. W. 1972. Permian Rotliegendes of N.W. Europe interpreted in the light of modern desert sedimentation studies. *Bull. Am. Ass. Petrol. Geol.*, **56**, 1048–71.

Jankowsky, W. 1970. Empirical investigation of some factors affecting elastic wave velocities in carbonate rocks. *Geophys. Prospect.*, **18**, 103–118.

Marr, J. D. 1971. Seismic stratigraphic exploration—Part III. *Geophysics*, **36**, 676–89.

Rhys, G. H. (Compiler) 1974. A proposed standard lithostratigraphic nomenclature for the southern North Sea and an outline structural nomenclature for the whole of the

(U.K.) North Sea. A report of the joint Oil Industry–Institute of Geological Sciences Committee on North Sea Nomenclature. *Rep. Inst. geol. Sci.*, **74/8,** 14 pp.

Stadler, P. J. 1973. Influence of crystallographic habit and aggregate structure of authigenic clay minerals on sandstone permeability. *Geologie. Mijnb.*, **52,** 217–20.

DISCUSSION

R. Byramjee (Compagnie Française des Petroles): Diagenesis in sandstone reservoirs consists mainly of dissolution and re-deposition resulting in the obstruction of pore-space. If we follow Professor Dunham's early gas generation, the pore-space could have been filled with gas in early stages, preventing any further diagenesis. The implication of this is that there could be, to a certain extent, a correlation between gas-accumulation and areas of low diagenesis.

J. P. P. Marie: The exact time of gas-migration in the Rotliegendes is not known yet with precision. We suppose that it could have taken place during the late Jurassic/early Cretaceous at which time the Rotliegendes was more deeply buried. If it is so, the early diagenesis (due to burial) and the later diagenesis (due to basin inversion after the late Mesozoic tectonic phase) would have started, and then the process indeed might have been stopped by gas-migration; but we do not think there was earlier gas accumulation. Therefore I do agree with Mr Byramjee that there is to a certain extent a correlation between gas-accumulation and areas of low diagenesis, which areas have been shown to be also dependent on the environment, burial and tectonic.

J. B. Blanche (National Coal Board, Exploration): (1) Did Dr Marie during his depth of burial study calculate a porosity gradient for the Rotliegendes Sandstone?

(2) Could Dr Marie elaborate on the role of authigenic clay minerals in the reduction of poroperm in the Rotliegendes Sandstone?

J. P. P. Marie: (1) At that stage of our study it seemed impractical to calculate a porosity gradient for the entire basin, as the diagenesis was not uniformly destructive. Nevertheless I do agree that such calculations could be made for specific areas, and especially for those where the history of burial and later inversion is known to have been severe.

(2) The role of authigenic clay minerals is indeed extremely important in the reduction of the porosity/permeability of the sandstones and this is why I have attempted to show their distribution on the environment maps. So far it is difficult to estimate their effect quantitatively; one can only say that in those areas in which we suppose there was little or no diagenesis the porosity of the pure aeolian sandstones is of the order of 20 to 30%, whilst the porosity of the "dirty sands" is lower (15 to 20%).

14

The West Sole Gas-Field

By J. B. BUTLER

(BP Petroleum Development Limited)

SUMMARY

The West Sole gas-field is located 35 miles (56km) east of Spurn Head. It was the first discovery in the North Sea and was brought to production in early 1967.

The gas is derived from underlying Coal Measures and has accumulated in Rotliegendes sandstone in a faulted anticline 19km long by 5km wide, trending north-west–south-east. An offset pillow of Zechstein salt overlies the field and forms a cap to the reservoir.

The gas-in-place is of the order of 67.97×10^9 m³ (2.4×10^{12} scf) with an initial formation pressure of 4250psig.

Production from the field is *via* 19 production wells which with the exception of two platform proving wells have been drilled from 3 fixed platforms.

INTRODUCTION

Evaluation of the hydrocarbon potential of the North Sea was encouraged by the discovery in 1959 of the giant Groningen gas-field in the Netherlands. At about this time the Geneva Convention on the continental shelf was under discussion. Following international agreement the British Government commenced in 1964 to offer blocks for licence. Exact definition of the international boundaries was agreed by 1966 (Kent, 1967b). Geophysical surveys were commenced by the oil industry in 1962 and the first depth map of the West Sole structure became available during 1965. The first North Sea discovery to be made was the West Sole Gas-Field in block 48/6 located 35 miles off Spurn Head (Fig 1). Water depths average 27–30m. The discovery well was drilled by the ill-fated Sea Gem and had been tested, plugged and abandoned when the jack-up barge capsized in December 1965 whilst preparing to move off location.

During the following year an appraisal well was drilled 1250m to the west-south-west by the new semi-submersible rig Sea Quest in a bottom supported position and simultaneously development drilling commenced from a fixed platform located 975m to the south-east of the discovery well. The appraisal well was subsequently produced from an individual production tower. Sea Quest also drilled two appraisal wells along the south-western margin of the central lobe and one of these (48/6–10) located the gas–water contact. These two wells were abandoned.

The field is produced from 19 wells most of which have been drilled from three fixed platforms.

GEOLOGICAL SUCCESSION

The West Sole Field (Hornabrook, *in press*) is situated in the south-western part of the English Zechstein Basin. It is an elongated and faulted dome 19km long and 5km wide trending north-west–south-east and falls within the area of halokinesis which extends eastwards into Holland and Germany (Kent, 1968).

Gas production is from the Rotliegendes sandstone of lower Permian age. The culmination of the structure is at 2680m and the gas–water contact is at 2940m subsea. The gas–water contact encompasses an area of 65 square kilometres excluding the unproven northern lobe.

The Rotliegendes sandstones were deposited unconformably on a gently undulating Carboniferous surface (Fig 2). Oxidation and other weathering features affect the Carboniferous mudstones, siltstones and sandstones to depths which may amount to several tens of metres. It is generally agreed that the gas which is almost pure methane is derived from the devolatilization of the underlying Westphalian coal seams.

The reservoir is the Lower Sandstone Member of the Rotliegendes and is on average 125m thick and is capped by 37m of reddish silty and sandy mudstones or Siltstone Member which separates it from 23m of gas-bearing but unproductive sandstone, termed the Upper Sandstone Member.

The Upper Sandstone Member is overlain conformably by the 1m-thick Kupferschiefer and 40m of Z1 and Z2 carbonates and anhydrite. These beds are succeeded by thick halites which include, particularly in the lower part, beds of polyhalite. The halite forms the flank of a large pillow which thickens considerably in a north-easterly direction. The basal clays, carbonates and anhydrites of the Z3 cycle are well developed over much of the field and are succeeded by halites, potassic salts and marls of the Z3 and Z4 cycles (Brunstrom and Walmsley, 1969). In three wells in the northern part of the field and one well in the south, the Z3 carbonates are missing and are presumed to have been removed by rafting.

A full sequence of Triassic follows the Zechstein.

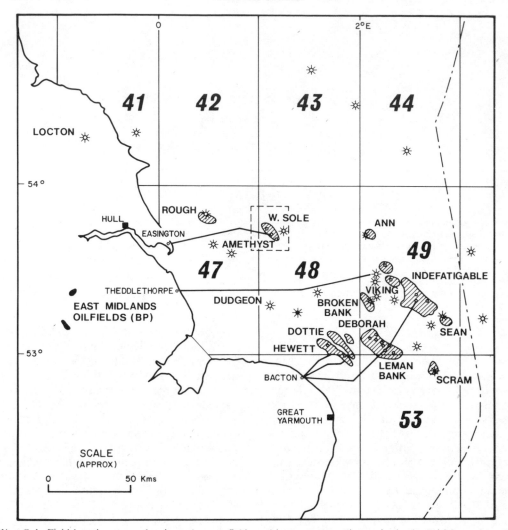

FIG 1.   West Sole Field location map, showing other gas fields and important gas discoveries in the British southern North Sea.

The Lower Bunter commences with a basal siltstone facies equivalent to the Brockelschiefer, which grades into a monotonous sequence of red mudstone. Oolith bands (Rogenstein) occur in the upper part and form a useful marker.

The Middle Bunter is developed in its typical sandstone facies. The top of the sandstone is erosional and this surface has been correlated with the Hardegsen Discordanz in Germany. The sandstones are overlain by the Upper Bunter which comprises anhydritic and dolomitic mudstones and one massive halite bed equivalent to the Rotsalinar.

The Muschelkalk mudstones are also anhydritic, and are characterized by numerous thin dolomite beds and one massive halite bed.

The Keuper mudstones are silty and anhydritic and contain halite beds. The red mudstones become mottled green and finally greenish grey in the upper part (Tea-green Marl).

The Rhaetic is a transgressive facies represented by marine sandstones and dark shales (Westbury Beds) grading into reddish marl (Cotham Beds).

As is the case with the Triassic formations the

Lias is also found to have a thicker and more complete sequence than in most UK land areas. Thin limestones and calcareous mudstones occur at the base and are equivalent to the Hydraulic Limestone. Micropalaeontological evidence shows that there is a fairly complete sequence of Lias amounting to a total of 500m. Middle Jurassic shales, sandstones and oolitic limestones occur in the two appraisal wells drilled on the western margin of the field. The section in the two wells is approximately 300m thick and ranges up to the Corallian/Cornbrash.

A thin cover (c. 15m) of Boulder Clay completes the succession.

THE ROTLIEGENDES FORMATION

The reservoir rock is termed the Lower Sandstone Member and averages 125m gross thickness (Fig 3). Nett pay calculated on a 10 per cent porosity cut-off averages 80m. Porosities within the pay rarely exceed 20 per cent and average about 15 per cent. Permeabilities as measured in cores are usually low

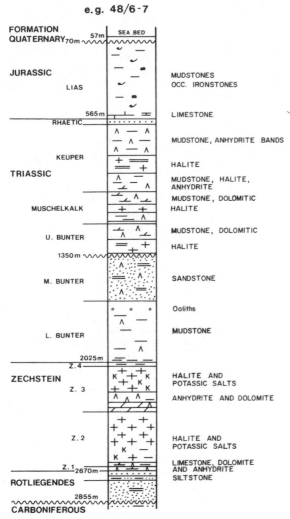

FIG 2.   General geological section of West Sole Field.

pebbles or granules of mudstone or rarely volcanic material at the base of the unit. Planar cross-bedding and irregular cross-bedding similar to festoon bedding is quite common. Porosities and permeabilities within the lower unit are generally below average especially near the base. The sandstones of the lower unit are interpreted as aeolian dune sands which have in part been reworked by flowing water, resulting in an interdigitation of dune, wadi-flood and alluvial deposits.

The middle unit also contains sandstones which are similar to those of the lower unit and in addition a sandstone type which appears to be restricted to this unit. This sandstone is pale grey to yellowish/brown, fine- to medium-grained; cementation is argillaceous and dolomitic and there are scattered blebs of highly cemented sand. Structures include laminae of coarse rounded grains concentrated in the troughs. The latter structures may be adhesion ripples indicating deposition in an interdune area close to the water

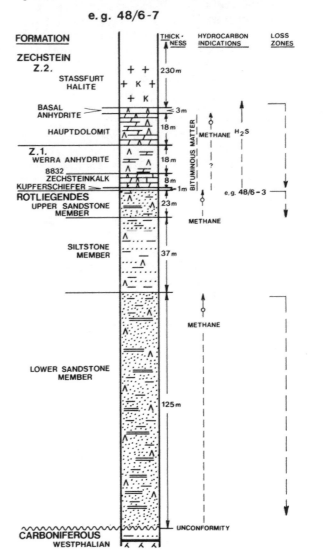

FIG 3.   Section of the Rotliegendes Sandstone and cap-rock in the West Sole Field.

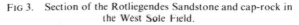

and rarely exceed 1md. Production is therefore dependent upon hair-line fractures and locally distributed zones of slightly higher permeability. This improvement results from a better degree of sorting within individual sandstone beds and also from the development of a fibrous type of cementation. Cementation is described in more detail later.

The Lower Sandstone Member can be sub-divided into three units which are transitional one to the other and appear to represent a progressive change in depositional environment from aeolian, partly reworked by water, through to shallow water (Blanche, 1973; Glennie, 1970, 1972; Smith, 1972).

The lower unit is the most variable both in thickness and log character, and this renders its correlation across the field difficult. The sandstone is variably pale grey to greyish red, fine- to medium-grained, with frequent laminae of coarse, rounded and frosted grains. Sorting is poor and the grain size distribution is often bimodal. Cementation is frequently strongly developed. Sedimentary units include graded units up to 1m thick, often with table. The sandstones are in part relatively well sorted, consequently permeabilities are above average and the middle unit is therefore the most productive.

The upper unit contains sandstones with interbedded siltstones and mudstones and sedimentation is frequently cyclic. A complete cycle is about 1m thick and commences with a basal layer of medium-grained sandstone which may contain shale clasts; the sandstone fines upwards into a thin silt or shale and a thin black shale may complete the cycle. Log correlation between wells is good, particularly in the upper part. Sedimentation of the upper unit evidently took place in a shallow water environment. Glauconite and tests of the marine foraminifera *Spirillina* have been noted in thin sections from two wells, indicating marine incursions in the upper unit.

Regionally the Lower Sandstone Member is thickest along a NW–SE line which passes through the north-east corner of block 48/6. It gradually thins across the field and considerable attenuation occurs beyond the south-east corner of the block.

The Siltstone Member conformably overlies the Lower Sandstone Member. It comprises a graded series of reddish brown dense siltstones and mudstones which contain varying proportions of very poorly sorted quartz grains which range from coarse rounded to very fine grains. Sedimentation is again cyclic and log correlation is very good, extending to a bed by bed correlation within each cycle. The Siltstone Member maintains a fairly constant thickness of about 37m across the field and is interpreted to be a distal wadi-flood deposit. Regionally the depositional trend is north-west–south-east. The Siltstone Member is absent outside the south-west corner of block 48/6 and gradually thickens across the block attaining a thickness of more than 100m beyond the north-east corner of the block.

The Upper Sandstone Member is transitional to the Siltstone Member and is about 23m thick. The basal 8m is a grey variable silty sandstone and is dense and hard. The upper 15m of sandstone is greyish white, fine-grained, dense and friable. It is cemented with white argillaceous material which is also anhydritic and dolomitic. The sandstone also contains pyrite and lenses of grey mudstone and therefore appears to have been deposited under reducing conditions and hence is transitional to the Zechstein. The contact with the overlying Kupferschiefer is conformable or slightly undulose and sharp. Although gas-bearing the high degree of cementation precludes any production.

Regionally the Upper Sandstone Member is not recognised to the south-west of block 48/6 due to the absence of the intervening Siltstone Member in the area. It reaches its maximum thickness along a line roughly coinciding with the axis of the field, although it rapidly disappears along the north-westerly continuation of this axis. Its development is most persistent in a north-easterly direction.

CEMENTATION WITHIN THE ROTLIEGENDES

The low poroperm quality of the Rotliegendes sandstone is due to cementation compounded by a poor degree of sorting of much of the sandstone.

Four main types of cement are present: (i) Clay, usually illite, mixed with varying proportions of red hematite. The red coloration of the sandstones appears to be random in its distribution and its origin is problematical. The hematite occurs as fine grains of minute crystallites which may have crystallized in place, suggesting periods of *in situ* oxidation. (ii) Dolomite, commonly in poikilitic habit. (iii) Limited authigenic silica. (iv) Anhydrite is particularly well developed along with dolomite near the base of the Rotliegendes and in and near to mineralizing faults and fractures.

An interesting sequence of cementation is the formation of a thin rim of illite with the crystallites wrapped around the detrital grains to form a thin pellicle. Iron-silicate crystallites have then grown normal to the illite pellicle. This fibrous cementation is associated with better porosity, which may have resulted from the inhibition of later dolomitic or anhydritic cementation.

STRUCTURE

The Lower Permian structure is overlain by Zechstein salt which has undergone halokinesis giving rise to a salt pillow offset to the north-east (Fig 4).

The sub-salt anticlinal structure (Fig 5) containing the gas reservoir is practically confined to block 48/6. It is an elongated faulted dome 19km long and 5km wide extending diagonally across the block from the south-eastern corner to the north-western corner (Kent and Walmsley, 1970). The growth of the structure is considered to be fault-controlled and due to late epeirogenic movement along a Hercynian fault system. This rejuvenation probably occurred in the Jurassic in association with isostatic readjustment.

The reservoir structure can be effectively divided into two areas: (a) a north-western area, where the axes of the central and northern lobes of the field and system of longitudinal faults are parallel to the NW–SE diagonal of the block; (b) a south-eastern area, where the structure of the southern lobe is more complex, the faults trend WSW–ENE and the anticlinal axis appears to have been displaced *en echelon* by a series of small sinistral movements.

A limited amount of faulting has been detected in wells, where sections of up to 30m of Rotliegendes are cut out.

The axis of the Zechstein salt pillow is parallel to the Rotliegendes structure but lies 5km to the north-east. Folding above the salt is virtually parallel, although some thinning of beds over the crest suggests that salt movement commenced during

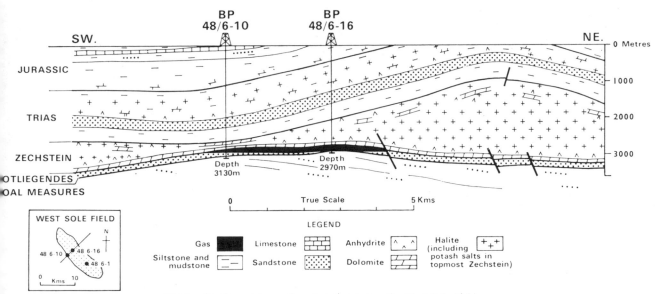

FIG 4. South-west to north-east section across the West Sole Field.

the Bunter. Gravity-type strike-slip faults occur on the flanks of the post-salt anticline. In addition a narrow graben has formed along the crest due to collapse into the salt following tension.

The differential velocity effect caused by the variable thickness of salt has made the seismic mapping of the Rotliegendes structure extremely difficult and sophisticated interpretation techniques have had to be employed (Hornabrook, *in press*). Poor data precludes detailed mapping of the structure in the Carboniferous.

PRODUCTION HISTORY

Production from the field commenced early in 1967 from 4 wells deviated from Platform 'A' into the southern lobe. Two additional wells were deviated from a jack-up barge and hooked into this Platform in 1974. The gas from the Platform 'A' wells is processed on a satellite production platform and then fed into the 16in sea-line.

Platform 'B' is located 2.7km north-west of Platform 'A' and was brought onto production in October 1967. Five wells drain the southern lobe and one well drains the central lobe of the field. The appraisal well 48/6–3 drilled in the southern lobe by Sea Quest in 1967 is served by the production tower 'E1' and the gas piped to Platform 'B'.

A platform proving well was drilled 4.5km north-north-west of Platform 'B'; this well was hooked into Platform 'C'. Four wells were then deviated into the central lobe and one well deviated into the southern lobe. Production began in January 1971. Thus a total of 20 producing wells have been drilled in the field, of which one has been abandoned, 13 drain the southern lobe and 6 drain the central lobe.

On completion the wells are fractured, resulting in increases in flow rates by a factor of 2–3 times

and with continued production the wells clean up and continue to improve. Individual wells are capable of producing between 565×10^3 m^3 to 990×10^3 m^3 (20 to 35 MMscf) of gas per day, although several wells have lower rates than this.

Analysis shows that the gas is practically pure methane as the following table of average values shows. This is consistent with a derivation from coal.

Carbon dioxide	0.5
Nitrogen	1.0
Methane	94.5
Ethane	3.0
Propane	0.5
Butane/Pentane	0.5
	100.0

Problems with corrosion have been encountered. The 4in tubing near the surface and the well-heads have suffered severe corrosion from dissolved CO_2 gas. The wells of Platforms 'A' and 'B' have been worked over and stainless well-heads and uppermost tubing joints installed.

After initial processing, gas from Platform 'C' flows to Platform 'B' for further processing and inlet to the main sea-line. Processing offshore consists of natural separation and glycol absorption to ensure that no free water is present at any time in the sea-line. The sea-line is 16 inches in diameter and flows to Easington where on arrival a pressure of at least 1050 psig is required. The gas is transferred at Easington into the trunk-line operated by the Gas Council.

The initial reservoir pressure was approximately 4250psig. For an average nett pay thickness of 80m with average values of 15 per cent porosity, 50 per cent water saturation and 5md reservoir permeability, the gas in place has been calculated to be 67.97×10^9 m^3 (2.40×10^{12} scf), comprising

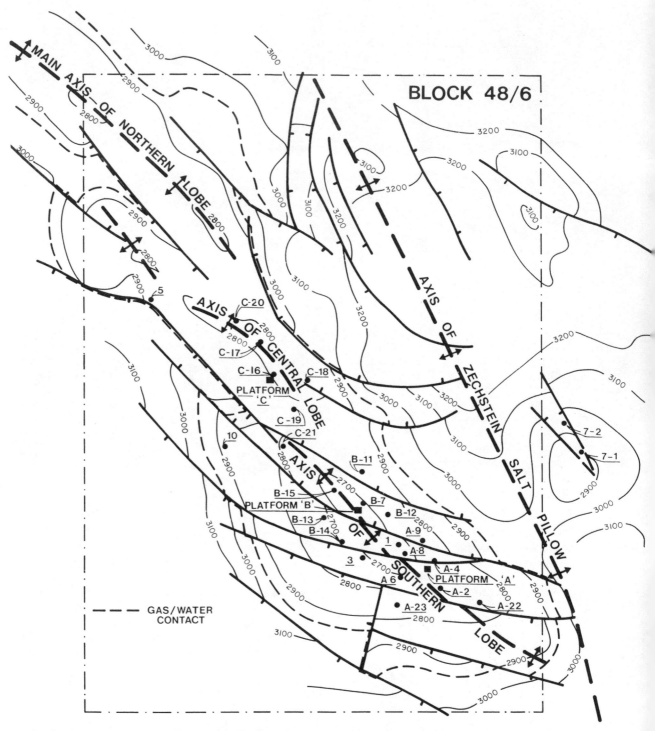

FIG 5.   Structural contours on the top of the Rotliegendes Lower Sandstone Member in the West Sole Field. (After D. R. Hughes and P. J. Moulds, internal BP report.)

55.23×10^9 m³ $(1.95 \times 10^{12}$ scf$)$ in the southern lobe and 12.74×10^9 m³ $(0.45$ scf $\times 10^{12})$ in the central lobe. The minimum water saturation in the most porous and structurally highest part of the reservoir is 25 per cent. Calculated reservoir permeabilities vary between 1.0 and 39md and are highest in the southern lobe.

The northern lobe is structurally lower than the remainder of the field. Its structure has not been confirmed by drilling, therefore, no reserves have been assigned to this portion of the field.

Cumulative production for the West Sole Field from the commencement of production up to the end of September 1974 was 12.03×10^9 m³ $(0.425 \times 10^{12}$ scf$)$ or 17.7 per cent of the original gas-in-place.

ACKNOWLEDGEMENTS

The author is indebted to his many colleagues whose unpublished reports have provided much of the information and to the Management and Board of the British Petroleum Company for permission to publish this paper.

REFERENCES

Blanche, G. B. 1973. The Rotliegendes Sandstone formation of the UK Sector of the Southern North Sea Basin. *IMM Section B Trans*, **82** (801) 85–9.

Brunstrom, R. G. W. 1974. Oil and Gas. *In Geology and Mineral Resources of Yorkshire. D. H. Rayner and J. E. Hemingway (Editors)*, 385–91, Yorkshire Geological Society.

Brunstrom, R. G. W. and Walmsley, P. J. 1969. Permian Evaporites in the North Sea Basin. *Bull. Am. Ass. Petrol. Geol.*, **53**, 870–83.

Glennie, K. W. 1970 *Desert sedimentary environments. Developments in Sedimentology 14*. Elsevier, Amsterdam, 222 pp.

Glennie, K. W. 1972. Permian Rotliegendes of northwest Europe interpreted in the light of modern sedimentation studies. *Bull. Am. Ass. Petrol. Geol.*, **56**, 1048–1071.

Hornabrook, J. T. (*in press*). Seismic interpretation of the West Sole gas field. *Proc. Bergen North Sea Conference, Norge, Geol. Unders.*

Kent, P. E. 1967a. Outline geology of the southern North Sea basin. *Proc. Yorks. geol. Soc.*, **36**, 1–22.

Kent, P. E. 1967b. North Sea exploration—a case history. *Geogrl Jl.*, **133**, 289–301.

Kent, P. E. 1968. Geological problems in North Sea exploration. *In Geology of Shelf Seas by D. T. Donovan (Ed.)*, 73–86, Oliver & Boyd, London.

Kent, P. E. and Walmsley, P. J. 1970. North Sea progress. *Bull. Am. Ass. Petrol. Geol.*, **54**, 168–181.

Smith, D. B. 1972. The Lower Permian in the British Isles. *In Rotliegend. Essays on European Lower Permian, G. H. Falke (Ed.)* Brill, Leiden.

DISCUSSION

H. A. van Adrichem Boogoert (Netherlands Geological Survey): Could Mr Butler give some information on the depositional environment of the Upper Sandstone Member of the Rotliegendes in the West Sole Field, and why this Member is not productive?

J. B. Butler: The Upper Sandstone Member differs from the Lower Sandstone Member by virtue of its light grey colour and higher degree of cementation; it also appears to be less indurated and may not therefore have developed the hair-line fracture system of the Lower Sandstone Member. It contains clasts of grey shale, disseminated pyrite and anhydritic, dolomitic and white clay cement. I regard it as having been deposited in a reducing environment and therefore a local precursor of the Zechstein transgression.

15

Geology of the Leman Gas-Field

By F. R. van VEEN

(Shell UK Exploration and Production Limited)

SUMMARY

In April 1966 Shell/Esso well 49/26-1 discovered the Leman Field in the Southern North Sea. Extensions into adjacent licence blocks were subsequently proved by the Amoco group (49/27), the Arpet Group (49/28), and Mobil (53/2) in the period 1966–1968. Appraisal drilling by the four licence operators established a recoverable gas reserve in the order of 10.5×10^{12} scf. Development drilling started in July 1967 and continues to the present. To date 118 production wells have been drilled from 10 platforms.

The field is a north-west–south-east trending anticlinal structure, dissected by numerous, mainly north-west–south-east running faults. The Rotliegendes reservoir, 600–900ft (180–270m) thick, consists of wadi, aeolian, and water-laid sandstones each with typical reservoir characteristics. The reservoir sands overlie the eroded Carboniferous surface and are in turn overlain by the thin Kuperschiefer and some 1500ft (460m) of Zechstein evaporites, which form an effective seal.

The deepest continuous seismic reflector, which has been used to construct the top reservoir map by adding a basal Zechstein isopach, occurs at the Base Upper Zechstein salt. The Zechstein is overlain by several thousand feet of Triassic clastics and evaporites. Thin Jurassic and Cretaceous sediments are present along the flanks of the structure, and the whole sequence is overlain by a Tertiary and Quaternary cover.

INTRODUCTION

The search for gas in the southern North Sea was triggered by the discovery in 1959 of the giant Groningen gas-field in The Netherlands, with reserves of 58×10^{12} cu ft in Lower Permian Rotliegendes Sandstones (Stauble and Milius, 1970).

The Leman Field, which is situated in the southern part of the UK sector of the North Sea, some 30 miles (48km) from the East Anglian coast (Fig 1), was discovered in April 1966 by Shell/Esso well 49/26-1 which proved the Rotliegendes Sandstones to be gas bearing. The inferred outline of the field after the drilling of the discovery well is shown on Fig 2. Extensions into adjacent licence blocks were subsequently proved by Amoco/Gas Council well 49/27-1 in September 1966, Arpet well 49/28-1 in October 1967 and Mobil 53/2-1 in January 1968. After extensive appraisal drilling (Table I) and seismic detailing, the field is now proved to extend considerably more to the south-east than originally assumed (Fig 3). It forms at top Rotliegendes level a north-west to south-east trending domal structure cut by numerous faults. The long axis of the field measures 18 miles (28.8km), the short axis 8 miles (12.8km). A generalized cross-section of the field is shown on Fig 4. The Rotliegendes reservoir is 600–900ft (180–270m) thick with a gas-water contact at 6700ft (2042m) subsea. The maximum gross gas-column is 800ft (244m). To date 118 production wells have been drilled from 10 platforms (Table II) and recoverable gas reserves are estimated at some 10.5×10^{12} scf of which 2.2×10^{12} scf have been produced.

TECTONIC FRAMEWORK AND STRATIGRAPHY

A generalized stratigraphic column is presented in Fig 5.

Carboniferous

Carboniferous sediments consisting predominantly of shales with intercalations of sandstones and coal layers were encountered in several wells. The coals and shales are generally thought to be the source of the Leman gas. Sedimentation was terminated by the Variscan orogeny which resulted in block-faulting and gentle folding of the present Southern North Sea Basin. The area was subsequently peneplained except for the basin margins where considerable topography was maintained. The Southern North Sea Basin has persisted since the end of the Carboniferous as a broad intracratonic basin.

Permian Rotliegendes (Leman Formation)

Rotliegendes sedimentation commenced during the Permian in a desert environment (Fig 6). Three facies units can be distinguished in the Leman Formation (Fig 7).

(i) *Lower unit: wadi deposits.* The lowermost 100–200ft (30–60m) of the Leman Formation overlying the truncated Carboniferous is characterized by wadi deposits, interbedded with

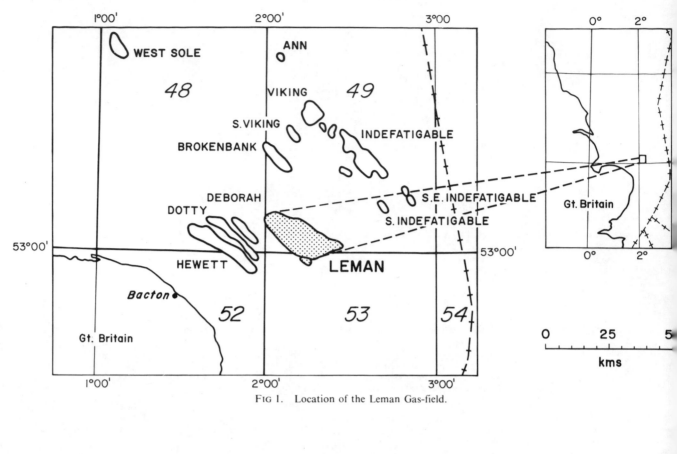

Fig 1. Location of the Leman Gas-field.

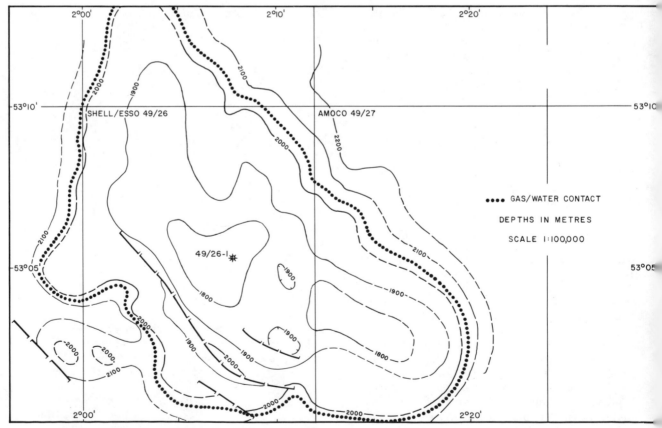

Fig 2. Initial outline of Leman Gas-field, showing contours on top of Rotliegendes Formation in metres below sea-level.

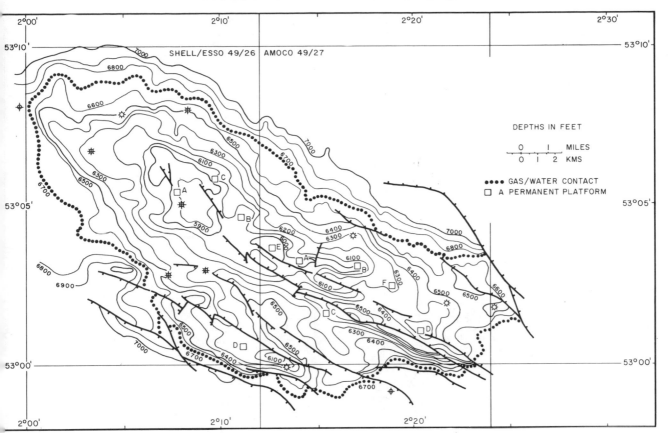

FIG 3.   Present interpretation of Leman Gas-field, showing contours on top of Rotliegendes in feet below sea-level.

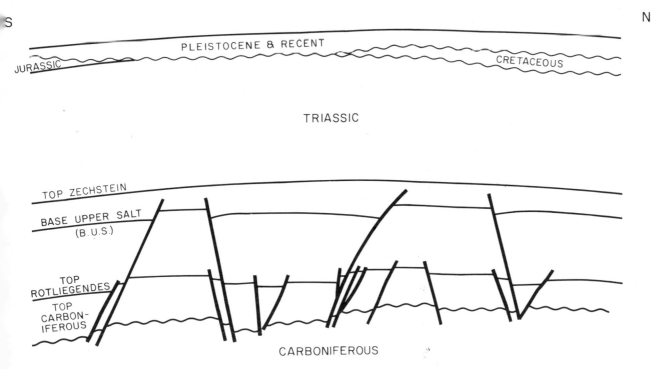

FIG 4.   Schematic cross-section through Leman Gas-field.

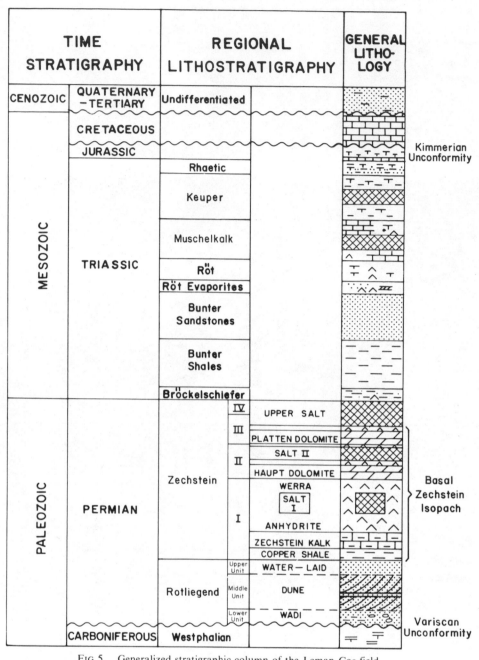

FIG 5.   Generalized stratigraphic column of the Leman Gas-field.

TABLE I
Exploration and appraisal of Leman Gas-field

Regional seismic surveys for lease selection		1962–1964
Detailed seismic surveys (Blocks 49/21 and 49/26) with Signal		1965
Shell/Esso Exploration well	49/26-1—Discovery	December 1965–April 1966
Shell/Esso Appraisal well	49/26-2	April 1966–July 1966
Shell/Esso Appraisal well	49/26-3	May 1966–July 1966
Amoco Appraisal well	49/27-1 (now A1)	June 1966–September 1966
Shell/Esso Appraisal well	49/26-4	August 1966–October 1966
Amoco Appraisal well	49/27-2 (now C1)	December 1966–February 1967
Shell/Esso Appraisal well	49/26-5	December 1966–April 1967
Amoco Appraisal well	49/27-3	January 1967–March 1967
Amoco Appraisal well	49/27-4	April 1967–June 1967
Shell/Esso Appraisal well	49/26-25	June 1970–August 1970

Additional detailed seismic surveys in 1966, 1967 and end 1970.

TABLE II
Development of Leman Gas-field

Shell Platform AD/AD II	wells 49/26–A6 to A24	(14) July 1967–September 1970
Amoco Platform A	plus 49/27–A2 to A12	(12) September 1967–March 1970
(over 49/27-1)		
Amoco Platform B	49/27–B1 to B12	(12) June 1968–December 1969
Amoco Platform C	plus 49/27–C2 to C12	(12) July 1968–February 1970
(over 47/27-2)		
Shell Platform B	49/26–B210 to B295	(16) September 1969–November 1971
Shell Platform C	49/26–C300 to C375	(16) June 1970–June 1972
Shell Platform D	plus 49/26–D400 to D455	(10) February 1972–End 1974
(over 49/26-25)		
Amoco Platform D	49/27–D1 to D12	(12) March 1972–February 1973
Amoco Platform E	49/27–E1 to E12	(12) September 1973–(drilling)
Amoco Platform F	49/27–F1 to F12	(12) January 1974–(drilling)

aeolian sands. The wadi deposits consist of very fine to coarse, water-laid sands, frequently containing large clay pebbles (up to 10cm) with intercalations of conglomerates and some discontinuous shale streaks. The poorly sorted aeolian sands have been deposited in thin cycles having a sub-horizontal base, and increasing foreset dips upwards.

This lower unit of the Leman formation has generally poor reservoir characteristics and is mainly below the present field gas/water contact.

(ii) *Middle unit: aeolian deposits.* The middle unit which measures between 450–650ft (140–200m) forms the main reservoir and consists entirely of aeolian dune sands. The sandstones are very fine to medium, locally coarse and clay intercalations are entirely absent.

Porosities range from 10–20 percent (average porosity 14 percent) and permeabilities from 4md to 1d (average 60md). Steeply dipping foresets (Plate I) which give excellent sedimentary dips on the dipmeter logs are frequent (Fig 8, average dip

25°). Alternating laminae have different grain size but within individual laminae the sorting is good to very good. A succession of large-sized dune crossbed sets with an average thickness of 15ft (4.5m) can be recognized whereby each dune is truncated by its successor. The resulting shape of the individual crossbed set is a flat oval disc. The dunes were of the transverse type with sinuous crests oriented perpendicular to the wind direction (Glennie, 1972), which was persistently from the east during the deposition of the aeolian beds. Detailed log-correlations have indicated that the average length of the crossbed sets in the wind direction is about 200 times the average thickness while the width is about half of this value. Aerial photographs of recent dunes of this type give also an impression of the shape of the units (Wilson, 1972).

Sand grains were transported over the windward dune slopes and slid down along the leeward dune slip-face, whereby discrete coarser and finer sand streaks were formed. Bagnold (1941) has described

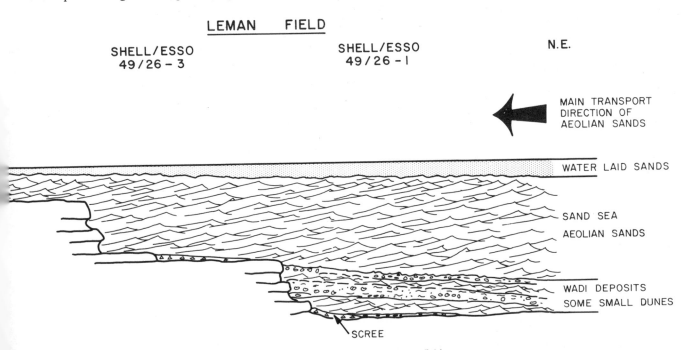

FIG 6.   Rotliegendes of the Leman Gas-field.

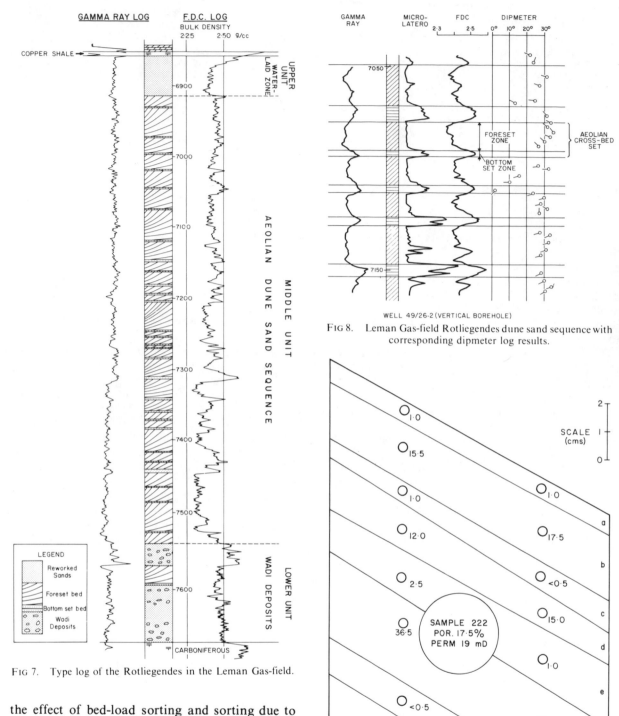

FIG 7.   Type log of the Rotliegendes in the Leman Gas-field.

FIG 8.   Leman Gas-field Rotliegendes dune sand sequence with corresponding dipmeter log results.

FIG 9.   Permeability distribution in aeolian cross-bed set; Well No. 49/26-A6, interval 6829ft 3in—6829ft 7in (2081.55—2081.66m).

the effect of bed-load sorting and sorting due to gravitational sliding. The differences in grain-size and sorting of the individual sand-streaks can be quite large resulting in considerable permeability differences of adjacent sand laminae (Fig 9). At the base of the leeward slope of a dune a mixture of sand grains rolling down the slope and fine particles suspended in the air and carried downwards by turbulence behind the dune crest is deposited. This mixture is poorly sorted and forms a finely laminated, low permeable layer beneath the zone of foreset laminae. This layer is called the bottom-set bed. The foreset laminae merge tangentially with the bottom-set laminae.

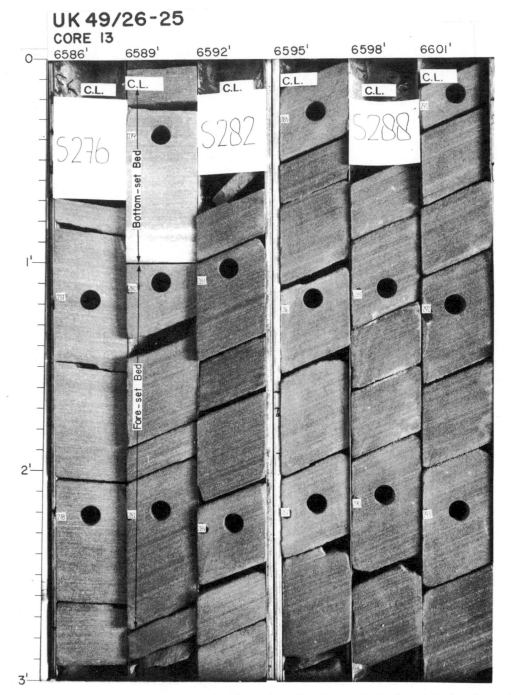

PLATE I.   A typical example of Bottom-set/Fore-set bedding in the aeolian dune sands (Rotliegendes) of the Leman Gas-field.

In the Leman Field cores the bottom sets usually contain a higher percentage of carbonate cement than the foreset laminae. The poor sorting and the carbonate cement cause a low porosity and high grain density which makes it possible to identify the bottom-sets in the FDC logs. Moreover the relatively higher clay content of the bottom sets result in peaks on the gamma-ray logs (Fig 8).

(iii) *Upper unit: waterlaid deposits.* The upper unit, 20–100ft (6–30m) thick, consists of water-laid sands which are generally fine to medium grained and grey coloured. They contain occasional clay pebbles and are moderately to poorly sorted. The sands are mostly structureless but show occasional faint irregular laminations or slump-structures.

These homogenized, originally aeolian sands, were reworked as a result of the Zechstein transgression which ended the desert conditions in the Leman area. Log-correlations indicate that the contact between the aeolian and waterlaid units is a sub-horizontal erosional surface. The differences in thickness of the waterlaid units are caused by erosional relief at the top rather than at the base of the unit.

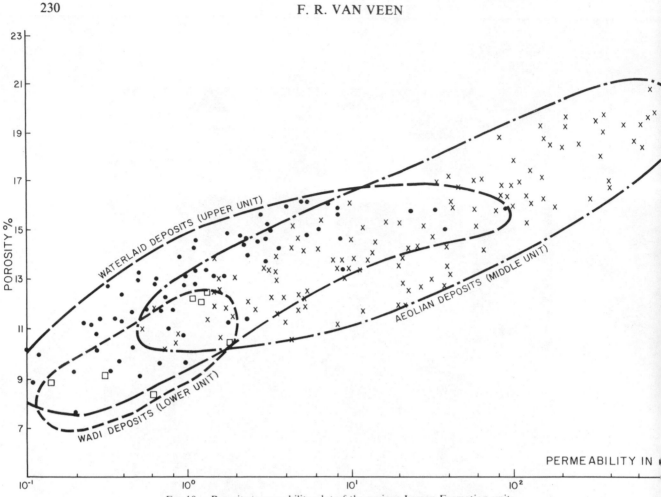

FIG 10.   Porosity/permeability plot of the various Leman Formation units.

The reservoir characteristics of the waterlaid unit are poorer (porosities between 8 percent and 15 percent, permeability 0.1md to 100md) than those of the aeolian unit as shown on a porosity–permeability plot of some representative samples of the three units (Fig 10).

Permian Zechstein

The Upper Permian Zechstein sequence, which measures up to 1500ft (460m) consists predominantly of evaporites, and forms an effective seal to the Leman gas accumulation.

Sedimentation began with a marine transgression resulting in the deposition of the 4 to 5ft (1.2 to 1.5m) thick black bituminous Kupferschiefer (Copper Shale), which overlies the waterlaid deposits of the Leman Formation. The transgression initiated four evaporite cycles of the Zechstein.

Zechstein I. Zechsteinkalk which reaches a maximum thickness of 26ft (7.9m), overlies the Kupferschiefer and is followed by an evaporite (mainly anhydrite) sequence, which varies in thickness from 350ft (105m) to 650ft (200m). In the northern part of the field a salt layer interfingers the anhydrite sequence.

Zechstein II. This cycle is characterized by dolomite, anhydrite and salt. The dolomite interval has been correlated with the well-known Hauptdolomit. The cycle ends with a layer of roof anhydrite about 5–10ft (1.5–3.04m) thick. The total thickness of this cycle varies from 200 to 400ft (60–120m).

Zechstein III. This sequence which measures 250–500ft (75–150m) began with a marine transgression depositing a grey salty clay of about 2–4ft (0.6–1.2m) thick, followed by a dolomite (equivalent to Plattendolomit) and an anhydrite member. The latter is related to the deepest continuous seismic reflector in the Leman field area. The cycle ended with the deposition of salt, rich in potassium. From this deepest seismic reflector, called the Base Upper Salt (B.U.S.) the top Rotliegendes structure map has been obtained by adding a "basal Zechstein isopach" to the Base Upper Salt contour map (Moore, 1973). This "basal Zechstein isopach" varies in thickness from 750 to 1300ft (230 to 400m).

Zechstein IV. The last cycle of the Zechstein sequence consists predominantly of potassium salts. In several wells the base is formed by an anhydrite layer of maximum 10ft (3m). In the absence of the anhydrite, no clear subdivision can be made between Salt III and Salt IV Zechstein. If present, the thickness of Zechstein IV can be measured, varying between 100ft (30m) and 300ft (90m).

Triassic-Jurassic

The Triassic occurs in Germanic facies and the three lithostratigraphic subdivisions equivalent to the Bunter, Muschelkalk and Keuper can be recognized. The thickness varies from 4000ft (1200m) in the south-west of the field to 5000ft (1500m) in the north-east.

Variable amounts of Keuper equivalents have been removed during the late-Mesozoic inversion tectonics. Jurassic sediments have only been preserved in a few wells off the crest of the structure outside the field limits.

Cretaceous

Upper Cretaceous Chalk of some 150ft (45m) thickness has been preserved on the flanks of the field. It is unconformably overlain by Tertiary and Quaternary sediments.

ACKNOWLEDGEMENTS

The author wishes to acknowledge the contribution of K. J. Weber to the sedimentological investigation of the Leman formation. He is also indebted to Shell UK Exploration and Production Co. Ltd., and Esso Europe Inc. for permission to publish this paper.

REFERENCES

Bagnold, R. A. 1941. *The physics of blown sands and desert dunes.* (Reprinted 1960). Methuen, London, 265 pp.

Glennie, K. W. 1972. Permian Rotliegendes of Northwest Europe interpreted in light of modern desert sedimentation studies. *Bull. Am. Ass. Petrol. geol.*, **56**, 1048–71.

Moore, J. W. 1973. Top reservoir mapping problems in the United Kingdom Sector of the North Sea, AIME paper 4308.

Stauble, A. J. and Milius, G. 1970. Geology of Groningen Gas Field Netherlands. *In* M. T. Halbouty (Ed.) Geology of Giant Petroleum Fields. *Am. Ass. Petrol. geol. Mem.*, **14**, 359–69.

Wilson, I. G. 1972. Aeolian bedforms—their development and origins. *Sedimentology*, **19**, 173–210.

16

The Geology of the Indefatigable Gas-Field

By D. S. FRANCE

(Amoco (UK) Exploration Company)

SUMMARY

The Indefatigable Field, discovered by the Gas Council/Amoco Group in 1966, is the second largest producing gas-field in the North Sea. The field is located 55 miles (88km) north-east of Great Yarmouth, England in about 100ft (30m) of water, and is jointly owned by the Gas Council/Amoco Group and Shell/Esso. The reservoir is the Lower Permian Rotliegendes Sandstone at a depth of about 8000ft (2438m). The trap is a complexly faulted horst, with three separate gas-bearing structures, which have closures above the gas–water contact of up to about 1300ft (396m), but with pay thicknesses ranging up to only approximately 420ft (128m). The field covers just over 63 sq miles (163km²) and contains an estimated recoverable reserve of about 4.5×10^{12} scf of gas. The field came on production in September 1971 and is currently producing 560MMcfd.

INTRODUCTION

The Indefatigable Field is the second largest producing gas-field in the North Sea, and is located 55 miles (88km) north-east of Great Yarmouth and 14 miles (23km) north-east of the Leman Field (Fig 1). It was discovered in June 1966 when the Gas Council/Amoco Group, drilled well 49/18-1 and established gas production from the Rotliegendes Sandstone at a depth of about 8000ft (2438m).

Subsequent appraisal drilling of thirteen wells showed that the field covers an area of approximately 63 sq miles (163km²) extending into the Gas Council/Amoco Group's licence block 49/23 and the Shell/Esso Group blocks 49/19 and 49/24. The participating companies in the Indefatigable Field are as follows:

Gas Council (Exploration) Ltd, Amoco (UK) Exploration Company, Amerada Petroleum Corporation of the United Kingdom Ltd, and North Sea Inc., comprise the Gas Council/Amoco Group.
Shell (UK) Exploration and Production Ltd, and Esso Exploration and Production UK Inc., comprise the Shell/Esso Group.

Average water depth is 100ft (30m) and the sea-floor is firm flat sand.

DEVELOPMENT

Between 1970 and 1972 three development platforms ("A", "J", "K") were installed with a total of 23 wells; these are now producing gas at an average rate of 560MMcfd through a 30in pipe system to the Bacton Terminal *via* Leman Field.

The Gas Council/Amoco Group is presently drilling the tenth and last well on their second platform while plans are being made for yet a third platform. The Shell/Esso Group has drilled a location proving well for their third platform.

STRATIGRAPHY

The stratigraphy of the area is, in broad outline, the standard one for the southern North Sea and has been governed by three phases of earth movements: the Hercynian orogeny in late Carboniferous times, the Cimmerian phase during the Mesozoic and the Laramide phase in late Cretaceous to early Tertiary times. The deepest drilled formations are of typical Upper Carboniferous deltaic facies with occasional coal seams up to 8ft thick and are early Westphalian in age. The coals are generally considered to be the source of the gas.

Resting unconformably on the Carboniferous post-Hercynian erosion surface lies the Lower Permian Rotliegendes Sandstone which forms the reservoir of the field. The Rotliegendes Sandstone varies from entirely grey to entirely red-brown, though usually the top 50 to 100ft (15 to 30m) are uniformly grey. Except at the base, where there are locally thin conglomerates or breccias with clasts of Carboniferous shales, and the occasional thin shale streak, the lithology is sandstone, generally fine- to medium-grained. In individual bands or laminae the sandstone can be silty or coarse. It varies from poorly to well sorted.

Quartz is the main detrital component with rare chert and quartzite fragments in the coarser layers. Orthoclase, plagioclase, muscovite, graphic granite fragments, and heavy minerals such as tourmaline and apatite also occur, all suggesting a mainly

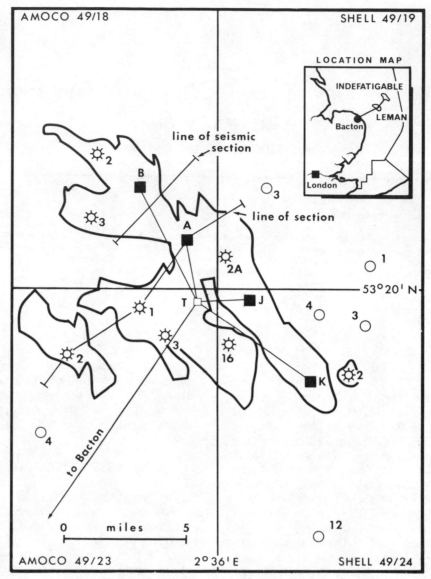

FIG 1.   Location map of Indefatigable Field.

granitic provenance. In the red-brown sandstone the grains are partially coated with hematite whereas in the grey zones the coating is diagenetic kaolinite. Authigenic anhydrite and carbonates occur as a groundmass or as a cement. Some of the sand grains have a secondary silica coating which encloses the encrusting hematite. Otherwise the matrix is detrital clay and the decomposition products of unstable minerals such as biotite which have been partially recrystallized to sericite, kaolinite, illite and leucoxene.

The sandstone is generally massively bedded or displays ripple structures, and only occasionally has interbedded light and dark grey bands or is thinly laminated with good sorting in the laminae. The latter type usually shows cross-bedding. The general lack of character seen on the electric logs confirms observations made on cores, which show that the sandstone as a whole is more uniform than in Leman Field, with only a few aeolian dune units,

and is composed predominantly of reworked aeolian sands which have been homogenized by flowing water. The decoloration of the hematite sand-coating could be due to the later percolation of reducing groundwater from the transgressive, euxinic Zechstein sea.

Overlying the Rotliegendes Sandstone, apparently conformably, and sealing migrating hydrocarbons is the Upper Permian Zechstein (Fig 2) carbonate-evaporite succession beginning with the ubiquitous Kupferschiefer, a thin organic shale. The four Zechstein cycles recognized in Germany are developed in the Indefatigable Field area. The salient features of the succession are as follows:

(i) The Kupferschiefer, Zechsteinkalk, Werra-anhydrit of Cycle Z1, together with the Hauptdolomit of Cycle Z2 form a fairly uniform thin sequence, the top of which is the deepest seismically mappable horizon in the area (Fig 3).

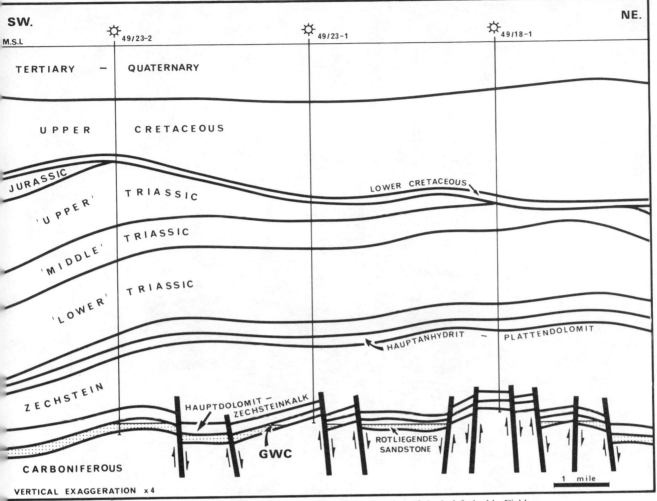

FIG 2.   South-west—north-east structural cross-section of the Indefatigable Field.

Thus in order to construct the Top Rotliegendes structure map it is necessary to relate the isopachs of this interval to the Top-Hauptdolomit seismic structure map. This basal Zechstein interval remains fairly constant over the field area at between 163 to 225ft (50–69m).

(ii) The Cycle Z2 halite unit comprised of the Basalanhydrit equivalent, which in this area is represented mainly by halite with occasional anhydrite and polyhalite stringers, together with the Stassfurt Halite, varies considerably in thickness from 772 to 2173ft (235–662m).

(iii) The Deckanhydrit of Cycle Z2, when developed, and the Cycle Z3 Grauer Saltzton, Plattendolomit and Hauptanhydrit together form a mainly carbonate–anhydrite unit which varies from 250 to 382ft (76–116m) in thickness over the area.

(iv) The Aller and Leine Halites of Cycles Z3 and Z4, both of which have variable developments of potassium salts are virtually one salt body, being separated only by the thin Röter Saltzton and a poorly developed Pegmatitanhydrit. The whole of this salt interval varies considerably in thickness from 65 to 528ft (20 to 161m).

Conformable with the Zechstein are the Triassic rocks, which are developed in the threefold Germanic lithostratigraphic subdivisions of Bunter, Muschelkalk and Keuper.

The Bunter, which has about 100ft (30m) of anhydritic siltstone—the Bröckelschiefer—at the base, consists principally of two red-brown shale members with a red-brown sandstone separating them. The Röt Evaporite, comprising approximately 100 ft (30m) of halite or anhydrite occurs in the Upper Shale section. The maximum thickness of the Bunter in the Indefatigable Field is about 2100ft (640m). The Bunter is overlain by a mainly red-brown shale section with dolomite interbeds in the Muschelkalk, and anhydrite interbeds in the Keuper. Both contain some halite; however the Keuper Halite is only locally developed. The maximum thickness of the Muschelkalk is 450ft (137m) and of the Keuper 1550ft (472m).

Because of post-Cimmerian erosion, the upper part of the Keuper sequence, the thin Rhaetian sandstone and grey shale sequence, and the conformable Lower Jurassic marine calcareous shales and limestones are present only in the

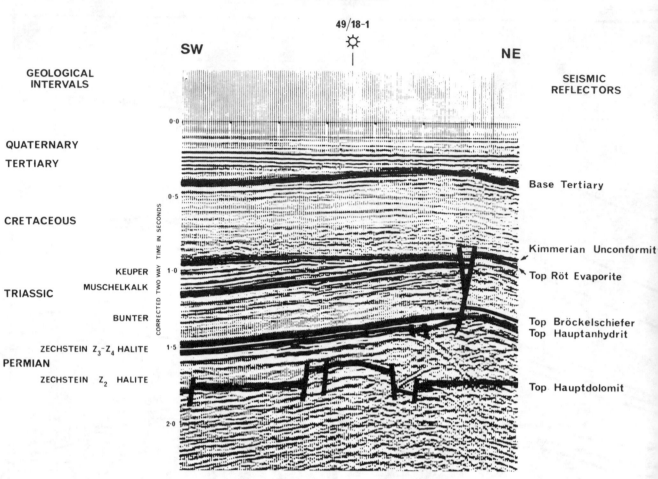

GEOLOGICAL
INTERVALS

SEISMIC
REFLECTORS

49/18-1

SW   NE

QUATERNARY

TERTIARY

Base Tertiary

CRETACEOUS

Kimmerian Unconformit

KEUPER

Top Röt Evaporite

MUSCHELKALK

TRIASSIC

BUNTER

Top Bröckelschiefer
Top Hauptanhydrit

ZECHSTEIN Z₃⁻Z₄ HALITE

PERMIAN

ZECHSTEIN Z₂ HALITE

Top Hauptdolomit

CORRECTED TWO WAY TIME IN SECONDS

1 mile

FIG 3.   Seismic section across Indefatigable Field.

western part of the area. The whole of the foregoing sequence was folded and faulted during the Cimmerian phase of earth movements, then subsequently uplifted and eroded. Subsidence was renewed and Lower Cretaceous shales with glauconitic sandstone interbeds were deposited on an uneven Cimmerian erosion surface.

Lower Cretaceous rocks overstep onto Lower Jurassic rocks in the western part of the area, then towards the east they rest on progressively older Triassic formations. Overlap also occurs since the basal beds are variously Barremian and Aptian in age. Conditions became uniform with the onset of Upper Cretaceous chalk deposition.

In late Cretaceous to early Tertiary times Laramide movement caused local uplift and partial erosion of the Chalk. Subsequent deposition of Upper Palaeocene marine shales was locally conformable with Danian Chalk; otherwise they rest on Maastrichtian Chalk. Marine conditions continued through to the Pliocene, with the Indefatigable area lying on the southern flank of the Northern Tertiary Basin. This situation was further complicated by mid-Tertiary uplift to the north of the field following salt movement and resulted in a thinning of the Tertiary succession.

The sedimentary sequence is completed by poorly consolidated Pleistocene glacial outwash sands and gravels with thin clay interbeds which in turn are overlain by Holocene post-glacial sands.

STRUCTURE

The field appears to be comprised of three gas-bearing structures which have a maximum closure above water of the order of 1300ft (396m), but with a pay thickness ranging up to approximately 420ft (128m). The three areas (Fig 4) are referred to as follows:

(i) the Main area, (ii) the 49/23-2 area, (iii) the 49/24-2 area.

(i) The Main area can be divided on structural characteristics into two sub-areas, the main horst and the main flank area. The main horst, trending north-west–south-east, is 15 miles (24km) long and 2 miles (3km) wide at its widest part. Its form is anticlinal with flanks dipping towards the bounding faults. The axis plunges gently away from the crest which is at an approximate depth of 7550ft (2301m) subsea and lies between the Gas Council/Amoco Group's A and B platforms. The

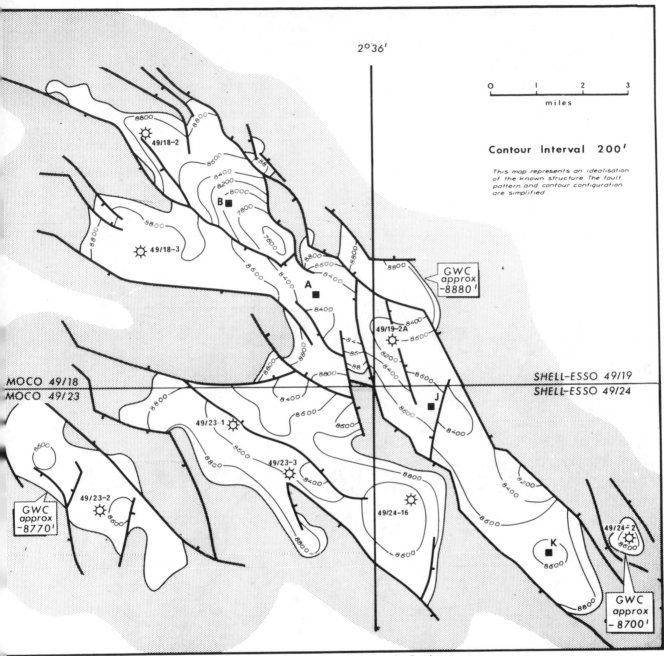

FIG 4.   Indefatigable Field—idealized top Rotliegendes Sandstone map.

horst is divided into several blocks by approximately north–south trending faults. The gas–water contact lies at approximately 8880ft (2707m) below sea level thus giving a maximum closure above water in excess of 1300ft (396m). The main flank area is a lower lying tract to the west and south-west of the main horst, which appears to have the same gas–water contact and is thus in communication with the main horst. The area has a highest point of approximately 8350ft (2545m) below sea-level, but consists of several low amplitude domal highs, all with approximately the same closure of about 420ft (128m). This area is partially fault-bounded, and is complexly faulted.

(ii) The 49/23-2 area is similar to the main flank area in being of low amplitude. It is periclinal, with its longest axis, 5 miles (8km) long, trending north-west–south-east and has several culminations, the highest being in the region of 8570ft (2612m) subsea. The gas–water contact lies approximately at 8770ft (2673m) subsea, some 110ft (34m) higher than in the main area.

(iii) The 49/24-2 area is a small domal horst to the south-east of the main horst, from which it is isolated by a narrow graben. Its crest is found around 8570ft (2673m) subsea and it has a gas–water contact at around 8700ft (2651m) subsea some 180ft (55m) higher than the main area.

FAULTING

The whole area is closely block-faulted into
generally elongated horsts and grabens (Fig 4).
There are two principal fault sets. The dominant set
trends north-west–south-east occasionally swing-
ing to east–west while the other set trends north-
north-west–south-south-east and occasionally
north-north-east–south-south-west. The latter set
tend to break up the relatively simple horst and
graben configuration of the dominant faults to give
a complex diamond pattern of block-faulting. This
effect is particularly seen in the main horst area.
The displacement varies considerably along the
faults and a maximum throw of 1300ft (396m) is

reached on one of the bounding faults of the main
horst.

DEFORMATION HISTORY

The Rotliegendes structures described above were
formed principally during late Jurassic times in the
late Cimmerian phase of earth-movements.
Throughout this phase regional tension caused the
north-west–south-east trending block-faulting of
the sub-Zechstein rocks. By this time sufficient
overburden to the Zechstein had been deposited for
plastic flow of the salt to be triggered by the
extensional faulting of the Rotliegendes and older
beds. The salt-flowage resulted in pillowing of the

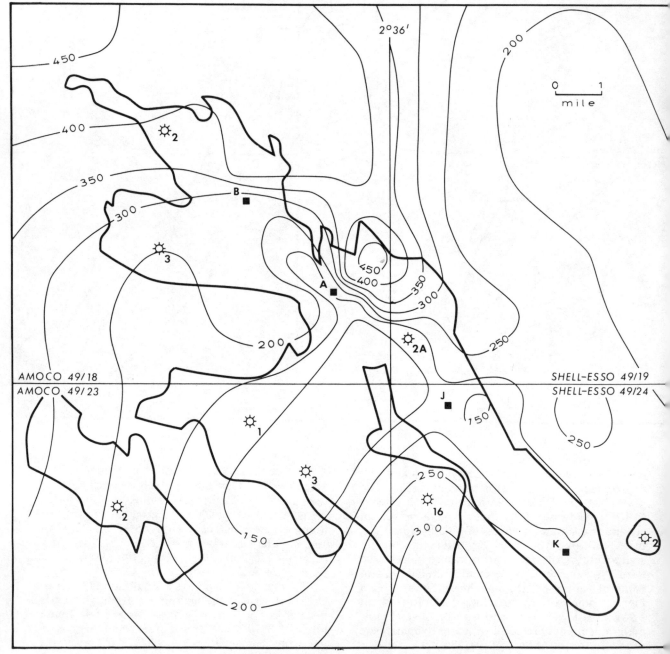

FIG 5.   Indefatigable Field—Rotliegendes Sandstone isopach map.

Zechstein sequence and the entire Triassic to Jurassic succession accommodated this differential vertical movement by flexuring and faulting. The flexures are long wave-length low amplitude north-west–south-east trending structures. The faults which displace Triassic formations (post-Cimmerian erosion having removed the Jurassic sequence), appear to die out downwards into Zechstein salt. They also trend north-west–south-east, thus confirming probable initiation during the Cimmerian phase. These faults often displace Cretaceous strata as well. This is probably due to reactivation and growth of the faults with continuing salt movement during the Cretaceous.

The main effect of the Laramide movements at the close of the Cretaceous and in early Tertiary times, was localized uplift with erosion of the Chalk and consequent local unconformity. Tertiary sediments were affected by uplift as a result of continuing salt movement, and it is possible that further fault displacement also took place then.

RESERVOIR PARAMETERS

(i) The reservoir sandstones vary from 115 to 419ft (35 to 128m) in thickness with the net pay varying from 50 to 419ft (15 to 128m). The average net pay over the Indefatigable Field is 197ft (60m). As can be seen from the isopach map (Fig 5), there is an approximate areal correspondence between the thinning of the sandstone and structural highs, particularly over the main horst, the main flank and the 49/23-4 areas. This is possibly related to Carboniferous palaeotopography.

(ii) The average porosities range from 9.3 to 21.6 percent with a field mean of about 15 percent.

(iii) Core permeabilities range from 0.1 to 2000md.

(iv) The field is at hydrostatic equilibrium.

(v) Gas saturations in the field range up to 80 percent.

(vi) Gas gravity is 0.614.

(vii) Type of gas—dry.

(viii) Initial condensate/gas ratio was 2bbls/MMcf.

(ix) BTU Rating—1050BTU per cubic foot of dry gas.

(x) Area of field is approximately 40 000 acres (163km²).

GAS RESERVES

Gas reserves were calculated in the following way:

1. Rotliegendes net pay Isopach and porosity distribution maps were required to calculate gas-in-place.
2. These were combined to give ϕ/h values at selected points.
3. Use of standard gas saturation curves for differing porosities and height above the gas–water contact resulted in a hydrocarbon pore volume map, which when planimetered gave the gas-in-place.
4. By applying a suitable recovery factor the reserves have been calculated at around 4.5×10^{12} scf.

FUTURE PLANS

The Gas Council/Amoco Group B platform should be fully developed by the end of 1974. A third 12-well platform is projected for installation in 1976. This will be situated on the main flank area.

The Shell/Esso Group plan their third four-well platform to start production in 1976; this will also be located in the main flank area at the 49/24-16 well location.

17

Viking Gas-Field

By I. GRAY

(Conoco Europe Ltd, London)

SUMMARY

To confine the Vikings' description to the brief one required here, North Viking alone has been considered, and only from the top of the Zechstein downwards to the Rotliegendes.

The Vikings are now seen as two interconnected anticlines *en echelon* with parallel north-west–south-east trending axes. North Viking has the simpler structure of the two. It is mapped as an asymmetric anticline, the south-western flank being deeper than the north-eastern, with a crestal graben running the length of the structure. Difficulties have been encountered in making depth conversions and in mapping gas–water contacts. A new three-dimensional perspective technique for modelling the reservoir is described. A conclusion is drawn that any closed Rotliegendes structure or fault-block can have an independent GWC and be a prospective exploratory location.

INTRODUCTION

The area covered by the Viking gas-fields now comprises the southern half of UK Block 49/12 and the northern half of Blocks 49/16 and 49/17. These licences are held equally by the National Coal Board (Exploration) Ltd and Conoco Ltd.

The discovery well on South Viking, 49/17-2 was drilled in May, 1968, and that on North Viking, 49/12-2 in March 1969. These two discoveries were developed with a three-platform complex on North Viking and a two-platform complex on South Viking. North Viking came on stream in October 1972 and South Viking in August 1973. A contract was signed with the British Gas Corporation to deliver initially 300MMcfpd, which is being stepped up annually to achieve a daily contract rate of 550MMcfpd by October, 1975. Contractually Conoco is also obligated to be capable of delivering up to 167 percent of this daily rate as a peak demand.

GEOLOGICAL SUCCESSION

The geological succession present is typically that seen throughout the UK southern North Sea and is illustrated in broad terms on the seismic section in Fig 1. The position of the line relative to the North Viking field can be seen from the "base of Zechstein" contour map, Fig 2.

Tertiary formations are usually present underlain by Upper Cretaceous Chalk with some Lower Cretaceous and Jurassic frequently occurring. The Jurassic is sometimes removed by Late Cimmerian erosion and occurs over North Viking as a "wedge" thickening to the north-east. The Triassic (Rhaetic to Lower Bunter) is normally the thickest sequence present and represents a fairly consistent 4000–4500ft (1219–1372m) of sediment.

Below the Triassic the Zechstein is present in greatly varying lithology and thickness. There is a high salt content in the Zechstein of the Vikings area and this provides an effective seal for the gas accumulation. Dolomites and anhydrites occur extensively within the salt but are frequently discontinuous. The Upper and Lower Magnesian Limestones occur typically as discrete layers within the salt but can thicken greatly over short geographical distances. Large "rafts" of dolomite are often discerned "floating" in diapiric salt structures.

Below the Zechstein lies the Rotliegendes Sandstone gas reservoir resting upon Carboniferous Westphalian strata from which the gas was derived.

STRUCTURE

The Viking gas-fields are a series of fault-bounded structures, each with its own gas–water contact and bearing a complex relationship with each other. In the main there are two anticlines at the Rotliegendes Sandstone level with their axes trending north-west–south-east and lying *en echelon.*

The South Viking Field is a broad anticline with a flat crest. In the centre, however, the structure is complicated by a fault-bounded upthrown block only 0.5 sq miles (1.3km²) in area which rises to over 1000ft (305m) above the general relief of the anticline. To describe this complicated area is beyond the scope of this paper; consequently the simpler North Viking field will be more fully discussed.

The North Viking field is an asymmetric

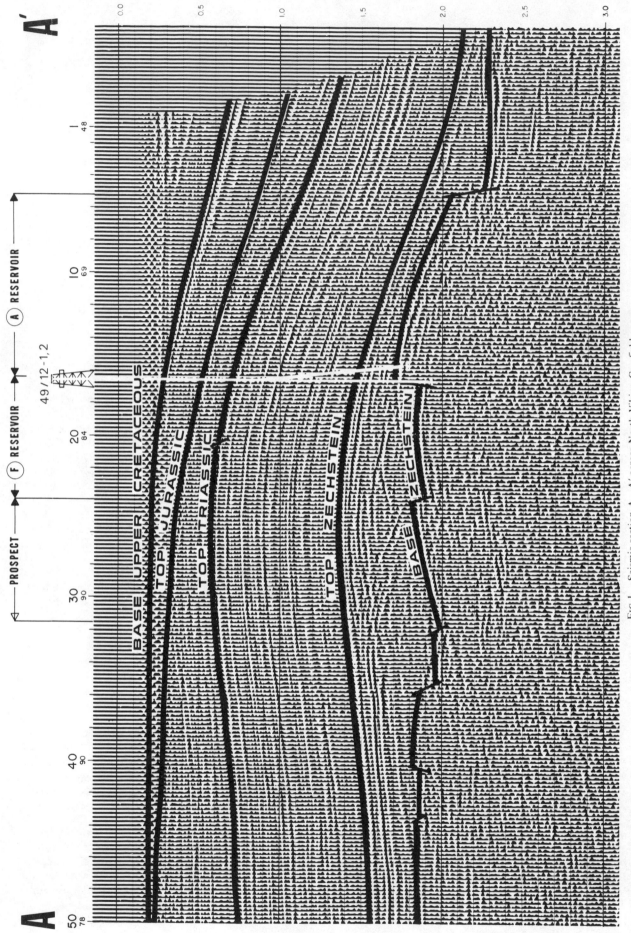

FIG 1.   Seismic section A—A' across North Viking Gas-field.

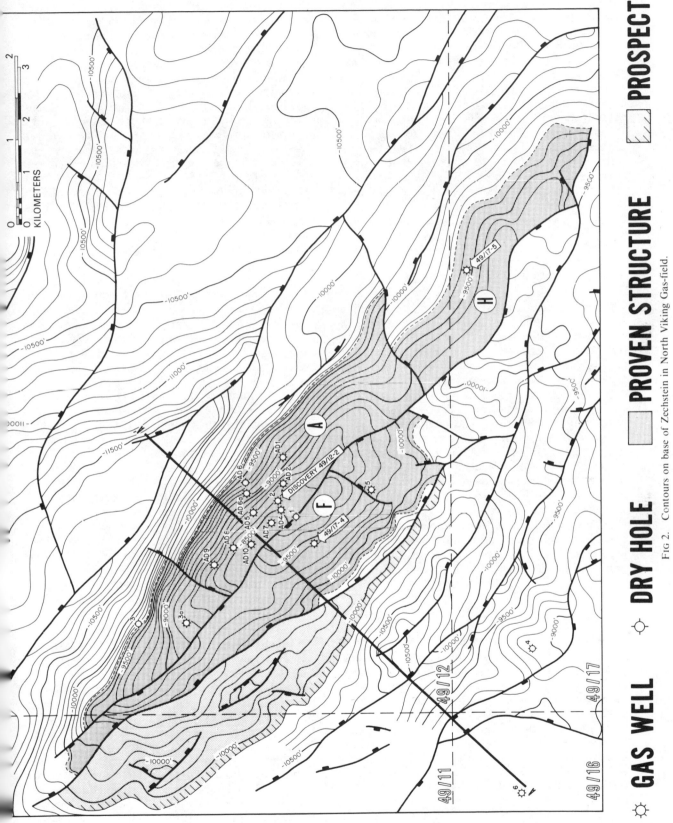

Fig 2. Contours on base of Zechstein in North Viking Gas-field.

☼ GAS WELL ◇ DRY HOLE ▨ PROVEN STRUCTURE ▨ PROSPECT

anticline about 10 miles (16km) long by 3 miles (5km) wide (Fig 2) lying to the north-west of South Viking on a parallel north-west–south-east trend. The north-eastern flank of North Viking is steeply dipping from about —8000ft (—2438m) to almost —11 000ft (—3353m). The south-western flank dips from —9500ft (—2896m) to over —10 500ft (—3200m) and there is a crestal graben running the length of the structure which dips in both directions axially along the structure from —9200ft (—2804m) to over —10 300ft (—3139m).

The fault bounding the north-eastern side of the graben is a major one with a throw to the south-west of up to 1000ft (305m) in the central area, decreasing to 100ft (30m) or so at each end.

The fault on the south-west has a maximum throw of 400ft (122m) to the north-east. This throw, as with the northern fault, decreases axially along the graben in both directions.

RESERVOIR CHARACTERISTICS

The Rotliegendes Sandstone of the North Viking Field is about 500ft (152m) thick on average and consists of aeolian and fluviatile deposits. The net-pay thickness encountered varies from 330–450ft (101–137m). The porosities in the reservoir vary from 14.5 to 18.5 percent and the permeability range is between 50 and 80md. The main north-eastern flank of the anticline, the "A" reservoir, is fairly uniform in porosity and permeability.

The top 200ft (61m) of Rotliegendes is frequently fairly "tight" and permeability is restricted, due perhaps to a fluviatile mode of deposition. In other parts of the Vikings section zones of very tight sand have also been encountered, caused mainly by secondary cementation. These zones probably coincided with the inter-dunal areas at the time of deposition and are almost impossible to predict before drilling as their presence seems to be quite unrelated to the structure as it is now seen.

GEOPHYSICS

The quality of data over the North Viking structure is very good for most horizons. The event usually picked to define the reservoir is that of the "Base Zechstein". The Rotliegendes itself, as typically for the Southern North Sea Basin, is rarely if ever seen, so the top of the Lower Magnesian Limestone is the event normally followed in the Vikings' area. This lithological unit is of fairly uniform thickness in North Viking, varying from 150 to 190ft (43–58m) so a constant thickness can be added to the top of the unit to derive the depth to the reservoir. Where this basal unit thickens appreciably, errors in the depth determination are introduced as the thickening can rarely be detected on the seismic sections.

Depth determinations are usually carried out by the "layer-cake" method. The interval velocities are determined from well surveys for the major units in the succession and summing the calculated isopachs to determine the depth to the reservoir.

Problems arise on the flanks of the structures as steep velocity gradients occur in beds with appreciable dip. This gradient is difficult to determine accurately since few wells are drilled far down the flanks of the producing structures in the gas-field, thus denying us the necessary velocity control. Consequently the depth to the crest of a structure can now be reasonably well derived, but the true areal extent and thickness of the reservoir as defined by the position of the gas/water contact on the flanks still presents serious problems. Seismically derived velocities have not proved very satisfactory in the Vikings. A typical problem in the method is to determine the Jurassic velocity in the wedge over the "A" reservoir. Derived velocities give values typically 40 percent too high for this low velocity layer.

Migration of the dip-lines has been undertaken but has not proved satisfactory. Empirical methods of relating the changes in velocity to changes in the time measured from the seismic section have been developed. These are an attempt to apply a linear relationship to the effects of compaction on the velocities used. The typical gradient for the Triassic as a whole, for example over North Viking, varies from 10 500ft/sec (3200m/sec) at the crest to 13 500ft/sec (4115m/sec) down the flank, a distance of less than 1 mile (1.61km). The sudden changes are presumed to be an effect of compaction. The method now utilized to encompass these variations simply uses the interval velocity directly derived at the well as a starting point and varies the velocity linearly on a scale relating to the mid-point of the formation as measured in time from the seismic section. This system has enabled a structural interpretation to be developed which gives reserve figures in close agreement with those derived from well-pressure studies. It may not be applicable, however, to the entire gas area.

PROBLEMS OF FIELD EVALUATION

"A" Reservoir

The first well to be drilled in North Viking was the 49/12-1. This accurately located the largest fault in the area by drilling through it and, after a brief encounter with the Zechstein Lower Magnesian Limestone, passed directly into Carboniferous sands. The well was then sidetracked about 500ft (152m) to the north and the main North Viking gas accumulation was discovered. The downthrow of the fault at the 12-1 location is about 800ft (244m) to the south-west.

The thickness of Rotliegendes Sandstone determined by the early wells on the North Viking crest was a fairly consistent 500ft (152m). A gas/water contact was not established until the 49/17-5 well was drilled. Later drilling from the AD platform demonstrated substantial discrepancies with the seismic depth maps while confirming the gas/water contact at a depth of 9680ft (2950m) subsea.

Platform drilling indicated a smaller reservoir than the original interpretation. This was confirmed by the production–pressure relationship, therefore the figures for reserves were reduced.

A reinterpretation of the seismic data produced a map that indicated that the 49/17-5 well drilled on the south-easterly extension of the field was separated from the "A" reservoir by a north-east–south-west trending fault. This fault had a downthrow to the south-east which varies from 370ft (113m) to over 500ft (152m). The Rotliegendes Sandstone thickness in this well is about 500ft (152m) so it is obviously not completely separated from the "A" reservoir although the gas/water contact is the same in both units. It had been suspected for some time that full separation of two adjacent sand formations by fault throw may not be necessary for a fault to be a sealing one. The Rotliegendes in the area is very susceptible to tight sections of greatly reduced porosity. These have probably been produced largely by secondary cementation and it is suspected that these zones occur frequently along fault planes so that the sealing propensity of a fault may well be larger than its presently seen throw would indicate.

The 49/17-5 well was re-opened in 1974 and pressure tested. The pressures were found to be those of the original measurements thus indicating that the area around the 17-5 well (now designated the "H" reservoir) is not in fact linked to the "A" reservoir. The portion of the structure drained by this well is calculated to contain substantial reserves.

Continuous velocity analysis of most of the seismic lines over the structure enabled a satisfactory velocity gradient to be calculated, after much trial and error, and the seismically calculated depth picture is now in good accord with the well results.

"F" Reservoir

When a closed structure was interpreted in the central graben it was assumed that at least 250ft (76m) of sand would be established above the —9680ft (—2950m) water level although doubts were expressed about the viability of the reservoir on the downside of a major fault.

When the 49/12-4 well was drilled in June 1973, the top 200ft (61m) of Rotliegendes Sandstone was found to be tight, as had been feared. The gas/water contact however was not encountered at —9680ft (—2950m) and was perhaps not really established in the 560ft (171m) of Rotliegendes penetrated. A water flow was obtained at the top of

the Carboniferous at —10 140ft (—3091m) but whether this is a genuine gas/water level or whether the water exists as a discrete layer between the Permian and Carboniferous formations is far from clear. The next well to be drilled in the graben, 49/12-5, also established a full gas-column and encountered water again at the base of the Permian at the same depth. The graben has now been designated the "F" reservoir.

The south-western flank of the anticline is yet undrilled but has obviously become a prime prospect. It has closing contours from —9600ft (—2926m) down to —10 200ft (3109m) and is separated from the graben by a fault which has a throw to the north-east of only 150ft (46m) in places. A gas/water contact very similar to the "F" reservoir is thus a reasonable expectation.

THREE-DIMENSIONAL ISOMETRICS

In an effort to model the North Viking reservoir effectively, a technique has been developed to plot on a two-dimensional plane a three-dimensional illustration of any selected formation or group of formations. This is accomplished by a computer plot of existing seismic sections each given an identical angular offset from the horizontal so that a series of them can be viewed simultaneously. They are, however, plotted as time-isometrics and the vertical time scale has been preserved so that measurements can be made from any section in the display just as from a normal format.

When these isometric diagrams are presented in colour with amplitude modulations, similar patterns of amplitude variation can be seen to be repeated on successive sections and also to be present where lines intersect. This indicates that these amplitude variations are not random and, although their true meaning is frequently not clear, especially in such a complex evaporite area as the southern gas-fields, a logical explanation for them must exist.

In the two examples shown here the models use the top of the Zechstein as a datum and extend downwards into the Carboniferous. The top of the Rotliegendes reservoir and the major faults affecting it have been outlined in the simpler of the two models (Fig 3). In the other model (Fig 4) more strike-lines have been added, thus aiding the definition of the fault-planes and lithological boundaries. Both models in this case are viewed from the south so that the structure is viewed from the same aspect as on the contour map in Fig 2.

The models can be simply prepared for viewing from any direction, in 30° steps, thus permitting an inspection of the reservoir from a wide range of viewpoints. As any formation can be examined in this manner this technique obviously will have a wide application in the building of detailed reservoir models from existing seismic sections.

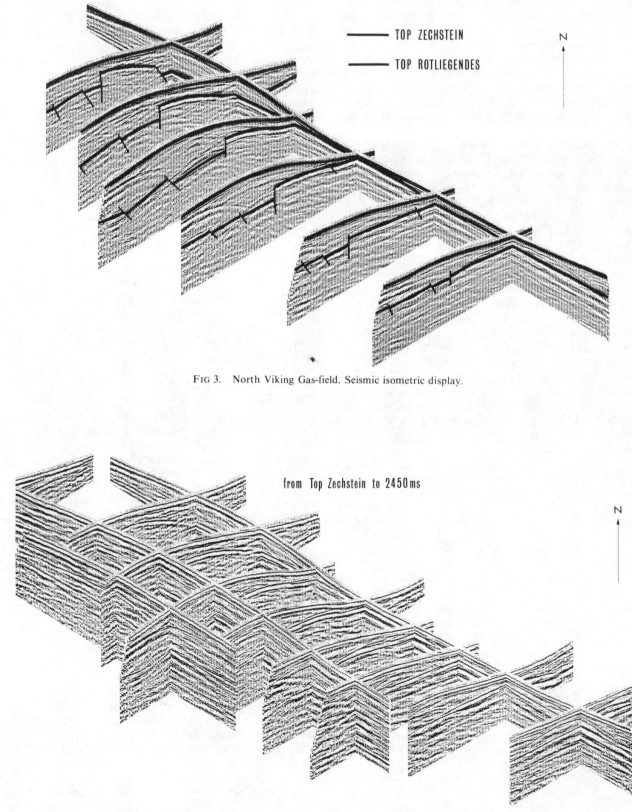

TOP ZECHSTEIN

TOP ROTLIEGENDES

N

Fɪɢ 3.   North Viking Gas-field. Seismic isometric display.

from Top Zechstein to 2450 ms

N

Fɪɢ 4.   North Viking Gas-field. Seismic isometric display from top of Zechstein to 2450m.

CONCLUSIONS

Six different gas/water contacts have been established in the Viking gas-fields varying from —9060ft (—2761m) to —10 140ft (—3091m), separated in some cases by faults of varying throw and in other cases separated by no physical discontinuity which can be discerned from the seismic data. The size of a fault needed to provide a seal has not been determined.

It would be a brave or foolish interpreter who would condemn as a future prospect any structure, of adequate area, in the Vikings region which can be demonstrated to exist as a fault-bounded closure.

ACKNOWLEDGEMENTS

The author would like to thank those friends and colleagues whose work and efforts in the gas field have made this paper possible. In particular I would like to thank Nigel Anstey of Seiscom Ltd for his help and advice and for producing the isometric displays illustrated in the above.

18

Zechstein of the English Sector of the Southern North Sea Basin

By J. C. M. TAYLOR

(V. C. Illing and Partners for British Petroleum Company Limited)

and V. S. COLTER

(British Gas Corporation)

SUMMARY

Major carbonate and anhydrite accumulations represent the early stages in each of the first three Zechstein cycles under the southern North Sea and adjacent areas of Yorkshire and Lincolnshire. They form prograding wedges, each reaching its maximum thickness farther from the margin than its predecessor before thinning towards the centre of the basin. The succeeding halite stages of the second, third, and fourth cycles appear to have built inwards in like manner, filling of the central depression being delayed till the end of the second cycle or probably later. Zechstein carbonate and evaporite formations therefore exhibit both platform (including sabkha) and true basin facies. Oolitic dolomites characterize the early platform facies of the first two cycles, with reef development in the first; sheets of *Calcinema* stems typify the third. Towards the end of the first cycle thin carbonate/anhydrite laminites formed beneath more than 100m of water in the basin whilst thick deposits of purer anhydrite were forming in shallow or emergent conditions around the margins. The incomplete succession flanking the Mid North Sea High, implying erratic vertical movement and periodic emergence, suggests a basin barrier, but further restrictions probably existed to the north.

INTRODUCTION

The Zechstein sequence in the southern North Sea has been described in outline by Kent (1967 a and b) and, with special reference to the evaporites, by Brunstrom and Walmsley (1969). Smith and Pattison (1972) and Smith (1974) have provided summaries of the Permian in Durham and Yorkshire, whilst much useful subsurface information relating to eastern England and the North Sea is given by the papers on correlation and nomenclature by Smith *et al.* (1974) and Rhys (1974). The purpose of the present paper is to give further details of Zechstein evaporites and especially the carbonates penetrated in exploratory wells onshore and in the English sector of the southern North Sea, mainly since 1964. We hope that a better appreciation of the basin-wide geometry and facies changes in this complex suite will help to put the specialist contributions of earlier writers into truer perspective, and clarify possible misunderstandings about evaporite genesis. On a more immediately practical level we believe it may assist seismic interpretation.

Methods

Correlation and gross lithology were established from wireline logs. Detailed lithology was determined by examination of material from 100 wells; about 1000m of core, almost entirely through carbonate and anhydrite rocks, were photographed and studied. Ditch cuttings from the carbonate and anhydrite units, hand-picked where necessary, were embedded in plastic and sliced to give surfaces that could be etched and stained, revealing textural details otherwise overlooked. Lithological logs were prepared using a system which enabled mineralogy, crystallinity—and in the carbonate units grain-size, type, and packing—to be recorded semi-quantitatively. This detailed characterization proved invaluable for comparing and distinguishing these rocks, which are typically diagenetically altered and unfossiliferous. Intervals used on the isopach maps (Figs 3 and 4) were confirmed by interpretation of wireline logs from a further 130 wells.

The type of evidence available varies; generally speaking the older wells onshore suffered from more primitive wireline logs but had better quality cuttings than is now common. More extensive coring through the platform carbonates and anhydrites in north Yorkshire tends to be balanced in offshore gas fields in the south-east of the study area by closer well control, providing more convincing gamma ray correlation.

The succession (Fig 1) is described in terms of major evaporative cycles, which correspond with those of Germany (Richter Bernburg, 1959) and Holland (Brueren, 1959). Generalized isopach and lithofacies maps are provided for the carbonates and evaporites of each cycle, with an additional pair for the complex carbonates of the second cycle, and an extra map for the terminal minor fifth cycle (Figs 2 and 3).

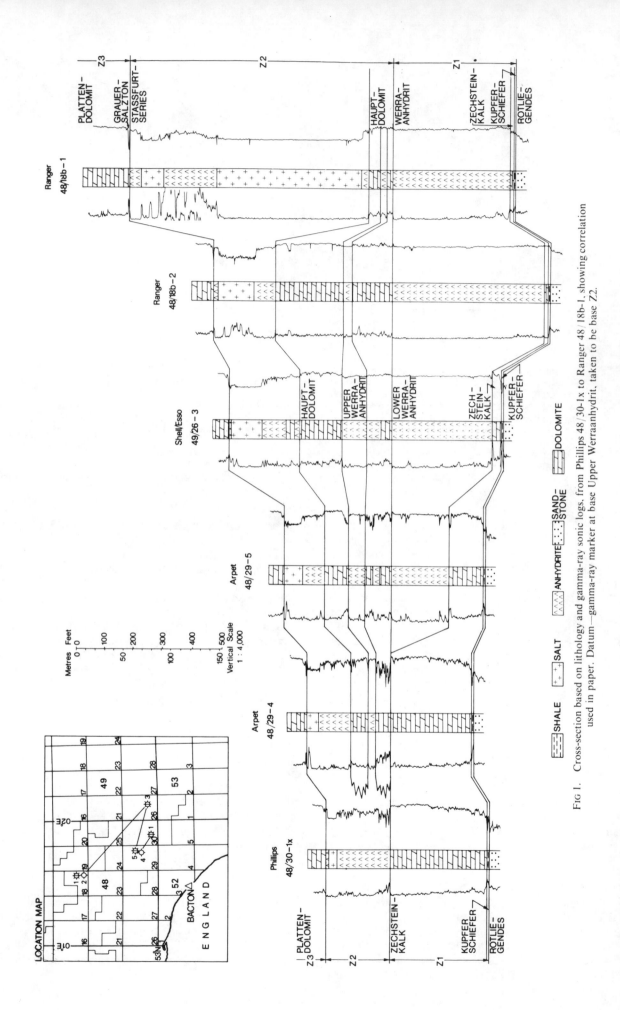

FIG 1. Cross-section based on lithology and gamma-ray sonic logs, from Phillips 48/30-1x to Ranger 48/18b-1, showing correlation used in paper. Datum—gamma-ray marker at base Upper Werraanhydrit, taken to be base Z2.

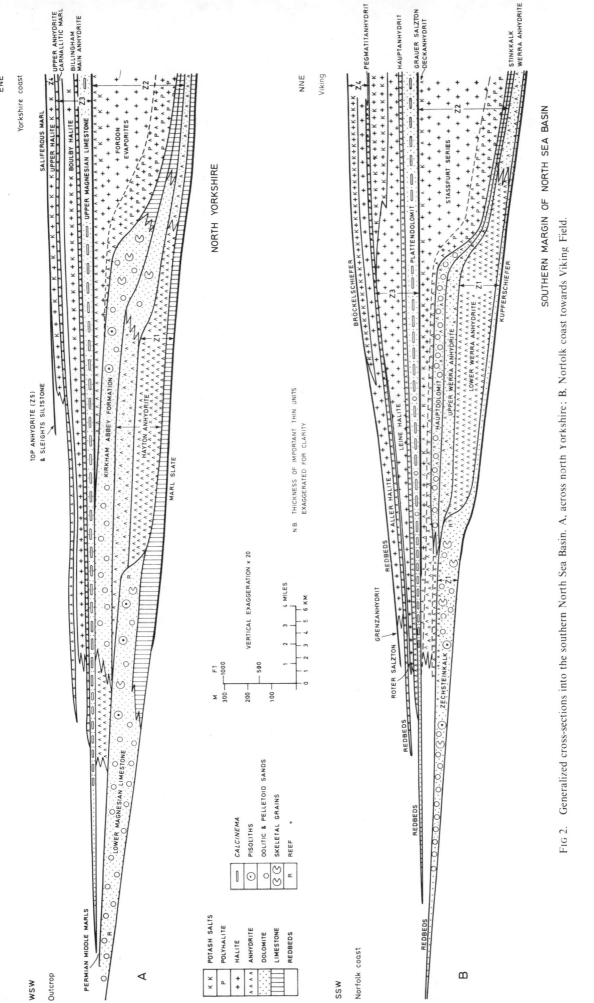

FIG 2. Generalized cross-sections into the southern North Sea Basin. A, across north Yorkshire; B, Norfolk coast towards Viking Field.

Nomenclature adheres so far as possible to the recommendations of Rhys (1974) and Smith *et al.* (1974). Two slightly generalized scale cross-sections, one into the North Sea in north Yorkshire and one from the Norfolk coast towards the Viking Field illustrate the time relations of the different facies and summarize the similarities and differences in these two areas (Fig 2).

FIRST CYCLE

Carbonates

Under most of the southern North Sea and north-east England, the Upper Permian Zechstein deposits commence with the thin sapropelic, illitic shale known in Durham as the Marl Slate, equivalent to the Kupferschiefer in Germany. Its high radioactivity provides an easily recognized log marker. It passes rapidly upwards into marine carbonates of the first cycle; these overlap it towards the edges of the basin, suggesting a further rise in relative sea-level after the initial transgression. The carbonates, corresponding to the Zechsteinkalk of Germany, are well known in outcrop as the Lower and Middle Magnesian Limestone of Durham and the Lower Magnesian Limestone in Yorkshire, Derbyshire and Nottinghamshire. Boreholes show that they thicken towards the basin to a maximum of about 120m before thinning, often abruptly, to between 15 and 30m, as shown on Fig 2A and B. They then thin progressively to as little as 3m over much of the basin floor.

The main body of the carbonates is composed of a monotonous series of greyish brown variably dolomitized lime muds, with occasional foraminifera and plant remains. Clay and pyrite are common near the base, but diminish in abundance upwards, whereas the proportion of dolomite increases. Under western Lincolnshire and again off the north-east Norfolk coast these beds pass laterally into an estuarine facies richer in clay, plant material and fossils (especially *Lingula credneri*) resembling the Permian Lower Marl at outcrop in Nottinghamshire.

The dolomitized lime muds pass upwards in most areas without a sharp break into a shallower facies averaging 30m thick. This consists of two main sub-facies: carbonate sands (predominantly oolitic and dolomitized) and reefs. This 30m-group, nearly always porous and permeable, extends over a belt more than 30km wide that surrounds the basin (Fig 3A and B). It seems likely that it is built of two or more diachronous lenses, represented by the upper and lower subdivisions of the Lower Magnesian Limestone at outcrop.

The sands are composed of a variety of carbonate grains. The commonest are fine to very fine and almost perfectly spherical; they rarely show much internal structure, but occasional concentric laminae suggest an oolitic origin.

Larger, distinctly laminated, grains occur locally. Though sometimes spherical, these tend towards irregular shapes. They are often flattened in contact with other grains as though flaccid at the time of burial. Very similar varieties are illustrated by Kerkmann (1964) and Füchtbauer (1968) from the first Zechstein cycle of Germany; they are interpreted as algal ooliths, pisoliths and oncoliths. Shells and bioclastic debris also occur locally. Centres of all types of grain were often leached during dolomitization. The sands thin under Lincolnshire against the London–Brabant Platform, passing into quartz sands and conglomerates cemented by coarse dolomite crystals. At many localities away from outcrop the top of the carbonate consists of finely oolitic sand cemented and partly replaced by anhydrite in what we interpret as a regressive sabkha sequence.

Study of outcropping patch reefs (algal and bryozoan mud-mounds) in the lower subdivision of the Lower Magnesian Limestone has shown that skeletal debris is micritized, abraded, rounded and oolitically coated until indistinguishable when dolomitized from the other pelletoid grains, all within a few kilometres of its source. Consequently the recognition of abundant fresh or partially altered skeletal material in cores or cuttings suggests close proximity to an area of prolific organic activity. In particular, intact bryozoan stems (Plate I, L) and carbonate mud bound by algal tubes are strong indicators of nearby reef construction. Such indicators have been noted in a number of wells, as shown on Fig 3B. Whereas at outcrop in Yorkshire, Derbyshire and Nottinghamshire reefs are restricted to the Lower Subdivision of the Lower Magnesian Limestone, as they are traced basinwards in the subsurface they occur progressively higher in the sequence, reaching the top of the Lower Magnesian Limestone where it is thickest. Wireline log correlation in areas of close control in the southern North Sea, for example between Arpet wells 48/29-4 and 5 (Fig 1) shows that the thick carbonates give way abruptly to the Werraanhydrit basinwards. The relationship appears to be similar to that between the Middle Magnesian Limestone and the Hartlepool Anhydrite in Durham, and we speculate that a reef scarp may extend round the basin.

The 3m-thick basin carbonates are dark, pyritic, very argillaceous (especially at the top and base) and sometimes silty or sandy. In several wells the uppermost argillaceous unit contains distinctive pisoliths up to 5mm across (Plate I, K) accompanied by a marine fauna of foraminifera and sometimes crinoids. The pisoliths appear to be algal, and if autochthonous set a depth limit to the basin water where they occur of perhaps 100–150m. A turbidite origin cannot be ruled out, but none of the appropriate sedimentary structures have been observed in cores. On the Mid North Sea High up to 3m of argillaceous dolomite has been

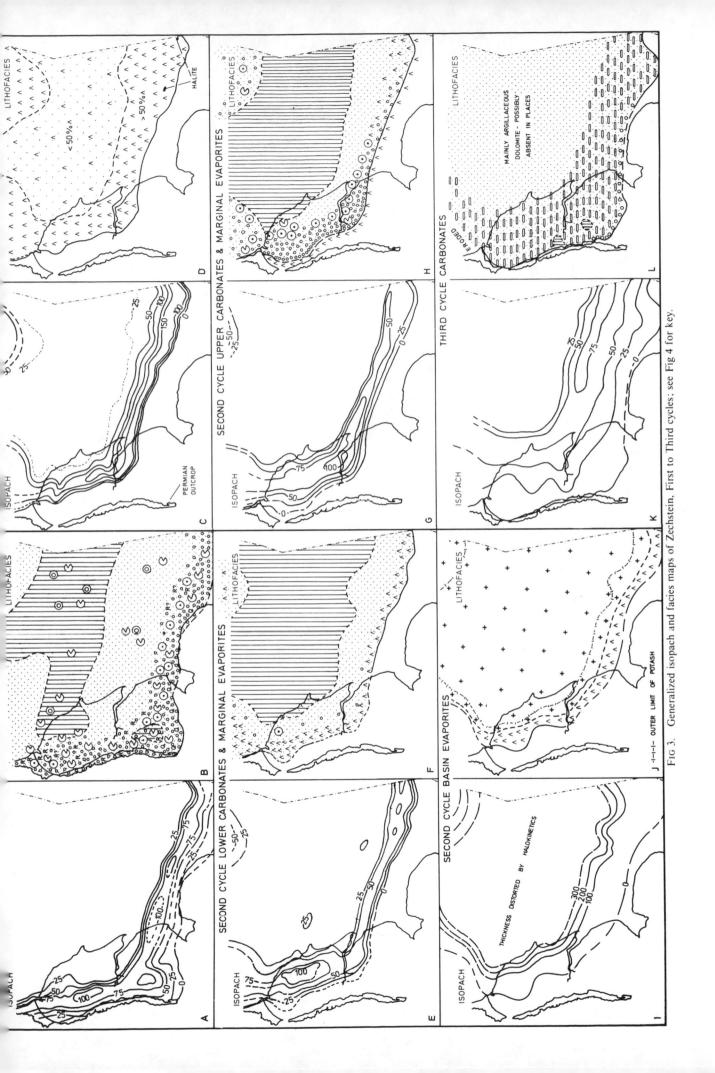

FIG 3. Generalized isopach and facies maps of Zechstein, First to Third cycles; see Fig 4 for key.

recorded, but in some wells both carbonate and Kupferschiefer are missing.

Evaporites

A sharp break separates the basin carbonates from the overlying Werraanhydrit or Hayton Anhydrite (Plate II, F), evidently representing a sudden increase in salinity. This anhydrite, contained for the most part within the encircling lens of the Lower Magnesian Limestone or Zechsteinkalk, thickens basinwards for 15 to 25km to form a nearly flat platform up to 200m thick before declining to typical basin-floor values of 20m (Fig 2). For reasons given later, we suggest that only the lower part of the thick platform succession should be included in the first cycle.

The thin basin-succession is finely and evenly laminated and consists of roughly equal proportions of anhydrite and dark grey-brown carbonate, the latter including microcrystalline dolomite and very coarsely crystalline calcite. The series has many of the features described by Shearman and Fuller (1969) in the laminites of the Middle Devonian Winnipegosis Formation of Saskatchewan. Four sub-cycles can usually be recognized; anhydrite maxima consisting of disruptive nodules (Plates I, J; II, D, E) and enterolithic bands pass gradually up into laminated carbonates, which tend to give way abruptly to the next anhydrite-rich layer. Laminae of sub-equant microcrystalline anhydrite, apparently primary, are common in the fourth sub-cycle. This, too, passes transitionally but rapidly into the argillaceous laminated carbonates of the second Zechstein cycle. The repeated gradational upward decrease in calcium sulphate is the reverse of observations in modern sabkhas.

The thicker platform succession contains a greater proportion of anhydrite as nodules, enterolithic bands, large cumulous masses and replacement crystals associated with varying amounts of brownish micro-crystalline dolomite (Plates I, I; II, F). The fine flat lamination of the basin is absent; instead the dolomite shows irregular dark laminae of algal aspect. Apart from a vaguely pelleted fabric in places, neither the Hayton Anhydrite nor the Werraanhydrit shows evidence of fauna, in striking contrast with the preceding carbonates. We suggest that the anhydrite is a shallow-water to subaerial sabkha deposit that rimmed the standing body of saline water from which the thinner basin equivalents were forming, presumably 150m deep at the foot of the platform and more at the basin centre. A number of anhydrite intervals are present in places on the Mid North Sea High, but their correlation is doubtful.

The only known halite in the first cycle occurs in the southern North Sea at the top of the Lower Werraanhydrit platform. A 13-m band is shown on the log of Shell/Esso 49/26-4 (Rhys, 1974). It probably formed in a marginal sabkha or salt pan.

SECOND CYCLE

Carbonates

Across the central basin-floor eastwards from the Yorkshire coast, the thin Werraanhydrit passes rapidly upwards into dark, bituminous laminated carbonates (Stinkdolomit or Stink-kalk), presumably as a result of the introduction of less concentrated water. The spacing and density of clay-rich laminae, evidently related to rates of carbonate deposition, produce a characteristic gamma ray signature. When traced shorewards into the thickened platform facies, this compound peak appears to run along the base of the Upper Werraanhydrit rather than over it (Figs 1, 2B). The platform Upper Werra is therefore the time equivalent of the basin carbonates. Roll (1969, Fig 5) illustrates a similar relationship between the Werraanhydrit and the Hauptdolomit in north-west Germany. If the boundary between Z1 and Z2 is defined as the top of the anhydrite in the basin it will be seen that the Upper Werra platform facies must be regarded as Z2, or vice versa.

In the Hewett field the marker can be traced over the top of the first cycle carbonates (Figs 1, 2B) passing eventually into shales of the bordering continental clastics. Correlation is less secure in the west; under central Yorkshire (Market Weighton axis) the marker probably terminates against the basinward edge (reef?) of the Lower Magnesian Limestone, but in north Yorkshire and south Durham it may again lie above it.

The Upper Werraanhydrit (upper part of the Hayton Anhydrite or lower unit of the Z2 carbonates) resembles the underlying platform facies regarding the predominantly displacive and partly replacive fabrics of its anhydrite. The proportion of dolomite, however, is greater. The crinkly laminations are more strongly reminiscent of modern strand-line algal mats, and at Egton Moor in Yorkshire bands of algal oncoliths have been found, resembling those in the succeeding dolomitic Kirkham Abbey Formation (Plate I, E, F). Alternating well-aerated shallow hypersaline and emergent sabkha conditions are implied for this unit. Its facies boundaries (Fig 3E, F) are not strictly parallel with its isopachs; areas richer in anhydrite project basinwards further in some places than others, presumably in response to local shoaling.

Across the platform provided by the Upper Werra, the Kirkham Abbey Formation or Hauptdolomit consists of 40–100m of shallow water to intertidal dolomite sands (Plate II, B). They interdigitate landwards with sabkha anhydrites and eventually the continental mudstones (Fig 2A, B) of the Middle Marls. Grains are typically very fine ooliths and pelletoids with leached centres (Plate I, D); larger pisoliths (Plate I, E) and algal sheets are common near the break in slope into the basin. Shells are rare, and no reefs are yet known. Along the western rim of the basin and in places on the Mid North Sea High the sands are porous, but from the Humber eastwards along the southern margin plugging by anhydrite (void-filling and replacive) and halite becomes prevalent (Plate I, C).

Basinwards of the platform edge the facies changes first to dolomitized lime muds, often pelleted and burrowed, with ostracodes, foraminifera and occasional lamellibranchs. On the basin floor the equivalent sequence is barely distinguishable from that below—except for a gamma ray baseline shift—being dark brown, foetid, barren, with close, even, flat laminations (Plate I, H). Limestone formed of a mosaic of crystals up to 5mm across predominates over microcrystalline dolomite.

An interesting transition zone occurs where this basin facies meets the surrounding dolomite muds. Rounded concretions of large calcite crystals grew during the onset of dolomitization and before compaction, finally merging and forming continuous limestone bands (Plates I, G; II, G).

Evaporites

The succeeding evaporites are the first to reach their thickest development in the *centre* of the basin. Here, a mixed zone consisting mainly of anhydrite, polyhalite and halite makes up the lowest 50–100m, and is followed by 300–600m or more of halite. Accurate figures for its original thickness cannot be determined, as this salt appears to have been the main source of Zechstein halokinetics. The same mobility frustrates attempts to explain sporadic intervals of anhydrite, potassium and magnesium minerals.

Conditions of deposition are difficult to establish because of lack of cores, but some clues are provided by the geometry of the marginal deposits. At the edges of the salt basin the first evaporite to be encountered is anhydrite. In the south this develops from the peripheral red beds (presumably a sabkha facies) and is difficult to separate from the underlying landward equivalents of the Hauptdolomit (Fig 2B). In the west, anhydrite deposits lap against the probably emergent Kirkham Abbey Formation (Fig 2A); cores from the nearby basin slope show a bedded rather than nodular structure. In both areas, as the thickness of the evaporites increases basinward, halite is found first in a single unit near the top of the section, and then as a series of intercalations soon accompanied by polyhalite (Plate I, B) until, when the evaporites have reached a thickness of about 300m, the complex series described by Stewart (1963) develops. Correlation between recent wells suggests that the 146m-section with halite, polyhalite and anhydrite recorded by Stewart (1963, pl 4) from 1968 to 2114m in Fordon No. 1 (and a similar thickness in other wells along the strike) thins basinwards to about 100m. At the same time the anhydrite from 1841 to 1851m at Fordon drops in the section basinwards, whilst

additional salt beds appear between it and the overlying Upper Magnesian Limestone. Thus it appears that the foresetting established in the previous Zechstein carbonate and anhydrite units may continue in the Z2 evaporites.

No halite or potash is known on the Mid North Sea High.

A widespread bed of carnallite or sylvite up to 10m thick occurs 10m from the top of the Z2 halite. The evaporites are capped almost everywhere by about 1m of anhydrite (Deckanhydrit). Where seen in cores (Plate II, C) it consists of spindle-shaped microcrystalline anhydrite crystals aligned with the bedding, possibly a sedimented solution-residue from the underlying halite, formed during the third cycle transgression.

THIRD CYCLE

Carbonates

The Upper Magnesian Limestone (Plattendolomit) has been extensively cored under Yorkshire and the North Sea. It possesses greater lateral uniformity than the carbonates of earlier cycles. The dark grey basal shale (equivalent to the Grauer Salzton) about a metre thick, is illitic with subsidiary chlorite and kaolinite, and forms a prominent gamma ray peak. It fails to reach as far west under eastern England as the Kupferschiefer; halfway between outcrop and the present coast higher beds of the Upper Magnesian Limestone overlap it directly onto the Kirkham Abbey Formation.

Near the southern margin of the formation, under Lincolnshire, a coarsely crystalline dolomitized oolite and quartz sand facies is developed, similar to that at outcrop south of the Don Valley.

An illuminating core from Home Oil's Lockton No. 3 in Yorkshire shows a rapid transition by interlamination from the Deckanhydrit of Z2 into the Grauer Salzton, and upwards into shaly dolomite. The break or breaks representing the true base of the third cycle occur a few metres lower, where contemporaneous solution of halite has produced weird structures (Plate II, C).

The Plattendolomit appears to have accumulated in decreasing depth of water. The dark shaly basal beds grade up into lighter grey-brown dolomite, though rounded concretions and bands of limestone sometimes occur in the middle and lower parts—an interesting and sometimes confusing parallel with Z2. The first fauna of stunted thin-shelled lamellibranchs, sometimes in bands crowded with individuals, appears 6–8m above the base. The characteristic tubular calcareous alga *Calcinema permiana* comes in a little higher, increases in abundance upwards, and continues to the top of the formation. It tends to occur in layers a few millimetres to a few centimetres thick consisting almost entirely of

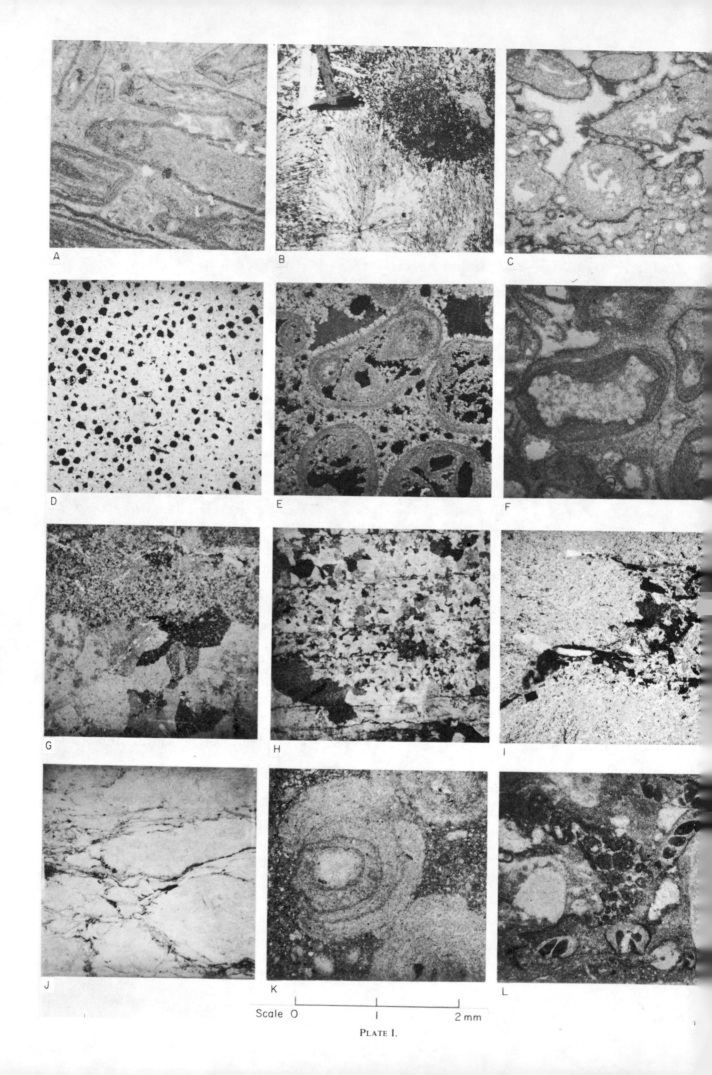

Scale O   I   2 mm

PLATE I.

PLATE II.

horizontal stems (Plate I, A), alternating with thicker layers of microdolomite with scattered fragments. Near the top current-bedded oolites often occur, and dark irregular laminations resembling blue-green algal mats are common. In Yorkshire and off the Norfolk coast (Plate II, A) one or more bands of nodular anhydrite are interspersed with the uppermost dolomites, resembling modern sabkha deposits and suggesting that the formation had built to sea level. The uppermost beds are shaly in the south (see Shell Esso 49/26-4 in Rhys, 1974), but in places replacement of carbonate by anhydrite confuses the boundary between the Plattendolomit and the overlying Hauptanhydrit (Billingham Main Anhydrite in Durham and East Yorkshire).

Calcinema and other fossils die out basinwards (Fig 3L). About half way to the centre, where the formation is at its thickest (90m), it consists of barren, tight, dark brown laminated micro-dolomite with abundant darker laths of replacement anhydrite. One or more thick anhydrite units are recorded on logs of some wells; cores from BP 48/6-1 have shown this to be a grey, dolomitic nodular anhydrite, strikingly similar to the Werra. Nearer still to the basin centre a thinner anhydritic dolomite is recorded by some companies as the Hauptanhydrit or un-differentiated Plattendolomit/Hauptanhydrit, and in some central wells no carbonate unit is detectable. Interpretation is difficult since halokinetics is known to distort, repeat, invert and even eliminate the Plattendolomit in places. We surmise that the anhydritic beds bear the same relation to the surrounding Plattendolomit lens that the Werra anhydrite does to the Z1 carbonate lens—that it is younger and was formed from more concentrated brines. There is insufficient evidence to show whether it was contemporaneous with the Hauptanhydrit of the surrounding shelf area, or whether the whole of the Plattendolomit represents more than one sub-cycle, divided by largely unrecognizable unconformities. Our maps (Fig 3, K, L) adopt the simpler alternative.

Only erosional remnants of the Plattendolomit (Seaham Formation in Durham) remain north of the Hartlepool fault zone on shore, and it is problematical whether it occurs on the Mid North Sea High.

Evaporites

The succeeding evaporites (Fig 4A and B) have been extensively cored in north-east Yorkshire and detailed accounts are available in the literature reviewed by Smith (1971 and 1973), who has revised the stratigraphy of the former Middle and Upper Evaporites, introducing new names and lettered subzones. Very shallow to emergent conditions are deduced for the entire sequence. Practically no cores have been cut in the North Sea. Wireline logs show that the Boulby Halite (probably equivalent to both the Riedel and

Ronnenberg Steinsalz) thickens from 30m near its margin towards the centre of the basin where halokinetics confuse the picture; an original value of between 100 and 150m seems probable.

Near the Yorkshire coast and off North Norfolk the potash zone (Smith's sub-zone "C") occupies 5–15m near the top of the Boulby Halite, separated from the overlying Carnallitic Marl by up to 3m of clean halite (sub-zone "D"). But over a large area in the centre of the basin the upper part of the third cycle is represented by a thick complex sequence of halite, potash salts (probably carnallite and polyhalite), and possibly mudstone. Accurate identification is difficult. This zone coincides with the area of maximum Zechstein salt thickness and encompasses the principal salt structures; problems of cause and effect arise.

A red salty clay or mudstone (poorly represented in ditch cuttings) occupying the position of the Carnallitic Marl of East Yorkshire between the third and fourth evaporite cycles, can be followed southwards into Lincolnshire and eastwards off Norfolk to the median line. Cores show a rapid transition from the underlying halite. This member appears to correspond with the Roter Salzton of Germany (Smith, 1970), but if this correlation is adopted it must be recognized that the Anhydritmittelzone at the base of the Riedel Group in Germany appears to have no counterpart in the English side of the basin (Brunstrom and Walmsley, 1969, 1970). Log correlation across the basin is difficult, for with increasing distance from the margin the clay becomes less easy to recognize, probably becoming saltier and losing its clay content, and its identity is finally lost in the thick series of impure potassic salt beds penetrated near the centre of the basin.

FOURTH AND FIFTH CYCLES: TOP OF THE ZECHSTEIN

Fourth Cycle

The fourth cycle (Fig 4C, D) is a subdued edition of the third. Its thin basal carbonate in Yorkshire and Teesside, the Upgang Formation (Smith, 1971) is not detectable under the North Sea, but the Upper Anhydrite (Pegmatitanhydrit) can be recognized over a wide area, especially round the margins of the basin. Towards the centre its identity becomes confused by intercalations of polyhalite and halite.

The Upper (Aller) Halite occupies a smaller area, first appearing where the distance between the Top Anhydrite and Upper Anhydrite is about 20m. Smith's almost pure "B" unit is continuous across the basin with little change in thickness (c20m), but his "C" and "D" units are more variable. Potash beds (mainly sylvite and carnallite) accompanied by red mudstone are developed at successively higher horizons into the basin, finally reaching the top of a greatly thickened "D" unit. This is

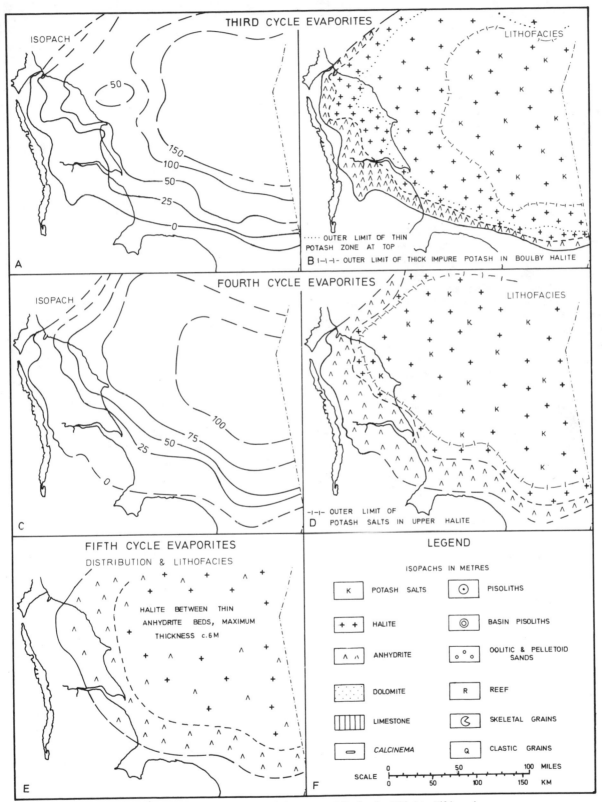

FIG 4.   Generalized isopach and facies maps of Zechstein, Third to Fifth cycles.

generally succeeded by a thin relatively pure halite (unit "E") which reaches a maximum thickness of about 7m under the North Sea, but passes laterally into anhydrite at its edges.

Fifth Cycle

The Top Anhydrite, separated under Teesside and the Whitby area by the red Sleights Siltstone, is regarded by Smith (1970) as a rudimentary fifth cycle, correlatable with the Grenzanhydrit of Germany. The mudstone/anhydrite couplet forms an easily recognized log marker which can be followed south across the Humber into Lincolnshire. Eastwards off the Norfolk coast it is liable to be confused with other thin anhydrite bands.

Cursory examination of wells from the centre of the basin gives the impression that the Top Anhydrite and Sleights Siltstone have disappeared and that the Bröckelschiefer or Upper Permian Marl rests directly on halite of the fourth cycle. However, when sequences disturbed by halokinetics are discarded, careful correlation suggests that the Top Anhydrite may swell basinward and split to embrace a thin halite member. As this thickens to its maximum of about 6m the anhydrites diminish to less than a metre each, close to the limit of wireline log resolution. The underlying Sleights Mudstone becomes salty at the same time; carnallite has been detected on the logs of some wells at this level.

The overlying Bröckelschiefer (Geiger and Hopping, 1968) in its typical dark red silty and slightly dolomitic facies is roughly co-extensive with the Top Anhydrite; outside that limit thinning and coarsening make it indistinguishable from other Permo-Triassic red beds. It contains thin anhydrite bands but no halite, and its base forms a convenient upper limit to the Zechstein.

CONCLUSIONS

The Zechstein ranks amongst the world's major evaporite basins. North Sea drilling has confirmed the link between successions in England and Germany, and shown close similarities in their lithologies and in the geometric relations between units. Perhaps more remarkable are the parallels which could be drawn with the Permian of the Delaware Basin, the Devonian of north-west Canada, and the Silurian of the Michigan Basin, particularly in the calcitic and anhydritic basin laminites. Whether these deposits formed under deep, shallow or emergent conditions is controversial; evidence from the North Sea Basin can contribute towards a solution of such problems.

Conditions of Deposition

The form and facies relationships of the units suggest the following adaptation of the classical barred basin model. The Southern North Sea Basin, one of a series of inter-communicating depressions which developed in the post-Variscan desert landscape, began subsiding during the Lower Permian, and may already have been a hundred metres below ocean level when sea water was admitted at the beginning of the Zechstein. In this nearly tideless, almost current-free inland sea, carbonate (and later evaporite) sedimentation was concentrated near the areas of most prolific production, namely the shallow, warm, aerated margins. This resulted in geologically rapid progradation of each unit inwards from the sides of the basin until the rate of infill was no longer able to exceed the rate of subsidence. From this critical zone the sediments thin landwards because their thickness was restricted by sea-level. They thin into the basin because the rate of sedimentation decreased with increasing water depth. But they extend circumferentially around the basin with more uniform thickness, because this thickness was controlled primarily by the rate of sediment production, which depended mainly on regional factors such as temperature and salinity.

With continuing subsidence, water depths in the centre of the basin increased with time except when communication with the ocean was restricted either by a eustatic drop in level or by the movement of intervening barriers. At each stage, marginal complexes of shoal, sabkha, salina, or salt-pan tended to produce minerals of somewhat higher evaporite grade and in greater profusion than was appropriate to the basin water concentration.

The incomplete succession on the Mid North Sea High suggests that it formed an intermittent barrier, but the presence of evaporites in the Northern North Sea Basin indicates that further restrictions existed to the north. The first major interruption raised salinities enough to precipitate anhydrite transiently in the basin, copiously around the margins. Despite the sabkha-like nodular and enterolithic anhydrite of the basin laminites—and evidence for some shallowing—the marginal thickening of the Werra downgrades the possibility that the deposits near the centre of the basin could have formed under emergent conditions. This dilemma can be resolved if magnesium-rich brines resulting from the evaporative precipitation of calcium sulphate round the basin margins sank to the bottom and dolomitized lime muds on the floor according to the well-known equation:

$$2CaCO_3 + Mg^{2+} + SO_4^{2-} \longrightarrow CaMg(CO_3)_2 + CaSO_4$$

The relative proportions of dolomite and anhydrite appear to accord with this equation which results in a net volume increase of 50–60 percent, probably sufficient to account for the displacement structures of the anhydrite.

Dilution by marine waters accompanied the transgressions of the Upper Werra and the

Hauptdolomit, but we suspect that the dense stagnant brines below effective sill depth remained there, influencing basin chemistry until much later. They may be a factor in the typical early diagenetic coarse calcite of the basin; if so, perhaps the concretionary facies of the slope deposits of the second cycle marks the pycnocline.

The second and subsequent major restrictions to circulation continued to the point at which halite was freely precipitated. If, however, infilling advanced centripetally from the margins as in the previous carbonate and anhydrite phases, it is possible that the centre of the basin was not completely filled until quite late in the Zechstein, perhaps explaining anomalies in the central deposits of the third and fourth cycles.

Hydrocarbons

In view of the established gas and oil production from the Zechstein in Germany and the Netherlands, exploration in the English sector of the basin has been disappointing. Limited commercial gas production has been obtained at Eskdale and Lockton in Yorkshire but other finds have not yet proved worth exploiting.

Good porosity and permeability occur in carbonate sands and reefs of the first cycle except near the Z1 anhydrite, and to a lesser degree in the carbonate sands of the second cycle away from the southern side of the basin where evaporite plugging prevails. Failure to obtain production from these potential reservoirs must be attributed to other variables such as proximity to source beds, timing and amount of structural uplift, lack of cap rocks, or the subsequent erosion of overburden. Careful search may yet find areas where these factors are favourable.

ACKNOWLEDGEMENTS

This communication is based on studies carried out on behalf of British Petroleum Company Limited between 1964 and 1972, and also for the Gas Council between 1967 and 1970, by V. C. Illing & Partners. It also relies on log correlation carried out by the British Gas Corporation since 1970. We are indebted to the management of British Petroleum and the British Gas Corporation for permission to publish these results, and to over twenty oil companies whose cooperation assisted the study. We are particularly grateful to Burmah Oil (North Sea) Ltd, Hamilton Bros. Petroleum (UK) Ltd, Home Oil of Canada Ltd, Mobil Producing North Sea Ltd, and Total Oil Marine Ltd for permission to reproduce the photographs in Plates I and II; and to Arco Oil Producing Inc., Phillips Petroleum UK Ltd, Ranger Oil (Canada) Ltd, and Shell UK Exploration and Production Ltd for the use of the logs in Fig 1.

We gratefully acknowledge the help and support of colleagues past and present who have shared the labour of recording the details of which this paper is a synthesis.

EXPLANATION OF PLATES

Plate I (all at same magnification)

A. Evenly oriented tubular *Calcinema* stems in microcrystalline dolomite; white patches, replacement anhydrite. Upper Magnesian Limestone (Z3, Plattendolomit). Plane polarized light. BP Eskdale No. 12 (East Yorkshire) 4071ft (1240.8m).

B. Polyhalite (spherulitic) and anhydrite (laths) from faintly bedded sequence near base of Fordon evaporites (Z2, Stassfurt series). Crossed polars. Home Oil Lockton No. 5 (East Yorkshire) 5960ft 8in (1816.8m).

C. Dolomite oncoliths with partly leached centres and inter-grain porosity plugged by clear halite. Z2 carbonate (Hauptdolomit). Plane polarized light. Burmah 47/18-1 (Southern North Sea) 6972ft (2125.1m).

D. Typical very fine pelletoid sand preserved in microcrystalline dolomite, with leached centres. Kirkham Abbey Formation (Z2, Hauptdolomit equivalent). Crossed polars. Home Oil Ralph Cross No. 1 (Yorkshire) 3514ft (1071.1m).

E. Dolomitized algal pisoliths; leached centres and inter-grain pores partly filled by clear anhydrite, partly by single calcite crystals, partly unfilled. Kirkham Abbey Formation (Upper Z2 carbonate Hauptdolomit equivalent). Crossed polars. BP Egton Moor No. 1 (Yorkshire) 4261ft 3in (1298.8m).

F. Algal pisoliths from thin dolomite band in nodular anhydrite, upper part of Hayton Anhydrite—actually a sabkha equivalent of the lower Z2 carbonates. Note similarity with E. Void filling and replacement anhydrite after early fringing carbonate cement. Plane polarized light. BP Egton Moor No. 1 (Yorkshire) 4506ft (1373.4m).

G. Concretionary zone of Z2 carbonates (Kirkham Abbey Formation) at junction between basin and basin-slope facies. Coarse calcite crystals forming a concretion (lower half) replace lime mud rich in microfossil and bivalve debris. Surrounding lime mud in upper half replaced by microcrystalline dolomite. Crossed polars. Home Oil Lockton No. 6 (Yorkshire) 5765ft (1757m).

H. Closely laminated coarsely crystalline basin limestone from upper part of Z2 carbonate. Similar beds are common down to the bottom of the Z1 (Werra) "anhydrite" and across the entire basin floor. Note the fine, even nature of the primary laminations. These are often accentuated though distorted by micro-stylolites. Crossed polars. BP Robin Hood's Bay No. 1 (East Yorkshire) 4740ft (1444.8m).

I. Anhydrite nodules showing typical fabric. Platform facies of the Hayton Anhydrite. Crossed polars. BP Hayton No. 1 (Yorkshire) 3152ft (960.7m).

J. Anhydrite nodules from basin facies of Hayton (Werra) Anhydrite, surrounded by relics of the carbonate, clay and organic matter in which they grew by displacement. Plane polarized light. BP Robin Hood's Bay No. 1 5031ft (1533.4m).

K. "Quiet water" algal pisolith and fine skeletal debris in silty argillaceous basin limestone near top of Z1. Total 44/21-1 (Southern North Sea) 12652ft (3856.3m).

L. Dolomitized skeltal packstone from Z1 carbonate platform (Zechsteinkalk); with crinoid ossicle (right). Large fresh fragments of bryozoan stem like those in centre are seldom found far from reefs. Mobil 53/1-2 (Southern North Sea) 5546ft (1690.4m).

Plate II

Photographs of slabbed cores. All continuous sequences except F, and all to same scale.

A. Regressive sequence at top of Upper Magnesian Limestone (Plattendolomit), southern North Sea; 2036–2040.3m in Burmah 47/18-1. Dolomitized "lagoonal" lime muds with bands of Schizodus and occasional nodules of anhydrite in slabs (5) and (6). Algal mats becoming more important upward from (5). Dome-like stromatolite in (2). Bands of grey microcrystalline anhydrite and algal mats in (1).

B. Cross-bedded oolites near top of Hauptdolomit, southern North Sea; 2144–2146.4m in Burmah 47/18-1. Cream microcrystalline dolomite. Bedding picked out by dark replacement and void-filling anhydrite.

C. Passage from Fordon (Stassfurt) Evaporites upwards into Upper Magnesian Limestone (Plattendolomit), east Yorkshire; 1535.4–1551.7m in Home Oil Lockton No. 3. Coarse halite in slabs (7) and (12), alternating with white bands of anhydrite in (6) to (11), the latter consisting of anhydrite "ooliths" like those figured by Stewart (1963, pls 1 and 3) in Fordon No. 1. Complex zone from top of (11) through (9) resulting from dissolution. Anhydrite (Deckanhydrit) in (5) grades imperceptibly into dark grey shale (Grauersaltzton) which in (3) grades into dark, pyritic, shaly, flat-bedded micro-crystalline dolomite (Plattendolomit) (1 and 2).

D. Upper part of Werraanhydrit near centre of basin; 3382.0–3390.8m in Hamilton–BP 44/11-1. White microcrystalline anhydrite interlaminated with dark microcrystalline dolomite and finely crystalline calcite. Flat lamination, 2–3 per mm, except where distorted by growth of anhydrite nodules, particularly in slabs (6), (7) and (8). Each of these disturbances dies out rapidly upwards, showing that the nodules grew a short distance below the depositional surface.

E. Upper part of Werra near western edge of basin; 1584.0–1588.6m in Total A 339/1-2. Note similarity to "D" from 160km away, and contrast to "F" 30km further west. Enterolithic anhydrite bands in slabs (3), (4) and (5), small anhydrite pods in (4) and (5).

F. Selected slabs from platform facies of Hayton Anhydrite (Lower Werraanhydrit), Yorkshire, resting on basin-edge Lower Magnesian Limestone (Zechsteinkalk), all from Home Oil Rosedale No. 1. Slabs (1) to (5) show a variety of distinct and diffuse nodules of microcrystalline to finely crystalline anhydrite (light grey) in disrupted bands of darker microcrystalline dolomite. In (6) note sharp contact with dark underlying dolomite, which has shrinkage cracks filled with sandy argillaceous dolomite. Sharp-edged white anhydrite nodules in dolomite (7) die out downwards. Wavy bedding in limestone containing foraminifera beneath (8) typical of first cycle. Top of slab (1) c. 1217.1m, (2) 1245.4m, (3) 1248.2m, slabs (4) to (6) 1251.1–1253.6m; top of (7) 1254.2m, (8) 1255.7m.

G. Concretionary zone in Kirkham Abbey Formation (Hauptdolomit); part of the transition between dolomitized flat-bedded, fossiliferous and bioturbated lime muds of the basin slope—slabs (1) and (2)—to dark coarsely crystalline early diagenetic calcite similar to that of basin limestones in (2) to (5). The light patches at bottom of (2) and (3) are salt encrustations, not anhydrite. Home Oil Lockton No. 5, 1853.8–1858.4m.

REFERENCES

Brueren, J. W. R. 1959. The stratigraphy of the Upper Permian "Zechstein" Formation in the eastern Netherlands. I Giacimenti Gassiferi dell'Europa Occidentale; *Atti del Convegno di Milano 1957*, **1**, 243–74.

Brunstrom, R. G. W. and Walmsley, P. J. 1969. Permian evaporites in North Sea basin. *Bull. Am. Ass. Petrol. Geol.*, **53**, 870–83.

Brunstrom, R. G. W. and Walmsley, P. J. 1970. Permian evaporites in North Sea basin. reply *Bull. Am. Ass. Petrol. Geol.*, **54**, 664.

Füchtbauer, H. 1968. Carbonate sedimentation and subsidence in the Zechstein Basin (northern Germany). *In: Muller, G. and Friedman, G. M. Recent developments in carbonate sedimentology in Central Europe.* Springer Verlag, Berlin, 196–204.

Geiger, M. E. and Hopping, C. A. 1968. Triassic stratigraphy of the southern North Sea basin. *Phil. Trans. R. Soc. Lond.*, *Series B*, **254**, 1–36.

Kent, P. E. 1967a. Outline geology of the southern North Sea basin. *Proc. Yorks. geol. Soc.*, **36**, 1–22.

Kent, P. E. 1967b. Progress of exploration in North Sea. *Bull. Am. Ass. Petrol. Geol.*, **51**, 731–41.

Kent, P. E. and Walmsley, P. J. 1970. North Sea Progress. *Bull. Am. Ass. Petrol. Geol.*, **54**, 168–81.

Kerkmann, K. 1964. Zur Kenntnis der Rittbildungen in der Werraserie des thüringischen Zechsteins. *Freiberger Forschhft*, Sonderdruck aus Heft C., 213.

Rhys, G. H. 1974. A proposed standard lithostratigraphic nomenclature for the Southern North Sea and an outline structural nomenclature for the whole of the (UK) North Sea. A report of the joint Oil Industry/Institute of Geological Sciences Committee on North Sea Nomenclature. *Rep. Inst. Geol. Sci.*, **74/8**, 14 pp.

Richter Bernburg, G. 1959. Zür Palaeogeographie der Zechstein. I Giacimenti Gassiferi dell'Europa Occidentale *Atti del Convegno di Milano 1957*, **1**, 88–99.

Roll, A. 1969. Recent development in German exploration for oil and gas. In: *Hepple, P. (Ed.), The exploration for petroleum in Europe and North Africa.* Institute of Petroleum; Elsevier, Amsterdam, 221–30.

Shearman, D. J. and Fuller, J. G. C. M. 1969. Anhydrite diagenesis, calcitization, and organic laminites, Winnipegosis Formation, Middle Devonian, Saskatchewan. *Can. Jl. Petrol. Geol.*, **17**, 496–525.

Smith, D. B. 1970. Permian evaporites in North Sea Basin: Discussion. *Bull. Am. Ass. Petrol. Geol.*, **54**, 662–4.

Smith, D. B. 1971. Possible displacive halite in the Permian Upper Evaporite group of northeast Yorkshire. *Sedimentology*, **17**, 221–32.

Smith, D. B. 1973. The origin of the Permian Middle and Upper Potash deposits of Yorkshire, England: an alternative hypothesis. *Proc. Yorks. geol. Soc.*, **39**, 327–46.

Smith, D. B. 1974. Permian. In: D. H. Rayner and J. E. Hemmingway (Eds.), The Geology and Mineral Resources of Yorkshire. *Yorks. geol. Soc.*, 115–43.

Smith, D. B. and Pattison, J. 1972. Permian and Trias. *In:* G. Hickling (Ed.), The Geology of Durham County. *Trans. nat. Hist. Soc. Northumb.*, **41**, 66–91.

Smith, D. B., Brunstrom, R. G. W., Manning, P. I., Simpson, S. and Shotton, F. W. 1974. A Correlation of Permian rocks in the British Isles. *Jl. geol. Soc. Lond.*, **130**, 1–45.

Stewart, F. H. 1963. The Permian Lower Evaporites of Fordon in Yorkshire. *Proc. Yorks. geol. Soc.*, **34**, 1–44.

19

Undershooting Salt-domes in the North Sea

By T. KREY and R. MARSCHALL

(Prakla-Seismos GmbH, Hanover, West Germany)

SUMMARY

Sometimes rather simple geological structures, which are important for oil and gas exploration, lie below complicated heterogeneous bodies. These deep structures very often cannot be resolved by normal seismic surveys because of the rapid velocity changes within these bodies. A typical example is a salt-dome and its immediate environment. Here normally, a reliable interpretation of the Zechstein Base is not possible below the salt-dome zone. To cope with this difficulty the "undershooting" technique was developed. The principal idea consists in using reflection ray-paths which are adapted to the geological problem, *i.e.* the ray-paths have to traverse the overburden where smooth layering is encountered and thereby avoid the disturbing body. The offsets involved in the "undershooting" technique make it necessary to use a wave-front method for the depth presentation of the results.

This technique, which has been widely used in Germany for several years has recently been extended to marine seismics. In marine work two ships have to be used because of the offsets involved. The navigational problems could be solved by special additional controls as for example water-breaks, mini-streamer recording, etc. The results of the technique in marine seismics are good and supplement the results of conventional seismic surveys in salt-dome areas, sometimes in a decisive manner.

INTRODUCTION AND METHOD

In most of the southern part of the North Sea basin the deeper Zechstein and Rotliegendes are major gas-bearing layers. These geological formations exhibit a rather smooth layering with some faulting but only minor to moderate dips. However, they are overlain by Tertiary, Mesozoic and upper Zechstein layers which have suffered complicated halokinetic movements. Figure 1 may remind you of the many salt-domes present in the Southern North Sea Basin. As a consequence large parts of the overburden of interesting layers have become more or less opaque for seismic rays. This fact can well be recognized in Fig 2 where reflections from horizon Z are missing in the region beneath the domes. The strongest velocity contrasts are encountered at the upper parts of the salt-dome flanks, and here the wave-fronts are most severely distorted, whereas the velocities do not differ essentially at the lower parts of the flanks where mainly Triassic layers are in contact with the salt. Therefore we tried to get the desired information from underneath the opaque bodies by placing shots and geophones at locations such that the ray-paths reflected by lower Zechstein and deeper layers avoid the upper salt-dome flanks. In Fig 3 the most opaque flank zones are marked by horizontal hatching, but very often the whole salt-dome is an unfavourable body for reflection seismics. In such a case ray-path 3, where shot, and geophone situated on opposite sides of the dome may yield the best reflections. A survey which uses such ray-paths is called total undershooting. But ray-paths 1 and 2, where only one of either the up-

or downgoing ray avoids the salt-dome, mostly result in useful reflections too. The corresponding surveys are called flank undershootings.

More details on the undershooting technique have been published by Bading *et al.* (1974). Here we should only like to encompass the main advantages of this method. "Undershooting" provides one to three or even five different CRP sections, each belonging to a certain in-line offset of say 3 or 8km or even up to 10km. These sections can be tied to each other and tied to the sections of normal seismic reflection lines by a certain shifting in the time direction. Sometimes minor tilting has to be added on account of differing velocities on both sides of the salt-dome. Thus we get information on the continuity of sub-salt horizons, especially on the base of the Zechstein. Faults and other structural features of minor extension, *e.g.* reefs, become recognizable as can be seen in Fig 4. However, the migrated depth presentation cannot start from such zero-offset seismogram sections, because geology and velocities under the midpoints between shot and geophone differ essentially from those along the up- and down-going ray-paths. Instead, we determine the reflecting points by using wave-fronts and constructing the ray-paths starting at the shot-points and those arriving at the geophone points, taking into account additionally the reflection times. For this purpose unstacked records without moveout corrections can be used or CRP sections stacked for two or three standard-shot-geophone distances which are encountered within the observation range. The velocities and the boundaries of the layers traversed by the rays are pretty well known in general, *e.g.* by normal

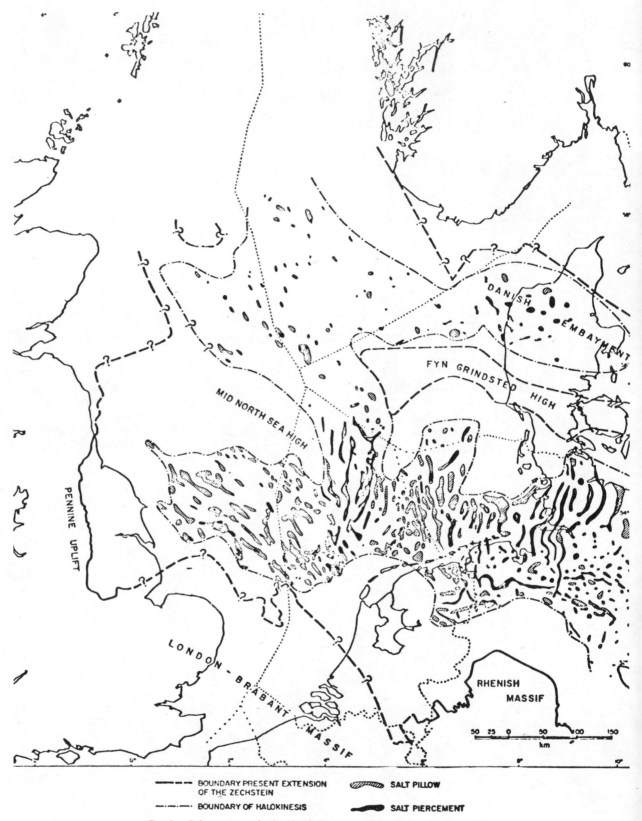

FIG 1.   Salt-structures in the North Sea area (after Heybroek *et al.*, 1967).

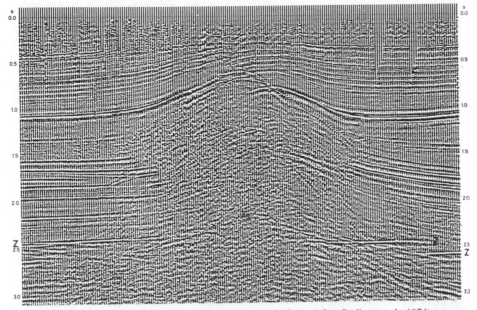

FIG 2.   Conventional seismic section across a salt-dome (after Bading *et al.*, 1974).

seismic lines and some well velocity surveys. Thus, the resulting depths normally fit well to those obtained by normal shooting. If deviations are observed, *i.e.* if there are steps between the horizons derived by normal shooting and those derived by flank undershooting or between the latter ones and that derived by total undershooting, the velocities applied can very often successfully be improved or information on anisotropy can be derived. This is a very important additional advantage of the "undershooting" technique because the structural closure in the deeper Zechstein and in the Rotliegendes formations is generally very small as compared to the depth

variations in the overlying Mesozoic and Tertiary layers.

APPLICATION

Salt-dome undershooting has successfully been applied onshore in north-west Germany and the Netherlands during the last decade. The method was soon extended to the tidal regions. In the meantime it was also proved by well shooting that the reflection time of undershooting corresponds to the sum of the transmission times from the supposed reflection point in Zechstein 2 to shot- and geophone-points.

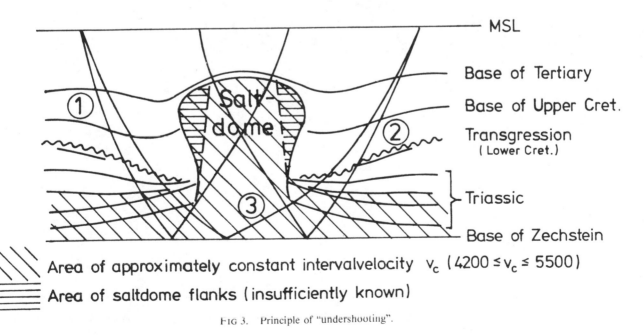

FIG 3.   Principle of "undershooting".

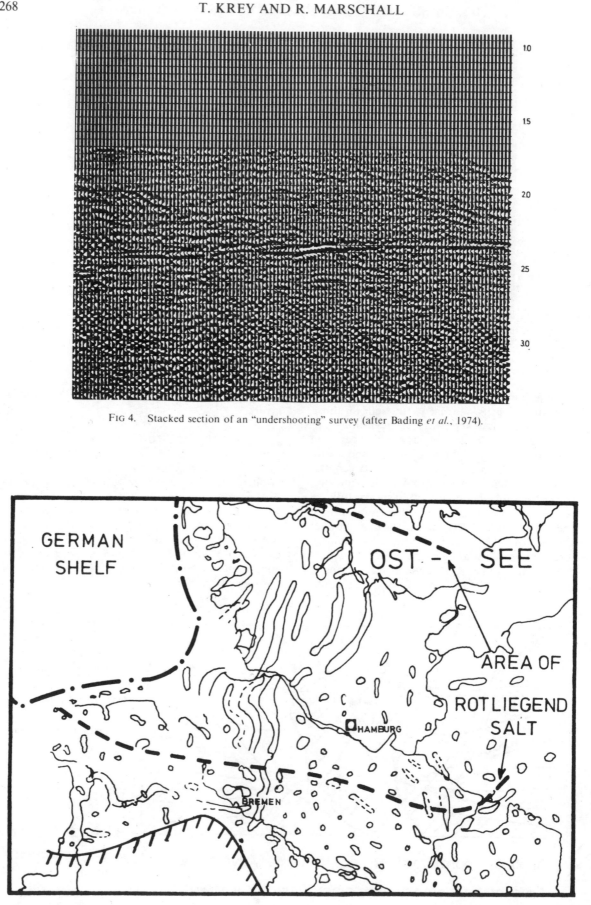

FIG 4. Stacked section of an "undershooting" survey (after Bading *et al.*, 1974).

FIG 5. Salt-structures in north-west Germany (after Trusheim, 1957; Plumhoff, 1966; Richter-Bernburg, 1953).

Offshore the "undershooting" method was accepted only recently. Due to the large in-line offset of 3 to 10km two ships have to be used instead of one. This involves additional costs per kilometre and some careful planning. But good cooperation between clients and contractors has resulted in field-work being carried out without major trouble using the two boats "Seismic Explorer" of Seismic Engineering Company for recording and "Explora" or "Prospecta" of Prakla-Seismos GmbH for shooting. Air guns proved to be the better energy source for our purpose as compared to gas exploders. The lines on which undershooting was to be carried out were run two or four times with either of one or two distances between the two boats, corresponding respectively to total and flank undershooting. Twelve- to 24-fold coverage was applied using a 2400m streamer. Running to and fro resulted in a doubling of the coverage for each kind of under-shooting. Onshore, where shot-hole drilling is necessary we carefully restrict our undershooting surveys to such shot and geophone locations whose reflection ray-paths are undisturbed by highly inhomogeneous regions. Offshore, it is advisable not to omit any shots which might yield useful information when the boats have to sail anyway. Usually up to one hour of continuous shooting is needed to perform one crossing of a salt-dome. One to 1.5 hours have to be added to each crossing for turn-around time. Therefore one complete undershooting, *i.e.* total and flank undershooting with four crossings of the salt-dome takes one Decca-day. Both flanks of the salt-dome can of course be undershot in one run. In order to reduce costs both boats may shoot other conventional lines during night time by using satellite navigation.

There is yet another way to render "undershooting" techniques more economical, *i.e.* by crossing more than one salt-dome in one run. This should easily be possible in case of elongated parallel salt walls as encountered for instance in western Holstein and at the mouth of the Elbe river (see Fig 5). Crossing four salt-domes in one run would result in a time reduction of 25 to 35 percent because three turnaround times of 3 to 4.5 hours can be replaced by three rather short straight runs of a total duration of say one hour.

In all kinds of marine geophysical surveys positioning is a very important problem, and this is valid to a still higher degree for undershooting salt-domes for three reasons. The first is a geological one. The main goal of our undershooting surveys was not to find broad structural features, but to discover and outline faults, which needs a higher precision. Secondly, the surveys had to be tied to older surveys carried out with possibly less precise Decca positioning. Thirdly, the relative position of the two boats to each other must be precisely known. Their distance from each other should especially be kept as constant as possible in order to facilitate CRP-stacking in the data centre. In order to meet these requirements a very precise Hi-Fix chain was used with satellite navigation as backup.

To facilitate comparisons with older surveys the shooting boat recorded the undershooting pops by ministreamer, and the recording boat shot and recorded with normal offset along the last 2km ahead of the first undershooting recording location

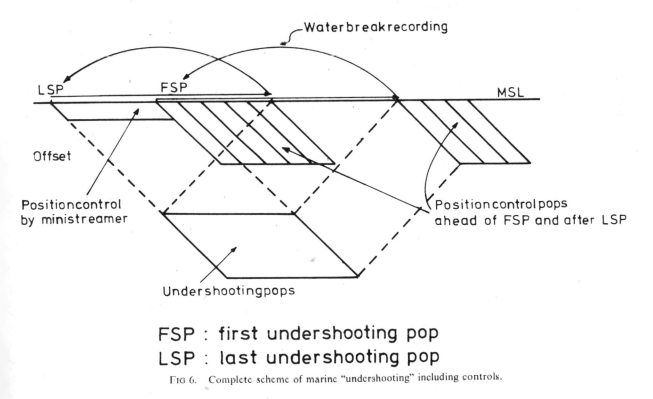

FSP : first undershooting pop
LSP : last undershooting pop

Fig 6. Complete scheme of marine "undershooting" including controls.

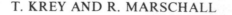

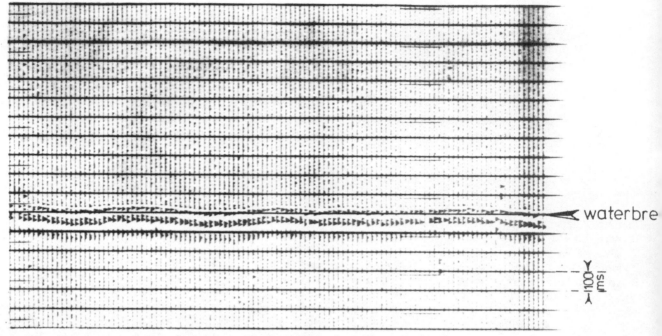

Fɪɢ 7.   Example of water-break record.

(FSP) and along the first 2km following the last undershooting recording location (LSP) (Fig 6). On account of the usually strong dips on the salt-dome flanks, positioning deviations relative to lines shot previously can easily be recognized.

Direct control of the distance between the two boats is offered by recording the waterbreaks of the undershooting pops at the recording boat. In spite of a distance of up to 10km the energy was sufficient. The waterbreaks were not only recorded at the recording boat but were automatically transmitted back by radio to the shooting boat, thus enabling this boat to continuously correct its speed. Figure 7 is an example of waterbreaks

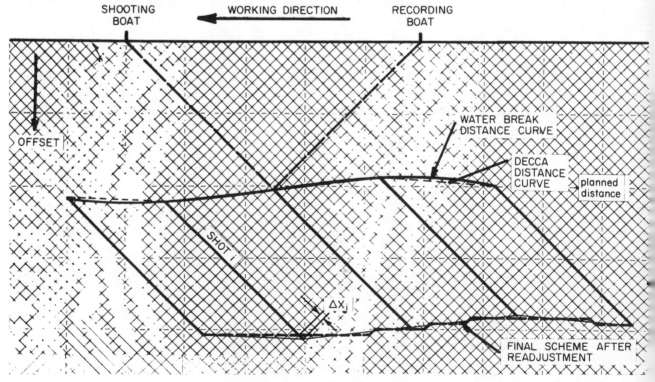

Fɪɢ 8.   Example of up-dated correlation scheme.

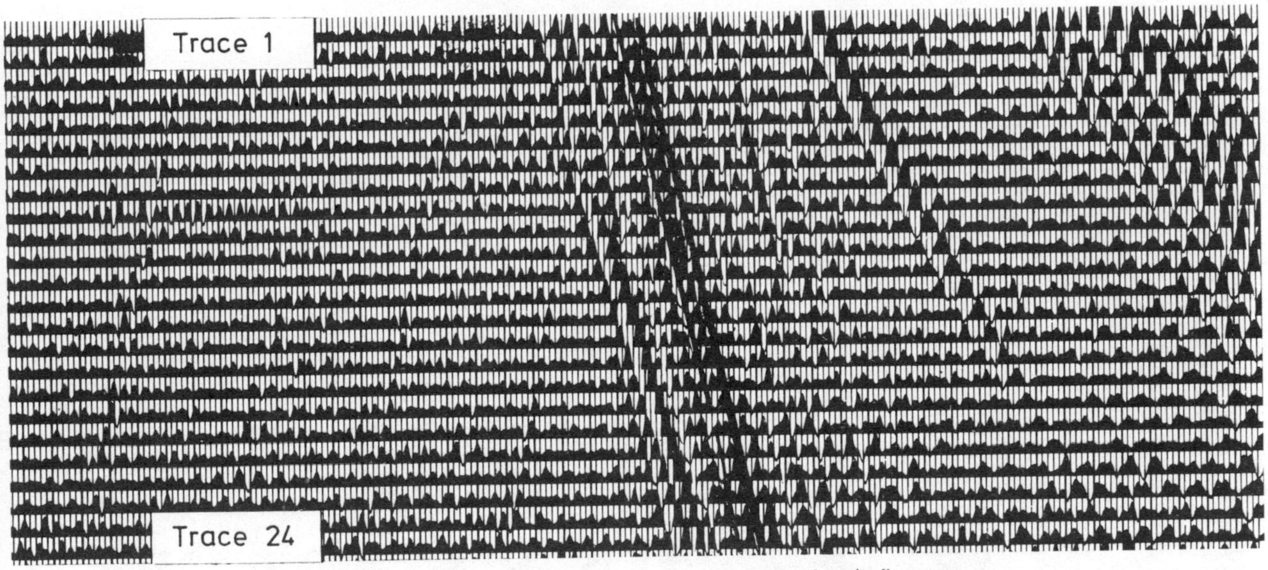

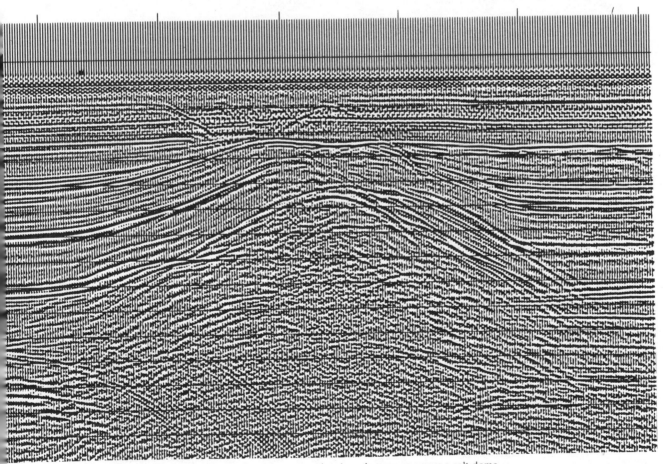

FIG 9.   Single shot display from a marine "undershooting" survey.

recorded at the shooting boats. The distance from line to line in this figure corresponds to 150m. During one favourable crossing of a salt-dome it was possible to keep the distance constant to within ±25m. When running against strong waves somewhat larger deviations were observed. In such cases the correlation scheme necessary for CRP-stacking was readjusted correspondingly after fitting the waterbreak recordings to the distances resulting from Hi-Fix (Fig 8).

A further distance control is provided by comparing the times of any seismic event of reversed traces where shot and geophone locations are interchanged. This uses the fact that the distance

FIG 10.   Section of a conventional marine survey across a salt-dome

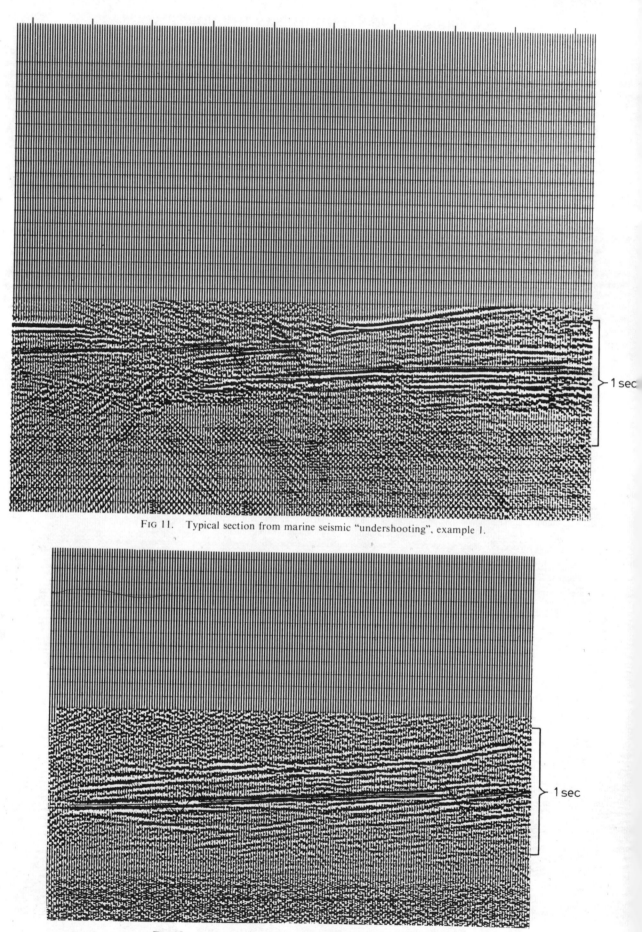

1 sec

FIG 11. Typical section from marine seismic "undershooting", example 1.

1 sec

FIG 12. Typical section from marine "undershooting", example 2.

between the positions of two hydrophone stations belonging to one shot are known very precisely, on account of the cable dimensions.

Besides positioning control there is another point to which attention must be paid. Figure 9 is a single shot from a marine seismic undershooting survey. The desired reflection is obvious and can be easily distinguished from undesirable direct and refraction noise waves by its smaller moveout. But the recognizability of the reflection might suffer seriously if it were to interfere with the preceding or following noise waves. Therefore, the inline offset should be well selected in undershooting surveys in order to let the desired reflection appear in a noisefree window, though velocity filtering before stacking may help to recover the desired reflections out of the interference pattern.

Figures 10–12 show some time sections obtained by offshore undershooting. Figure 10 shows a section of a conventional survey crossing a salt-dome in the North Sea area. The weak reflection quality of horizon Z beneath the salt-dome is obvious. In Fig. 11 the stacked section of the corresponding undershooting line is presented. The base of the Zechstein shows up very clearly, including two good faults. In Fig 12 another line is presented. The quality of this line is similar to the previous one. Two minor faults are indicated here too.

CONCLUSIONS

The undershooting technique has proved to be a very useful tool in modern seismic exploration, not only onshore but offshore too. With the positioning controls mentioned no trouble in navigation can occur. Records are fair to excellent and depth presentation proved to be reliable. Thus, the gaps left underneath the salt-domes by normal shooting can now successfully be filled in.

ACKNOWLEDGEMENTS

We would like to thank the various oil companies for permission to show the records presented here, and also Prakla-Seismos GmbH for the permission to publish this paper.

REFERENCES

Bading, R., Echterhoff, J., Krey, Th. and Marschall, R. 1974. New possibilities for reflection seismics by undershooting bodies with complicated tectonics. *Geophys. Prospect.*, **22**, 1–21.

Gill, W. D. 1967. The North Sea Basin. *Wld. Petrol. Congr. 7. Mexico, Proc.*, **2**, 211–19.

Glennie, K. W. 1972. Permian Rotliegendes of northwest Europe interpreted in light of modern desert sedimentation studies. *Bull.Am. Ass. Petrol. Geol.*, **56**, 1046–71.

K.H. and H.-P.B. 1974. Öl und Gas aus der Nordsee. *Oel-Z. Min. Hamburg.*, **12**, 150–7.

Heybroek, P., Haanstra, U. and Erdman, D. A. 1967. Observations on the geology of the North Sea area. *Wld. Petrol. Congr. 7, Mexico, Proc.* **2**, 905–16.

Krey, Th. and Marschall, R. 1969. Undershooting saltdomes. Paper presented on the 31st EAEG-Meeting, Venice, Italy.

Marschall, R. 1972. Parameter optimization of velocity functions of given form by the use of RMS-velocities. *Geophys. Prospect.*, **20**, 700–11.

Meissner, R. 1965. Multiple events in refraction shooting. *Geophys. Prospect.*, **13**, 617–58.

Plumhoff, F. 1966. Marines Ober-Rotliegendes (Perm) in Zentrum des nordwestdeutschen Rotliegende-Beckens. Neue Beweise und Folgerungen. *Erdöl Kohle*, **10**, 713–20.

Richter-Bernburg, G. 1953. Über salinare Sedimention. *Z. dt. geol. Ges.*, **105**, 593–645.

Richter-Bernburg, G. 1959. Die nordwestdeutschen Salzstöcke und ihre Bedeutung für die Bildung von Erdol-Lagerstatten. *Erdol Kohle*, **5**, 294–303.

Sanneman, D. 1963 Über Salztockfamilien in NW-Deutschland. *Erdöl-Z.*, **11**, 3–10.

Trusheim, F. 1957. Uber Halokinese und ihre Bedeutung für die strukturelle Entwicklung Norddeutschlands. *Z. dt. geol. Ges.*, **109**, 111–51.

20

The Auk Oil-Field

By T. P. BRENNAND and F. R. van VEEN

(Shell UK Exploration and Production Limited)

SUMMARY

Exploration in the south-eastern part of the Forth Approaches Basin had shown by 1970 that the main potential oil reservoirs then known (Tertiary sands and Chalk) were unfavourably developed in the Auk area.

Thick porous Rotliegendes sands and tight Zechstein carbonates were, however, proved in several wells. Seismic evidence showed that on the Auk horst, a pronounced westwards-tilted high on the western rim of the Central Graben, Rotliegendes sands should be in a structurally suitable position to trap oil which had been generated in deeply buried Jurassic shales in the Central Graben.

The Auk discovery well showed that the Rotliegendes, wet and partly tight but overlain by a thin Zechstein carbonate sequence with good reservoir characteristics, had been preserved under the Cimmerian unconformity.

The Zechstein carbonates were deposited in a lagoonal to supratidal environment. Exposure of the rocks to meteoric waters caused leaching which created a well-developed vugular porosity during Mesozoic erosional phases.

Three appraisal wells have established an expectation of reserves in the order of 50 million barrels which can be recovered from a centrally located platform.

GENERAL STRUCTURAL SETTING

The concept of drilling for oil at the margins of the Central Graben rested upon two important facts—the discovery of oil at Ekofisk and Josephine both within the graben, in 1969–1970, and the knowledge that the Permian north of the Mid North Sea High was developed in a facies comparable to that seen in the south. It seemed reasonable, therefore that oil generated within the graben might migrate into structures immediately bordering the graben edge, despite the fact that the anticipated reservoirs would be older than the Jurassic source rocks supplying the oil from a structurally lower position.

The western margin of the Central Graben is expressed as a series of major eastwards-hading faults which progressively offset one another towards the south-east (Fig 1). Fault displacement affects rocks of Cretaceous age and older along this system while younger rocks reflect only a syn-depositional effect of rapid eastwards thickening over the margin, with post-depositional flexuring. Older rocks dip generally westwards from the margin, whereas younger rocks generally rise westwards from the margin, the thickness increase being taken up by Zechstein salt and Triassic mudstones (Fig 2). This westwards-rising of the younger capping rocks is inherently unfavourable for the retention of hydrocarbons at the margin. However trap-forming conditions arise as a result of the off-setting of the major boundary faults.

As each fault dies out southwards it is taken over, as described above, by a similar eastwards-hading fault. In the overlap area there is a tendency for a minor "back basin" to be formed, closed to the south and open down-plunge towards the graben.

It is on the "back-basin" side of the new offset upthrown block that the all important counter regional west-hading faults can develop. Such faults provide a mechanism for counteracting the westwards migration of hydrocarbons.

THE AUK PROSPECT

Block 30/16 is situated at the northern end of an offset fault, trending south-eastwards through the block. This fault takes over from a major system which is mappable between Blocks 22/21 and 29/7, some 50km to the west, and dies out rapidly in a

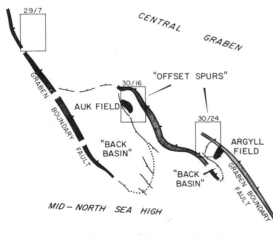

FIG 1. Structure of the Auk Oil-field.

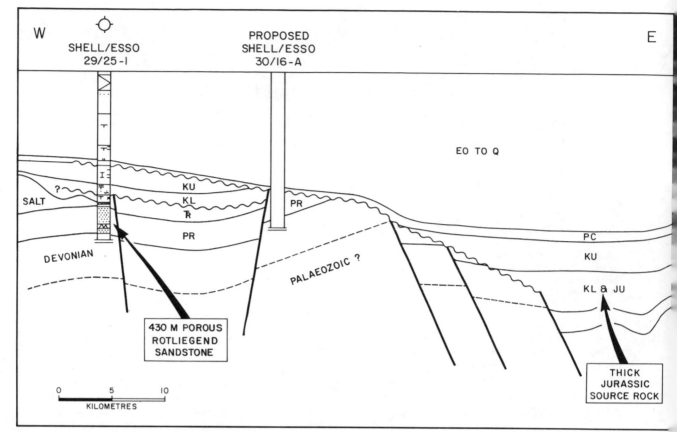

FIG 2.   Structural interpretation prior to Auk Field drilling of Mid North Sea High—Central Graben.

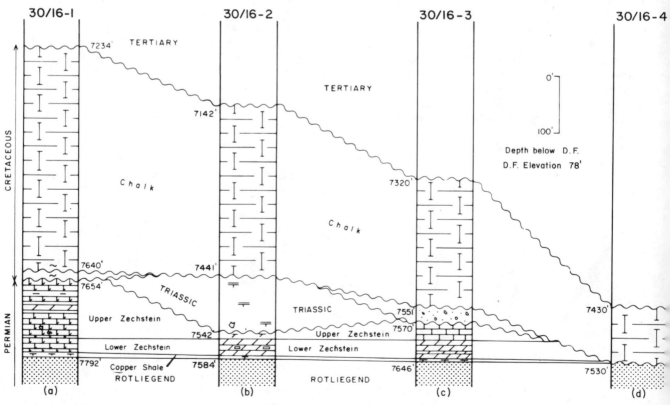

FIG 3.   Sections of wells in the Auk Field.

heavily faulted area to the south of 29/7. The "back basin" thus created is unusually wide, rather shallow and considerably fault-dissected. Stress resolution on its eastern side gave rise to westwards-hading normal faults, creating a prominent horst-like "offset spur" in Block 30/16 (Fig 1). Dip directions within the older, horst-block sequence, which were visible on seismic profiles and confirmed by drilling, are consistent with the regional westwards trend. Younger rocks drape over the spur, a sharp flexure overlying the major eastwards-hading fault, with a minor flexure over the westwards-hading fault.

As regards the potential reservoir, attention was initially focused upon the Permian, and in particular on the Rotliegendes sands which were known in several wells to be over 1000ft (300m) thick. Some of this sand could reasonably be expected to have survived the anticipated deep Mesozoic erosion on the spur.

The seismic evidence showed that the strong basal Tertiary reflector passes over the block as an unconformity down-cutting eastwards into the Cretaceous Chalk. The resulting thinness of the Chalk prevents any coherent information being obtained immediately below the regional Laramide unconformity.

In view of the westwards dip in the pre-Cimmerian and the eastwards Laramide downcutting the chance of any upper Permian carbonates being preserved was not rated as high.

While uncertainty prevailed over the thickness of the reservoir that might exist on the prospect, the vertical seal was less in doubt. Wells in the vicinity had shown by this time that the Tertiary was devoid of sand in this part of the basin, and that the Chalk was likely to be tight. Minimum closure was therefore established by the drape of the sealing Tertiary, and the hope was that this might be increased by fault-sealing.

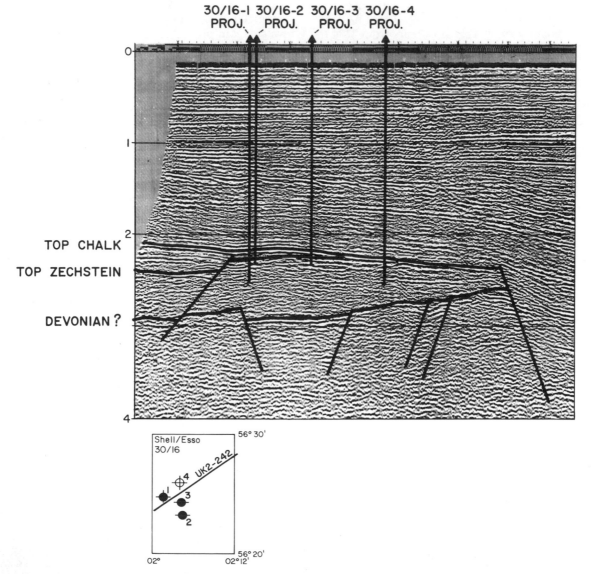

FIG 4.   Seismic line UK2-242 across Auk Field.

DISCOVERY AND APPRAISAL DRILLING RESULTS

The location of the first well was selected on the western edge of the horst where the chances seemed greatest for the preservation of a reservoir sequence.

After spudding in September 1970 the well encountered Rotliegendes clastics below a 2-ft (0.6-m) Kupferschiefer bed at 7790 to 7792ft (2374.4 to 2375m). These sands extended down to 9382ft (2860m) and below this possible Devonian sandstones were interpreted to be present down to TD at 9403ft (2866m). This clastic sequence showed good reservoir qualities but was water-bearing. Overlying the Kupferschiefer, hydro-carbon-bearing Zechstein carbonates with good vugular porosity were unexpectedly encountered, from 7790 to 7654ft (2374.4 to 2332.9m). Thin lower Cretaceous shales at 7654 to 7637ft (2332.9 to 2327.7m) overlie the unconformity and are in turn unconformably overlain by some 400ft (122m) of Chalk at 7637 to 7234ft (2327.7 to 2205m). Predominantly shaly Tertiary occurs from 7234 to 1630ft (2205 to 497m) (Fig 3a).

The well was tested at a maximum of 5900b/d of 37° API oil at a formation drawdown of only 20psi from the Zechstein interval.

Because the Zechstein reservoir could not be confidently distinguished on seismic sections due to its occurrence close under the strong Chalk-top reflector, the extent of the thin Zechstein could not be defined. The presence of two unconformities made the possibility of it being eroded at a short distance from the discovery well not improbable. If the Zechstein were to parallel a deeper Palaeozoic reflector, showing a pronounced south-western dipping flank, the reservoir would be truncated by the north-easterly dipping base Tertiary (Fig 4). But even if the Zechstein extended over a larger area, the well-developed vugular porosity could be a strictly local phenomenon, since tight Zechstein was present in nearby well 29/25-1 (Fig 5).

The first appraisal well 30/16-2 was therefore drilled some 4km to the south-east near the crest of the top Chalk structure to evaluate the Rotliegendes and possible Zechstein reservoirs in a structurally higher position. The well encountered the Zechstein carbonates from 7542 to 7584ft (2299 to 2311.6m), much reduced in thickness with respect to the first well. Underlying the Zechstein was a 1-ft (0.3-m) Kupferschiefer and below that some 500ft (150m) of Rotliegendes was penetrated which again proved non-productive.

Overlying the Zechstein occurred 104ft (31.7m) of grey to reddish brown mudstones which were dated on palynological evidence as Lower Trias (Scythian). The presence of a third, local sub-Triassic unconformity increased the possibility of the reservoir rock being eroded a short distance

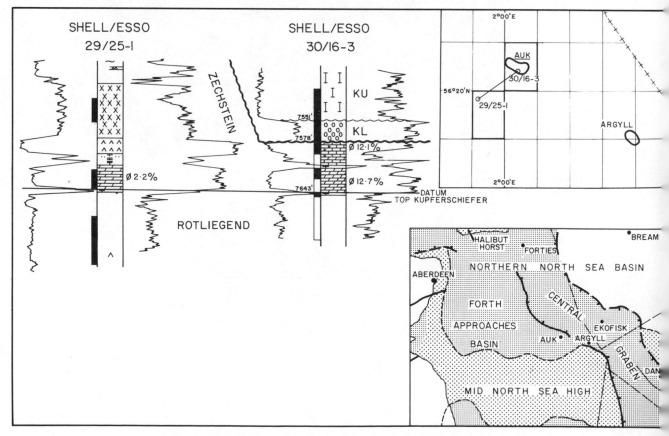

FIG 5. Auk Field: stratigraphic sections of wells 29/25-1 and 30/16-3.

from this location. The Chalk in the well measured 300ft (91.4m) compared to the 400ft (122m) found in the discovery well (Fig 3b).

A third well 30/16-3 was drilled 2.5km to the east of the discovery well to investigate the eastern extension of the field. The Zechstein carbonates were again encountered, displaying vuggy porosity and with a thickness of 69ft (21m), intermediate between that found in the first two wells. The Zechstein was unconformably overlain by a monomict conglomerate containing well-rounded cobble-sized dark green-grey volcanic rocks in a yellowish to red-brown calcareous sandstone matrix. This conglomerate was dated as Lower Cretaceous (Aptian/Albian) and measured 24ft (7.3m). The Chalk in this well was again reduced in thickness (230ft—70m) (Fig 3c).

The third well proved the minimum reserves of some 50×10^6 bbl required for economical field development. In order to locate the oil–water contact, a fourth well was drilled 3km north of well 30/16-3. This well, however, proved the Zechstein reservoir entirely absent due to erosion and found a 100-ft (30-m) Chalk immediately overlying structurally high, but wet Rotliegendes sandstones (Fig 3d).

A seismic line through the Auk structure shows the tilted pre-Cimmerian fault-block, which has been truncated by Mesozoic erosion during several phases (Fig 4). A seismic event, possible top-Rotliegendes, has been mapped, resulting in the outline of the Zechstein accumulation shown on Fig 6.

PLATE I.   Top: Zechstein dolomitic limestone, collapse breccia; bottom: Zechstein vugular dolomite.

SEDIMENTOLOGY AND DIAGENESIS OF THE ZECHSTEIN CARBONATES

The lowermost part of the Zechstein carbonates immediately overlying the Kupferschiefer is rich in insoluble residue, but carbonate purity increases rapidly towards the top. Cores show a pelletoidal–oncolitic fabric with stromatolitic laminations which together with abundant replacement anhydrite, recognizable from crystal moulds, points to a high intertidal to supratidal sabkha environment. The lack of marine organisms is interpreted to indicate an abnormally high water-salinity leading to growth of early replacement gypsum and anhydrite. Early dolomitization is indicated by the small crystal size of the dolomite and the good preservation locally of the original sedimentary fabric.

A conspicuous mudstone layer, some 4ft (1.2m) in thickness, which appears to be highly organic on the gamma-ray log, separates the lower Zechstein dolomite from the upper Zechstein dolomitic limestones which show strong brecciation.

Exposure of the Zechstein carbonates to meteoric water during Mesozoic erosional phases caused leaching of calcium sulphate creating a highly porous and permeable rock with abundant

vugs (Plate I). It is thought that the brecciation of the upper part of the Zechstein resulted from collapse after the vuggy and cavernous rock became mechanically unstable.

The matrix of this solution-collapse breccia originates partly from the Zechstein dolomite itself, but contains also coarse clastic and Lower Cretaceous limestone fragments indicating a persistent leaching and reworking process on the high.

The fact that vugular porosity did not develop in the Zechstein carbonates in 29/25-1 can be explained by the preservation of overlying evaporites which prevented exposure to leaching agents (Fig 5).

PRODUCTION PLANS

Initially the Auk Field will be developed by six platform-drilled wells with deviations up to 7000ft (2130m) at reservoir depth. The platform which is centrally located between the three Auk oil wells was floated out during August 1974.

A natural water drive is not expected to develop and water injection will probably be required

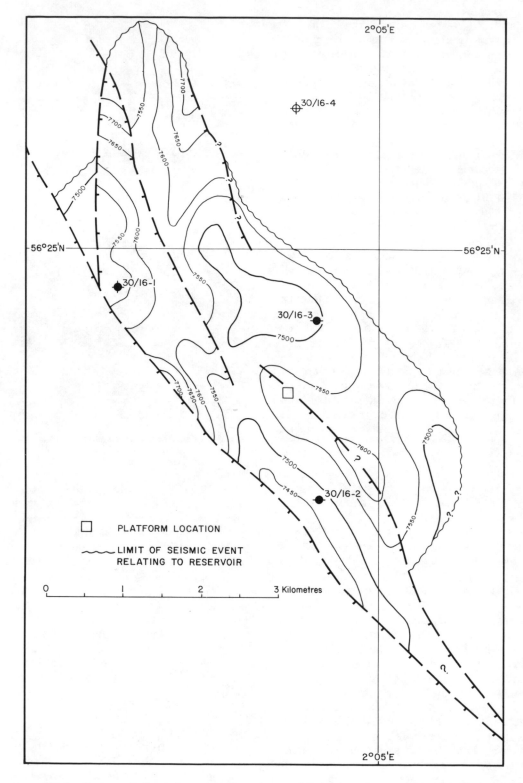

FIG 6.   Contours (in feet below sea level) on the top of Zechstein carbonates in the Auk Field.

within a few weeks after first production to prevent a reservoir pressure drop of more than 1000psi. The final annual average plateau rate for the field will be 40 000bopd. Provisions are made in the development plan to drill up to six additional platform wells and two subsea completed wells, should reservoir characteristics require them.

No offshore crude storage is planned for Auk, and the field will stay closed-in when tanker-loading operations are not taking place. Relatively few problems are foreseen with this production method.

ACKNOWLEDGEMENTS

The authors would like to acknowledge their indebtedness to their colleagues in Shell UK Exploration and Production Co. Ltd and Koninklijke Shell Exploratie en Productie Laboratorium Rijswijk, who have been involved in the discovery and interpretation of the Auk field.

They also thank Shell UK Exploration and Production Co. Ltd, and Esso Europe Incorporated for permission to publish this account.

DISCUSSION

James B. Blanche (NCB Exploration Ltd): Could the authors elaborate on the facies-types and probable environments of deposition of the Rotliegendes in the Auk Field, and how do these facies-types resemble those in the Rotliegendes in the Southern North Sea Basin.

Dr T. P. Brennand: The Auk Rotliegendes has much in common with the same formation's clastic fringe in the Southern North Sea Basin. Less is known of the details, however, but overall, the sandstones are comparable in grain-size and colour. Waterlaid, aeolian and wadi beds have also been identified. No subdivision has been attempted as there are not enough sections available.

G. H. Rhys (Institute of Geological Sciences) commented on the reservoirs encountered in the Auk and Argyll Fields. Apparently the reservoir characteristics of the Zechstein carbonates are the result of solution beneath an early unconformity which, had it cut down a further 100ft or so, would have removed the carbonate entirely. That such a reservoir should have been found seemed fortuitous; that two such occurrences had been found seemed to indicate something more than good fortune. If another should be found at some future date would it be by design or would it be yet another happy accident?

Dr T. P. Brennand: The Permian carbonate erosion relicts on highs bordering the western rim of the graben seem indeed to be chance survivors from Cimmerian or later erosional periods. However, the discovery once made in one locality is understandably repeated elsewhere owing to the extreme topographic uniformity that prevailed in pre-Jurassic times outside subsiding troughs. It is quite probable there are other patches of thin Zechstein carbonate preserved beneath the Cimmerian unconformity at other localities with similar post-Triassic tectonic history.

J. J. Pennington (author of Paper 21): During the early stages of North Sea exploration the nature of the stratigraphic section in many areas was one of the risks which companies, such as Hamilton, were willing to undertake. If companies had been able to accurately predict the stratigraphy and reservoir-characterization in advance of drilling, there would have been a marked reduction in the number of expensive dry holes.

In the case of Argyll, the initial well was drilled in search of Palaeocene sands and porous Danian carbonates. Both objectives proved to be devoid of any reservoir potential but shows of oil were found in the Zechstein. An offset well was subsequently drilled to investigate these shows and to test the Rotliegendes sandstones. No doubt other fields of this nature will be found, but whether such discoveries are the result of astute geology and geophysics, luck or a combination of the two may be difficult to ascertain.

Dr D. R. Whitbread (Cluff Oil Ltd): Has the geochemical analyses of Auk and Argyll crude oils included any sulphur isotope determinations? These have proved extremely useful in determining the age of Middle East crudes.

Dr T. P. Brennand: We have not detected any correlation between the characteristics of Palaeocene source-rock extracts and the Argyll crude. The Auk, Argyll, Forties and Ekofisk crudes have not exhibited any significant geochemical differences which would point to different sources, in any of the analyses we in Shell have made so far. The geochemical studies we have made have not included sulphur isotope determinations.

Kjell G. Finstad (Saga Petroleum): Has it been possible to detect any seismic reflector corresponding to the marine Devonian?

J. J. Pennington: There is a strong seismic reflector from the Middle Devonian carbonates which can be traced over most of the area south-west of the large fault bordering the Central Graben. This reflector appears to be completely faulted but the majority of the faults seem to die out in the Upper Devonian/Rotliegendes interval. Indications of what has been interpreted as the same seismic event have been noted in the Auk field area and in parts of the mid North Sea High.

R. Byramjee (Compagnie Française des Pétroles): (1) Has any geochemical analysis been carried out on Palaeocene shales and if so how does it match with the Argyll crude characteristics?

(2) Are the Auk and Argyll crudes similar and how do they compare with the crudes of Ekofisk and Forties?

J. J. Pennington: (1) We have not run a geochemical analysis of the Palaeocene shales in the Argyll field area; such work has been done in adjacent areas but no attempt has been made to match these results with Argyll crude.

(2) All of these crudes are quite similar in character.

21

The Geology of the Argyll Field

By J. J. PENNINGTON

(Hamilton Brothers Oil and Gas, Ltd)

SUMMARY

The Argyll Field was discovered in 1971 and is located in the British sector of the North Sea 200 miles (322km) east-south-east of Aberdeen, Scotland. Oil has been found in Zechstein dolomite and in Rotliegendes sandstone at depths ranging from 8700 to 9400ft (2652–2865m). One well tested oil from a sandstone of probable Jurassic age. The proven productive area is in excess of 3000 acres (13km²) and is contained within a horst near the crest of a much larger tilted fault block. Four wells are being completed on the seafloor in an average water depth of 250ft (76m). Field production is expected to average between 35 000 and 40 000 barrels per day.

INTRODUCTION

The Argyll Field is located within the United Kingdom sector of the North Sea 200 miles (322km) east-south-east of Aberdeen, Scotland (Fig 1). In 1969 the Hamilton Group drilled well 30/24-1 to evaluate a seismic structure. Although this well was plugged and abandoned, shows of oil were present at total depth in a Zechstein dolomite. The discovery well 30/24-2 was drilled in 1971 to investigate these shows and test the Rotliegendes.

Four additional wells have been drilled to determine the productive limits of the field. Three of these were successful and the other found only residual shows of oil in a sandstone of probable Jurassic age. The productive area covers 3200 acres (13km²) but has not been fully established. The two primary reservoirs are the Zechstein and the Rotliegendes which are oil-saturated at depths ranging from —8733 to —9373ft (—2662 to —2857m). One well tested oil from a thin sandstone of probable Jurassic age.

The Hamilton Group plans to bring the field on production in the early part of 1975. Four wells are being completed on the seafloor and a semi-submersible drilling rig will be utilized as a production platform. Oil will be loaded into tankers from a single point mooring system. No provision has been made for storage facilities and the well will be "shut-in" about 30 percent of the time when sea conditions are too severe to permit tanker loading. The field production rate is expected to average about 35 000 barrels per day.

STRATIGRAPHY

Cenozoic

A generalized stratigraphic column for the Argyll Field is shown in Fig 2. The total thickness of the Cenozoic is about 8800ft (2682m) and all series are represented. The Tertiary is comprised predominantly of marine mudstones and clays and is devoid of potential reservoirs. The clays have a tendency to heave and drilling was accomplished with mud weights of 12 to 12.5 pounds per gallon.

In the lower part of the Tertiary the Danian is a rather dense chalky limestone with a thickness of 131 to 248ft (40–76m). The high porosities found in the Ekofisk area 30 miles (48km) to the north-east are not present.

Mesozoic

The Danian is conformable with the Maastrichtian and the lithologies of the two are similar. Seismic evidence shows over 2000ft (610m) of Upper Cretaceous onlapping the northern portion of the Argyll structure and a similar thickness is present to the north-east on the down-thrown block of a major fault. In the field area only the youngest part of the Upper Cretaceous is present. In well 30/24-3 (Fig 3) 37ft (11m) of Maastrichtian rests directly on the Permian.

Outpost well 30/24-4 contains the most complete Mesozoic section with 43ft (13m) of Albian/Aptian marl beneath the Maastrichtian. This overlies 120ft (37m) of shallow marine argillaceous and glauconitic sandstone with residual oil shows in the top 40ft (12m). This sandstone is unfossiliferous but its stratigraphic position and relation to other wells would indicate a probable Jurassic age. Beneath this sandstone lies 128ft (39m) of typical Triassic red-brown waxy shale and silty mudstone resting on Zechstein.

Log correlations within the Mesozoic are fairly reliable as illustrated in Fig 3. Variations in the thickness of these sediments are due to erosion (Cimmerian unconformity) of the Jurassic and Triassic and to onlap of the unconformity surface by the Lower Cretaceous near the apex of the structure.

Well 30/24-6 tested oil at the rate of 1200 barrels

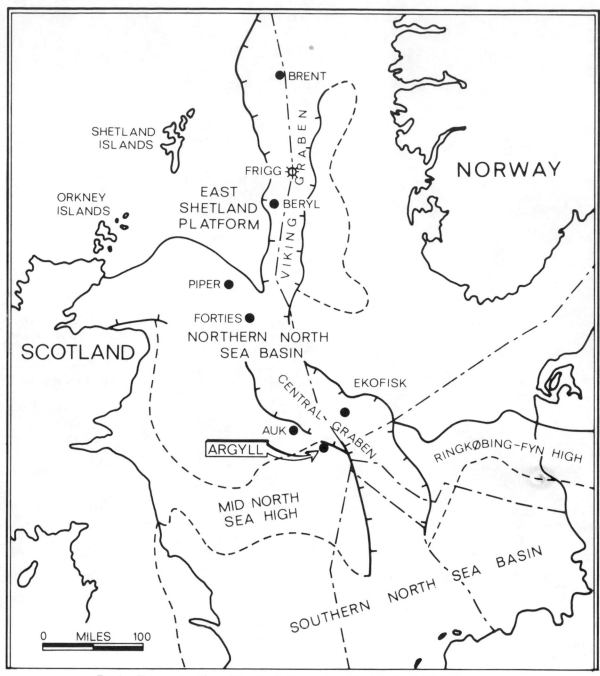

FIG 1.   Structural setting of the North Sea, and position of Argyll Oil-field.

per day on a restricted choke from 4ft (1.2m) of perforations in a 24ft (7m) Jurassic sand. Further drilling would be required to determine the areal distribution of this reservoir.

Palaeozoic

Zechstein. The Zechstein (Fig 3) has been dated as Upper Permian on the basis of flora from a dark grey argillaceous dolomite found in a conventional core from well 30/24-5. Core and electric log data show the Zechstein to be composed largely of a dense grey dolomite which varies in thickness from 33 to 103ft (10–31m). Conventional cores of the dolomite from well 30/24-2 are vuggy and highly

fractured and many of the vugs and fractures are partially filled with crystals of pyrite. Salt deposited during the Zechstein evaporitic cycles dissolved causing the dolomite to collapse. It has not been established to which cycle of the Zechstein the dolomite should be assigned and possibly more than one cycle is represented.

To date the position of the water-level within the Zechstein has not been established more accurately than somewhere between —8916 and —9646ft (—2718 to —2940m). Average log porosities range from 18.2 percent in well 30/24-6 to 8.2 percent in well 30/24-3. Core porosities in well 30/24-2 averaged only 5.1 percent but this figure is not

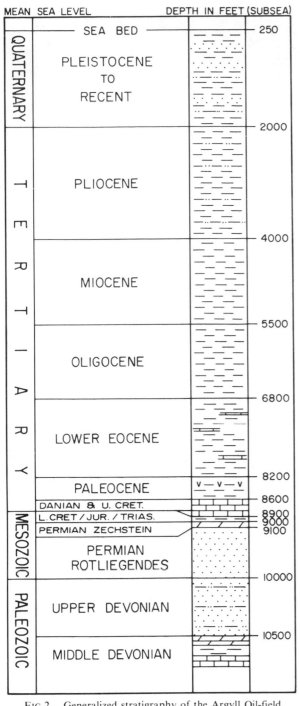

FIG 2.   Generalized stratigraphy of the Argyll Oil-field.

results are interpreted as showing an alternating sequence of fluviatile and aeolian sandstones largely derived from a source to the south-west. Well 30/24-2 penetrated 1490ft (454m) of Rotliegendes and a detailed petrographic study was conducted on 62 sidewall cores recovered from the upper 600ft (183m). The sandstones were found to be remarkably uniform, soft, friable, fine- to very fine-grained with a variable amount of clay matrix. The basic sandstone type is a subarkose with lesser amounts of orthoquartzite and subgreywacke. Minor amounts of calcareous and ferruginous cement are present but the interstitial clay also contributes to the cohesion of the rock. Two of the cores contained highly altered basalt, which may indicate the presence of a thin flow or intrusive body.

Precise dating has not been possible due to the paucity of fossils. A floral assemblage was recovered from a thin shale cored in well 30/24-3. This shale occurs at a depth of —8788ft (—2679m) about 28ft (8.5m) below the base of the Zechstein. If the structural correlations referred to in the following paragraph are correct, the stratigraphic position of the shale would be near the base of the Rotliegendes. According to Paleoservices Ltd the spore assemblage is typical of the Zechstein of Western Europe. This conflicts with the usual tendency to regard the Rotliegendes as Lower Permian. Some recent workers in palynology believe the German Upper Permian contains both a Zechstein and a Rotliegendes facies and that the North Sea Rotliegendes should be assigned to the Upper Permian.

Detailed log correlations within the Rotliegendes are extremely poor due to the type of deposition and the distances between wells. Correlations of gross log parameters, detailed sample logs and dipmeter results all support the existence of a major angular unconformity between the Zechstein and Rotliegendes. Between wells 30/24-2 and 30/24-3 about 1350ft (411m) of Rotliegendes are truncated by this late Variscan unconformity.

A total of 144ft (44m) of conventional core was recovered from the Rotliegendes in wells 30/24-4 and 30/24-5. Core porosities average 14.8 percent and the average permeability is 51.9md. Variations in permeability are largely related to the amount of interstitial clay.

The gross thickness of the oil column within the Rotliegendes varies from 131ft (40m) in well 30/24-3 to 367ft (112m) in well 30/24-2. Residual oil shows are often present below the saturated zone and water levels are difficult to establish from electric logs. Each well in the productive area encountered either a tight zone or a water level at depths varying from —8881ft (—2707m) in well 30/24-3, to —9373ft (—2857m) in well 30/24-2. Although the saturated interval of the Rotliegendes becomes progressively deeper along the axis of the structure in a south-westerly

considered to be representative as the plugs were cut from the more massive pieces of the core. Three wells have tested oil from the Zechstein with very little pressure drawdown at the limit of the test facilities. Flow rates in excess of 10 000 barrels per well per day are anticipated.

Rotliegendes. The Rotliegendes facies is comprised of a thick sequence of non-marine sandstones and thin interbedded shales lithologically similar to the Rotliegendes of the Southern North Sea Basin. Core and dipmeter

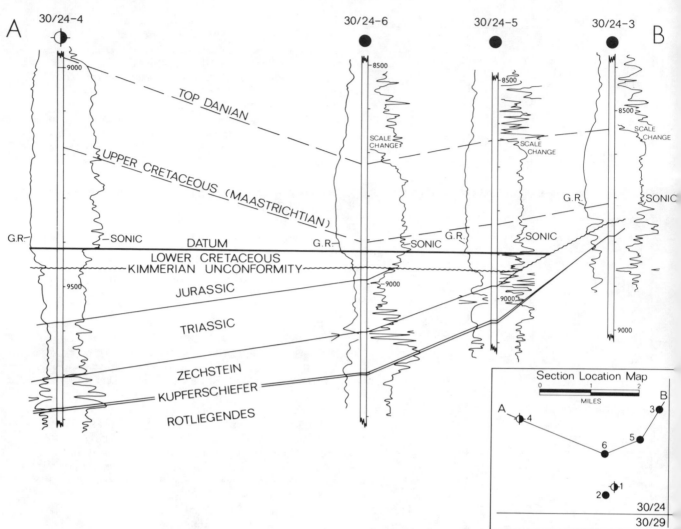

FIG 3.   Cross-section through wells of the Argyll Oil-field.

direction, the hydrodynamic conditions necessary for a tilted water-table are not evident.

It is postulated that a number of vertical permeability barriers are present between various sandstone members of the Rotliegendes and oil was transmitted to these truncated reservoirs from the overlying fractured dolomite. The depth to which the oil migrated could be determined by slight variations in the physical characteristics of the sandstones.

Devonian. No evidence has been found to support the existence of Carboniferous and the Rotliegendes is essentially parallel to the underlying Devonian. The most complete section of Devonian was encountered in well 30/24-3 (Fig 4) where the interpreted boundary between the Rotliegendes and the Upper Devonian at —8888ft (—2709m) is marked by a change in the lithology with an associated increase in the sonic velocity. The dipmeter also changes from the patterns typical of fluviatile and aeolian sandstones to a uniform structural dip of 10°.

The base of the Upper Devonian is closely established by palaeontology and the total thickness is about 800ft (244m). Thinly interbedded anhydritic siltstones, silty fine-grained sandstones and shales predominate. The only fossils recovered are Frasnian spores from the bottom 50ft (15m).

Middle Devonian occurs at a depth of —9698ft (—2956m) and is marked by a marine zone of thinly interbedded calcareous dolomites and shales containing ostracodes. At —9968ft (—3038m) the well encountered a thick fossiliferous limestone containing both rugose and tabulate corals. Drilling was suspended at a depth of —10 080ft (—3072m) and a drillstem test of the limestone flowed water at the rate of 4900 barrels per day with a chloride salinity of 32 000ppm. This limestone is the source of the deepest consistent seismic reflector.

STRUCTURE

The Argyll Field is located in the extreme southern part of the Northern North Sea Basin (Fig 1) near the crest of a large tilted fault-block paralleling the

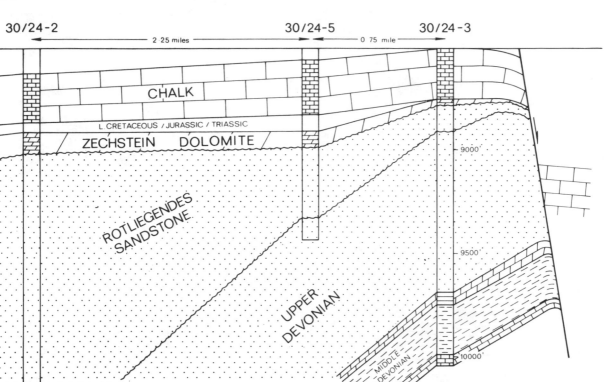

FIG 4. Structural section across Argyll Oil-field.

south-western margin of the Central Graben. The known productive area of the field is confined to a local horst-block striking at right angles to the Central Graben.

The first apparent large scale folding occurred during the Upper Permian and the Middle Devonian seismic reflector provides a reliable guide to its magnitude. Over 6000ft (1829m) of relief is present between the crest of the structure and a low area 5 miles (8km) to the south-west in the direction of the Mid North Sea High. The highest portion of this older structure was centred to the north of the field where the Middle Devonian is intersected by the unconformity between the Zechstein and the Rotliegendes.

Stable conditions prevailed during the Triassic and the Jurassic as both periods are represented by thin sediments which are essentially parallel to the Zechstein. Kimmeridgian folding at the end of the Jurassic created the north-east to south-west trending axis on which the field is located (Fig 5). During Cimmerian time the highest areas were eroded (Fig 4) and vugs were formed within the dolomite. Meteoric waters may have invaded the Rotliegendes, since connate waters recovered from several tests of the Rotliegendes have an average chloride salinity of only 35 000ppm.

No significant structural movements can be attributed to the Lower Cretaceous, but with the beginning of the Upper Cretaceous the area to the north of the field again tilted to the north.

The hinge for this downwarp also coincides with the field axis. During the Tertiary the entire area subsided about 8000ft (2438m) and this downwarp is superimposed on all of the older structure.

Faulting played an important role in both the accumulation and areal distribution of the oil.

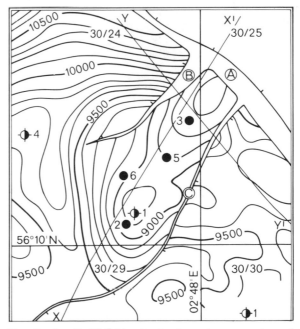

FIG 5. Argyll Oil-field: structural contours on top of Zechstein. Area shown is about 5¼ × 4¾ miles.

SEISMIC PROFILE SW — NE

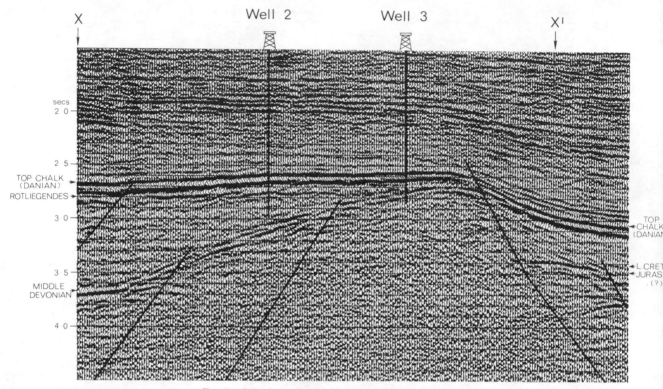

FIG 6.   Seismic profile across the Argyll Oil-field (SW—NE).

Fault "A" (Fig 5) is one of the boundary faults of the Central Graben and limits the productive area to the north-east. The displacement at the top of the Danian is about 200ft (61m), but the Jurassic is probably offset by 4000 to 5000ft (1219 to 1524m). Fault "B" bounds the productive area to the north-west. This scissors fault has dip in opposite directions on the upthrown and downthrown fault blocks. Fault "C" forms the south-eastern margin of the horst block. Its displacement is unlikely to be more than 200 to 300ft (61 to 91m) at the level of the Zechstein. A structural closure exists on the down-thrown block of this fault which may warrant further exploration.

GEOPHYSICS

The strongest and most continuous reflector stems from the top of the Danian (Fig 6). This reflector has two prominent cycles which mask other reflectors for a distance of about 450ft (137m) below the top of the Danian. Within the field area the interval from the top of the Danian to the base of the Zechstein is less than 450ft (137m) and the attitude of the Zechstein and the top of the Rotliegendes cannot be determined solely by reflection seismic data. The map shown in Fig 3 is based on well control in the centre of the field and reflectors of poor quality in offstructure areas

where the Danian to Zechstein thickness exceeds 450ft (137m).

ORIGIN OF THE OIL

There was not an effective seal for the reservoirs until Danian time and it is unlikely the oil stemmed from Zechstein or older source beds. Most of the evidence would suggest a Palaeocene shale source in fault communication with the reservoirs. Seismic data is interpreted to show erosion along fault plane "B" and the Palaeocene shales may be in direct contact with the Permian. The Zechstein could have served as a "pipeline" for distribution of the oil to the Rotliegendes sandstones.

From production tests the pressure gradient for the Zechstein has been established as 0.59psi/ft. This abnormal pressure was superimposed on the reservoir after an effective cap had been established. The Palaeocene and younger shales are also over-pressured and are therefore the most likely source for the excessive reservoir pressures.

CHARACTERISTICS OF THE CRUDE

The oil is classified as an intermediate paraffin type of light crude with low viscosity and a pour point of plus 12°C. Its sulphur content is only 0.21 percent and the wax content is 6 percent. Zechstein crude has a gravity of around 38° API and a gas/oil ratio of 210/1 whilst that from the Rotliegendes is about 34° API gravity with a gas/oil ratio of 110/1.

favourable relationship between source rock and potential reservoirs. In many areas the thin Zechstein to Upper Cretaceous interval, which contains two or more unconformities, will continue to present problems in reflection identification.

CONCLUSIONS

The search for Permian oil fields in the northern North Sea Basin is still in an early stage. An appreciation of the stratigraphic and structural history is essential in locating prospects with a

ACKNOWLEDGEMENTS

My thanks are extended to the following who have helped in the production of this paper: Francis J. Brophy, Eric J. Loughead, David J. Warwick, Jo-Ann Bates, and Anna Gedroyc.

DISCUSSION

For discussion relating to this paper see T. P. Brennand and F. R. Van Veen, Paper 20.

22

The Triassic of the North Sea

By T. P. BRENNAND

(Shell UK Exploration and Production Limited)

SUMMARY

Complete sections of Triassic have so far been proved only in the southern part of the North Sea. Here a widespread phase of red shale and later coarse clastic deposition was replaced in mid-Triassic times by conditions in which cyclic repetitions of halite, red marls and claystones, with dolomite or anhydrite, were laid down. At each level, deposition throughout the Triassic was regionally uniform and log correlations are possible over wide areas.

To the north of the Mid North Sea High, on which Triassic sediments are mainly absent, evidence is fragmentary owing to subsequent erosion. Age-dating is uncertain, but evidence suggests that erosion was deepest in the west where only thin red shales are preserved.

In the northern North Sea, thick sections of red silts and mudstones with thin sandstones are widely present, for which sparse dating evidence points to an upper Triassic age.

Synchronous block-faulting probably gives rise to considerable thickness variations between Triassic sequences on and off the main highs.

INTRODUCTION

The last review of the Triassic of the North Sea was a paper by Geiger and Hopping (1968) in which they presented the first integrated picture of the stratigraphy of the Southern North Sea Basin, and showed the value of palynology in correlating what had hitherto been regarded as an unfossiliferous, mainly continental series. Since that paper many wells have been drilled both in the south and in the north and we are today in possession of a picture of some complexity. Drilling since 1968 has confirmed the conclusions of the above paper as regards the south and, despite the fact that the ranges of some of the diagnostic flora have been extended as more material has been studied, no major revisions of the time and rock correlations have resulted. In the south, therefore, the later drilling has filled out the previous picture concerning the areal distribution of rock units. In the north, it has revealed the existence of basins not previously known and as yet undescribed in published form.

This paper will concentrate therefore on a brief review of the main aspects of each of these basins, in the south and in the north, indicating the approximate limits, nature and age of the lithologies encountered. No discussion of the problems of local rock correlation, or of palynological zonation is undertaken. In the case of the former the author has relied on published reviews (Audley-Charles, 1970a, b; Geiger and Hopping, 1968), and for the latter on the considerable assistance of Mr T. Schroeder who has identified most of the palynomorphs mentioned in the text. It will be noted that most of the supporting detail applies to UK areas, a fact which indicates only that UK data were the most readily accessible to the author.

SETTING OF THE BASINS

The Triassic of the North Sea at the present day lies in several distinct basins (Fig 1), some of which were only separated by late or post-Triassic events. The deposits vary considerably from basin to basin as regards thickness, lithology and the ages of the sequences preserved. Knowledge of the complete succession is available in certain parts of the southern North Sea, while in central and northern parts of the area the picture is usually incomplete. This is due to post-Triassic erosion, or poor age control in the largely unfossiliferous rocks, or to the incomplete penetration of the section in deeper parts of central and northern basins. The isopach map (Fig 1) indicates in a broad way the relative thicknesses of Triassic sediments preserved today in the various basins. In the southern basins a maximum of 5000ft (1500m) accumulated in the Anglo-Dutch basin, and 2500ft (750m) in the South-West Netherlands Trough. In the south Central Graben (Netherlands offshore) a total of some 6000ft (1800m) has been preserved, while south-eastwards in the North German Basin rather greater thicknesses were laid down.

In the west Central Graben, north of the Mid North Sea High no more than 3000ft (900m) of Triassic are now preserved, while in the Danish Embayment thicknesses exceeding 10 000ft (3050m) were laid down and have been variably eroded.

In the Moray Firth Basin and the Viking Graben, where tectonic control of sedimentation

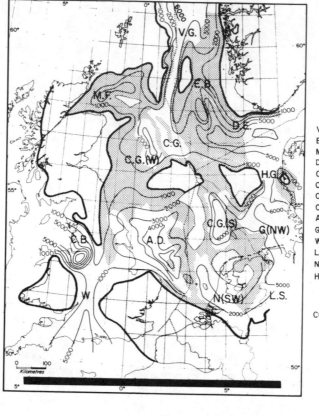

V.G.   VIKING GRABEN
E.B.   EGERSUND BASIN
M.F.   MORAY FIRTH BASIN
D.E.   DANISH EMBAYMENT
C.G.   CENTRAL GRABEN
C.G.(W) WEST CENTRAL GRABEN
C.G.(S) SOUTH CENTRAL GRABEN
C.B.   CHESHIRE BASIN
A.D.   ANGLO-DUTCH BASIN
G(NW)   NW GERMAN BASIN
W   WORCESTER GRABEN
L.S.   LOWER SAXONY BASIN
N(SW)   SOUTH WEST NETHERLANDS BASIN
H.G.   HORN GRABEN

CONTOURS IN FEET

NON-DEPOSITION AREAS IN TRIASSIC TIME.

LATE KIMMERIAN UPLIFTS. NO TRIASSIC PRESERVED.

PARTIALLY ERODED TRIASSIC.

FIG 1.   Generalized isopach map of preserved North Sea Triassic.

was most marked, maximum thicknesses of about 3000ft (900m) and 6000ft (1800m) respectively are known from seismic evidence and drilling.

Despite the disparity in the thickness and nature of each basin-fill a few generalizations in respect of the North Sea Triassic are permissible.

Firstly, the Triassic sections encountered, with very few exceptions, lie directly and without evidence of tectonism or major time hiatus, upon Upper Permian rocks.

Secondly, the Triassic displays over the whole North Sea area an overall continental depositional aspect, with definite marine influences confined only to parts of the middle Triassic in southern areas and to the uppermost Rhaetian passage beds into the Jurassic. Possible marine influences have also been recorded in the lowermost Scythian in the Central Graben area.

South of the Mid North Sea and Ringkøbing–Fyn highs the basins show a continuance of the uniformity of deposition that was established in pre-Triassic, post-Hercynian times and typifies the underlying Permian.

Middle and upper Triassic evaporites are now mainly restricted to areas south of the Mid North

Sea High and it is unlikely that they ever extended much farther north.

Palynological occurrences (Fig 2) have been referred to Alpine stages according to the compiled range charts currently in use in Shell Companies. However, as Visscher (1974) has pointed out, an inherent weakness in the Triassic stratigraphic subdivision widely used today is that palynologically defined time intervals are being expressed in terms of Alpine "standard" stages before the pollen ranges in the Alpine type sections have been fully worked out. Only the Carnian and Rhaetian are yet published.

THE SOUTHERN NORTH SEA BASIN

The broadly east-south-east–west-north-west trending post-Hercynian basin system extends from Germany to England bounded on the southern rim by the Rhenish and London–Brabant massifs. The main basin terminates to the west against the Pennine High, with a narrow connection maintained through the English Midlands to the deep fault-bounded Cheshire and

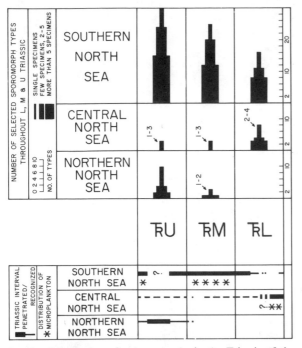

FIG 2. Distribution of palynomorphs in the Triassic of the North Sea.

Worcester grabens. To the north, the basin system is bounded by the highs extending from Northern Denmark to Northumberland. Post-Hercynian/early Cimmerian tectonic trends are superimposed in the Horn and south Central Graben areas and in the Cheshire–Worcester basin subsidence.

Parts of the southern North Sea basins have suffered considerable post-Triassic uplift and denudation (Fig 1) notably in those areas marginal to the later Mesozoic graben, such as the Cleaver Bank and Texel–IJsselmeer highs. Thus the original, probably continuous southern North Sea Triassic basin is now broken up into several sub-basins—the North-west German Basin, the southern Central Graben basin (Netherlands offshore), the Broad Fourteens and South-west Netherlands basins, and the Anglo-Dutch Basin.

The sediment-fill varied with time, but great lateral uniformity prevailed, except at the margins (Fig 5). The initial phase (equivalent to Lower and Middle Bunter) saw the deposition of mainly fine-grained terrigenous clastics, which become increasingly coarse and cyclical in the eastern half of the region. The subsequent phase of deposition (Upper Bunter, Muschelkalk and Keuper) was dominated by fine-grained clays alternating with evaporites. Marine carbonates were important early on, particularly in south-eastern areas, but progressively less so towards the UK. Coarse clastic deposition persisted longer in marginal areas, such as the UK onshore and southern Netherlands.

In the southern basin, therefore, deposition throughout the Triassic took place in a broad,

remarkably flat basin inherited from late Permian time, though reduced in area. In contrast to the Upper Permian, however, marine influences very rarely reached the basin, and the sediments contain a higher proportion of coarse clastic material. Seen as a whole the Triassic in the south constitutes an overall fining-upwards sequence culminating in the first sustained fully marine incursion in the Rhaetian.

The correlation section (Fig 3) extends 250km in a south-east to north-west direction, more or less axial to the Triassic basin itself.

Cuttings and sidewall samples have provided the only direct lithological evidence, and this has been augmented by petrophysical logs, sonic and gamma ray logs being the most commonly available.

In the southern North Sea, the Triassic is clearly divisible into two lithostratigraphic groups, a lower one characterized by alternations of mudstones and sandstones and an upper one in which cyclic repetitions of clay and evaporites with minor carbonate streaks prevail (Figs 3 and 5). The traditional four-fold subdivision of Bunter, Muschelkalk, Keuper and Rhaetic which was, until recently, extended from the German type areas westwards to the UK by virtue of excellent log-marker correlation, has been revised in the UK sector of the North Sea to accommodate the phasing out of the Muschelkalk, towards the west. In UK waters a three-fold subdivision has been proposed by the joint Oil Industry–Institute of Geological Science committee (Rhys, 1974) and is employed in this paper.

The relationship between the old and the new rock terminology is as follows (see also Fig 3):

Proposed new terms (UK offshore)	German subdivision (Netherlands, Germany, Denmark)
Winterton Formation	Rhaetic
Haisborough Group	Keuper Muschelkalk Upper Buntsandstein
Bacton Group	Middle and Lower Buntsandstein

The Triassic of the southern North Sea will be reviewed in two phases:

(1) an initial phase of clastic deposition persistent throughout the Lower and Middle Scythian, and represented by red mudstones and sandstones (Bacton Group; Lower and Middle Buntsandstein);

(2) a subsequent phase of fine-grained cyclical deposits of muds and evaporites, ranging from Upper Scythian to Rhaetian (Haisborough Group and Winterton Formation; Upper Buntsandstein, Muschelkalk, Keuper and Rhaetic).

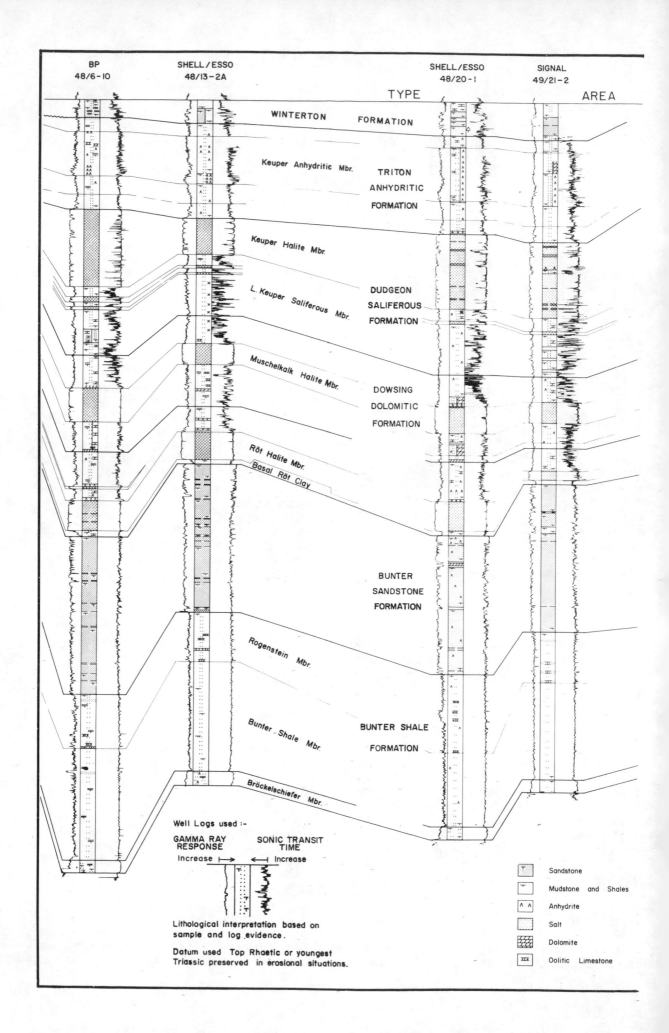

BP
48/6-10

SHELL/ESSO
48/13-2A

SHELL/ESSO
48/20-1

SIGNAL
49/21-2

TYPE   AREA

WINTERTON   FORMATION

Keuper Anhydritic Mbr.   TRITON
                                ANHYDRITIC
                                FORMATION

Keuper Halite Mbr.

                                DUDGEON
L. Keuper Saliferous Mbr.   SALIFEROUS
                                FORMATION

Muschelkalk Halite Mbr.
                                DOWSING
                                DOLOMITIC
                                FORMATION

Röt Halite Mbr.
Basal Röt Clay

                                BUNTER
                                SANDSTONE
                                FORMATION

Rogenstein Mbr.

Bunter Shale Mbr.   BUNTER SHALE
                                FORMATION

Bröckelschiefer Mbr.

Well Logs used :-

GAMMA RAY   SONIC TRANSIT
RESPONSE   TIME

Increase ⟶   ⟵ Increase

Lithological interpretation based on
sample and log evidence.

Datum used Top Rhaetic or youngest
Triassic preserved in erosional situations.

	Sandstone
	Mudstone and Shales
	Anhydrite
	Salt
	Dolomite
	Oolitic Limestone

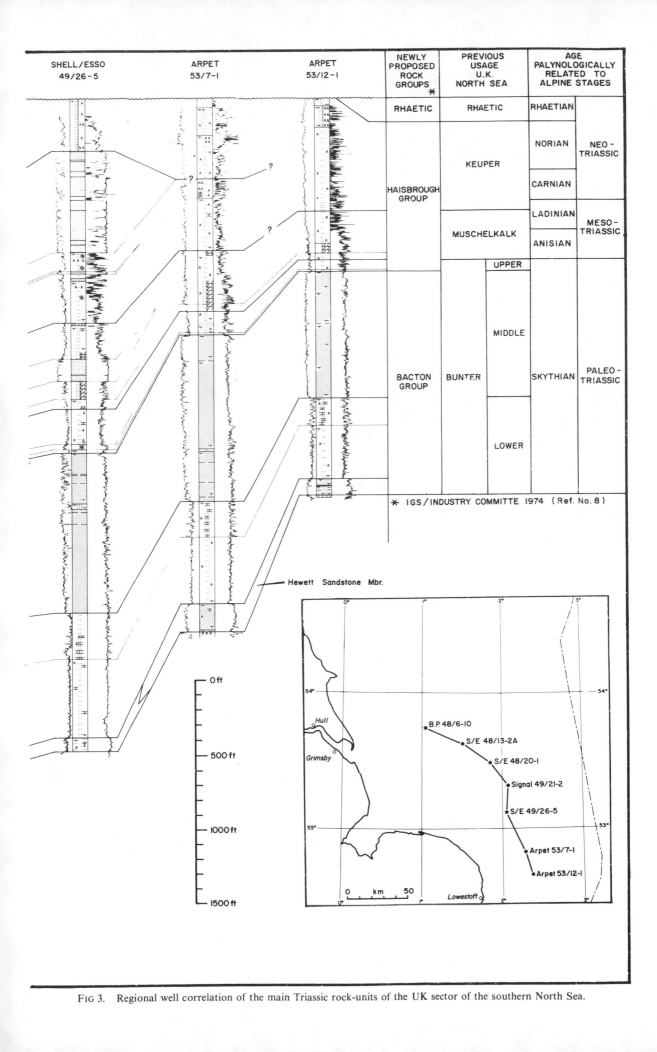

SHELL/ESSO 49/26-5	ARPET 53/7-1	ARPET 53/12-1	NEWLY PROPOSED ROCK GROUPS *	PREVIOUS USAGE U.K. NORTH SEA	AGE PALYNOLOGICALLY RELATED TO ALPINE STAGES	
			RHAETIC	RHAETIC	RHAETIAN	
			HAISBROUGH GROUP	KEUPER	NORIAN	NEO-TRIASSIC
					CARNIAN	
				MUSCHELKALK	LADINIAN	MESO-TRIASSIC
					ANISIAN	
			BACTON GROUP	BUNTER UPPER	SKYTHIAN	PALEO-TRIASSIC
				MIDDLE		
				LOWER		

* IGS / INDUSTRY COMMITTE 1974 (Ref. No.8)

Hewett Sandstone Mbr.

0 ft

500 ft

1000 ft

1500 ft

Hull
Grimsby
B.P. 48/6-10
S/E 48/13-2A
S/E 48/20-1
Signal 49/21-2
S/E 49/26-5
Arpet 53/7-1
Arpet 53/12-1
Lowestoft

0   km   50

FIG 3.   Regional well correlation of the main Triassic rock-units of the UK sector of the southern North Sea.

Bacton Group: Lower and Middle Buntsandstein

The Anglo-Dutch Basin (the Bacton Group). At the end of the Permian the basin margins were probably covered with a thin but extensive cover of fine-grained sabkha deposits. The onset of Triassic conditions was signalled by an influx of clastic deposits, coarsest at the margin but very rapidly fining into the centre of the basin. The first sign of the withdrawal of the sea and the increased energy of erosion and deposition is remarkably distinctive and widespread throughout the basin. In north-east England the Permian Saliferous Marls pass upwards into the Lower Mottled Sandstone, the base of which must be regarded as the first indication of the regressive phase. Time stratigraphical control is, however, obscure owing to the rarity of fossil evidence, and the Palaeozoic/Mesozoic boundary is not easily defined. There is some slender evidence for a lowermost Scythian age in the Bröckelschiefer of the West Sole gas-field and in wells of the quadrant 38 area.

Two formations are distinguished in the Bacton Group, the Bunter Shale Formation and the overlying Bunter Sandstone Formation. The basal unit of the Bunter Shale is the Bröckelschiefer Member, which is represented by characteristically argillaceous sandstones with varying amounts of clay, dolomite or anhydrite. Towards the south and west of the basin fine- to coarse-grained Hewett sandstones are present in the upper part of the Bröckelschiefer, which remains argillaceous above and below, while retaining local intercalations of anhydrite or dolomite.

Distinguishing the Bröckelschiefer Member from the overlying Bunter Shale and the underlying Permian evaporites is usually clear on well-logs, gamma ray response and acoustic travel time both decreasing characteristically from the uniform Bunter Shale values (Fig 3).

The succeeding Bunter Shale is a well-defined unit which can be traced throughout the entire Southern North Sea Basin, and its lithology remains constant from well to well. Typically it comprises thick red-brown and some greenish shales and clays and mudstones, which are variably anhydritic. There are frequent thin intercalations of calcareous micaceous siltstones and the shales and clays are generally somewhat silty. In the upper part there are thin streaks of silty or sandy limestone or dolomite within which ooliths with a ferruginous coating are prominent. The streaks are usually about 5ft (1.5m) or less in thickness and several may occur over a total interval of 400ft (122m) in the thickest part of the Trias. This sandy oolith-bearing interval is clearly the equivalent of the "Rogenstein" in the Lower Buntsandstein of Lower Saxony, and it has been retained as a second member of the Bunter Shale Formation in the UK sector of the North Sea.

As with the underlying Bröckelschiefer the Bunter Shale itself displays an astonishing uniformity of lithology and thickness testifying to the existence of a very flat basin surrounded by featureless and low-lying source areas. The fineness and extreme redness of the sediments suggest that much of the source area was covered by Permian and that the cause of the Triassic regression was initially less due to major earth movements uplifting basin margins than to gentle fault controlled subsidence within the basin, which failed to keep pace with a continent-wide withdrawal of the sea at the end of the Palaeozoic.

The marginal equivalent of the Bunter Shale and Bröckelschiefer could well be the Lower Mottled Sandstone of north-east England. No direct evidence for the age of the Bunter Shale has yet been obtained in the Southern North Sea Basin, where it is everywhere barren of sporomorphs. North of the Mid North Sea High comparable lithologies have, however, yielded a lowermost Scythian flora in several wells. This will be discussed later.

In the Anglo-Dutch Basin the next unit above the Bunter Shale is a thick sheet-sand complex which has been named the Bunter Sandstone Formation. Though the lower part is more argillaceous than the upper, the onset of coarse deposition was abrupt and from log-evidence apparently simultaneous across the basin. Such an influx of coarse material was probably caused by uplift of the source areas. Around the margins of the basin were spread laterally equivalent wadi deposits including coarse pebbly fans such as the Bunter Pebble Beds of the English Midlands.

In the basin itself, the sands which make up the bulk of this unit are generally fine- to medium-grained in the main depositional area, and are variably grey to red-brown in colour. Calcareous cement is present, forming thin intercalations throughout the sand. Locally, anhydrite may develop as an important cementing mineral particularly in the upper part of the unit. Interstitial reddish clay and silt are frequently present, but mica is only occasionally reported.

Numerous thin argillaceous beds occur throughout, the great majority attaining a thickness of 5ft (1.5m) or less. They are reddish-brown to greenish-grey silty mudstones and the frequency of their occurrence is shown by the serrated nature of the gamma ray curves in most wells. As can be seen from Fig 3, the cleanest sands appear in the upper part of the complex, but this is accompanied by a greater frequency of interbedded claystones, at least one of which is of basinwide correlative value. It is in these clay interbeds that the first characteristic Upper Scythian Triassic flora appear. (*Spinotriletes echinoides, Taeniasporites noviaulensis* (Geiger and Hopping, 1968).)

The Bunter Sandstone correlates eastwards with the Middle Buntsandstein of Germany which will be referred to below. In Germany and Holland the unit contains a break in the sequence, the

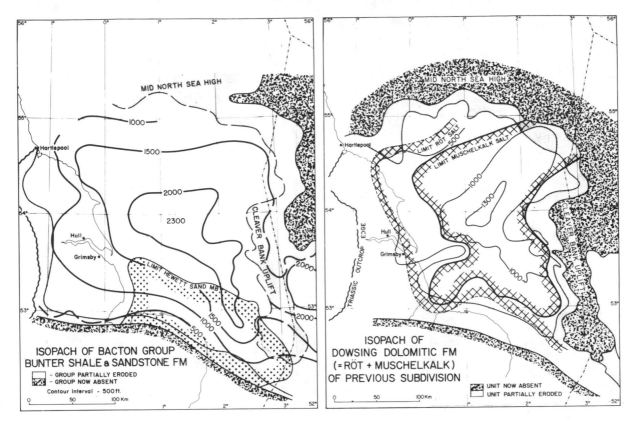

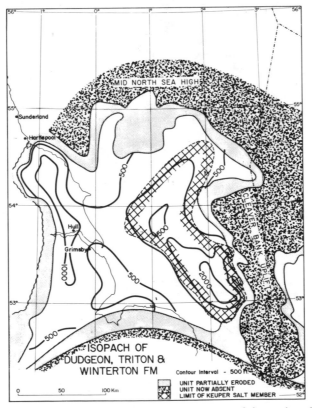

FIG 4.   Thickness variations in the Triassic of the UK sector of the southern North Sea.

Hardegsen unconformity, which can be demonstrated to be erosional in certain areas marginal to the basin and at some of the sites of later Cimmerian uplifts. In British waters it is not recognized except near late Cimmerian highs, and this may be due to the difficulty of detecting a break

in the much sandier sequence of the Bunter Sandstone.

The maximum thickness of 2300ft (700m) of Bacton Group sediments was deposited in the centre of the basin some 80km offshore from Britain as the regional isopachs (Fig 4) show. Sand

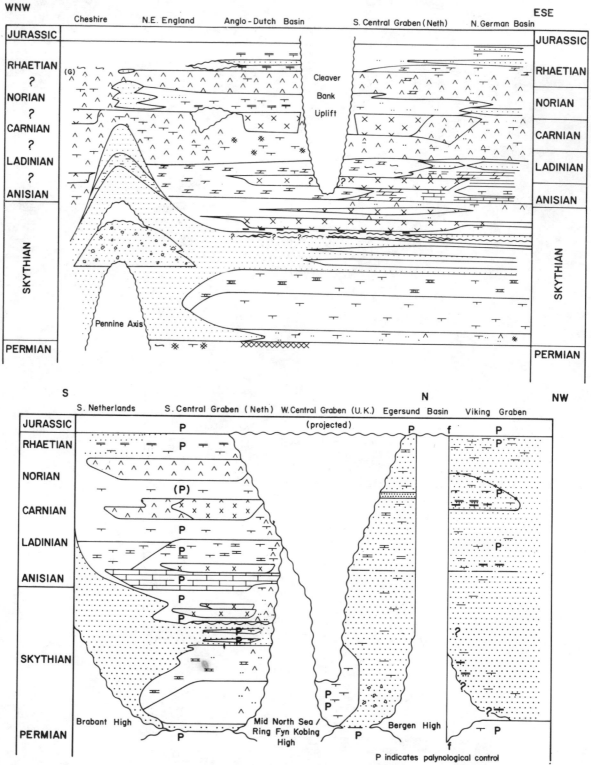

FIG 5. Generalized time-facies diagrams for the Triassic of the North Sea: upper section, east—west from Central England to north Germany; lower section, south—north from Netherlands to Viking Graben.

accumulation itself exceeded 1000ft (300m) in areas east of Spurn Head and Bacton.

The Eastern basins—Netherlands, Germany and Denmark (Lower and Middle Buntsandstein). During the initial phase of the Triassic very similar conditions to those in UK waters prevailed in on- and off-shore Netherlands and in Germany. In the marginal areas in the south Netherlands difficulty again arises in defining the Permian/Triassic boundary in the entirely sandy facies overlying the Carboniferous. As in UK, the red shale facies of the Lower Buntsandstein is absent at the fringe, but in the basin it is clearly the product of identical depositional conditions as existed in the UK offshore. The basal siltstone member, the red mudstones and ferruginous oolites are indistinguishable from their counterparts in the western area. In the upper sandier beds of the Middle Buntsandstein, however, differences become noticeable east of the Cleaver Bank uplift and the Texel–IJsselmeer High. What in the UK area is largely a homogeneous sheet-sand is split up into four cyclic repetitions of sand and mudstone which can be traced to the classic German exposures where they are designated, in upwards succession, the Volpriehausen, Detfurth, Hardegsen and Solling cycles respectively (Fig 5).

Between the cyclic basinal Middle Buntsandstein and the coarse clastic fringe there is no equivalent of the British sheet-sand complex. This could imply a less extensive sand source area supplying the eastern parts of the Southern North Sea Basin. The cycles presumably indicate water-level fluctuations in a very flat basin and their boundaries should be close to time lines. Sand distribution studies in the Netherlands indicate that the clastic supply originated from the west and the south.

No age diagnostic flora has been published for the Middle Buntsandstein of the Netherlands, the Scythian age being attributed by correlation from the well-dated cycles in Saxony (Geiger and Hopping, 1968). Unpublished confirmation of this has since been obtained in many later wells.

The northern flank of the southern basin lies in north Netherlands, Germany and Danish waters, and the deposits wedge out by erosion northwards against the Ringkøbing–Fyn High. A thin sequence of 55ft (17m) of unfossiliferous red-brown claystone with some anhydrite and a basal siltstone was penetrated in Dansk Nordsø well B-1 (Rasmussen, 1974). Some 420ft (128m) of similar material was encountered in Dansk Nordsø well D-1 and both sequences are most probably thin Bunter Shale equivalents conformably overlying Permian Zechstein.

A minor development of thin brownish sands with clay interbeds is present in well B-1 between the red claystone and the first dated Jurassic; there are possible feather-edge equivalents of the Middle Buntsandstein preserved beneath the early Cimmerian unconformity.

Haisborough Group and Winterton Formation: Upper Buntsandstein, Muschelkalk, Keuper and Rhaetic

Except in the extreme fringe areas the sand supply was abruptly curtailed towards the end of the Scythian. Throughout the remaining Triassic a regime of fine-grained, and at times marine, deposition prevailed. Periods of evaporitic deposition alternated with distal floodplain and coastal sabkha conditions. The relief in the source areas induced by post-Permian movements and sea withdrawal was by now peneplained, and the region must have been entirely flat over immense areas, across which marine influences could be rapidly established and as rapidly removed by only minor fluctuations of sea-level. Transitional areas between marine and non-marine conditions must have been very wide. Thus log correlations of quite minor beds can be carried great distances as, for example, with the basal bed, the thin transgressive Röt clay (Fig 3).

The Anglo-Dutch Basin (Haisborough Group and Winterton Formation). There were three phases of salt deposition in the Anglo-Dutch Basin, each accompanied by a progressive reduction in area of the salt-lakes (Fig 4). The earliest, that of the Röt Salt, was probably continuous over the entire Southern Basin and was reminiscent of earlier, Zechstein salt, deposition. The later two (the Muschelkalk and Keuper salts) were apparently restricted to smaller areas by early movements and were later affected by more severe early Cimmerian movements.

The middle salt-cycle equates with the marine Muschelkalk phase, which is fully developed only in the east. There are dolomite stringers both above and below the Muschelkalk Salt but no boundary based on the presence of such stringers can usefully be drawn between the first and second cycles.

The boundary between the 2nd and 3rd cycles is taken at a sonic log marker indicative of a regionally transgressive clay.

The lithological characteristics of the three cycles in the western area are summarized in the table on p. 304.

The lower two cycles attain comparable thicknesses of some 500ft (150m) apiece, and the uppermost cycle can exceed 1000ft (300m).

The age of the lowest cycle is not determined in the UK area owing to the lack of short-ranging pollen at this level, but it is probably at least Upper Scythian in age.

The second cycle cannot be younger than Anisian in the lower part (on the basis of *Aequitriradites minor* (Geiger and Hopping, 1968; Madler, 1964a, 1964b; Visscher and Commissaris, 1968)) but probably ranges up into the Ladinian above the salt member (*Illinites chitonoides, Striatoabietites balmei* and *Microsachryidites sittleri* (Döring, 1966; Schultz, 1964; and others)). This interval is of interest in that it contains

Cycles	Name (Rhys, 1974)	Lithology	Base
3.	Dudgeon Saliferous Fm. (= previous Lower Keuper)	Red silty clays with abundant salt streaks, passing up into the Keuper Halite Member at the top	Transgressive clay, regional sonic log marker
2.	Dowsing Dolomitic Fm., upper part (= previous Muschelkalk)	Red silty clays with abundant anhydrite or dolomite streaks; some minor greyish mudstones; the clays sandwich the Muschelkalk salt	Arbitrary dolomite streak, at less well defined regional sonic log minimum
1.	Dowsing Dolomitic Fm., lower part (= previous Upper Buntsandstein, Röt)	Red silty clays with minor anhydrite and dolomite streaks, sandwiching one major and one minor salt	Transgressive clay, good log marker

acritarchs, indicative of at least semi-marine conditions.

The third cycle could well be mainly of Carnian age owing to the presence of *Retisulcites perforatus* in the lower part and *Camerosporites secatus* in the upper (Döring, 1966; Madler, 1964b; Scheuring, 1970).

The correlation of the three cycles with the English marginal facies is possible owing to the presence of a flora restricted to the Anisian in parts of the Waterstones (Fisher, 1972; Smith and Warrington, 1971) thereby implying that this formation is the marginal equivalent of the most marine influence yet noted in the basinal later Trias.

The Keuper Halite Member is the thickest salt body in the UK North Sea Triassic, but it is more restricted in area than the earlier salts. The abrupt pinchouts between wells 49/26-5 and 53/7-1 and between 48/6-10 and 47/8-1 may in part be due to erosion and overstep by younger sediments at the basin margins.

The penultimate non-marine Triassic is represented by reddish mudstones with interbedded anhydrite and minor dolomite. The tendency for anhydrite and dolomite to be well developed in the upper part enables a distinction to be made between a lower claystone member and an upper Keuper Anhydritic Member, both easily recognizable on log records (Fig 3), the former directly overlying the Keuper Halite. The unit can be correlated towards the basin margins where on log evidence it appears to overstep the underlying bed at an erosional contact.

Only long-ranging pollen has been found, but the evidence would not conflict with a Norian age.

Except in the south-east, the Rhaetic sands and clays (Winterton Formation) which succeed the Haisborough Group have little in common with the Triassic already described. Above a basal clay, the sandstones are grey-green to whitish grey, fine grained and are interbedded with greyish green to dark grey mudstones. Plant debris has been noted, and there is evidence of an influx of new sporomorphs at the base of the unit.

In the north-west, log correlation within the underlying Keuper anhydrites shows the Rhaetic basal clay transgressively overstepping partly eroded Haisborough Group. Similar cut-outs are widespread, marginal to localities of more severe erosion in late Jurassic time, and are indicated on Figs 1 and 4.

The Eastern Basins—Netherlands, Germany and Denmark (Upper Buntsandstein, Muschelkalk, Keuper and Rhaetic). The same rock succession can be traced eastwards through the Netherlands into the German offshore with modifications developing, firstly in the lowermost salt cycle, in which the minor salt becomes increasingly important, giving rise to an upper and lower Röt halite, and secondly in the Muschelkalk itself, in which carbonate stringers coalesce eastwards into more massive limestones and dolomites of the classic development. In Netherlands waters all the salt cycles are developed in the Broad Fourteen's and Southern Central Graben Basins, though the Keuper Halite is restricted mainly to the latter; none is present in the marginal Central Netherlands Basin. The full suite reappears in the Northern German Basin, near the Dutch land border. In north Germany, Triassic salts have largely wedged out, and in the basin the successive areal shrinkage of each salt lake matches the picture in the Anglo-Dutch Basin.

Infill well data-points have confirmed the age allocations to the units made in 1968 by Geiger and Hopping (1968), with the exception that the Upper Muschelkalk is now interpreted, as in the UK, to be Ladinian in age, in this case on the uncertain basis of the occurrence of *Triadispora spp.*

At the southern margin of the basin in the south Netherlands, conditions were similar to those in north-east England, with sands persisting from the Lower Trias into the Anisian, and being replaced by anhydritic siltstones in the Ladinian.

Here also post-Ladinian/Pre-Rhaetian uplifts must account for the removal of the Keuper over much of the Netherlands land area, since Rhaetic sands and clays lie directly on Muschelkalk over large areas.

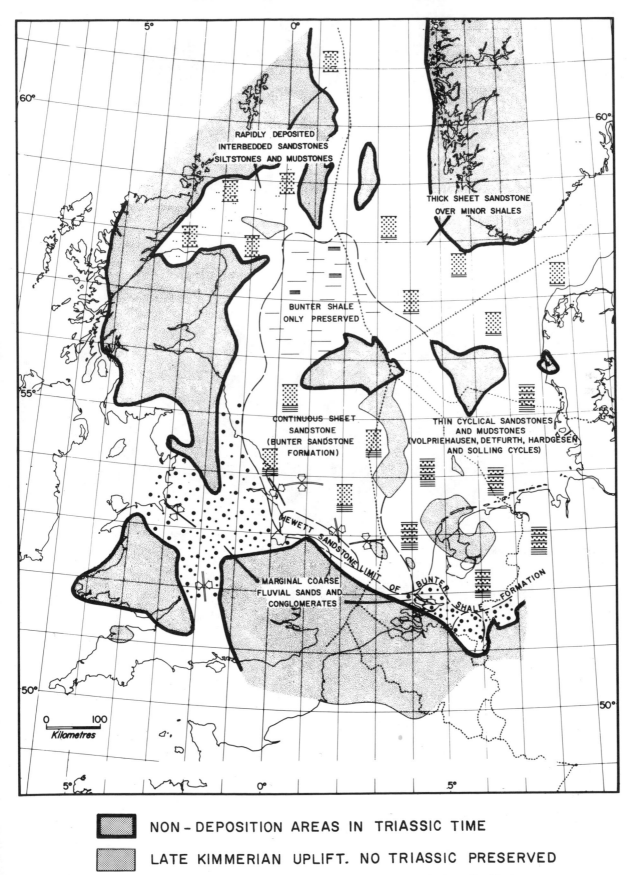

RAPIDLY DEPOSITED
INTERBEDDED SANDSTONES
SILTSTONES AND MUDSTONES

THICK SHEET SANDSTONE
OVER MINOR SHALES

BUNTER SHALE
ONLY PRESERVED

CONTINUOUS SHEET
SANDSTONE
(BUNTER SANDSTONE
FORMATION)

THIN CYCLICAL SANDSTONES
AND MUDSTONES
(VOLPRIEHAUSEN, DETFURTH, HARDGESEN
AND SOLLING CYCLES)

HEWETT SANDSTONE LIMIT OF BUNTER SHALE FORMATION

MARGINAL COARSE
FLUVIAL SANDS AND
CONGLOMERATES

0   100
Kilometres

NON-DEPOSITION AREAS IN TRIASSIC TIME

LATE KIMMERIAN UPLIFT. NO TRIASSIC PRESERVED

FIG 6.   Regional lithofacies in the Triassic of the North Sea: early phase (Scythian).

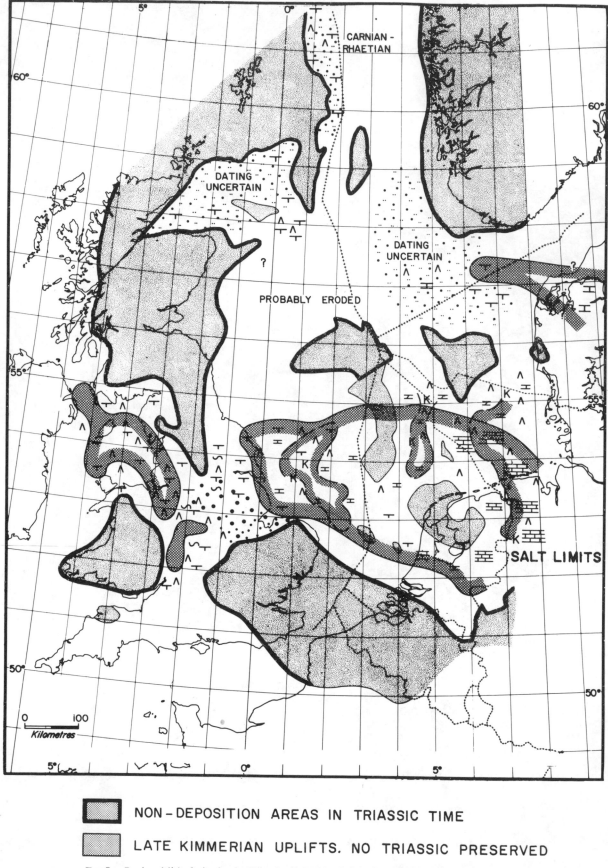

NON – DEPOSITION AREAS IN TRIASSIC TIME

LATE KIMMERIAN UPLIFTS. NO TRIASSIC PRESERVED

FIG 7.   Regional lithofacies in the Triassic of the North Sea: late phase (Anisian—Rhaetic).

To the north-east, against the Ringkøbing–Fyn High, Muschelkalk and Keuper sediments are developed in a semi-marginal facies, with limestones alternating with anhydrite clays, passing up through anhydritic mudstones into a sandy and occasionally dolomitic facies near graben margins. Clay with some anhydrite terminates the succession in more basinal areas.

No palynological age-data for the sections in German and Danish waters are available for inclusion in this paper.

The regional lithofacies for the southern basin systems has been summarized in map form (Figs 6 and 7) and in a time-facies diagram (Fig 5).

THE NORTHERN NORTH SEA BASINS

The Central Graben (West)

Most of the wells drilled in the Central Graben and Forth Approaches area contain some Triassic (Fig 8). No complete section has however been found in this part of the North Sea. Within the area indicated on Fig 1 the Triassic is preserved only in a brown-red to brick-red mudstone facies very similar to that of the Bunter Shale Formation, though lithologies characterizing the basal siltstones (the Bröckelschiefer Member) and the Rogenstein oolites are not present. The red mudstone in this area lies directly upon Permian Zechstein, equally upon basinal and shelf facies, without itself indicating any variation in facies.

The upper boundary of the red mudstones is always erosional or non-sequential, the overlying sequences varying in age from probable mid-Jurassic to late-Cretaceous. As a result thicknesses vary greatly from well to well, ranging from near zero to some 2000ft (600m), and in undrilled synclines apparently reaching 3000 to 4000ft (900 to 1200m). The erosion that caused this wide thickness variation was intensified by local uplifts near the graben margin or by local post-Triassic salt movements in the extensive Zechstein basinal area.

As may be expected, the time-stratigraphic evidence in this oxidized sequence is extremely sparse. However, the mudstone has yielded a few samples of age-diagnostic flora, all of which point to a Lower Triassic age in the Forth Approaches and Central Graben (UK) areas. In Shell/Esso 21/26-1D, 21/11-1, 29/25-1 and 30/16-2 there is an assemblage of flora known to range from Permian into lower Scythian, containing *Gnetaceaepollenites* cf *steevesi* (in all 4 wells), *Endosporites papillatus* and *Taeniaesporites sp.* (Jansonius, 1962; McGregor, 1965; Visscher, 1971). In the absence of any flora restricted to the Permian and in view of the lithology, the age of the formation is interpreted to be Lower Scythian, and equivalent to that of the Bunter Shale of the south. The spores and pollen have been found low in sections 2000ft (600m) thick which pass up without

lithological change; acritarchs were also found at the base.

The youngest age indications are poor and rare, a single specimen of *Porcellispora longdonensis* was found in Shell/Esso 29/25-1 suggesting that this uniform red mudstone facies persists into the Anisian or even later (Geiger and Hopping, 1968; Visscher and Commissaris, 1968). *Chordasporites* occurs near the top of the 900ft (274m) of red mudstones in Shell/Esso 21/26-1D, pointing to an Upper Scythian or later age (Smith and Warrington, 1971).

The floral ages obtained would, if correct, indicate that this part of the North Sea remained a flat, featureless, non-marine reception area for fine red muds and dusts, immune from the incursions of the Middle Bunter clastics or the Muschelkalk sea. Source areas for the deposits preserved must have been distant and were probably thickly covered with fine Permian muds and carbonates.

The Danish Embayment/Egersund Basin

As in the southern basin Cimmerian movements severely reduce the thickness of the Triassic rocks particularly near the Central Graben margins, both west and east. The underlying Zechstein salts thicken enormously in north Danish and southern Norwegian waters, and are associated with major halokinesis. Nevertheless, great thicknesses of Triassic have been penetrated in wells north of the Ringkøbing–Fyn High, and these indicate that the energy of sediment supply was far greater in the vicinity of the Scandinavian shield, than in the western area described above.

The sediments accumulated in great thickness in an east–west trending trough, the Danish Embayment/Egersund Basin, lying between the Ringkøbing–Fyn High and the Scandinavian Shield (Fig 6). Isopachs in Fig 1 indicate a preserved thickness of some 12 000ft (3650m) or more in the deepest part of this trough, which is comparable in magnitude and appearance to the Cheshire Basin. The surviving fill of the Danish Embayment is however mainly of clastics, with sands and siltstones predominant (Fig 5). There is evidence, regionally, for an initial short-lived phase of clay-deposition with a maximum drilled thickness of about 500ft (150m), before sands and silty mudstones poured into the basin, some 3500ft (1065m) being preserved locally beneath the regional late-Cimmerian unconformity.

Unlike the Cheshire Basin where mudstones and thick evaporites provided the later fill, salts have been penetrated only in the easternmost part. Conceivably these salts could originally have been extensive over the trough, but were later removed by the post-Triassic erosion, evidence for which is present everywhere in this part of the North Sea.

Westwards from the axis of the Egersund Basin, the depth of post-Triassic erosion progressively attenuates the thickness of preserved Triassic towards the Central Graben.

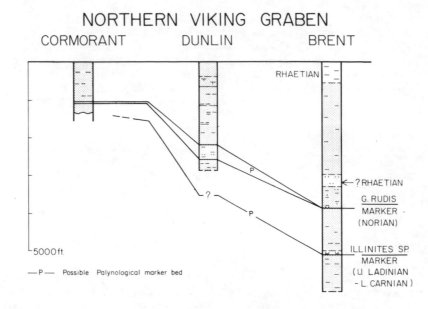

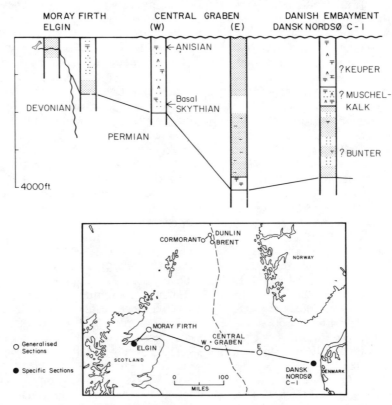

FIG 8.   Triassic sections in the northern North Sea.

The Moray Firth Basin

A succession of fine- to medium-grained white to reddish grey sandstones passing upwards into a sequence in which the argillaceous and silty component is uniformly more prevalent, has been penetrated in this area without, unfortunately, any age diagnosis being obtained. Interposed between definite Permian and Jurassic, however, it is highly likely the sequence is entirely Triassic in age and is of marginal facies (Fig 8). Close by, on land at Elgin, dune sands have yielded a reptilian fauna which is argued by Walker (1973) to favour a Triassic age. This would support the above facies interpretation that the Moray Firth was situated close to the margin of the Triassic basin. Up to 1500ft (450m) have been penetrated in the area, but seismic evidence indicates that even greater thicknesses are present.

Viking Graben

Substantial thicknesses of Triassic rocks have been drilled in the northernmost part of the North Sea, mainly in the greater Brent area. Only a few wells penetrate into beds older than Triassic, however, and nowhere is there a single section in which the complete Triassic has been proven. Evidence of age is extremely sparse, but rare occurrences of fairly fossiliferous beds are sufficient to show that large thicknesses of Upper Triassic were deposited in certain places (Fig 8).

As might be expected from the tectonic domain in which these sections are situated, the basin fill is dominantly of the marginal clastic type, comprising sandstones, red siltstones and mudstones which rapidly infilled a subsiding fault-bounded depression. No salts have been encountered and, as in the Moray Firth, there is an absence of any marine indications.

Block-faulting influenced the rate of subsidence in different parts of the province during the Upper Triassic, thicknesses varying greatly from fault-block to fault-block. However, no attendant facies changes have been noted.

Throughout the province the succession is composed of rapidly alternating sandstones, siltstones and mudstones in which individual bed-thickness is generally 10ft (3m) or less. Occasionally sandstones thicken into larger groups—at Dunlin they average 20ft (6m) thick or more, the maximum for the province rarely amounting to hundreds of feet. Clay interbeds show the same pattern.

The sands are uniformly white to pink, and only rarely red-brown, while the characteristic reddening is carried by the finer clastics, silts and clays. There are rare examples of oxidized mudstone.

A tentative correlation of the Viking Graben Triassic can be put forward on the basis of two possible pollen-marker intervals which are lithologically characterized by a high proportion of claystone to siltstone and sandstone.

The upper of the two, a clay-dominant interval some 300ft (90m) thick in several wells (e.g. 211/23-1, 211/29-5), contains a distinctive flora. In four wells (211/29-4, 5, 211/23-1 and 211/26-2) an assemblage containing *Ovalipollis sp.* and *Granuloperculatipollis rudis* was obtained, and would indicate an Upper Triassic Norian to Lower Rhaetian age. Confirmatory evidence is the additional presence of *Circulina meyeriana* in 211/29-5 and in 211/23-1 (Döring *et al.*, 1966; Rasmussen, 1974; Schultz, 1967).

In the Cormorant well 211/26-1 where an attenuated Triassic onlaps Caledonian basement, the same floral assemblage was obtained in a thin clay-bed. The apparently singular occurrence of this pollen-rich clay interval in each of these wells, and its apparent province-wide distribution would argue for a unique relative water-level rise in Norian times, resulting in a temporary reduction of the inflow of sands into the subsiding graben.

Some 500–1200ft (150–365m) below this probable Norian marker, occurs another pollen-bearing clay-rich interval. This typically contains *Illinites sp.*, which would not be expected to range above lower Carnian. Also present is the longer ranging *P. longdonensis* (Anisian, Norian). No older flora has been obtained above the Permian.

A rhaetian age has been assigned to the highest beds containing flora, on the basis of *Ricciisporites tuberculatus, Cinguilizonates rhaeticus* (Heybroek, 1974), *Rhaetipollis germanicus* and *Striomonosaccites sp.* (Fisher, 1972; Madler, 1964; Schulz, 1967).

There are no satisfactory log correlations possible within the northernmost Triassic. The haphazard distribution of fine to medium sand, silt and clay is consistent with a sediment source in a rather flat hinterland, and deposition in a narrow, rapidly subsiding trough. The major fault-system which controlled the later development of the Jurassic tilted truncation-traps was, on the basis of the above evidence, clearly in existence in Upper Triassic times.

North-East Greenland

Correlation northwards into north-east Greenland provides an interesting partial return to the overall pattern visible in the southern North Sea.

The time–facies interpretation of the Scoresbyland Triassic by Perch-Nielsen (1974) shows that a transition occurs from marine to continental deposition during a time-interval ranging from the Permian to the Scythian. A flood of coarse clastics (Pingo Dal Formation) in the later Scythian is strongly reminiscent of the Bunter Sandstone phase of the southern North Sea. Fine-grained, occasionally semi-marine conditions then returned in the middle Triassic giving rise to a succession (The Gipsdalen Formation, with *Myalina* limestones), not unlike the UK Muschelkalk. The upper part of the succession (the Fleming Fjord

Formation) however resembles more the sandy
Upper Triassic of the Brent Province. No
palynological support is however put forward to
justify these correlations.

ACKNOWLEDGEMENTS

The author wishes to acknowledge the help given
by many colleagues in Shell for contributions and
discussions. He also wishes to thank Shell UK
Exploration & Production Co. Ltd and Esso
Exploration & Production (UK) Inc. for
permission to publish this review. Thanks are also
due to Arco Oil Producing Inc., Signal Oil and Gas
Co. Ltd, and BP Petroleum Development Ltd for
permission to publish Triassic data from wells.

REFERENCES

Audley-Charles, M. G. 1970a. Stratigraphical correlation of the Triassic rocks of the British Isles. *Q. Jl. geol. Soc. Lond.*, **126**, 19–47.

Audley-Charles, M. G. 1970b. Triassic palaeogeography of the British Isles. *Q. Jl. geol. Soc. Lond.*, **126**, 49–89.

Döring, H. 1966. Erläuterungen zu den sporenstratigraphischen Tabellen vom Zechstein bis zum Oligozän. *Abh. Zentr. Geol. Inst.*, **8**, 61–78.

Fisher, M. J. 1972. A record of palynomorphs from the Waterstones (Triassic) of Liverpool. *Geol Jl.*, **8**, 17–22.

Geiger, M. E. and Hopping, C. A. 1968. Triassic stratigraphy of the Southern North Sea Basin. *Phil. Trans. R. Soc. Series B*, **254**, 1–36.

Heybroek, P. 1974. Explanation to tectonic maps of the Netherlands. *Geologie Mijnb.*, **53**, 43–50.

Jansonius, J. 1962. Palynology of Permian and Triassic sediments, Peace River area Western Canada. *Palaeontographica*, **110B**, 35–98.

Madler, K. 1964a. Die geologische Verbreitung von Sporen und Pollen in der deutschen Trias. *Beih. geol Jb.*, **65**, 1–147.

Madler, 1964b. Bemerkenswerte Sporenformen aus dem Keuper und unteren Lias. *Fortschr. Geol. Rheinld. Westf.*, **12**, 169–200.

McGregor, D. C. 1965. Triassic, Jurassic, Lower Cretaceous spores and pollen of Arctic Canada. *Geol. Surv. Pap. Canada*, **64/55**, 1–32.

Perch-Nielsen, K., Birkenmajer, K., Birlelund, T. and Aellen, M. 1974. Revision of Triassic stratigraphy of the Scoresby Land and Jameson Land Region, East Greenland. *Bull. Grønlands. geol. Unders.*, **109**, 51.

Rasmussen, L. B. 1974. Some geological results from the first five Danish exploration wells in the North Sea. *Danm. geol. Unders., 42 Series* (III).

Rhys, G. H. (Compiler) 1974. A proposed standard lithostratigraphic nomenclature for the southern North Sea and an outline structural nomenclature for the whole of the (UK) North Sea. A report of the joint Oil Industry–Institute of Geological Sciences Committee on North Sea Nomenclature. *Rep. Inst. Geol. Sci.*, **74/8**, 14 pp.

Schulz, E. 1964. Sporen und Pollen aus dem Mittleren Buntsandstein des Germanischen Beckens. *Mber. dt. Akad. Wiss. Berl.*, **6**, 597–606.

Schulz, E. 1967. Sporenpaläontologische Untersuchungen rätoliassischer Schichten im Zentralteil des Germanischen Beckens. *Paläont. Abh. B.*, **3**, 544–633.

Smith, E. G. and Warrington, G. 1971. The age and relationships of the Triassic rocks assigned to the Lower part of the Keuper in North Nottinghamshire, North West Lincolnshire and South Yorkshire. *Proc. Yorks. geol. Soc.*, **38**, 201–27.

Venkatachala, B. S. and Goozan, F. 1964. The spore pollen flora of the Hungarian "Koessen Facies". *Acta. geol. Hung.*, **8**, 203–28.

Visscher, H. 1971. The Permian and Triassic of the Kingscourt outlier, Ireland: a palynological investigation related to regional stratigraphical problems in the Permian and Triassic of western Europe. *Geol. Surv. Ireland Spec. Pap. 1*, 114 pp.

Visscher, H. 1974. The impact of palynology on Permian and Triassic Stratigraphy in western Europe. *Rev. Palaeobot. and Palyn.*, **17**, 5–19.

Visscher, H. and Commissaris, A. L. T. M. 1968. Middle Triassic pollen spores from the Lower Muschelkalk of Winterswijk (the Netherlands), *Pollen Spores*, **10**, 161–76.

Walker, A. D. 1973. The age of the Cutties Hillock Sandstone (Permo-Triassic) of the Elgin area. *Scott. Jl. Geol.*, **9**, 177–83.

DISCUSSION

D. G. Roberts (Institute of Oceanographic Sciences): Would Dr Brennand comment on the relative importance of Triassic and post-Jurassic faulting in the Viking Graben area?

Professor Arthur Whiteman: Dr Brennand mentioned that in the northern North Sea area near Brent there is a fault-system which must have a downthrow of over 6000ft. Immediately east of Statfjord Field two closely adjacent faults between them displace the Dogger and Lias sands by over 6500ft. These are enormous down-throw-faults within the Viking Graben system and there are others. Could Dr Brennand give us the Shell view of how these faults were formed, because such fundamental structures must continue deep into the crust?

Dr T. P. Brennand: There is insufficient data as yet to determine the relative importance of Triassic, intra-Jurassic and post-Jurassic faulting. A fair idea of the magnitude of post-Jurassic faulting is given by regional seismic mapping at the level of the late Cimmerian unconformity, where the base of the Lower Cretaceous envelops, by onlap, topographic features several thousands of feet high. Absolute fault-throws will not be established until wells have been drilled on the down-thrown side of such faults and established the base of the Jurassic there.

Seismic profiles are of no help in regard to Triassic fault-movements, since the Caledonian basement reflector is not regionally mappable. Drilling on high points has up to now shown that subsidence during Triassic varied from block to block, the relative differences for the Upper Triassic being, again, in the order of thousands of feet. If Triassic sedimentation began in fault-bounded depressions, developing in a largely peneplained landmass, the throws of the bounding fault may well have reached 10 000ft in places.

The origin of the faults must be linked with the development of the Viking Graben itself. The association of tilted fault-blocks and graben-formation is well documented, as in the Red Sea case for example. However, while it is not difficult to visualize the break-up and foundering of the crust within the graben as it forms, the origin of the graben itself is still not resolved.

Robert E. King (Consultant): Dr D. B. Smith has shown that onshore in Yorkshire the Upper Permian evaporite marls are diachronous with the Bunter, and that the Permian-Triassic contact may be within the Bunter. Mr W. H. Ziegler has suggested that the Permian-Triassic boundary may actually be at the Hardegsen unconformity. Dr Brennand does not show the Hardegsen unconformity on his section in the UK sector of the North Sea. How much significance does he attach to this unconformity, and does he think the Permian-Triassic boundary may be within the Bunter?

Dr T. P. Brennand: The Hardegsen unconformity does in fact feature in UK waters as a minor discontinuity representing a local cut-out on the western flank of the Cleaver Bank high (see Fig 4). It can be detected here because of the presence of shale breaks within the Bunter Sandstone Formation which can be correlated with reasonable certainty. In the basin itself no such cut-out has been mapped (Fig 2). I believe the importance of the unconformity diminishes westwards from Germany, and its main expression in the Netherlands or eastern UK waters are minor cut-outs in the upper part of the Middle Bunter, usually in the vicinity of pre-existing or late Cimmerian highs.

As is indicated in Fig 4 and discussed in the text, I believe the Permian-Triassic boundary could well lie near the base of the Bröckelschiefer in the offshore basin, on the basis of poor palynological evidence, and by palaeogeographic inference, near the base of the Lower Mottled Sandstone in the UK onshore.

23

The Geology and Development of the Hewett Gas-Field[1]

By A. D. CUMMING

(Atlantic Richfield Company, Los Angeles, USA)

and C. L. WYNDHAM

(Phillips Petroleum Company, Lima, Peru)

SUMMARY

The Hewett gas-field is located 15 miles off the Norfolk coast. Initial recoverable gas was estimated at 3.5×10^{12} scf contained in 2 Triassic sandstones. The field was discovered late in 1966 and placed on production in July, 1969. By that time 22 wells had been drilled, permanent offshore and onshore facilities installed, unitization negotiations concluded and market secured. The contract with the purchaser called for a gradual increase in production to an average rate of 600 MMcfd by 1974. The 2 reservoirs are largely coextensive, and the field, roughly elliptical in outline, has a length of 18 miles and a maximum width of 3 miles. Average depth to the Bunter Sandstone is 3000ft and to the Hewett Sandstone is 4150ft. Maximum observed gross pay thicknesses are 323ft (Bunter) and 202ft (Hewett). Both reservoirs have excellent porosity and permeability. The Hewett field structure is apparent on seismic profiles and at both reservoir levels a fault-bounded, north-west–south-east trending anticline is present. The gas in the Hewett Sandstone differs from that in the Bunter Sandstone in that it is free of hydrogen sulphide, but whether this implies separate sources has not been demonstrated conclusively. The Hewett Sandstone has a limited distribution in the North Sea area. Since the discovery of the Hewett field several North Sea wells have found gas in the Bunter Sandstone, but follow-up wells have been unsuccessful. The Hewett field may remain unique.

INTRODUCTION

The Rotliegendes sandstone of Lower Permian age was undoubtedly the prime objective during the initial stages of petroleum exploration in the North Sea area. With vast reserves of gas already established at that level in the Groningen field on the Dutch mainland, this was hardly surprising, and seemed the more justified when early exploratory efforts in the British sector of the North Sea were rewarded by the discovery of major gas accumulations in the Rotliegendes at West Sole and at Leman. These, and more recently discovered fields, are shown in Fig 1.

Operators nonetheless entertained the hope that production would not prove to be confined to the Lower Permian, but that sediments of various ages, known to have equally attractive reservoir characteristics, would prove to be hydrocarbon-bearing. This hope was first realized with the drilling of the Arpet 48/29-1 Hewett field discovery well in the latter half of 1966.

The general, post-Carboniferous stratigraphy of the Hewett area is shown in Fig 2. The discovery well, bottomed at 7281ft (2219m) in the Carboniferous, found gas in the Triassic Bunter Sandstone, in the Triassic Hewett Sandstone and in the Upper Permian Hauptdolomit. The Bunter Sandstone came in at 3034ft (925m) and on test of the uppermost 15ft (4.6m) yielded gas at the

rate of 18.1 MMcf/day. The Hewett Sandstone, encountered at 4178ft (1273m), yielded gas at the rate of 23.4 MMcf/day from a 10ft (3m) perforated interval. The maximum flow rate recorded from the Hauptdolomit was a comparatively modest 1.52 MMcf/day.

A follow-up well, Arpet 48/29-2, shown in Fig 3, located 1.5 miles (2.4km) south-south-east of the discovery found gas in the forementioned three zones and in addition yielded a small quantity of light gravity oil from the Hauptdolomit. It was the drilling of the Phillips 52/5-1 well, however, some 9 miles (14.5km) south-east of Arpet 48/29-1 which gave final confirmation that a field of significant magnitude had been discovered. This well proved to be productive in both Triassic sandstones, testing gas at the rate of 52.5 MMcf/day from the Bunter, and at 23.9 MMcf/day from the Hewett. Further testing yielded gas from both Plattendolomit and Hauptdolomit sections, but at rates which again demonstrated the relatively minor importance of these two carbonate reservoirs. Though apparently capable of limited production from fracture porosity, neither reservoir has appreciable matrix porosity or permeability. Both are excluded from further consideration in this paper.

As delineated by subsequent drilling and seismic mapping, the Hewett field (so named from its proximity to the sea-bed feature known as the

---

[1] Paper presented at the Conference by J. L. Martin, ARCO Oil Producing, Inc.

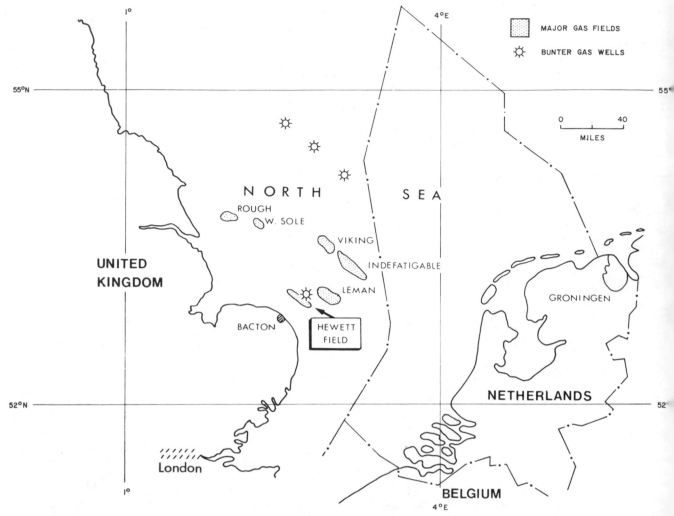

Fɪɢ 1.   General location map of North Sea gas-fields.

Hewett Ledges) has the configuration of an attenuated ellipse with its long axis aligned north-west–south-east. As will be seen from Fig 3, the two Triassic reservoirs are largely co-extensive. Considering both reservoirs the Hewett field is some 18 miles (30km) long and has a maximum width of 3 miles, (4.8km). Located approximately 15 miles (24km) north-east of the Norfolk coast, Hewett is the closest producing field to shore in the North Sea. Water depth does not exceed 130ft (40m).

The maximum measured gross pay thickness in the Bunter Sandstone is 323ft (98.5m) and in the Hewett Sandstone is 202ft (61.5m). Porosity and permeability are excellent in both reservoirs; averaging 25.7 percent and 474md respectively in the Bunter Sandstone, and 21.4 percent and 1310md in the Hewett sand. Total recoverable reserves on the Trias are estimated at 3.5 trillion cu. ft., of which figure approximately 1.5 trillion cu. ft. are attributed to the Bunter Sandstone and 2.0 trillion cu. ft. to the Hewett Sand. The presence of a small but significant quantity of hydrogen sulphide distinguishes the Bunter from the Hewett Sandstone gas.

DEVELOPMENT OF THE FIELD

By March 1967, a total of seven wells had been drilled along the long axis of the Hewett field. In Arpet 48/29-3 only the Hewett Sandstone is gas-bearing, and in Arpet 48/28-1 both sands are wet. The north-western limit of the field was thus established, and the results obtained from the drilling of Phillips 52/5-3 showed this well to be located close to the south-eastern extremity.

Subsequent to the drilling of these seven wells a further twenty-three have been drilled from three fixed platforms within the Hewett field: eight each from Arpet Platforms 48/29-A and 48/29-B and seven from Phillips Platform 52/5-A, which was centred over the Phillips 52/5-1 well. Beneath each platform the deviated well bores are some 900ft (275m) distant horizontally from the centre of the cluster at Bunter Sandstone level and some 1500ft (460m) distant in the Hewett Sandstone.

Long before fixed platform drilling was begun in December 1967, however, the Arpet and Phillips Groups had been aware that unitization of their respective interests would do much to assure the efficient and economic development and

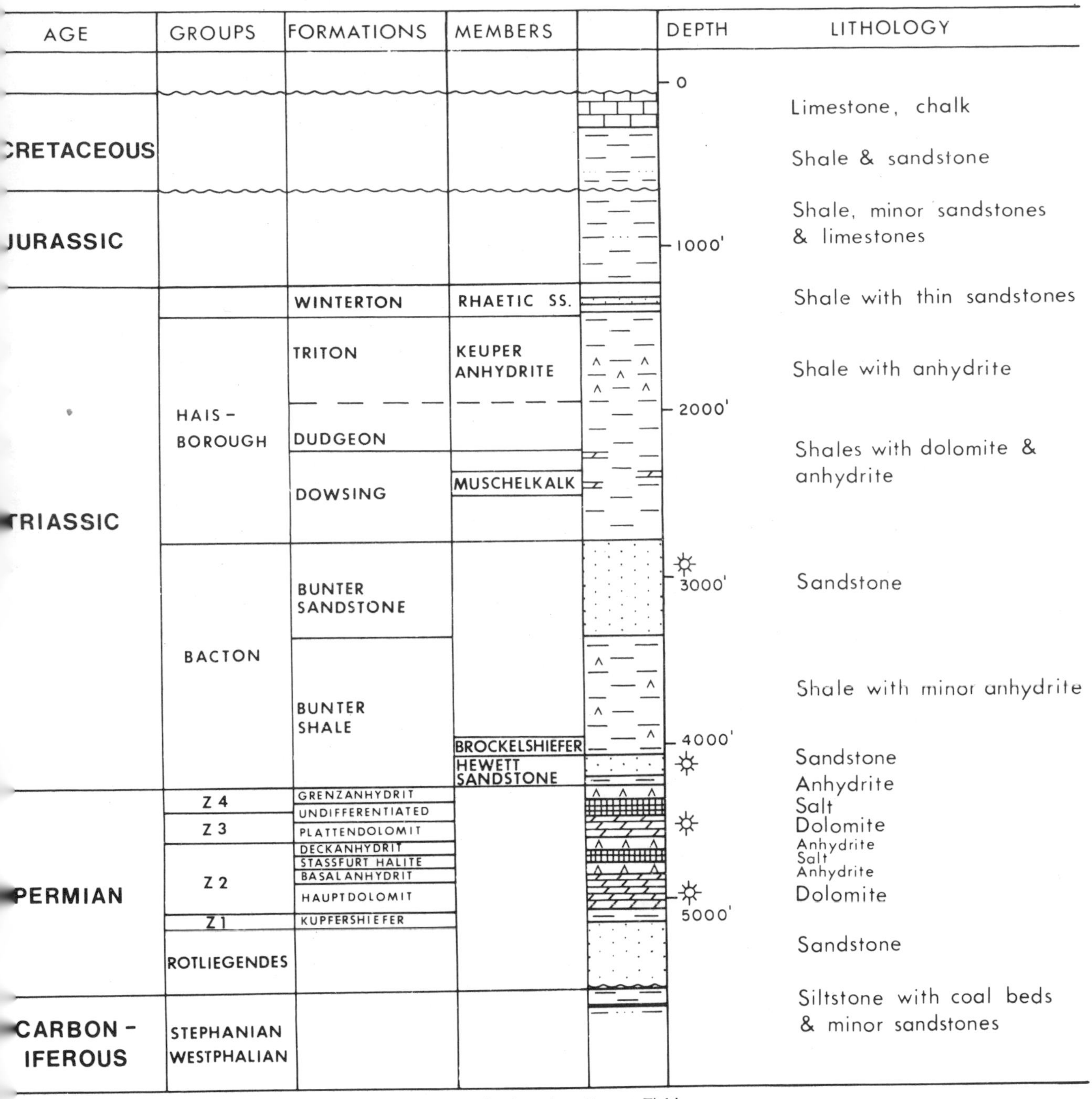

AGE	GROUPS	FORMATIONS	MEMBERS		DEPTH	LITHOLOGY
					─ 0	Limestone, chalk
CRETACEOUS						Shale & sandstone
JURASSIC					─ 1000'	Shale, minor sandstones & limestones
TRIASSIC	HAIS-BOROUGH	WINTERTON	RHAETIC SS.			Shale with thin sandstones
		TRITON	KEUPER ANHYDRITE			Shale with anhydrite
		DUDGEON			─ 2000'	Shales with dolomite & anhydrite
		DOWSING	MUSCHELKALK			
	BACTON	BUNTER SANDSTONE			─ 3000'	Sandstone
		BUNTER SHALE				Shale with minor anhydrite
			BROCKELSHIEFER		─ 4000'	Sandstone
			HEWETT SANDSTONE			Anhydrite
PERMIAN	Z4	GRENZANHYDRIT				Salt
		UNDIFFERENTIATED				Dolomite
	Z3	PLATTENDOLOMIT				Anhydrite
		DECKANHYDRIT				Salt
	Z2	STASSFURT HALITE				Anhydrite
		BASALANHYDRIT				Dolomite
		HAUPTDOLOMIT			─ 5000'	
	Z1	KUPFERSHIEFER				Sandstone
	ROTLIEGENDES					Siltstone with coal beds & minor sandstones
CARBON-IFEROUS	STEPHANIAN WESTPHALIAN					

FIG 2.   Generalized section, Hewett Field area.

exploitation of the new field. With this end in view, negotiations were initiated in April, 1967, which were to culminate in the signing of a Unit Agreement in April 1969. Out of these deliberations the Arpet equity in the single unit was fixed at 45.8 percent and the Phillips equity at 54.2 percent with no provision being made for redetermination.

While the utilization studies and negotiations were being conducted, other matters essential to the development of the field were receiving attention. In March 1968 the Phillips Group signed a sales contract with the Gas Council to supply their portion of the Hewett field gas. In July 1968, the Arpet Group signed an identical contract. In April 1968 the Arpet Group began the laying of the 19.5 miles (31.4km) of 30in trunkline which now links the Bacton Plant with the Field Terminal Platform set immediately south of Arpet 48/29-A. This project was completed in June 1968, in which month the Phillips Group began laying the 24in flowline connecting Phillips 52/5-A with the Field Terminal Platform. Construction of the Bacton Plant was begun by the Phillips Group in July 1968 and completed in June 1969.

With all necessary facilities installed and tested,

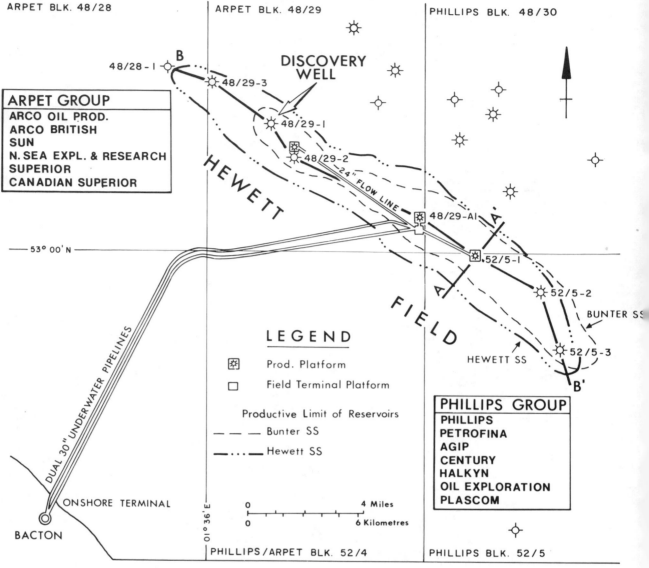

ARPET BLK. 48/28   ARPET BLK. 48/29   PHILLIPS BLK. 48/30

48/28-1

B

DISCOVERY
WELL

48/29-3

ARPET GROUP
ARCO OIL PROD.
ARCO BRITISH
SUN
N. SEA EXPL. & RESEARCH
SUPERIOR
CANADIAN SUPERIOR

48/29-1

48/29-2

HEWETT

24" FLOW LINE

53° 00′ N

48/29-A1

A

52/5-1

A

52/5-2

FIELD

BUNTER SS

LEGEND

52/5-3

HEWETT SS

B′

Prod. Platform

Field Terminal Platform

PHILLIPS GROUP
PHILLIPS
PETROFINA
AGIP
CENTURY
HALKYN
OIL EXPLORATION
PLASCOM

Productive Limit of Reservoirs

— — — Bunter SS

—·····— Hewett SS

ONSHORE TERMINAL

0   4 Miles

0   6 Kilometres

BACTON

PHILLIPS/ARPET BLK. 52/4   PHILLIPS BLK. 52/5

DUAL 30″ UNDERWATER PIPELINES

01° 36′ E

FIG 3.   Hewett Field location map.

the field unitized and a market secured, the Hewett Sandstone reservoir of the Hewett field was placed on production on July 12, 1969; less than three years from the spudding of the Arpet 48/29-1 discovery well.

To sustain production at the required level a third fixed platform 48/29-B, was installed in 1972, and eight Hewett sand wells were completed from it. Concurrently an additional 24in flowline to the Field Terminal Platform and a second 30in trunkline to Bacton Terminal onshore were laid. Desulphurization equipment at Bacton was completed in September 1973 and Bunter Sandstone sour gas production from recompleted wells on the 52/5-A platform commenced shortly thereafter. Onshore compression to maintain the gas supplied to the Gas Council distribution system at 1000psi pressure has been operational since August 1973. The Hewett Field is presently yielding an average of 800MMcf gas per day with

approximately 1800 barrels per day of associated condensate and 12 long tons of sulphur per day.

STRATIGRAPHY

Attention has been drawn to the typical section encountered in the Hewett field. Its relationship to certain aspects of the geological framework of the North Sea area may now be considered.

From Fig 4 it will be seen that the Hewett field is located relatively close to the East Anglia or London–Brabant Platform, which served as the southern boundary of the North Sea basin through much of Palaeozoic and Mesozoic time. It will also be noted that the field lies south of the area of discernible Permian salt movement, from which it may be concluded that the Hewett structure owes nothing to those halokinetic movements which are responsible for many of the structures found in

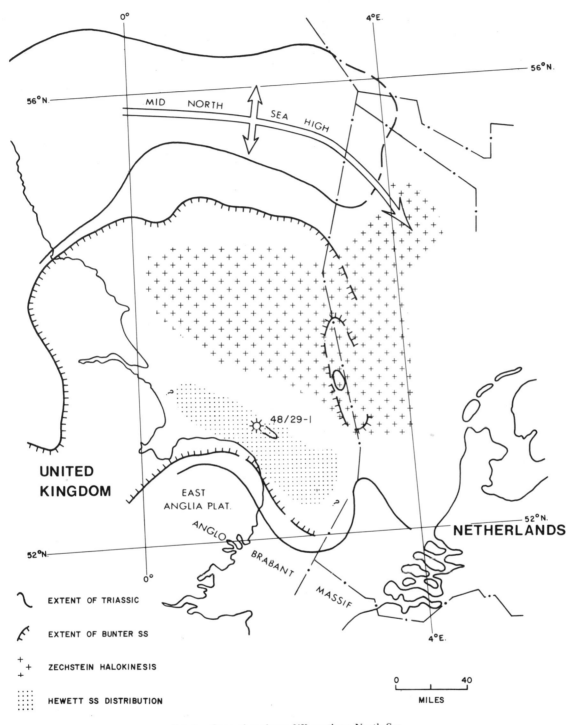

FIG 4. General geology: UK southern North Sea.

post-Permian beds elsewhere in the basin. Finally, it will be apparent that although some members are more restricted in distribution than others, Triassic sediments as a whole are widely developed throughout the North Sea area.

Geiger and Hopping (1968) have traced the German Triassic type section from outcrop in Lower Saxony westwards on subsurface information to outcrop on the English mainland. The correlation which these authors have established on lithological and palynological grounds makes it evident that the finely drawn subdivision of the Trias of the German basin cannot be carried with complete authority into the condensed, arenaceous, basin-margin facies of the English mainland. As far west as the Hewett field, however, the main elements of the German subdivision of the Trias are obviously valid. Keuper, Muschelkalk and Bunter are readily discernible, and on the basis of lithology the tripartite division of the German Buntsandstein may evidently be applied to the typical Hewett section, where the markedly arenaceous character of the middle unit emphasizes the appropriateness of such a subdivision.

Though there can be little doubt that the Bunter Sandstone of the Hewett area is correlative with the Middle Buntsandstein of the German succession, the precise equivalent of the Hewett Sandstone is less apparent. As is seen in Fig 4, this sandstone is severely restricted in distribution compared with other Triassic sediments. It is a local development, lacking regional stratigraphic significance, which probably owes its existence to erosion following some minor readjustment of the London–Brabant Platform. As such, and in the absence of definitive palynological evidence, it is extremely difficult to

fit the Hewett Sandstone precisely into the existing stratigraphic classification. Doubts regarding the true position of the Permo-Triassic boundary in the North Sea area compound the problem and are beyond the scope of this account. For present purposes, however, it may be stated that we refer the Hewett Sandstone in its entirety to the Lower Buntsandstein of the German sequence and recognize that it may be in part or even wholly equivalent to the Bröckelschiefer—a series of red, silty, anhydritic and dolomitic mudstones which in Germany are commonly accepted as marking the base of the Trias.

Lithologically, the Bunter Sandstone sequence of the Hewett field area is composed almost

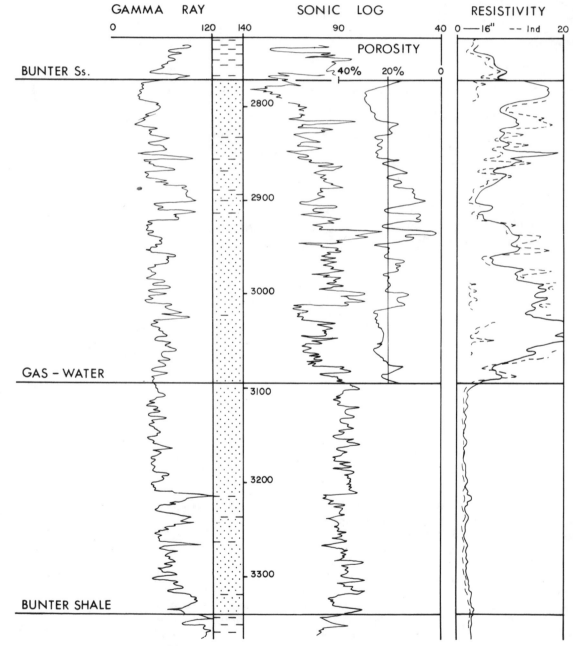

FIG 5.   Typical well logs of Hewett Field, Bunter Sandstone.

entirely of light grey to grey-brown, poorly-sorted, quartzose sandstone in which grains of microcline and plagioclase feldspar are a minor (10 percent) constituent. The sandstone is largely friable, but anhydrite and dolomite are present in varying amounts as cementing agents. A scarce, patchy clay matrix is observed. The sandstone is fine to medium grained; the grains subangular to subrounded. The heavy mineral suite is restricted to numerous granules of pyrite, small scattered concentrations of mica flakes and rare crystals of tourmaline. Occasional thin partings of grey and brick red shale are sporadically developed throughout the section.

As previously noted, the average porosity and permeability of the Bunter Sandstone pay-section in Hewett are 25.7 percent and 474md respectively. Values substantially greater than these have been recorded for certain intervals. The Bunter Sandstone as a whole increases in thickness across the field from west to east, measuring 355ft (108m) in the Arpet 48/28-1 well just beyond the western extremity of the field and 701ft (213.7m) in the Phillips 52/5-3 well on its eastern margin. The maximum observed gross pay-thickness of 323ft (98.5m) was recorded in the Phillips 52/5-1 well.

The response of various logs to a typical Hewett field Bunter Sandstone section is shown in Fig 5 in which some of the properties of the sandstone just described are apparent. Although not shown in the figure, it is worth noting that an overlay of the Laterolog and Microlaterolog curves of the

Bunter Sandstone gives a rapid and accurate indication of the thickness of gas-bearing section above water. The higher resistivities recorded by the deeper-reading Laterolog effectively separate the two curves above the gas/water contact. The method has proved to be of general application to gas-bearing sandstone sections in the North Sea area. Although not accurate enough to satisfy the log analyst it is a useful aid to the well-site geologist.

A maximum observed thickness of 202ft (61.5m) of Hewett Sandstone was penetrated in the Arpet 48/29-A2 platform well. Away from this central location the sandstone thins in both directions along the long axis of the field and is in the order of 70ft (21m) thick at the north-western and south-eastern margins. Drilling beyond the field boundaries has demonstrated a regional thinning to the north-east. Typically (Fig 6) the Hewett Sandstone may be divided into three parts, identified in field terminology as the Main Clean Sand (or Hewett Sand proper) sandwiched between the relatively thin Upper and Lower Shaly Sand which are probably equivalent to the Bröckelschiefer. The main unit is a red-brown, medium- to coarse-grained, moderately well-sorted, friable, quartzose sandstone. Feldspar content and clay matrix are present as described in the case of the Bunter Sandstone, but the degree of cementation by dolomite and anhydrite is less. Tourmaline is the main constituent of a rare heavy mineral suite. Rounded pebbles of metamorphic

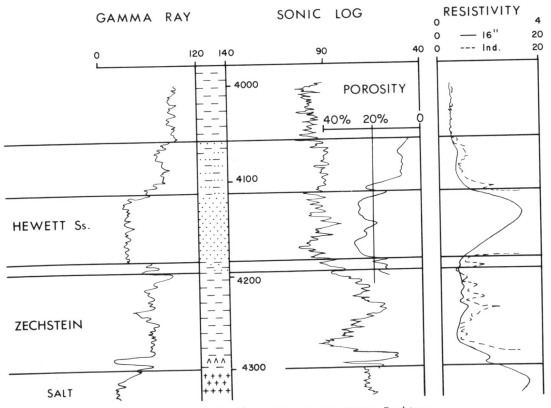

FIG 6. Typical well logs of Hewett Field, Hewett Sandstone.

and sedimentary origin occur sporadically through the sandstone but may be locally concentrated in thin, well-cemented, conglomeratic layers. The intervals above and below the Main Clean Sand are composed of laminated, red-brown, shales interbedded with thin, argillaceous, slightly dolomitic, fine-grained sandstones and siltstones. Porosity and permeability are best developed in the Main Clean Sand, but for the Hewett Sandstone as a whole the average values for the field are 21.4 percent and 1310md respectively.

STRUCTURE

The trapping mechanism in the Hewett field is structural. Both productive sands are contained in an anticline which is aligned north-west to south-east and is bounded by faults on its north-east and south-west sides.

Section A–A′ (Fig 7) is at right angles to the long axis of the field and runs north-east to south-west

through the Phillips 52/5-1 well. As previously noted, drilling thus far in the Hewett field has been confined to crestal positions on the structure. It is consequently all the more fortunate that seismic coverage of the area shows the structure to advantage. Some indication of the clarity of the seismic delineation may be gained from an examination of the record section which forms part of this illustration and is the basis of the structure section shown above it.

The two reflections, which for field mapping purposes are identified with the top of the Bunter Sandstone and the base of the Hewett Main Clean Sand respectively, have been identified on the record section. The anticlinal form at both horizons and the bounding faults of the north-east and south-west sides of the field are all readily visible. On this line of section the north-eastern fault is a major dislocation with downthrow to the north-east of the order of 600ft (183m). The depiction of the faulting at the south-west boundary of the field is here a debatable point. It

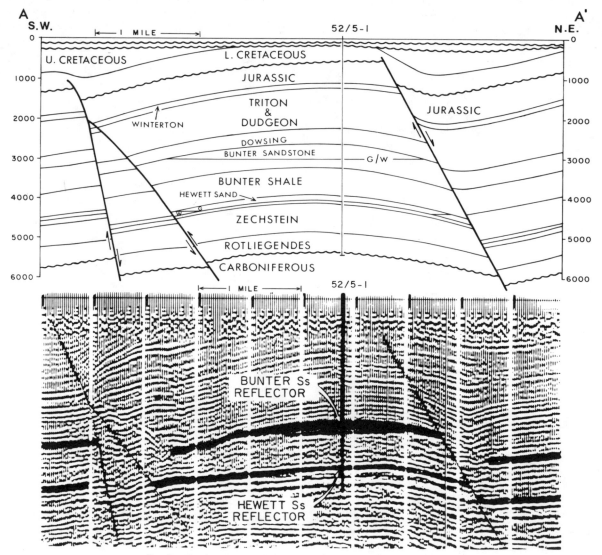

FIG 7.   Section A—A′ of Hewett Field; for line of section see Fig 3.

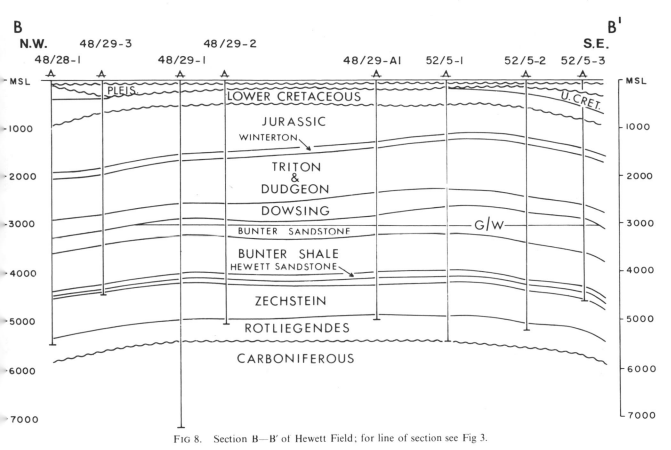

FIG 8.   Section B—B' of Hewett Field; for line of section see Fig 3.

might equally well have been shown as a single fault passing almost midway between the two planes which have been drawn in this illustration. In any event, the amount of movement involved is less than in the faulting on the north-east margin of the field, and the net effect is that strata within the field are thrown down with respect to those beyond its south-western edge.

In that none of the wells has cut either of these two bounding faults it is difficult to be precise regarding time of movement. The possible effect of other faulting in the area renders unreliable such interpretation as might be made of the thickening of sediment off the Hewett structure proper. From the record section shown in this Fig 7 and from other seismic control in the area it is evident that the faulting has affected beds younger than the Bunter Sandstone. Down-to-the-basin faulting may have occurred at different times in response to sediment load, but, allowing that the faults probably provided a route for the migration of the gas now found in the Triassic reservoirs, it seems likely that major movements were post-Triassic in time. To label them Jura-Cretaceous is perhaps the best estimate.

Before passing to other considerations it is of interest to draw attention here to the relatively flat-reflecting horizon apparent a short distance beneath the Bunter Sandstone reflection. One plausible interpretation which has been placed on this event is that it is caused by the gas/water contact in the Bunter Sandstone reservoir.

Structural cross-section B-B' (Fig 8) runs approximately north-west to south-east linking wells along the long axis of the field and utilizing seismic data in the spaces between wells. This section reveals undulations in the Bunter Sandstone surface and the overall eastwards thickening of that formation to which reference has already been made. As proven by drilling the Bunter Sandstone is structurally highest in the Phillips 52/5-1 well. Over the length of the field the Bunter Shale Formation displays a less pronounced structural variation than the Bunter Sandstone, but, in common with other strata shown, plunges rather steeply to the south-east beyond the Phillips 52/5-3 well. The Upper Cretaceous is absent over much of the field area, but this is of regional rather than local significance.

Figures 9 and 10 show structure contour maps on the top of the Bunter Sandstone and the base of the Hewett Main Clean Sand respectively. In the preparation of these maps velocity corrections were applied to the seismic data to effect a tie with well data. The fact that the horizon mapped in the case of the Hewett Sandstone lies within the pay-section necessitated further modification before the map could be used to determine the areal extent of that reservoir in calculating reserves. In each figure the productive limit of the reservoir

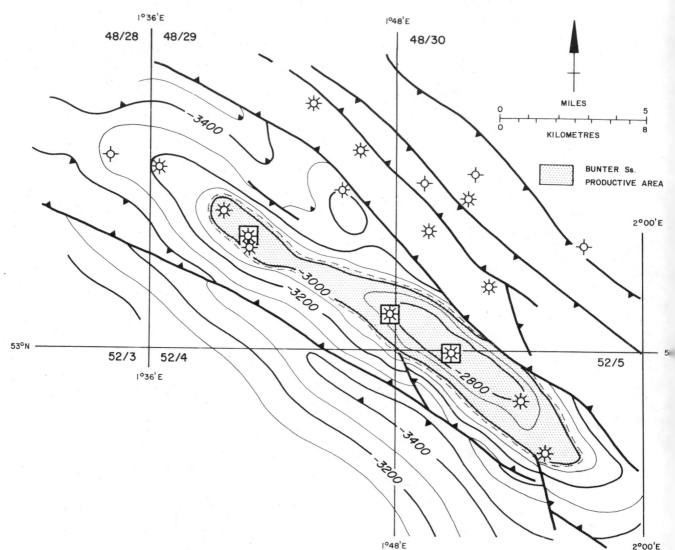

FIG 9. Structural contours on the top of the Bunter Sandstone, Hewett Field.

concerned is indicated. Both show the faults bounding the Hewett structure and include others present in the area north of the field.

Taken together, the figures serve to emphasize certain points already made or included without comment in other illustrations. The greater size of the Hewett Sandstone reservoir, the narrowness of the field at Bunter Sandstone level south-east of the Arpet 48/29-2 well and the south-eastwards plunge beyond the Phillips 52/5-3 well are aspects which are immediately apparent. The dip on the north-east and south-west flanks into the bounding faults is seen to be relatively steep.

The north-west to south-east alignment of the structure at both levels is of interest. This particular orientation is one which the Hewett structure has in common with other productive features in the southern portion of the North Sea area. Collectively they would appear to imply a dominant structural grain. At what time this was first impressed is uncertain, but it is perhaps significant that several authors have subscribed to

the view that the post-Silurian Caledonian movements in the area of Kent and East Anglia were related to those of the Mid-European belt which produced folds with a north-west–south-east alignment. It is possible that in these younger features we are seeing later manifestations of such an earlier structural trend. The folding associated with the Hewett structure may have occurred contemporaneously with the faulting in Jura-Cretaceous time.

NATURE AND ORIGIN OF HEWETT FIELD GASES*

The hypothesis that the gas now found in Lower Permian Rotliegendes reservoirs in the North Sea area owes its origin to the recoalification of Upper

* A number of the observations made in this section are based upon an unpublished report by D. Leythaeuser, on a geochemical study of hydrocarbon sources in the North Sea area.

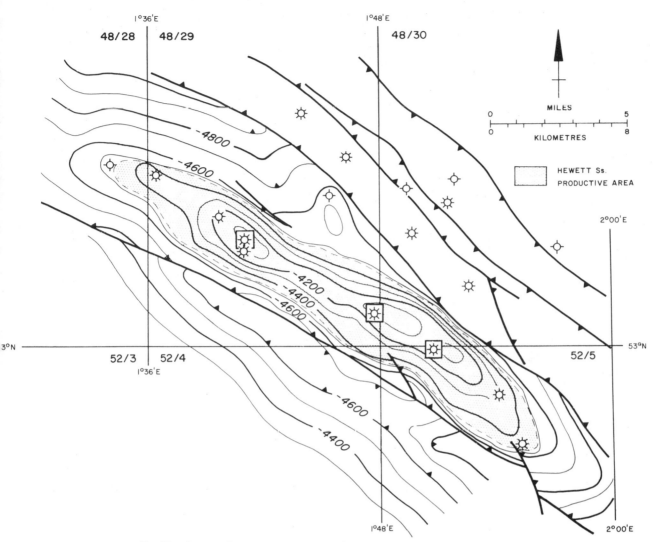

FIG 10.   Structural contours on the base of the Hewett Sandstone, Hewett Field.

Carboniferous coal has won general acceptance. In common with the Rotliegendes, neither the Hewett nor the Bunter Sandstone contains any obvious hydrocarbon source material. It seems most probable that the Carboniferous was at least the principal source of the gases contained in these two Triassic reservoirs, but certain aspects of the nature of the Bunter Sandstone gas indicate it to be of mixed derivation.

Excluding minor quantities of helium, the average composition of the gas as found in the Bunter Sandstone and Hewett reservoirs of the Hewett field is shown in Table I. With respect to the hydrocarbon content of the two gases perhaps the most noticeable difference in composition is seen in the percentage of methane present: 83.19 percent in the Bunter and 92.13 percent in the Hewett Sandstone. In that the Hewett reservoir lies some 1200ft (365m) below the Bunter Sandstone, this observed difference may be a function of depth of burial of reservoir, in which context it may be noted that the methane content for Leman field gas of average composition is 95.00

percent and that in Leman the Rotliegendes reservoir is approximately 1400ft (425m) deeper than the Hewett Sandstone in Hewett. Treating the remaining hydrocarbon constituents as a whole, the percentage content of ethane and its heavier homologues apparently decreases with depth from 8.31 percent in the Bunter Sandstone through 5.49 percent in the Hewett Sandstone of Hewett to 3.74 percent in the Rotliegendes of Leman.

Differences in composition of the two Hewett field gases are most pronounced in the case of the non-hydrocarbon constituents. The increase in nitrogen content from an average of 2.4 percent in the Hewett Sandstone reservoir to 8.4 percent in the Bunter is particularly striking. Although the process of recoalification is probably adequate to account for the hydrocarbon content of the gases found in the Permian and Trias of the North Sea area, it is not a mechanism which generates large amounts of nitrogen (Bokhoven and Theuwen, 1966). No single explanation of the relatively high nitrogen concentration in the Bunter Sandstone gas can be advanced with certainty, but certain

TABLE I
**Comparison of average gas analysis
Hewett Field**

Component	Bunter Sand	Hewett Sand
	(Values expressed in mol. fractions)	
Carbon dioxide	.000 8	.000 2
Nitrogen	.084 0	.023 6
Hydrogen sulphide	.000 2	
*Methane	.831 9	.921 3
Ethane	.053 2	.035 6
Propane	.021 4	.008 5
iso-Butane	.002 1	.001 6
n-Butane	.001 5	.002 2
iso-Pentane	.000 8	.001 0
n-Pentane		.000 8
Hexanes	.001 3	.001 3
Heptanes	.002 8	.003 9
	1.000 0	1.000 0

*Includes Helium.

factors which may have contributed to the observed end result can be suggested. First, it may be noted that nitrogen is a relatively inert gas and has what may be termed greater migrating ability than the hydrocarbon members of a gaseous mixture. Again, it is possible that the relative concentration of nitrogen may have been increased by preferential absorption of hydrocarbons during migration. Such absorption would be greatest on the surfaces of clay minerals concentrated in shales. Migration through the several hundred feet of Bunter Shale which separate the two reservoirs may therefore account in part for the difference in composition of the two gases. Finally, the Bunter Sandstone reservoir may have received a contribution of nitrogen from a source other than the Carboniferous; one which did not supply gas to the Hewett Sandstone. In this regard the sediments of the Haisborough group seem the most likely candidates.

The major distinction between the gas of the Bunter Sandstone reservoir and that of the Hewett Sandstone is the presence in the former of hydrogen sulphide, an occurrence which is of considerable academic interest and very real economic significance in terms of production costs. Classified as an acid gas, hydrogen sulphide as a constituent of natural gases in the North Sea area appears to be restricted to certain Zechstein carbonate reservoirs and to the Bunter Sandstone reservoir in Hewett.

With no trace of the gas apparent in the Hewett Sandstone, its presence in the Bunter Sandstone reservoir can only be accounted for in terms of derivation from a source which contributed solely to the younger reservoir. It is commonly supposed that hydrogen sulphide is most probably derived from the action of sulphate-reducing bacteria on anhydrite in the presence of hydrocarbons, with carbon dioxide liberated as a by-product of this process. In this regard it is noteworthy that the

average carbon dioxide content of the Bunter Sandstone gas in Hewett is four times greater than that in the Hewett Sandstone.

The likeliest source-beds to support the generation of hydrogen sulphide are to be found in the Haisborough Group sediments in which anhydrite is an important constituent. Significantly, faulting has served to bring these beds in juxtaposition with the Bunter Sandstone on the north side of the Hewett field. It may be supposed that the pyrite, which is now commonly found in the Bunter Sandstone, is a secondary mineral formed subsequent to the introduction of hydrogen sulphide to the sandstone.

An alternative hypothesis, that the hydrogen sulphide was produced *in situ* by bacterial action on anhydrite disseminated through the Bunter Sandstone is less attractive. Thus far the only sour gas found in Triassic sediments of the North Sea area is in the Bunter Sandstone of the Hewett area. Several other isolated discoveries of gas in the Bunter Sandstone have been made and although in all instances the formation is no less anhydritic than in Hewett, only in the case of the Phillips 48/30-1 well is hydrogen sulphide a constituent of the contained gas. The proximity of that well to the Hewett field proper suggests that a similar process of migration and subsequent introduction of hydrogen sulphide which was not operative in other areas may be adduced to account for the nature of the gas now found in the Bunter Sandstone.

CONCLUSIONS

While it has been demonstrated that gas is present in Triassic sediments in North Sea areas other than Hewett, it has yet to be shown that reservoirs of similar type and comparable magnitude exist. Where Bunter Sandstone discoveries have been made further north in the basin, follow-up drilling has either not been attempted or has been unsuccessful, suggesting that the reservoirs are of no great size. The search for further Hewett Sandstone accumulations is of necessity confined to the southern portion of the basin by the restricted nature of its distribution. By the same token, the area within which both sandstones might be found productive in the same well is subject to the same limitations. To date, despite reasonably intensive drilling, production from the Hewett Sandstones has not been established outside the Hewett field.

To be dogmatic in such matters is to invite being proved wrong, but it does seem probable that the combination of favourable facies, structure, migration processes and final entrapment found in Hewett are not duplicated elsewhere in the North Sea area.

REFERENCES

Bokhoven, C. and Theuwen, H. J. 1966. Determination of the abundance of carbon and nitrogen isotopes in Dutch coals and natural gas. *Nature, Lond.,* **211,** 927–9.

Geiger, M. E. and Hopping, C. A. 1968. Triassic stratigraphy of the southern North Sea basin. *Phil. Trans. R. Soc.* Series B, **254,** 1–36.

Rhys, G. H. 1974. A proposed standard lithostratigraphic nomenclature for the Southern North Sea and an outline structural nomenclature for the whole of the (U.K.) North Sea. A report of the joint Oil Industry–Institute of Geological Sciences Committee on North Sea Nomenclature. *Rep. Inst. Geol. Sci.,* **74/8,** 14 pp.

24

Structural Styles in North Sea Oil and Gas Fields

By D. G. BLAIR

(Esso Europe Inc., Exploration Studies, England)

SUMMARY

The overall structure of the North Sea Basin has largely resulted from a long tensional history dating back to the Permian.

There have been some variations on the basic tensional theme however, as evidenced by the salt-supported structures and wrench-related anticlines of the Southern North Sea Basin.

Almost all significant hydrocarbon discoveries to date have been located on closed structures, which can be assigned to four structural styles:

1. Basic tensional structures characterized by rotated fault blocks, horsts, and grabens. Hydrocarbon traps are developed in sub-unconformity sands and porous carbonates as in Cormorant, Brent and Auk fields.
2. Drape, compaction, or late movement structures as illustrated by Frigg field, where the depositional topography of Tertiary sands has been enhanced by compaction or late structuring over a deep rotated fault block.
3. Salt-supported structures illustrated by Ekofisk-type fields.
4. Polyphase structures such as Leman and Indefatigable fields, in which an early tensional phase was followed by a later compressive phase.

INTRODUCTION

The exploration for hydrocarbons in the North Sea has progressed rapidly in the relatively short time of less than ten years. During this time numerous exploratory wells have been drilled and thousands of miles of geophysical data have been gathered. From these data four general structural styles have been interpreted:

1. Basic tensional structures characterized by rotated fault-blocks, horsts, and grabens are common all over the North Sea Basin. Thus far these structures have proved most productive in the northern part of the North Sea, where hydrocarbon traps are developed in Jurassic sub-unconformity sands.
2. Drape, compaction, or late movement structures are generally developed on younger horizons overlying deep fault blocks and therefore have as wide a distribution as the basic tensional structures. To date, however, production in structures of this style is confined mainly to the north central part of the North Sea, where Lower Tertiary reservoir sands have a limited areal distribution.
3. Salt-supported structures are found predominantly in the central part of the North Sea, where salt movement has caused structuring and fracturing of porous Cretaceous and Lower Tertiary chalks.
4. Polyphase structures, which show evidence of more than one structural style, occur in the southern North Sea and extend onshore into the Netherlands and Germany.

Hydrocarbons in these structures are most commonly trapped in Permian and Triassic sands.

The Mesozoic–Cenozoic deformations which produced these structural styles were influenced by an old structural grain established within the underlying craton during the Palaeozoic. In Fig 1 a stress-strain ellipsoid showing the three directions of the Palaeozoic grain is superimposed on a tectonic fabric map of the North Sea. Two of the directions are conjugate shear-pairs—the north-east–south-west Great Glen Fault system and the north-west–south-east Tornquist Line. The third direction is the north–south extensional normal-fault direction (Ziegler, 1975).

Mesozoic–Cenozoic structural elements are aligned roughly parallel to these three pre-existing directions. For example, the bounding faults and fault-blocks of the Viking Graben are predominantly parallel to the north–south extensional direction. Structural alignment parallel to the two shear conjugates developed to a somewhat lesser extent as evidenced by north–south normal faults which in the northern part change strike to north-east–south-west. The bounding faults of the Central Graben follow the north-west–south-east Tornquist direction as well as the north–south direction. In the southern North Sea area a series of structural elements, *i.e.*, faults, anticlines, horsts and grabens, included within the Texel High–Anglo-Dutch Basin complex, parallel the Tornquist direction. These examples suggest that the old Palaeozoic structural grain has dictated fault directions and structural alignment of younger deformations.

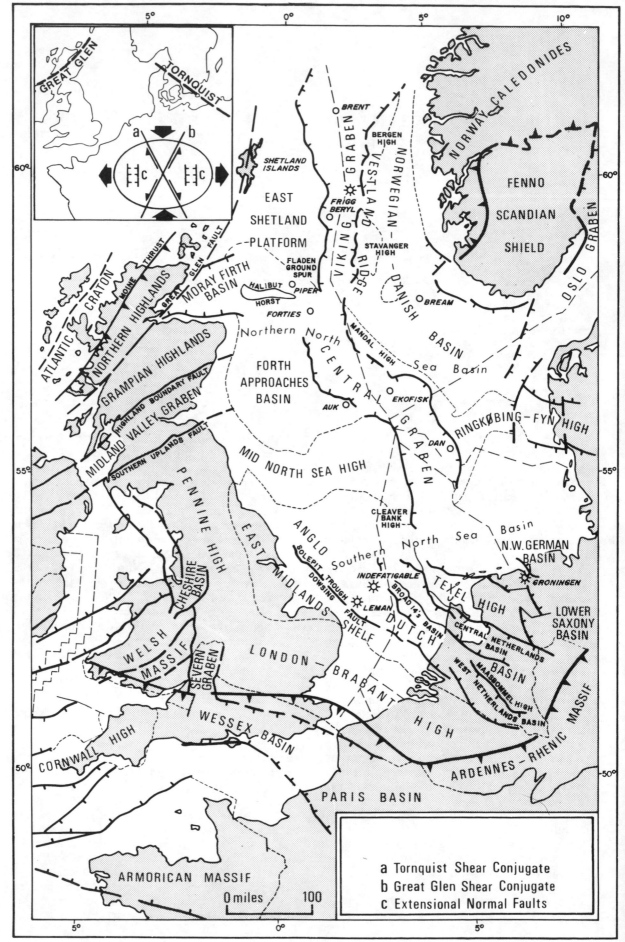

FIG 1.   Tectonic fabric map.

The dominant Mesozoic–Cenozoic stress regime has been tensional. At least two lines of evidence support this. First, the North Sea Basin is an intracratonic basin underlain by continental crust. Extension and thinning of the crust are necessary to allow an intracratonic basin to form. Second, if extension and thinning of the underlying crust occur, the passive overlying sedimentary layer should respond in normal faulting. This appears to be the case in that normal block faulting, that is, rotated fault blocks, horsts and grabens, is the dominant structural style. Most of the movement in the North Sea has been normal dip-slip, not strike-slip. In fact strike-slip and compression related structures appear to be more local phenomena, perhaps resulting from an uneven distribution of tensional stress.

BASIC TENSIONAL STRUCTURES

One of the most common structural styles in the North Sea is the basic tensional style typified at Brent Field, located mostly in UK block 211/29 (Fig 2). In an extensional stress regime normal faulting develops to the virtual exclusion of folding,

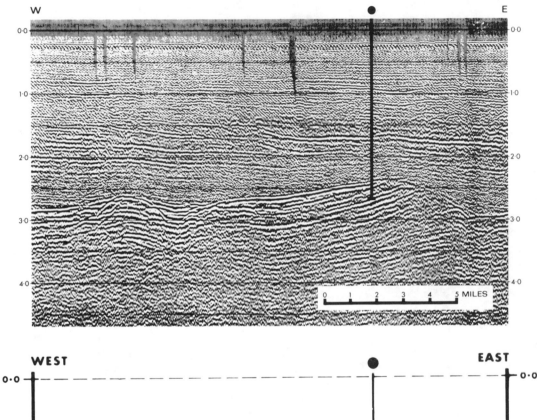

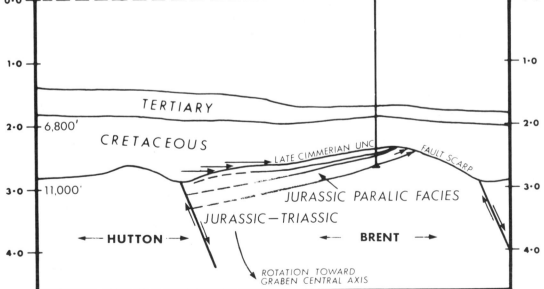

FIG 2.   West—east section through the Brent oil-field; based on interpretation of seismic profile above: sub-unconformity trap on a Cimmerian tilted fault-block. Production from Jurassic paralic sands.

and thus structure contour maps show fault block closures, not fold closures. Figure 2 is a west–east geophysical section across the Brent fault block. A fault scarp bounds the block on the east with the Late Cimmerian unconformity surface dipping west. On the left, west of the Brent block, the Hutton fault block has a similar aspect with east bounding normal fault and Late Cimmerian surface dipping west. In this style the dip on the fault blocks is often away from a central graben axis or "deep". As a graben system develops, extension causes normal faulting parallel to the graben axis. Under continuing extension, the fault-blocks rotate toward the central axis, so that individual blocks dip away from the axis (Fig 2).

In plan view the graben axis is to the east of Brent. The fault pattern is predominantly north–south but changes to the north-east–south-west to the north. Whereas the north–south faulting parallels the extensional normal fault direction of the stress-strain ellipsoid in Fig 1, the north-east–south-west alignment parallels the Great Glen shear direction, suggesting a pre-determined bias.

The production at Brent is from the sub-unconformity Jurassic paralic sands which in Fig 2 are shown truncated along the scarp face. These beds dip westwards at approximately 8°–10°, a slightly greater angle of dip than the late Cimmerian surface above. The major fault movement ranged from middle to late Jurassic. Renewed movement occurred in the middle Tertiary (Alpine orogeny) as evidenced by gentle arching of Lower Tertiary and Cretaceous beds over the crest of the deep structure.

One of the unique features of the tensional block fault structural style is the development of the hydrocarbon trap. In the Brent example normally flat lying beds were faulted, and rotated toward the graben centre, producing critical dip to the west. Up-dip, the tilted reservoir beds were truncated by the late Cimmerian unconformity. Finally the trap was sealed by early Cretaceous shales. The same relative structural attitudes have been maintained to the present day by rather uniform subsidence of the whole basin.

Preservation of the trap is of particular importance in assessing the hydrocarbon potential of this structural style. In this context uniform subsidence of the whole basin plays a key role. Consider for a moment this structural style as associated with a pull-apart continental margin, instead of an intracratonic basin. The early formation of the trap is the same, with fault blocks rotating initially toward the central axis and the establishment of critical dip toward the continent. But in the continental margin case the initial graben central axis becomes, through continued subsidence, an ocean basin. Inasmuch as the ocean basin subsides at a much greater rate than the continental margin, the critical dip toward the continent is ultimately destroyed, and the final dip

of the fault blocks is toward the ocean basin centre. Any initially trapped hydrocarbons may then escape up-dip toward the continent.

Oil and gas fields' in the basic tensional style category commonly have large reserves, because the fault blocks tend to have large trap volumes and generally are well located for entrapping hydrocarbons generated in the adjacent basin deep. Fields discovered to date include Auk, Cormorant, Hutton, Ninian, Dunlin and the newly discovered Statfjord to name a few. Exploration is in a comparatively youthful stage, and prospects for finding additional fields of this style are considered excellent.

DRAPE, COMPACTION, OR LATE MOVEMENT STRUCTURES

In this structural style the depositional topography on younger horizons has been enhanced by drape and/or differential compaction over deep fault-blocks. Simple drape, as defined here, is caused by rejuvenation of movement on old faults, resulting in uplift. Differential compaction assumes less lithostatic pressure on the structural crest than on the flanks, such that sediments in the flank position undergo greater compaction.

In Frigg Gas-Field (UK Block 10/1 and Norway Block 25/1) the depositional topography of a lower Eocene deep-water submarine fan has been structurally enhanced by drape and/or compaction over a Cimmerian rotated fault-block. Time of structural enhancement appears to have been post-Eocene and was probably associated with Alpine movements which caused rejuvenation on Jurassic faults controlling the deep structure. In Fig 3, a west–east geophysical section across Frigg Field, note that the topographic relief on the late Cimmerian unconformity surface was essentially filled by Lower Cretaceous onlap. Upper Cretaceous and Palaeocene have a fairly constant thickness over the deep structure; hence there appears to have been no deep structural expression during that time. The overlying Eocene Frigg Sand, on the other hand, had considerable topographic relief and local closure due to the moundy nature of its deposition. After deposition, drape and probably compaction as well, played the final role in shaping the present day structural configuration of the Frigg Sand reservoir. The amount of compaction is difficult to determine from the interpretation of the geophysical data. Drape, however, can be estimated by assuming that the top-of-Cretaceous reflector was essentially flat at the time of Frigg Sand deposition. The amount of present day reversal on the reflector is 200–500ft (60–150m). At least this much structural enhancement of the Frigg reservoir can be inferred.

Some of the other fields in this structural style are East Frigg, Maureen, Montrose and Forties. All of these sand reservoirs are either deepwater

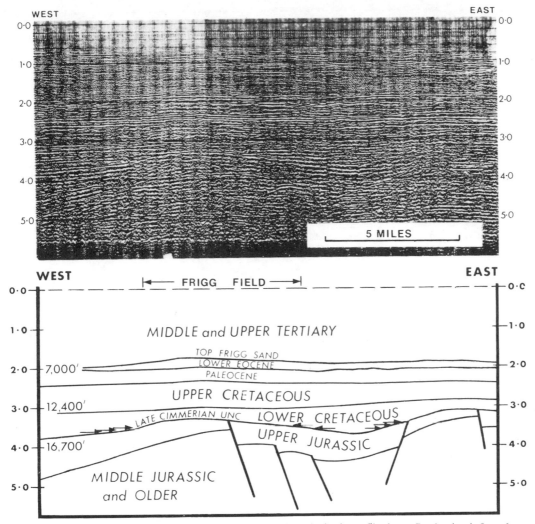

FIG 3.   West—east section from Frigg gas-field; based on interpretation of seismic profile above. Production is from Lower Eocene sands. Structural closure has been enhanced by drape over deep structure.

turbidites or submarine fans of Palaeocene and Eocene age, and as such have varying amounts of depositional topography. Further structuring of these sands by later movement of deep underlying fault blocks has enhanced their present day closure.

In the North Sea there is no lack of structuring in young beds overlying old fault-blocks, and undoubtedly there would be more fields in the drape category if more reservoir rock were available. Unfortunately, most of the post-Cimmerian deposition is dominated by shales and non-permeable carbonates, and Lower Tertiary reservoir-quality sands seem to have limited areal distribution in the north-central part of the North Sea. Therefore, future exploration for fields in this style has a somewhat limited potential.

SALT-SUPPORTED STRUCTURES

Salt movement is the dominant force causing structural closure in this structural style. Closure may be of two types: (1) drape or arching over a salt

swell or pillow; and (2) closure against a piercement dome. The drape or arching is similar to that produced by the rejuvenation of movement of deep fault blocks, except that here salt is the prime mover. Although piercement domes are common in the North Sea area, few have trapped sizeable hydrocarbon reserves.

An example of the salt-supported structural style is Ekofisk Field (Norway block 2/4) which, together with surrounding fields, forms one of the largest hydrocarbon accumulations found to date in the North Sea. This field is unique in that salt movement was not only responsible for the structural closure, but also produced an effective reservoir in the normally high porosity–low permeability North Sea chalk by fracturing and solution. Figure 4 is a series of diagrammatic restorations showing the evolution of Ekofisk-type structures based on current interpretation.

Stage I shows the deposition of Permian Zechstein salt and Lower Triassic clastics. Since the salt was originally deposited in a restricted environment formed by grabens and half grabens,

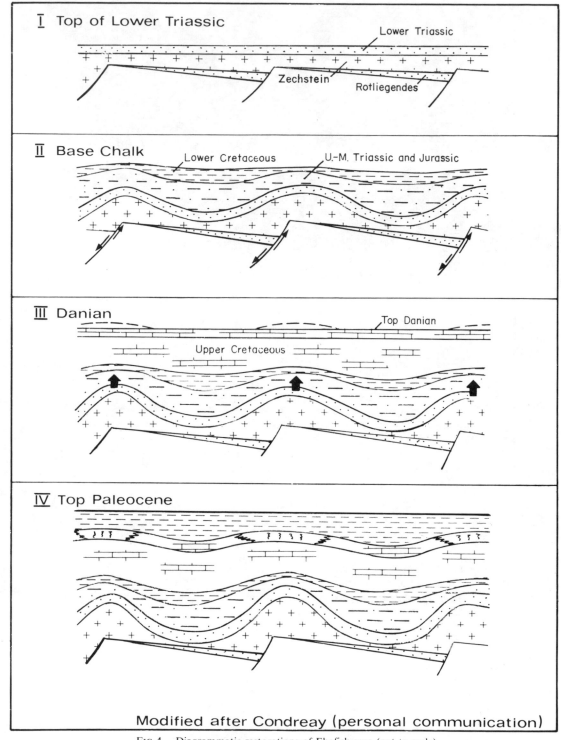

FIG 4. Diagrammatic restorations of Ekofisk area (not to scale).

it may well have received impetus for upward movement by continued movement on underlying faults. In this connection, movement of older fault blocks may be a common thread linking the drape style discussed previously with the salt-supported style.

In Stage II the salt probably began movement in the Triassic. Individual pillows and domes created local highs. Late Triassic, Jurassic and Lower Cretaceous sediments thinned by onlap onto these highs.

Stage III shows the deposition of the Upper Cretaceous and Danian Chalk. The lower part of the Upper Cretaceous onlapped Lower Cretaceous topography. Accelerated upward movement of the salt at the end of Danian time caused structuring and fracturing of the chalk. Topographic relief may have been great enough for subaerial exposure and solution weathering. Together or individually, these two events produced permeability, and thus reservoir potential in the initially non-reservoir chalk.

In Stage IV the reservoir was sealed by onlap of Lower Tertiary shales. Salt movement died out by mid-Tertiary. At this time a second episode of fracturing may have occurred. Overburden had by then become great enough to increase significantly the pore pressure in the chalk. As the pore pressure increased, the effective strength of the rock making up the most porous zones decreased, resulting in fracturing (Harper and Shaw, 1974). This further enhanced the permeability of the reservoir.

Although the Ekofisk example is a carbonate reservoir, this does not preclude salt-supported clastic reservoirs even though they are in the minority. Maureen Field, which was classified under the preceding style, could be considered salt-supported inasmuch as salt was drilled. In spite of this evidence, rejuvenation of movement of old fault-blocks was interpreted as the more important cause of structuring.

Although exploration of salt-supported structures is in a fairly mature stage, some potential remains for significant finds, particularly in the UK, Norwegian and German sectors.

POLYPHASE STRUCTURES

Polyphase structures are characterized by evidence of more than one structural style, and needless to say often have a complex structural history. In the Permian and Triassic gas-fields of the southern North Sea there are elements of tension and wrench-related structuring as well as salt tectonism. The old Palaeozoic north-west–south-east grain paralleling the Tornquist Line appears to have had a recurring influence on the deformation in that many of the structures—the Solepit Trough, the Dowsing Fault, the Broad 14's and Netherlands Basins, the Texel High, and individual anticlines align roughly with this direction. Description of one field example cannot serve as a general statement about all the fields in this style category, but perhaps this discussion will give some indication of the type of structural deformation involved.

Figure 5 is a south-west–north-east geophysical section through Indefatigable Field located in UK blocks 49/18, 19, 23 and 24. Gas production is from the Permian Rotliegendes Sand. The present structural configuration is an elongated, north-west–south-east trending anticline within the Anglo-Dutch Basin (Fig 1). Owing to the structural complexity of Indefatigable, not all the structural relationships can be seen on the geophysical section in Fig 5. For the complete structural synthesis, a regional seismic grid and well control were utilized.

There were at least three important phases in the structural development of Indefatigable. First, the

Post-Hercynian movement produced faulted basins north and south of the Mid North Sea—Ringkøbing-Fyn High (Ziegler, 1975). Although these movements basically had tensional manifestations, such as rifting and normal block-faulting, there undoubtedly was some associated strike-slip along the older Tornquist direction. Because data are lacking below the salt and the Rotliegendes is thin, these relationships cannot be seen clearly on geophysical sections such as Fig 5. For this reason, Rotliegendes structuring is interpreted from isopach "thicks" and "thins" based on well data.

On the other hand, aspects of the second phase of structural development can readily be seen on geophysical sections. Hardegsen movements at the end of the early Triassic probably initiated salt flowage. Where the salt was thick, large salt swells formed, and Upper Triassic sediments thinned over them. Not all the topography however, was produced by salt movement. On geophysical sections an early phase of wrench-related uplifts is contemporaneous with salt structures. The Hardegsen pulse probably rejuvenated wrench movements along the older north-west–south-east shear direction. Finally, the second phase culminated in the erosion of the late Cimmerian surface, which truncated and bevelled the topography produced by the Hardegsen and salt movements. In Fig 5 note the thin remnants of Jurassic left by the Cimmerian erosion.

Phase three in the structural development of Indefatigable began at the end of early Cretaceous and extended through the early Tertiary. Fig 5, shows a uniform thickness of Lower Cretaceous over the late Cimmerian unconformity suggesting deposition on a fairly flat surface. At the beginning of Upper Cretaceous deposition, further wrenching resulted in compression and uplift. Upper Cretaceous sediments onlapped the newly structured early Cretaceous surface. The north-east side of Indefatigable is bounded by a wrench-fault with interpreted left-lateral displacement bringing a thicker Triassic section south-west of the fault opposite a thinner section to the north-east (Fig 5). On other geophysical sections the wrenching also produced disharmonic folding, in which an Upper Cretaceous low is superimposed over a Zechstein–Rotliegendes high. Wrench movement continued through the late Cretaceous. In fact, it appears to be spasmodic, occurring at the beginning of late Cretaceous at some places and near the end of late Cretaceous elsewhere.

It is of interest to note that although the wrench zone generally parallels the old Tornquist shear conjugate, the interpreted late Cretaceous movement is left-lateral, opposite to the inferred Palaeozoic movement. This further suggests that the late Cretaceous stress field used a pre-existing structural grain direction for stress release.

A Rotliegendes isopach "thin", underlying the Indefatigable Field, suggests early structural relief.

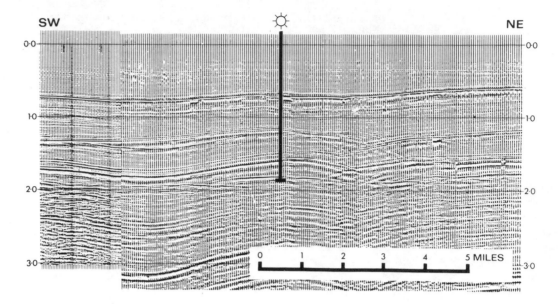

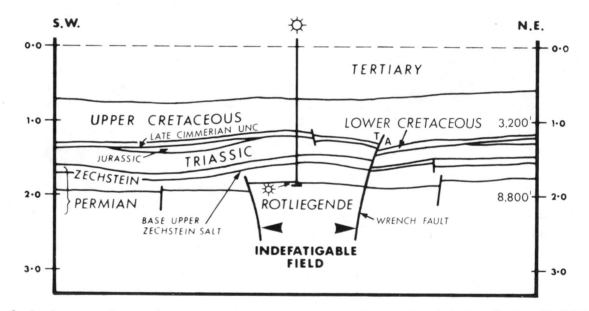

FIG 5. South-west—north-east section through Indefatigable Gas-field; based on interpretation of seismic profile above. The field is a large complex faulted anticline trending north-west—south-east. The reservoir is Permian Rotliegendes sand.

Following the various movements outlined above, the present day high fortuitously coincides with the old high.

The Rotliegendes reservoir in this structural category was one of the first North Sea exploration targets. Some of the gas-fields discovered to date include Leman, Hewett, Dotty, Deborah, Broken Bank, West Sole and Viking. In spite of the relative maturity of exploration good potential in this style exists in the UK and Netherlands sectors.

As the intensity of exploration in the North Sea continues to increase, more data will become available to broaden our knowledge of structural styles and their underlying causes. It is therefore anticipated that the rather general style

classification presented here will in time become more sharply defined and undoubtedly more complex. For the present this paper should be viewed as a preliminary sampling of structural styles interpreted from currently available geophysical and geological data.

ACKNOWLEDGEMENTS

The author gratefully acknowledges the contributions made to this paper by the geologists and geophysicists of Esso Europe's Exploration Studies Group. Special thanks are expressed to Brian Garner for preparing the illustrations.

REFERENCES

Harper, M. L. and Shaw, B. E. 1974. Cretaceous–Tertiary carbonate reservoirs in the North Sea. *In* Exploration geology and geophysics offshore North Sea. *Technology* *Conference and Exhibition, Stavanger, Norway, September 3–6, 1974.*

Ziegler, W. H. 1975. Outline of the geological history of the North Sea. *Paper 11, this volume.*

DISCUSSION

D. G. Roberts (Institute of Oceanographic Sciences): In your presentation you commented on the palimpsest control exerted by various structures. In fact the Great Glen Fault and Viking Graben are both truncated northwards by the continental margin between Shetland and Norway. Would you like to comment on the stress relationships between the Viking Graben and the margin?

D. G. Blair: In the paper three Palaeozoic structural grains are identified and illustrated in Fig 1. The name "Great Glen" is given to the north-east–south-west direction, as that is the prominent direction of the structural elements which parallel the Great Glen Fault in Scotland. The usual interpretation of the northwards extension of the Great Glen Fault connects the Scotland part with faults in the Shetland Islands as indicated by the dashed line in Fig 1. This interpretation causes the north-east–south-west trend of the Great Glen to swing to north–south. Perhaps this is puzzling inasmuch as the north-east–south-west structural direction is very prominent in the northern part of the Viking Graben. For example, as mentioned in the paper, the strike of a number of major faults changes from north-south to north-east–south-west to the north of Brent. Aeromagnetic data show rather strong north-east–south-west trends in this same area. North of 62°N there appears to be some evidence for a north-east–south-west trending graben; which parallels the north-west coast of Norway. This graben may join with the Viking Graben north of Brent, producing an offset or dog-leg to the north-east. Furthermore the continental margin between about 58°N to 64°N trends north-east–south-west. North of 64°N and south of 58°N the margin appears to trend north–south for about 2° of latitude. Since the north-east–south-west structural direction is so prominent in the northern North Sea I wonder if the Great Glen Fault does not continue in the north-east–south-west direction rather than turn northwards to connect with faults in the Shetland Islands, which could be more closely related to the north-south grain direction.

Several papers at the meeting discussed the various tectonic pulses which have contributed to the structure and stratigraphy in the North Sea (see papers by W. H. Ziegler and P. A. Ziegler). These pulses were related to interactions between the major tectonic plates of North America, Europe–Asia, and Africa. The Viking Graben had a good portion of its structural development in the Jurassic and early Cretaceous, whereas the continental margin between Shetland and Norway resulted from rifting and extension from early Cretaceous to Palaeocene. The stress regimes which produced these two structural features were probably not the same although their underlying causes, *i.e.* boundary interactions between the tectonic plates, may be the same. One of the major premises of my paper was that the three dominant directions of the North Sea tectonic fabric parallel major zones of weakness and/or fragment boundaries established within the craton during the Palaeozoic. Mesozoic–Cenozoic stress fields have utilized these old zones of weakness for stress-release. The break-off of Greenland from north-west Europe took place along the old north-east–south-west direction.

Dr D. A. Robson (Newcastle upon Tyne University): The structures within the North Sea Graben, described by Mr Blair, seem remarkably similar to those in the Gulf of Suez region of the African Rift. These structures of the so-called "Clysmic" Gulf of Suez, have been recorded by detailed surface mapping and from the logs of oil wells. Some of the principal features of the Clysmic Rift structures are: (1) The curving trend, in plan, of the great normal faults bounding the rift on either side which, at intervals along the strike, break up into several splay faults or are even entirely replaced by a steep, inward-facing monocline; (2) the consistently high angle of dip, about 70°, of all the rift faults; (3) the large number of faulted blocks which tilt away from (or are rotated towards) the centre of the rift—these blocks are terminated along their length either (a) by their bounding faults curving away (in plan) towards the outer margin of the rift, or (b) by the bounding fault being broken up into splay faults with ever diminishing throws; (4) the existence of occasional faulted blocks tilted towards (or rotated away from) the centre of the rift; (5) the rejuvenation of the boundary faults of the blocks, such that later sedimentation keeps pace with movement, producing a thick succession on the downthrow side but only a thin cover over the block; (6) continuation of the rejuvenation of the boundary faults of the blocks, producing (a) gentle or even (b) sharp flexures in the later sediments; and (7) persistent tensional movements throughout the history of rifting. Is Mr Blair able to cite evidence for similar patterns associated with the North Sea Graben?

D. G. Blair: All the features you list for the Gulf of Suez region have definite counterparts in the North Sea. The plan or map view of faulting trends is similar, although some interpretations may resolve the curving changes in strike of the major faults into straight-line segments. North Sea faults commonly have a high angle of dip similar to your example. The erosional fault-scarps dip about 10°. Fault-blocks usually are tilted away from the centre of the rift; however, it is not uncommon to find them tilted towards the centre. Horsts and grabens or half-grabens are often observed. Faults bounding individual structures are as you describe them and could be called "trap-door" faults. For example, many of the blocks are bounded by a major fault which parallels the central axis of the rift and by splay- or cross-faults which run at some angle to the major fault and have ever-diminishing throws going in the direction away from the major fault. The tilted fault-block assumes the appearance of a partially opened trap-door. The rejuvenation of boundary-faults such that later sedimentation keeps pace with movement is well documented on geophysical sections in the North Sea. The development of flexures in later sediments due to the continued rejuvenation of older faulting is so common that it was used as the basis for the drape structural style in the paper. Finally, concerning your last point, tensional movements certainly seem to have been dominant throughout much of the Mesozoic–Cenozoic history of the North Sea; however, episodic wrench movements are evident as described under "polyphase structures".

M. Ala (Seagull Exploration): I would like to direct two questions at Mr Blair:

(a) To what extent would he consider his "polyphase compressive structures" as having resulted from local resolutions of essentially epeirogenic movements rather than from true horizontal compression?

(b) In a number of his slides he has related salt-dome development to fault-blocks. While this is true of some salt-domes, many others are not associated with faulting as demonstrated by Christian (Amoco) in his contribution to the Brighton Conference in 1969.

D. G. Blair: (a) The term "polyphase structures" was used for structures with more than one style of deformation. In the Permian and Triassic gas-fields of the southern North Sea there is evidence for tension- and wrench-related structuring as well as salt-tectonism, hence the name "polyphase" for these structures. "Epeirogenic movements" imply broad uplift or subsidence of areas in which the strata is not folded but may be tilted or remain essentially horizontal. There is no suggestion of *cause* of uplift or subsidence in the definition. Cratonic blocks underlying these polyphase structures do not go up and down of their own volition. An explanation that addresses the cause of structuring is preferred. There are several ways of producing vertical movements. These include phase changes in the mantle, convective upwelling as in the initial arching stages of rifting, orogenesis, and wrench or strike-slip faulting. Of the four, the best explanation for the structuring of the southern North Sea gas-fields is wrench-related compression. This hypothesis is supported by the following:

(1) On geophysical sections the bounding faults of Indefatigable Field can be interpreted as high-angle reverse-faults or upthrusts which are commonly associated with wrench or strike-slip faulting. This has been confirmed by at least one well which has drilled a repeated section.

(2) On successive geophysical sections the same fault often shows normal and reverse dip-slip separation along its strike. Isopach "thicks" and "thins" are juxtaposed and appear to be out of place across the interpreted wrench-fault as shown in section view in Fig 5. These are common phenomena associated with lateral movement.

(3) An analysis of the directional properties of fractures, slickensides, and kinks in oriented Rotliegendes cores (14 wells) from the southern North Sea gas-area has shown that left-lateral strike-slip motion played a significant role in the structural development. As mentioned in the paper, the wrench-zone at Indefatigable (see also Fig 5) parallels the old Tornquist shear conjugate; however, the interpreted Late Cretaceous movement is left-lateral opposite to the inferred Palaeozoic movement. The interpretation is that the change over from right-lateral to left-lateral strike-slip movement probably took place during the late Cretaceous.

(4) Finally, the fault-patterns on structure maps of Indefatigable are compatible with left-lateral shear. Synthetic faulting is dominant with north-west–south-east strike. The antithetic faults, striking north-east–south-west, are developed to a lesser extent.

(b) In Fig 4 the faulting in Stage I is diagrammatic and was intended to portray the deposition of salt in a restricted environment of a block-faulted basin. I agree that faulting is not necessary to the formation of salt-domes, swells, or pillows. In fact modelling of the buoyant phenomena has shown that it can be self-initiating. In the Ekofisk area faults in the section below the salt can be interpreted on geophysical sections. In the Gulf Coast basin of the US a majority of salt-domes are associated with major down-to-basin growth-faults. (See also the next question.)

Dr D. H. Matthews (Cambridge University): Have you any explanation of the marked north–south orientation of the numerous elongated salt-domes in the southern North Sea that we have seen on diagrams by T. Krey and R. Marschall (Paper 19) and by J. B. Butler (Paper 14). If this elongation is due to formation of the domes along the line of a reactivated fault in the (Carboniferous) economic basement, as you have suggested for Ekofisk, it seems to imply a remarkable uniformity of fault-directions imposed during Hercynian movements. Or is the elongation a response to (post-depositional) stresses acting at the time of initiation of salt-flow?

D. G. Blair: Whether or not the north–south alignment of salt-domes in the southern North Sea is controlled by the old Palaeozoic structural grain is difficult to say. If the craton underlying the North Sea has been subjected to stresses which produced the conjugate shear pairs in Fig 1, then the extensional direction would be north–south. In the inset of Fig 1, the a, b, and c directions are possible fracture-directions within the craton. Cratonic fragments would have boundaries that followed these directions. Stresses imposed upon this mosaic of cratonic fragments would cause differential movements between the fragments, and the passive sedimentary layers would respond in the various structures observed. Salt-movement could have been initiated by movement along old basement fractures. In the Ekofisk example Fig 4, stage I, the faults shown were a diagrammatic indication of salt-deposition in a faulted basin. Faults can be interpreted underlying a number of salt-domes and swells in the North Sea.

The Gulf Coast Basin of the US is a tensional basin with down-to-basin faults which strike concentrically about the basin-deep to the south and, therefore, are parallel to the strike of the sedimentary fill as shown in Murray's Fig 4.1 (Murray, G. E. 1961. *Geology of the Atlantic and Gulf Coastal Provinces of North America;* Harper, New York, 692 pp.) The faulting in the sedimentary layers results from sediment creep into the subsiding basin deep. Tension vectors are at right angles to the fault-strike. The salt-domes or ridges generally follow the concentric fault-strike and are related to the faulting as shown in Murray's (1961) Figs 4.37 and 5.1. Presumably the faulting within the sedimentary section is controlled by the down-faulted basement underlying the 10 000m or more of fill in the Gulf basin. The prominent fault-pattern of the sedimentary section is a shallow expression of the deep basement structure. The alignment of Gulf Coast salt-domes parallel to the regional fracture pattern, particularly in the Texas–Louisiana area, suggests a cause and effect relationship. Perhaps the parallel does not extend to the North Sea; however, the Gulf Coast salt-dome province should be considered in possible explanations of the alignment of North Sea salt-features.

25

On the Structure of the Dutch Part of the Central North Sea Graben

By P. HEYBROEK

(Nederlandse Aardolie Maatschappij B.V.)

SUMMARY

The Dutch North Sea Graben is a north–south trending down-faulted feature, 200km long by 40km wide.

Relative subsidence of the graben apparently began with deposition of Upper Aalenian to Lower Bajocian age lacustrine sediments. By the Middle Bajocian marine conditions prevailed regionally, but before the end of the Middle Jurassic, the graben's southerly part was partially uplifted. Following this uplift, Oxfordian age paralic sediments were deposited. Marine sedimentation which had continued uninterruptedly to the north, had by Kimmeridgian time transgressed southwards over this paralic facies.

The area was regionally uplifted prior to the Cretaceous transgression which started to fill the graben and later encroached on to the bordering highs. During the Upper Cretaceous, the graben itself was uplifted (inversion movements) and at its centre deeply eroded.

By the late Palaeocene all differential movements ceased and the area was covered by a thick marine Tertiary sequence.

INTRODUCTION

The Central North Sea Graben in the Dutch offshore is a north–south trending major down-faulted feature. It extends for 190km between longitudes 4° and 5° E from the northern offshore boundary towards the south where it disappears before reaching the Dutch coast (Fig 1). It forms the southern portion of a much larger feature occupying an axial position in the North Sea, and reaching the Atlantic Ocean between the Shetland Islands and Norway (Ziegler, 1975). The southern portion of the graben, which will be described here, seems to constitute a distinct sub-unit, whose structural connection to the more northerly units falls outside the area discussed.

The graben appears to be a complicated block-faulted feature. Maximum subsidence has taken place at the centre, flanked by intermediate blocks which step down from the bordering *Mid North Sea High* and *Cleaver Bank High* in the west, the *Schill Grund High* in the east and the *Texel–IJsselmeer High* in the south (Fig 2).

Intensive halokinetic movements took place in and outside the graben originating in the thick sequence of Zechstein salt deposited in the area. These movements hamper direct and detailed structural analysis in three ways.

Firstly they render seismic identification of the pre-salt basement difficult, and its present configuration lacks detail. Only a generalized picture can be offered which delineates very approximately the area and depth of the present-day graben (Fig 2). Secondly the block-fault movements in the pre-salt basement are smoothed in the post-salt layers due to the plastic behaviour of the salt. The structural configuration above and below the salt may be very different. Thirdly, the post-salt beds may have suffered extreme local deformation, independent of "basement" tectonics, due to the flowage of salt towards the numerous salt piercements. This has occurred to such an extent that there is often little salt remaining in the interdomal areas (Fig 9).

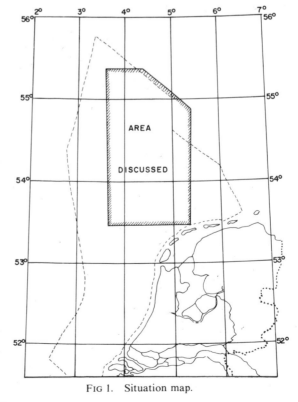

FIG 1. Situation map.

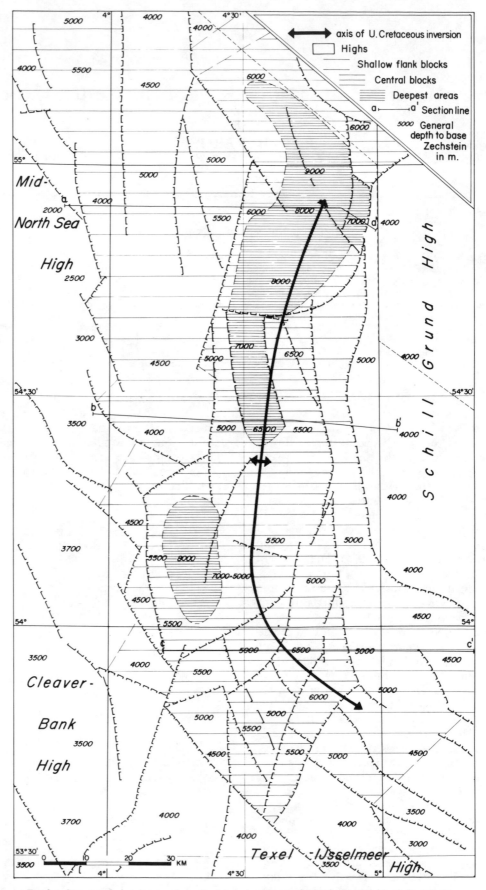

FIG 2.   Structural sketchmap of pre-Zechstein basement of Dutch Central North Sea Graben.

It is clear that, under these circumstances, thickness differences in post-salt layers have no significance in terms of regional tectonics. However, it appears that the Jurassic and Lower Cretaceous sediments which are present in the graben are absent and reduced respectively on the surrounding highs. The stratigraphical analysis of these sediments penetrated in wells provides the evidence for determining structural evolution of the graben and for defining the units for efficient seismic mapping. This permits the reconstruction of the order and direction of movement in time, but the detailed areal assessment will remain vague.

Thus the regional sedimentation and erosion in the graben forms the main topic of the following argument. It is schematically represented in Fig 3.

PRE MIDDLE JURASSIC SETTING

The area under consideration crosses at right angles the centre of the east–west trending north-west European Permo-Triassic basin, in which about 4000m of sediments were deposited unconformably on older beds (Heybroek, 1974; Heybroek et al., 1967). The pre-Permian basement has not been reached by drilling. It probably consists, for a large part, of Carboniferous strata of different ages, and possibly also some Devonian in the northern part of the Netherlands offshore.

Permo-Triassic deposition started with the

Rotliegendes, developed in a basinal facies of shale with some intercalated salt beds (Glennie, 1972). It is topped by the Zechstein in a fully basinal facies in which rocksalt forms more than 95 percent of the sequence. The depositional thickness may be estimated at around 1200 to 1500m, as it is elsewhere in the centre of the Zechstein basin (Christian, 1969). It is this salt which gave rise to the intensive halokinetic movements in the Permo-Triassic basin, in and outside the graben area (Trusheim, 1960; Heybroek et al., 1967). However, in the north-west the halokinetic movements stop abruptly at the eastern boundary fault of the Mid-North Sea High. On the high, west of the fault, the Zechstein is largely eroded but it reappears further west in a limestone–anhydrite facies with subordinate salt beds (Heybroek et al., 1967). This indicates that the Mid-North Sea High existed during Zechstein deposition and that the eastern boundary fault made a sudden step in the basin of deposition. The boundary fault continued to play a peripheral role in the later development of the graben.

The Trias was deposited on the Zechstein as a sequence which is closely comparable with the basinal sequence in northern Germany (Wolburg, 1969). It has a thickness of 1500 to 2500m. The variation is due to initial salt flowage which started in the Keuper, and possibly also to block movements in the pre-salt basement (Fig 9, section C). It is assumed that these early block movements

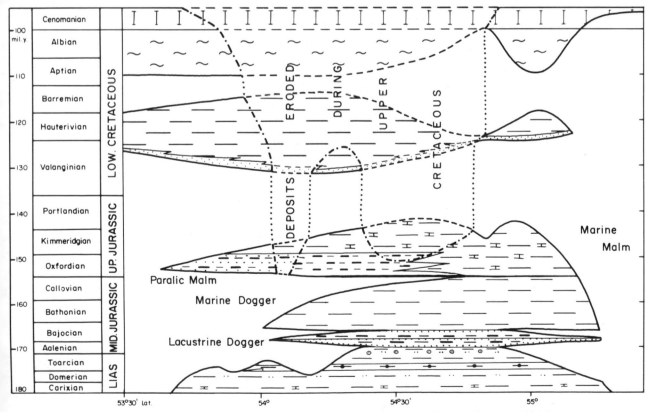

FIG 3.   Jurassic and Lower Cretaceous sedimentation and erosion along the axis of the Dutch Central North Sea Graben.

are connected with the south-west–north-east trending "Horn Graben-Off Holland Low" (Ziegler, 1975; Heybroek, 1974), a Permo-Triassic structural element which crosses obliquely the southern part of the Central Graben, a feature of later formation. The Lias was deposited conformably on the Trias. It is fully preserved in the centre of the graben only, and completely eroded on the surrounding highs (Fig 4). This sequence of black calcareous and sometimes bituminous shales is comparable in detail with the Lias sequences of the Netherlands, north-west Germany and eastern England (Wilson et al., 1934; Wilson, 1948). This uniform marine facies reflects the widespread transgression during the Lias. A thickness of around 1000m outside the halokinetic rim-synclines is not excessive for the north-west European basin, and cannot be taken as proof of relative subsidence of the Central North Sea Graben during the Lias (Sorgenfrei, 1969).

During the Lower Aalenian, sedimentation continued in a shallowing sea as evidenced by the intercalations of thin sandstones and iron oolites.

MIDDLE JURASSIC DEVELOPMENT

After the Lower Aalenian a regional uplift took place but the area of the graben lagged behind. In the centre of the graben sedimentation continued, but in a lacustrine facies, with the deposition of shales and sandstones with coals. The area of largest relative subsidence (least uplift) is found in the axial region between 54° 20′ and 55° N where the lacustrine Dogger beds rest, without apparent unconformity, on the Lower Aalenian.

Away from the central region the lacustrine Dogger rests unconformably on more or less deeply eroded Lias. This provides for the earliest clear evidence of relative subsidence of the graben in the Aalenian. Figure 4 shows the distribution of the Lias preserved after the Aalenian and later erosion periods, and the onlap of the Middle Jurassic. The lacustrine Dogger may reach a thickness in excess of 1000m and is inferred to be of Upper Aalenian–Lower Bajocian age (Dogger β and γ), from palaeontological evidence below and above the sequence. During this time a lake occupied the graben area. The sea must have retreated to the south where marine sequences of this age are known from the Netherlands (Haanstra, 1963) and north-west Germany (Brand and Hoffman, 1963) and partly also from Lincolnshire (Wilson, 1948). A lacustrine or terrestrial sequence in the Middle Jurassic is known also from Yorkshire (Wilson et al., 1934) and the North Jutland basin (Larsen, 1960). In Yorkshire and northern Lincolnshire this sequence (the Lower Estuarine series) rests unconformably on eroded Lias. In Yorkshire and north-west Germany clastic influx from the north is evident. These observations lead to the postulation of a

regional tilt in the North Sea basin; after the end of the Liassic the positive tendency was more pronounced in the northern part of the basin than in the south.

The lacustrine sedimentation in the graben was interrupted by a marine transgression which started in the Bajocian (Dogger δ). In this sea a monotonous shale sequence was deposited with a uniform thickness of about 300m. The lack of coarser clastics in the sequence suggests a large extension of this transgression over the surrounding highs, from which all traces have been removed by later erosion. The marine shale rests unconformably on the lacustrine beds near the edges of the basin. This need not necessarily imply uplift and erosion, for it could be due to the gradual infilling of the subsiding lake: the older sediments filled the border zone of the lake to base level, while the younger sediments were deposited only in the centre.

Towards the end of the Middle Jurassic, the axis of the graben was tilted. The part south of 54° 40′ N was uplifted above sea-level and progressively eroded towards the south. North of the hingeline, marine sedimentation continued. This tilt of the graben axis reflects most probably tectonic activity on a much larger scale in the North Sea basin, as may be deduced from the following evidence: a general uplift occurred within the Oxfordian in the Netherlands onshore (Haanstra, 1963) and north-west Germany (Boigk et al., 1963), while in Yorkshire the deltaic deposition of the Middle Jurassic ended with a Callovian transgression (Christian, 1969). This transgression is expressed in Lincolnshire and farther south by a more marine facies (Wilson, 1948). Apparently the North Sea basin subsided in the north and west, and was uplifted in the south-east, and thus in an opposite direction to the tilt postulated to have occurred during early Middle Jurassic.

The distribution of the Middle Jurassic which includes the lacustrine and the marine beds is given on Fig 5. The lacustrine beds are practically everywhere covered by marine beds, so that the present distribution of the Middle Jurassic coincides fairly well with the form of the early Dogger lake. Only in the north the lake may have extended far beyond the present subcrop.

The large formation thicknesses in the north are mainly due to differences in the depositional thickness of the lacustrine sequence, and much less to later erosion.

UPPER JURASSIC DEVELOPMENT

In the southern part of the graben, the Upper Jurassic occurs in a paralic facies and rests with pronounced unconformity on Middle Jurassic, Liassic and Triassic strata (Fig 4). North of the end-Middle Jurassic hingeline, no unconformity is apparent and the Upper Jurassic is developed in a

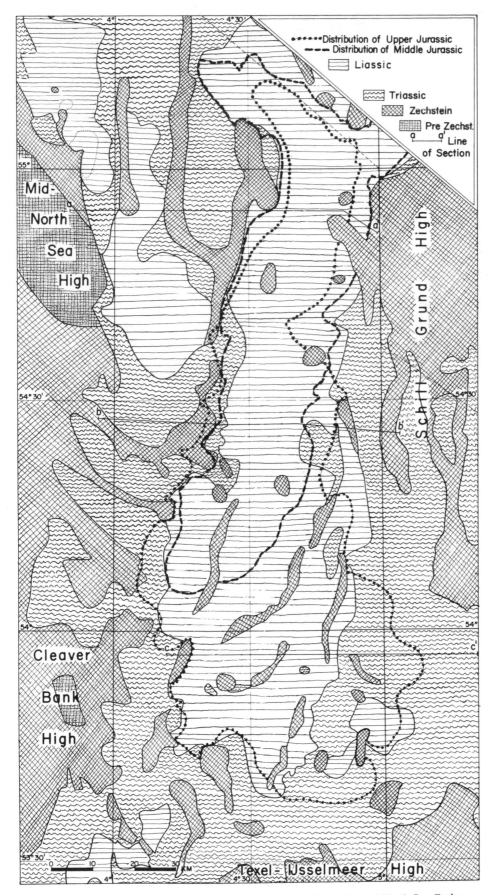

FIG 4. Geological map of pre Upper Aalenian formations; Dutch Central North Sea Graben.

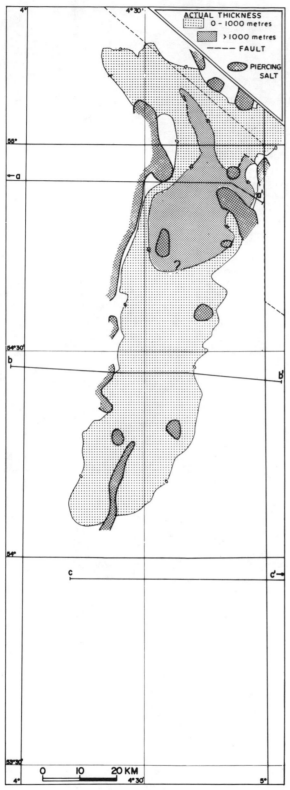

FIG 5. Distribution of the Middle Jurassic in the Dutch
Central North Sea Graben.

sequence consists only of an alternation of shale
and coal.

Overlying the paralic sequence a shale of
probably restricted marine facies was deposited, in
which ostracodes of Kimmeridge age are found. A

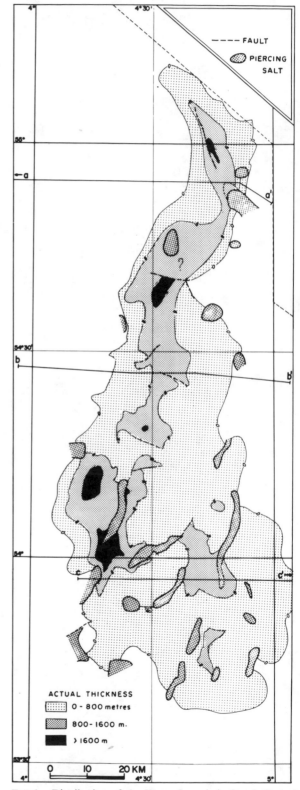

FIG 6. Distribution of the Upper Jurassic in Dutch Central
North Sea Graben.

fully marine facies of shales with thin limestone
bands of Oxfordian to Kimmeridge age. The
southern paralic facies consists of shales, with
irregular sandstone and coal intercalations. The
sandstones disappear towards the north where the

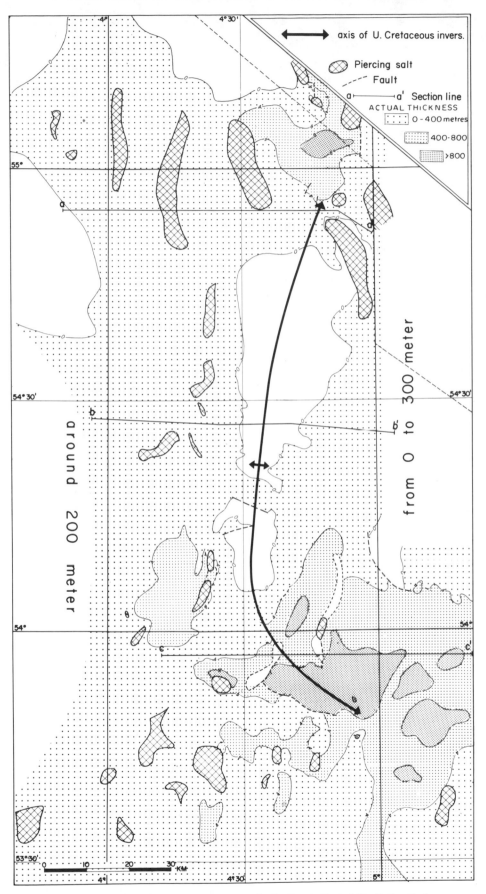

FIG 7. Distribution of the Lower Cretaceous in the Dutch Central North Sea Graben.

restricted marine incursion from the north apparently covered the paralic facies. The paralic Upper Jurassic sequence south of the hingeline can be compared with similar Upper Jurassic sediments found in the Netherlands onshore in elongated narrow basins, surrounded by highs which supplied the clastics found in the basin (Haanstra, 1963; Heybroek, 1974).

A large break in sedimentation is recognized between the Upper Jurassic and Lower Cretaceous. In the graben an appreciable amount of the deposited Upper Jurassic must have been eroded before the marine Lower Cretaceous transgressed over the area; on the surrounding highs all Jurassic deposits that may have been deposited were removed by erosion.

The distribution of the Upper Jurassic is given in Fig 6. Since a considerable amount was removed by pre-Lower Cretaceous and Upper Cretaceous erosion, this map is not indicative of the original distribution and thickness of the Upper Jurassic. However, the present distribution gives the approximate form of the subsiding graben at the time of the Lower Cretaceous transgression. During the Upper Jurassic notable extension and broadening of the graben towards the south took place, in contrast to the development of the Middle Jurassic graben (Fig 5).

CRETACEOUS DEVELOPMENT

After a period of uplift and erosion the Lower Cretaceous sea transgressed over the area. In the south, Middle to Upper Valanginian beds rest on the unconformity; in the north the oldest Lower Cretaceous found so far is of Hauterivian age. Apparently the transgression progressed from the south, filling first the graben depression and finally encroaching on the bordering highs.

The Lower Cretaceous sedimentation was regionally interrupted at the end of Barremian times. After a short period of erosion, the Aptian–Albian sea transgressed over the area leading the way for the general Upper Cretaceous transgression, which covered uniformly both the graben and the former high areas, and extended over large parts of north-western Europe. In this sea a thick monotonous sequence of Chalk was deposited. With this transgression, the differential subsidence in the graben came apparently to an end.

Tectonic activity did not stop, however, in the Dutch part of the graben. Some time in the Senonian, the graben area was affected by differential uplift, the axis of which consistently followed the earlier axis of largest relative subsidence.

These inversion movements in the Senonian are a common feature in several of the Jurassic–Lower Cretaceous troughs along the southern margin of the north-west European basin (Kölbel, 1956;

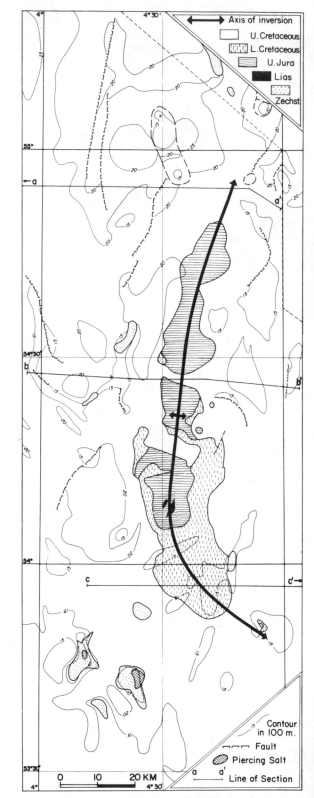

FIG 8.   Subcrop map of the base of the Tertiary in the Dutch Central North Sea Graben.

Voigt, 1962; Heybroek et al., 1967; Heybroek, 1974; Boigk, 1968). The uplift brought most of the Central Graben above sea-level, and in the axial area erosion removed the Upper Cretaceous Chalk, the Lower Cretaceous and part of the Upper

Jurassic. The inversion movement took place in two or more pulses each separated by temporary marine advances with the result that young Chalk beds may extend far over the eroded core of the graben.

Finally the late Palaeocene transgression covered the uplifted graben and the surrounding areas alike, and all differential movements ceased. The Lower Cretaceous has a quasi-general distribution over the area (Fig 7). Only the Mid North Sea High and part of the Schill Grund High escaped the general transgression. In the centre of the graben the Lower Cretaceous was removed by erosion during the Upper Cretaceous inversion movements, and the record of subsidence in this area is lost. However, in the north and south-east where the inversion movement tapers out, thicknesses of 600 to 800m are normal, not counting the extreme deviations due to halokinetic movements. This means a thickness of 400 or 600m more than on the surrounding highs. Noteworthy is the south-eastern extension of the area of subsidence, well outside the outline of the Jurassic graben.

The subcrop map of the base of the Tertiary (*i.e.* base of the Upper Palaeocene, Fig 8) shows the area of total erosion of the Upper Cretaceous Chalk which indicates the area of the maximum inversion movement in the graben. From a comparison of Chalk thicknesses, the axis of

inversion may be followed some distance north and south of the eroded area, before it fades away. It appears that the southern spur of the axis makes a bend to the south-east, following the trend of thick Lower Cretaceous deposits.

The Central North Sea Graben continues beyond the Dutch offshore border into the northern North Sea (Ziegler, 1975). The Upper Cretaceous inversion movements seem, however, to be confined to the Dutch part of the graben and taper out northward.

GEOLOGICAL SECTIONS

In Fig 9 three typical geological cross sections through the graben are presented. A palaeo-reconstruction of section B–B' in the central region is attempted in Fig 10. The palaeo-reconstruction shows in five steps the earlier deduced development of the graben. It lacks precision due to the uncertain timing and amount of salt flowage. Moreover the thickness of eroded strata has to be inferred. For this reconstruction an average salt thickness of 1200m has been assumed. Thickness variations in the Triassic and Liassic are attributed to initial salt flowage into pillows. Downwarp of the graben started after the Liassic deposition and diapirs came into existence. Differential subsidence was maintained both during the three

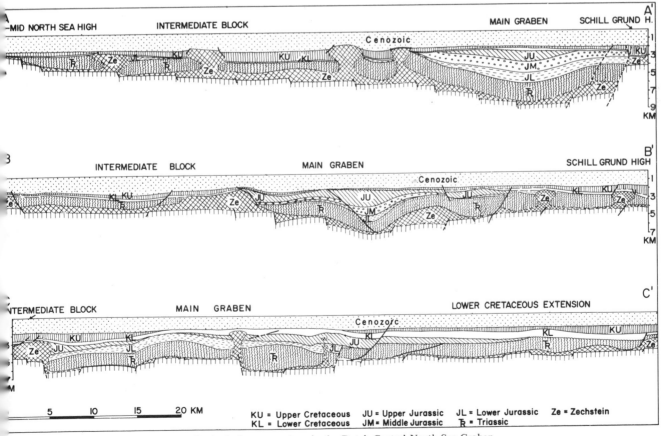

FIG 9. Geological cross-sections in the Dutch Central North Sea Graben.

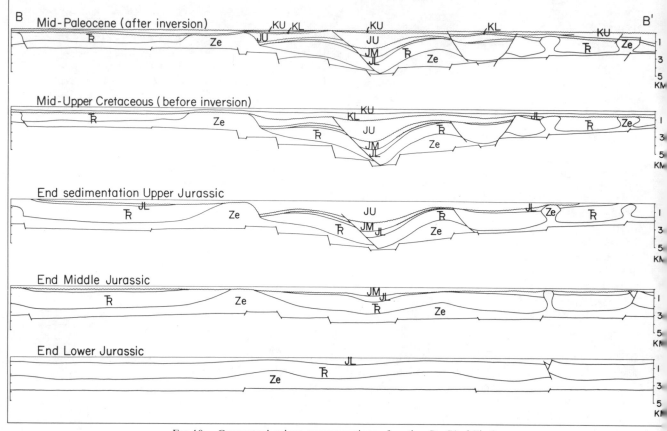

FIG 10.   Conceptual palaeo-reconstructions of section B—B' of Fig 9.

main periods of sedimentation in the Middle Jurassic, Upper Jurassic, and Lower Cretaceous, and the main periods of erosion in between. During the differential subsidence the salt in the graben was largely concentrated in saltstocks and later halokinetic movements were minor. The graben was deepest during the middle Upper Cretaceous. The subsequent inversion movements (in two pulses) resulted in an uplift in the centre of about 1500m. Finally an 1800m-thick Cenozoic sequence was deposited over the entire area.

The southern part of the graben (section C–C', Fig 9) is characterized by the absence of Middle Jurassic and by thick Lower Cretaceous which extends across the border of the Upper Jurassic graben.

Section A–A' across the northern part shows that Jurassic sedimentation in the main graben was hardly interrupted.

CONCLUSIONS

From the foregoing, the following general conclusions can be drawn on the evolution of the Dutch part of the Central North Sea Graben.

1. The relative subsidence of the graben started in the Middle Jurassic and persisted during alternating periods of regional subsidence and uplift until the end of the Lower Cretaceous. As a result, the graben contains the record of the regional tectonic evolution which is removed by erosion from the bordering highs.
2. The area of the graben progressively enlarged towards the south and south-east during its subsidence.
3. The marked inversion of the subsiding movement in the Upper Cretaceous is restricted to the Dutch part of the graben.
4. Detailed stratigraphical and facies analysis provides the main source of information for the structural evolution of the graben, where halokinetic movements obscure direct evidence.

ACKNOWLEDGEMENTS

The author is indebted to a large number of earth scientists of the Nederlandse Aardolie Maatschappij, who contributed to the stratigraphical and structural interpretation of the area discussed. Credit should especially go to H. Struwe and J. Tiemens who did the major part of the seismic interpretation. Permission of Shell Nederland B.V. and Esso Nederland B.V. as the parent companies of the Nederlandse Aardolie Maatschappij B.V. to publish this paper is gratefully acknowledged.

REFERENCES

Boigk, H. 1968. Gedanken zur Entwicklung des Niedersächsichen Tektogens. *Geol. Jb.*, **85**, 861–900.

Boigk, H., Hark, H. U. and Schott, W. 1963. Oil migration and accumulation at the northern border of the Lower Saxony Basin. *Wld. Petrol. Congr. 6, Frankfurt, Proc.* **1**, 435–56.

Brand, E. and Hoffman, K. 1963. Stratigraphy and facies of north-west German Jurassic and genesis of its oil deposits. *Wld. Petrol. Congr. 6, Frankfurt, Proc.* **1**, 223–46.

Christian, H. E. 1969. Some observations on the initiation of salt structures of the southern British North Sea. *In:* P. Hepple (Ed.), The Exploration for Petroleum in Europe and North Africa. Inst. of Petrol., London, 231–50.

Glennie, K. W. 1972. Permian Rotliegendes of northwest Europe interpreted in light of modern desert sedimentation studies. *Bull. Am. Ass. Petrol. Geol.*, **56**, 1048–71.

Haanstra, U. 1963. A review of Mesozoic geological history in the Netherlands. *Verh. kon. Nederlands. Geol. mijnb. Gen.*, **21**, 35–57.

Heybroek, P. 1974. Explanation to tectonic maps of the Netherlands. *Geologie en Mijnb.*, **53**, 43–50.

Heybroek, P., Haanstra, U. and Erdman, D. A. 1967. Observations on the geology of the North Sea area. *Wld. Petrol. Congr. 7, Mexico, Proc.* **2**, 905–16.

Kölbel, H. 1956. Uber wechselnde Tendenzen in der tektonischen Entwicklung Westmecklendung. *In:* Lotze, F. (Ed.) Geotektonisches Symposium zu ehren von Hans Stille, Stuttgart, Enke, 205–212.

Larsen, G. 1960. Rhaetic–Jurassic–Lower Cretaceous sediments in the Danish Embayment. *Geol. Surv. Denmark, II Series, No. 91*, Copenhagen.

Sorgenfrei, Th. 1969. Geological perspectives in the North Sea area. *Bull. Geol. Soc. of Denmark*, **19**, 160–96.

Trusheim, F. 1960. Mechanism of salt migration in Northern Germany. *Bull. Am. Ass. Petrol. Geol.*, **44**, 1519–40.

Voigt, E. 1962. Uber Randtroge von Schollenrandurn und ihre Bedentung im Gebiet der Mitteleuropaischen Senke und angrenzender Gebiete. *Z. dt. geol. Ges.*, **114**, 378–418.

Wilson, V. 1948. East Yorkshire and Lincolnshire. *British Regional Geology*. Geol. Surv., HMSO, 94 pp.

Wilson, V., Hemingway, J. E. and Black, M. 1934. A synopsis of the Jurassic rocks of Yorkshire. *Proc. Geol. Ass.*, **45**, 247–90.

Wolburg, J. 1969. Die Epirogenetischen Phasen der Muschelkalk und Keuper-Entwicklung Nordwest-Deutschlands, mit einen Rückblick auf den Buntsandstein. *Geotekt. Forsch*, **32**, 1–65.

Ziegler, P. A. 1975. North Sea basin history in the tectonic framework of north-western Europe. *Paper 9, this volume.*

DISCUSSION

T. D. Eames (Conoco Europe Ltd): (1) What evidence is there for the lacustrine facies in the Middle Jurassic?

(2) Can you say anything about the source of the clastics in the Middle and Upper Jurassic— whether they are shed off the east and west sides of the graben, or directed by sediment transport in a north–south direction along the axis of the basin?

D. P. Heybroek: (1) The fresh-water to paralic facies of deposition occurring in the Middle and Upper Jurassic is deduced from the occurrence of sizable coal beds, from the lack of marine fossils, from the limited extent of the sand intercalations and from the sedimentary structures found in cores. This interpretation is supported by the rapid vertical alternation of the three lithologies present. The palaeo-geographical setting of the Middle Jurassic fresh-water to paralic facies, separated by an inferred land area from the marine deposits of the same age to the south, make me think of a deposition in a lake.

(2) The distribution of sand in both intervals favours a nearby source of the clastics from the surrounding areas of erosion, rather than from a distance, with transport along the basin axis.

26

The Brent Oil-Field

By J. M. BOWEN

(Shell UK Exploration and Production Limited)

SUMMARY

The Brent Field lies in the extreme northern part of the UK sector of the North Sea, close to the median boundary with Norway. The field, which was discovered in July 1971 and declared commercial the following year, is situated mainly in licence block 211/29 some 88 miles (140km) north-east of Unst in the Shetlands. The average water depth is 460ft (140m). The trap is of the combination structural/stratigraphic truncation type comprising a westerly tilted and partially eroded fault-block (one of many within the Viking Graben) containing the reservoirs, overlain unconformably by sealing shales. There are two main reservoirs, both sandstone, one of Middle Jurassic, the other of early Jurassic to Rhaetian age, separated by Liassic shales and siltstones; a gross reservoir thickness of 1550ft (473m) is found below depths of 8000ft (2440m) in the crestal part of the structure. The overlying shales range from late Jurassic to late Cretaceous in age. The combined recoverable reserves of the field are estimated to be in the order of 2.0×10^9 barrels of oil, with some 3.5×10^{12} scf of associated gas.

REGIONAL SETTING

The Brent Field lies mainly in the Shell/Esso block 211/29 (61° N) some 5–10km west of the median line between the British and Norwegian sectors of North Sea; the southern limits of the field extend into the Texaco block 3/4.

In terms of the regional tectonic picture (Fig 1), the field is centrally situated in the northern, wider part of the Viking Graben, an area referred to here as the Brent Province where an extraordinary number of exploration successes have been scored, of which the Brent Field discovery was the first.

EXPLORATION BACKGROUND

The chief characteristic of the Brent Province, which became apparent as soon as the earliest seismic reconnaissance lines had been shot in 1966–67, is a thick tectonically undisturbed sedimentary fill of some 7000 to 10 000ft (2130 to 3050m), overlying and apparently onlapping an irregular erosional surface, giving an impression of a buried relief. This impression is enhanced by the shape of the buried highs which have a tendency to be elongated north–south and asymmetric east–west, giving a suggestion of dip slopes and scarps.

A very strong seismic reflector (Fig 2) is normally associated with this surface and, perhaps because of this, the early seismic profiles rarely showed any deeper reflections. It could be safely assumed that the overlying fill was of Tertiary and Cretaceous age since horizons such as top Palaeocene and top Chalk could be mapped by extrapolation from wells in the Central Graben area farther south. The age and nature of the rocks forming the buried features, however, was entirely open to conjecture; these could have been of any age from early Cretaceous to Precambrian.

By 1970, when a small number of blocks in the area were put on offer by the British Government, the seismic quality had improved sufficiently to permit the mapping of deeper reflectors in a few of the features, notably at Brent. Velocity analyses and aeromagnetic data also indicated that sedimentary sequences could be expected below the apparent unconformity; the question of their age and reservoir potential remained to be resolved. The general resemblance of the Viking Graben to the East Greenland basin, which in a pre-drift situation lay to the north of the Brent Province, also gave some encouragement, as Triassic and Jurassic sandstone reservoirs are present there in a similar block-faulted tectonic setting.

Following the allocation of blocks 211/29 and 30 to Shell/Esso in March 1970, the first well in the area (211/29-1) was spudded on May 10th, 1971. The well, in 467ft (142m) of water and 90 miles (145km) north-east of Shetland, lay 230 miles (370km) north of the nearest UK well control in the Forties Field, and was then the most northerly offshore well in the world.

The location chosen for the first well was 1km down-dip to the west of the crest of the buried high, with the intention of penetrating the most complete sequences, both immediately above and below the apparent major regional unconformity.

During the drilling, which took two months, the Fourth Round of concessions was announced (June 22nd) in which a large number of blocks in the vicinity were offered to the industry, some on a sealed bid basis. The consequent pressure on rig availability resulted in the curtailment of the

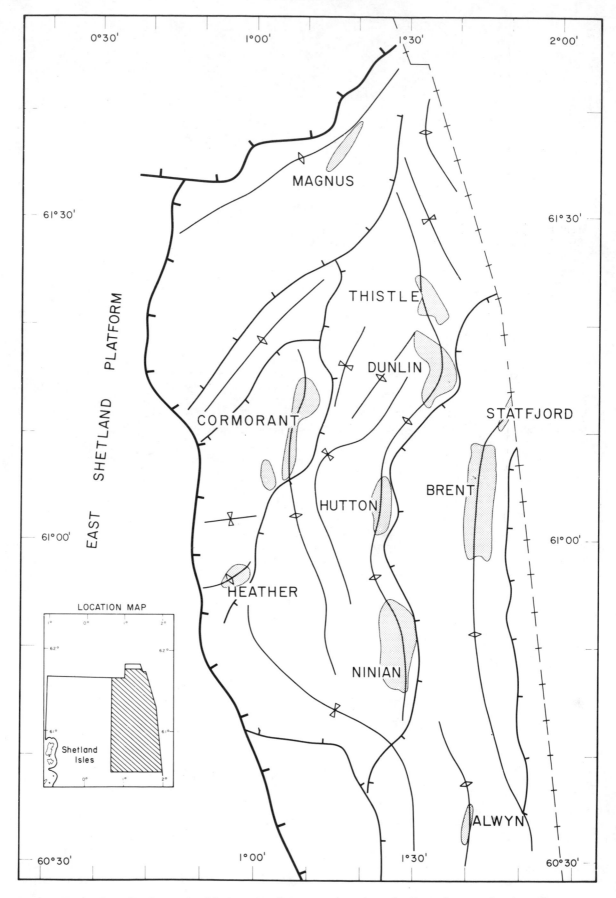

FIG 1.   Brent Province regional fault pattern and structural trends on the Cimmerian unconformity surface.

SHELL/ESSO
211/29−4
(proj.)

SHELL/ESSO
211/29−5

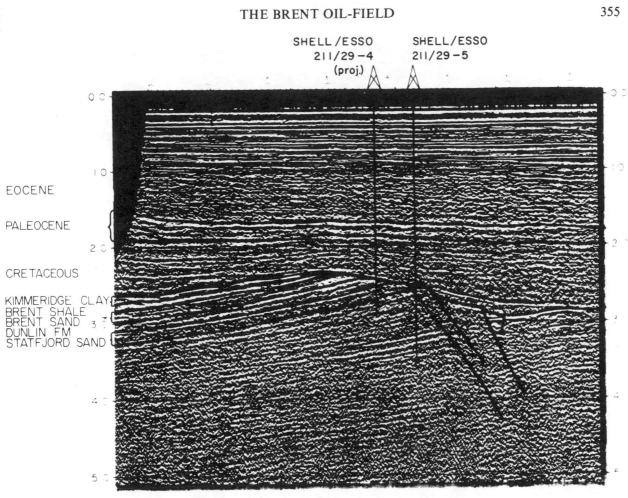

EOCENE

PALEOCENE

CRETACEOUS

KIMMERIDGE CLAY
BRENT SHALE
BRENT SAND
DUNLIN FM
STATFJORD SAND

FIG 2.   Seismic line U2-392.4 in Brent Province.

operations on 211/29-1 both in respect of the drilled depth and full evaluation.

STRUCTURE

The Late Cimmerian Brent structure below the flat-lying Tertiary and slightly draped Cretaceous fill consists of a north–south striking westerly gently dipping, partially eroded fault-block with a relief of 3000ft (910m) above the structurally deeper areas to east and west (Fig 3). A major eastwards hading fault bounding the Brent block to the east lies some 3.5km east of the crest of the buried escarpment; this is presumed to have a throw of many thousands of feet. At least one subsidiary fault with some 700ft (213m) of displacement, probably parallel to the main fault, occurs on the eastern slope of the structure (Fig 2). The total displacement of the Brent fault-block, relative to the next block to the east, probably exceeds 6000ft (1830m).

Dip-faulting is almost absent in the main part of the Brent structure, but is more common on the southern plunge, while an important north-eastwards hading fault occurs in the saddle which

separates Brent from the Statfjord structure to the north-east.

STRATIGRAPHY

The Brent discovery well 211/29-1 was drilled through a predominantly sandy Neogene section underlain by equally sandy Oligocene and Upper Eocene sections. Below, mudstones predominated with only thin-bedded sandstones and siltstones occurring until the base of the Tertiary was reached.

The Chalk, as known in the central and southern North Sea, was found to have been replaced by siltstones, claystones and marls with occasional thin beds of chalky limestone. Below, a 250-ft (76-m) section ascribed to the Lower Cretaceous again consists mainly of marls becoming reddish brown near the base where limestone is also recorded.

At this point, the well reached the prominent deep seismic reflector corresponding to the erosional surface and penetrated an important stratigraphic hiatus at which Aptian–Albian marls and limestone overlie 180ft (55m) of

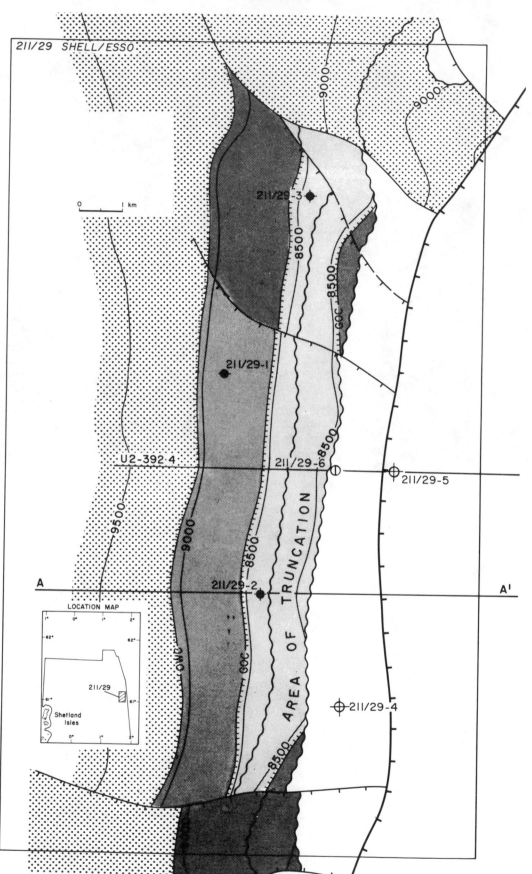

FIG 3.   Depth contour map of top of Brent Sand; contours in feet below sea-level.

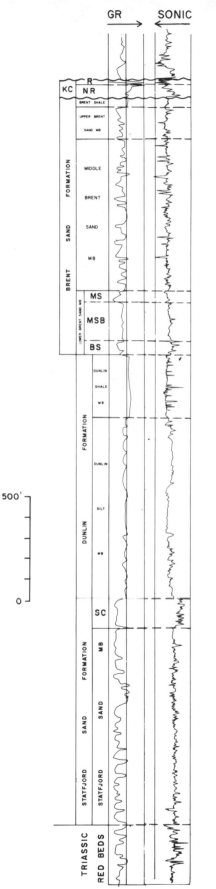

GR   SONIC

500'

0

FIG 4. Log of well 211/29-3 showing proposed lithostratigraphic units. KC: Kimmeridge Clay Formation; R: radioactive MB; NR: non-radioactive MB; MS: Massive sand bed; MSB: micaceous sand bed; BS: basal sand bed; SC: Statfjord calcareous MB.

ROCK STRATIGRAPHIC UNITS		TIME STRATIGRAPHIC UNITS	
FORMATION	MEMBER	STAGE	SERIES
			Lower Cretaceous
Kimmeridge Clay	Radioactive	Portlandian Kimmeridgian Oxfordian	Upper Jurassic
	Non-Radioactive	Callovian	
Brent Sand	Shale	Bathonian	Middle Jurassic
	Upper Sand	Bajocian	
	Middle Sand		
	Lower Sand		
Dunlin	Shale	Aalenian / Toarcian / Pliensbachian	Lower Jurassic
	Silt	Sinemurian / Hettangian	
Statfjord Sand	Calcareous		
	Sand	Rhaetian	Upper Triassic

FIG 5.   Rock and time stratigraphic units in the Jurassic of the Viking Graben.

richly organic, black radioactive shale of Kimmeridgian–Portlandian age, here referred to as the Kimmeridge Clay Formation; thus strata of earliest Cretaceous age were found to be absent. Some 200ft (60m) of dark grey non-radioactive shale of late Kimmeridgian age lies immediately below (Fig 5).

A second important unconformity was then penetrated, below which is an unbroken Middle–Lower Jurassic sequence. While the relationship of these two Cimmerian unconformities and the tectonic events which caused them have still to be fully elucidated, the remarkable persistence of the Kimmeridge Clay throughout the Brent Province is now well established. On the eastern erosional face of the Brent feature, however, the later erosion has cut out the Kimmeridge Clay entirely bringing the Lower Cretaceous into contact with Middle and Lower Jurassic rocks below (Fig 6).

Below the lower unconformity the discovery well penetrated an 800-ft (240-m) Middle Jurassic sand sequence of Bajocian age before reaching total depth in Lower Jurassic shales.

The Middle Jurassic section, proposed here as the Brent Sand Formation, consists of massive sands (482ft (147m) net in 211/29-1) deposited in a fluvio-deltaic environment. The formation has now been subdivided into three units on the basis of detailed sedimentological studies. The deepest unit overlying Lower Jurassic shales comprises massive sands which in turn can be subdivided into a thin basal coarse sand overlain by fine-grained micaceous coastal to shallow marine sheet-sands and followed by coarse-grained channel sands. The middle unit is more argillaceous with sandstones, shales and coals, interpreted as deltaic plain deposits, while the upper unit comprises massive coastal to shallow-marine sands.

211/29-1 discovered an oil column of 197ft (60m), 140ft (43m) net, in the Brent sands, below which the sands were water-bearing; only wireline testing was carried out.

211/29-2, drilled a year after the discovery well, was situated in a crestal position and encountered 845ft (257m) of Brent Sand Formation of which 680ft (207m) was wet sand; a 700-ft (213-m) hydrocarbon column including 170ft (52m) of gas, over 375ft (114m) of net oil-sand confirmed the presence of a major field.

Below the base of the Brent Sands, 210ft (64m) of shale was followed by a 600-ft (183-m) sequence consisting of shale, siltstones and a few thin sands. The Lower Jurassic succession below the base of the Brent Sands is proposed here as the Dunlin Formation with upper Shale and lower Siltstone members. Below the Dunlin Formation the well penetrated 90ft (27.4m) of lime-cemented sandstone and 175ft (53.3m) of another massive porous sandstone body before being terminated for mechanical reasons. This deeper sand encountered for the first time in this well, was wet.

It was not until two years after the discovery that the entire Jurassic sequence was penetrated (Fig 6). Well 211/29-3 (Fig 4), also in a crestal position, after drilling the Brent and Dunlin sequence as seen in 211/29-2, penetrated 80ft (24.4m) of lime-cemented sandstone overlying 585ft (178.3m) of massive sandstone before bottoming in typical Triassic sandstones and shales; again the deeper sand was water-bearing. This second major sandstone body, which was first fully penetrated in Conoco's well 211/24-1 to the north of Brent, is proposed here as the Statfjord Sand Formation; it is believed to range in age from early Jurassic (Hettangian–Sinemurian) to Rhaetian in the lower part.

The depositional environment of the Statfjord Sands is overall more continental than that of the Brent Sands. Again three units have been identified on a sedimentological basis; the lowermost and thickest unit consists of braided stream deposits whereas the upper two units comprise field-wide sheets of coastal barrier sands.

Subsequently a fourth well (24/29-4) was drilled to test the Statfjord Sands in an updip position and found them to contain a 610-ft (185.9m) hydrocarbon column with 180ft (54.9m) of gas overlying 370ft (112.8m) of oil.

The Triassic sequence underlying the Brent Field consists of several thousand feet of thin-bedded sandstones and red claystones, the upper part of which is of Rhaetian to Norian age.

RESERVOIR PROPERTIES

Except for the unconsolidated Eocene and younger sands, no reservoir rocks were encountered other than the Brent, Statfjord and Triassic sands.

Best reservoir properties occur in the Brent Sands in which porosities vary from 7 to 37 percent. Permeabilities range up to some 8000md. Despite these high figures, the heterogeneous nature of the reservoir is likely to have a negative effect upon recovery efficiency.

The Statfjord Sands properties are relatively inferior except in the lower unit of the blanket sands; porosities range from 10 to 26 percent and permeabilities up to some 5500md.

Reservoir properties in the Triassic sands underlying the Statfjord are very poor, due mainly to carbonate cementation.

TRAP DEVELOPMENT

The delineation of the Brent fault-block as a distinct structural unit between the north–south trending normal faults of the developing Viking Graben was established in the late Triassic, if not earlier. Differential movement between the Brent and adjacent fault-blocks continued from late Triassic until late Middle Jurassic times when the rate of westwards tilting increased and major erosion of the resulting fault-scarps took place.

Rapid subsidence and transgression followed. The tilting continued during the deposition of the deep-water euxinic Kimmeridge clays which were draped over the pre-existing, now drowned, relief. In late Portlandian times the rate of tilting again increased resulting in the complete erosion of the Kimmeridge and some of the underlying Middle Jurassic rocks from the scarp face prior to the deposition of Lower Cretaceous sediments. This latter minor erosion probably occurred in a submarine environment. At this stage movement between the fault-blocks essentially ceased and, as the whole area subsided, the remaining relief was engulfed by the onlap of flat-lying Lower Cretaceous clays and marls.

The development described above has given rise to a combination structural/stratigraphic trap (Fig 7). On the west flank the Brent and Statfjord sands are sealed above by the Brent Shale member and Dunlin Formation respectively. On the east the seal is the Kimmeridge organic shale where present and Lower Cretaceous marls. To the south, closure is provided by the plunge of the structure as a whole, while to the north closure is provided partly by a saddle at top Cimmerian level but mainly by closure across the north-eastwards hading fault already mentioned at both reservoir levels.

HYDROCARBONS

The bulk of the hydrocarbon reserves lie in the main fault-block in the Brent and Statfjord sands which are separate reservoirs. Recent drilling has indicated that some small additional reserves may remain to be found in the Statfjord Sand where it is

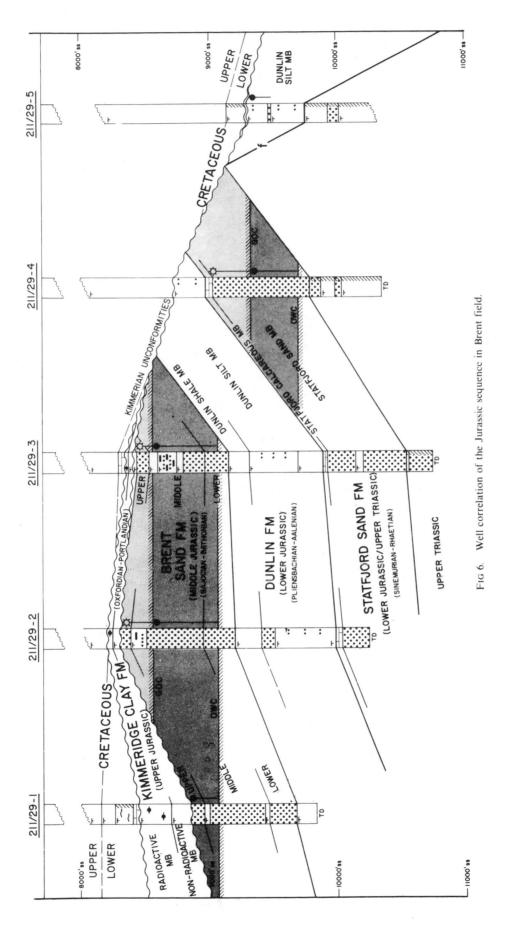

Fig 6. Well correlation of the Jurassic sequence in Brent field.

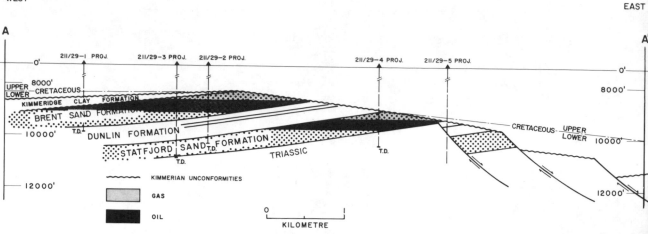

FIG 7.   Schematic cross-section through the Brent field; for line of section see Fig 3.

down-faulted along the eastern margin of the field. No hydrocarbons have been found in the underlying Triassic sands.

The Brent Sands reservoir has a gas-cap of approximately 260ft (79.2m) and a total hydrocarbon column in the order of 740ft (225.5m). Oil gravity is 36° API and the GOR varies from 1500 to 1700cu ft/bl. Sulphur content is 0.3 percent.

The Statfjord Sand reservoir, because of the geometry of the field, lies mainly to the east of the Brent Sand accumulation; it has a gas-cap of 465ft (141.7m) and a total hydrocarbon column of some 900ft (274m). Oil gravity is 38.5° API and the GOR varies from 1000 to 3000cu ft/bl. Sulphur content is 0.11 to 0.36 percent.

The oil is thought to have been generated mainly from Upper Jurassic shales deeply buried in the troughs flanking the Brent structure to the east and west, whereas the gas may have originated in part from the Brent sequence itself which is rather rich in plant material and small coal seams.

On the basis of present limited well data (three Brent Sand, three Statfjord Sand penetrations) reserve figures are necessarily tentative. However, total liquid oil reserves are likely to amount to

2000×10^6 bbls while the gas reserves may reach 3.5×10^{12} scf.

ACKNOWLEDGEMENTS

The author wishes to express his thanks to E. W. F. Berger for his help in preparing the paper and to acknowledge the contributions of many members of the Shell staff, in particular those of D. L. Batho, P. J. Ealey and P. Morgenroth, to the understanding of the depositional environments and biostratigraphy.

Permission to publish was given by Shell UK Exploration and Production Co. Ltd, and Esso Exploration and Production UK Inc.

BIBLIOGRAPHY

Haller, J. 1911. *Geology of the East Greenland Caledonides.* Interscience Publishers, London, 413 pp.

Rhys, G. H. (Compiler). 1974. A proposed standard lithostratigraphic nomenclature for the Southern North Sea and an outline structural nomenclature for the (U.K.) North Sea. A report of the joint Oil Industry–Institute of Geological Sciences Committee on North Sea Nomenclature. *Rep. Inst. Geol. Sci.*, **74/8,** 14 pp.

DISCUSSION

M. F. Osmaston: Dr Bowen's sections extend no deeper than the Triassic. Has he any evidence as to the existence and thickness of underlying sedimentary strata? His remarks regarding possible source-rocks for the Brent oil did not appear to entertain the possibility that the source was in the sub-Triassic strata. Could he please explain?

Dr J. M. Bowen: Several thousand feet of Triassic continental sediments were penetrated in one of the Brent Field wells. Although seismic evidence suggests the presence of a deeper sedimentary section nothing is known of its age or nature as yet. Although it is possible that deeper source-rocks may occur, particularly for gas, all the available evidence points to the Upper Jurassic section as the main source of the hydrocarbons.

Erich Brand (Wintershall): I would like to ask Mr Bowen about his ideas concerning source-rocks for the Brent field. As I understood it, the Triassic underlying the pay-zones is developed in a more paralic or lacustrine facies. Would Mr Bowen consider that the shaly Upper Cretaceous of the deep block could have been active in generation of hydrocarbons? Since it is not a big problem to distinguish oil being generated by marine or by non-marine source-rocks the provenance of the Brent oil should be easily solved.

Dr J. M. Bowen: The Triassic is developed in purely continental, mainly red-bed facies and is therefore unlikely to contain source-rocks. It is not impossible that Cretaceous shales in deeper fault-blocks have produced some hydrocarbons, but in our opinion they do not possess the excellent source-rock properties of the Upper Jurassic Kimmeridge Clay formation.

N. Morton (Birbeck College): (1) I would like to ask about the extent to which the Middle Jurassic in the northern areas are marine deposits.

(2) Do you have any information about the sediment-transport direction? In the Hebrides these seem to be parallel to the shore-line.

Dr J. M. Bowen: (1) In the Brent Field area the lower and upper parts of the Brent Sand Formation are marine whereas the middle part is considered to be of deltaic (lower coastal plain) origin.

(2) Judging by dipmeter results foresetting in the sands would appear to indicate a transport direction from between south and east. The position and alignment of the shorelines are as yet unknown.

27

The Piper Oil-Field, UK North Sea: a Fault-Block Structure With Upper Jurassic Beach-Bar Reservoir Sands

By J. J. WILLIAMS, D. C. CONNER and K. E. PETERSON

(Occidental of Britain, Inc., London)

Piper Field discovery well, 15/17-1a, drilled through 192ft (52m) of oil-bearing sandstone of Upper Jurassic age on December 22, 1972. Subsequent appraisal wells delineated approximately 8600 productive acres (3.5km²) with an estimated 1.55 billion barrels of 36° API gravity low sulphur oil in place and 650–900 million barrels recoverable. Reservoir sandstones are Oxfordian/Lower Kimmeridgian in age, of marine origin, and unconformably overlie a non-marine Middle Jurassic sedimentary sequence. The gross reservoir thickness averages 250ft (76m) over the field area and is comprised of several individual sand bodies 40–70ft (12–21m) in thickness. Within individual sand bodies the grain-size grades either upwards or downwards from very fine sandstone or siltstone into coarse-grained sandstone. The sandstones are generally well sorted, highly bioturbated, friable and have excellent porosity and permeability. Individual sand bodies record local regressions or transgressions. Regressive sands, accreting seaward as foreset beds, are generally thicker than transgressive sands.

Isopachs of Triassic through Cretaceous units suggest that the Piper structure grew intermittently before, during and after deposition of the reservoir sands. Post Albian–pre-Turonian/Coniacian faulting and erosion, plus contemporaneous deposition and faulting during pre-Maastrichtian Upper Cretaceous time formed a series of eroded and tilted fault blocks. The domal structure mapped at the level of the base of the Tertiary is not evident in the younger horizons due to masking effect of middle Tertiary deltaic deposits. Regional subsidence continued throughout the Tertiary and formed the North Sea Basin as we see it today.

INTRODUCTION

The Piper Oil Field is located in UK North Sea block 15/17. Geologically, the area is within the sub-basin referred to as either the eastern end of the Moray Firth Basin or the northern end of the Northern North Sea Basin (Fig 1). The discovery well, 15/17-1a, was the first to find commercial quantities of oil within this sub-basin. The nearest previous discovery, the Forties Field, 55 miles (88km) south-east of Piper, is a separate geologic setting, producing from a Palaeocene reservoir.

This paper was prepared incorporating results from the 15/17-1a discovery and seven follow-up wells including Burmah 15/12-1 and extensive seismic coverage (Fig 2). Three wells, 15/17-4, 5 and 6 were nearly continuously cored in the reservoir sands and detailed studies of these cores are in progress at the time of writing.

OCCIDENTAL'S NORTH SEA EXPLORATION HISTORY

Occidental Petroleum, a relative latecomer to the North Sea exploration scene, began regional seismic studies and investigating petroleum potential of onshore and offshore United Kingdom late in 1969 in anticipation of forthcoming licence application invitations by either the United Kingdom or Norway. Occidental began seismic shooting in December 1970 and by early 1971 it became apparent that the United Kingdom would be the next to offer production licences and our work was, therefore, concentrated in the UK sector. A consortium was formed, composed of Occidental as operator with a 36.5 percent share, Getty Oil International (England) Limited, 23.5 percent; Allied Chemical (North Sea) Limited, 20 percent; and Thomson Scottish Petroleum Limited, 20 percent. In the period January to August 1971, our group evaluated over 19 000 line miles of seismic data in the UK North Sea between 56° N and 62° N latitude.

In July, 1971, the Occidental group contracted with Ocean Drilling and Exploration Company to construct a semisubmersible rig, the Ocean Victory capable of northern North Sea drilling activity all year round. In addition to the Ocean Victory, which would be available late in 1972, the group also contracted for the drillship, Sonda I, in order

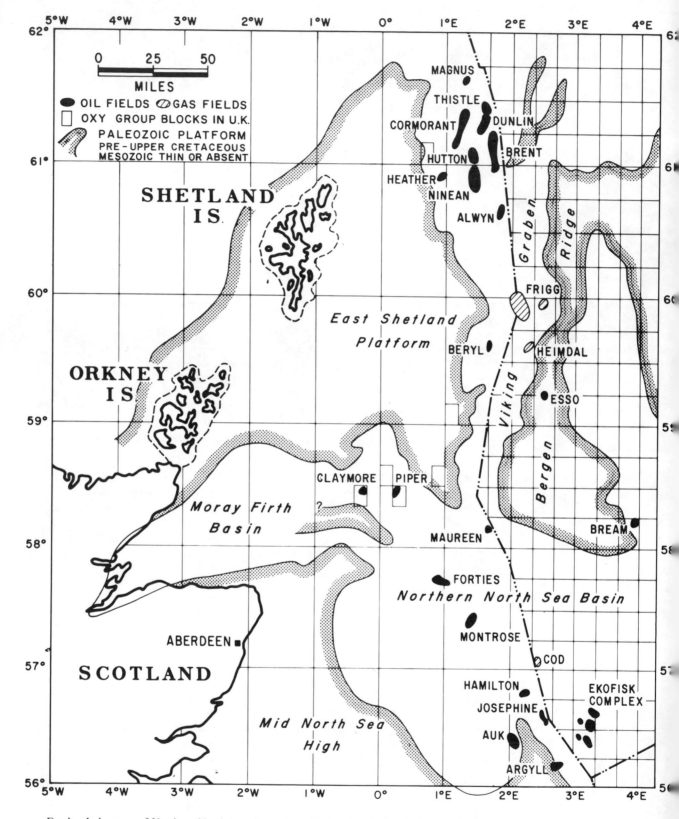

FIG 1. Index map of Northern North Sea oil province, offshore Scotland and Norway, showing main tectonic elements, Occidental Group's blocks and oil and gas fields discovered at the time of writing.

to take advantage of the more favourable summer drilling period, if licences were awarded in the spring of 1972. Both of the above mentioned rigs were contracted prior to our application, in anticipation of being awarded geologically favourable blocks which the group would hope to explore rapidly and aggressively.

Applications for petroleum production licences were submitted in August, 1971 and licences awarded by Her Majesty's Government in March 1972. Selection of blocks for which the Occidental group applied was based on geological analysis of seismically defined structures and the inferred geological framework within which these structures were located. Subsurface stratigraphic inferences were made based on the seismic models and meagre published and scouted information.

The Occidental group was granted a licence containing six blocks varying from 200–216km² each for a total of 1279.3km² or 315 987 acres (Fig 1). Three of these blocks, 14/19, 15/11 and 15/17 are within the same geological setting, at the north end of the Northern North Sea Basin and east end of the Moray Firth Basin. Each of the three blocks contains at least one large domal structure.

The Sonda I moved to the North Sea in April 1972 and spudded 15/11-1 on 3rd May, 1972, a little more than one month after the fourth round licences were awarded. The well, located near the common boundary of Occidental's 15/11-1 and Texaco's 15/16 blocks, tested a large faulted anticlinal structure as mapped on the "Base Tertiary" seismic horizon. The cost of drilling was shared jointly by both licence holders but, unfortunately, no oil was found. A second well, drilled near the crest of another large anticline located in block 14/19 tested a non-commercial quantity of oil from very thin sands which were considered to be either Jurassic or Lower Cretaceous in age. The Sonda I was then moved to 15/17-1, near the crest of a third large structure. Because of the proximity of British Petroleum's Forties Field, the primary objective of the first two tests was Palaeocene sands; a Mesozoic reservoir was a secondary objective. The 15/17-1 wildcat was primarily a Mesozoic sandstone test as the 15/11 and 14/19 wells indicated that basal Palaeocene sands were non-productive and geologically less attractive in this general area than had been originally anticipated. Seismically it appeared that the Mesozoic section was thicker over the 15/17 structure than at either the 15/11 or 14/19 crestal locations and encouragement was provided by oil shows in 14/19-1. The 15/17-1 was abandoned in September 1972 at a drilled depth of only 1500ft (457m) following several anchoring and shallow casing problems. Because of anticipated unfavourable winter weather, the drillship was then released to the contractor.

The semi-submersible Ocean Victory arrived at the 15/17-1a location in November 1972 and spudded in about one half mile east and slightly downdip from the original 15/17-1 location. On 22nd December, 1972, 192ft (59m) of highly porous and permeable oil sand of Jurassic age was encountered in the interval 7600ft (2316m) to 7806ft (2379m). Thus, after a disappointing first series of holes drilled during the summer of 1972, a very Merry Christmas resulted for all concerned.

Piper Field Appraisal and Drilling History

Production tests in the discovery well yielded 36° gravity, low sulphur oil at flow rates of 3629BOPD and 5266BOPD from perforated intervals 7780–7806ft and 7603–7635ft (2371–2379m and 2317–2327m). It was apparent to the Occidental group that a significant discovery had been made and it was named Piper Field.

The 15/17-2 well was drilled about two miles north-west of the discovery well and encountered a full oil column in the same reservoir sand at a depth of 7933–8192ft (2418–2497m). Production tests through a 2-in (51mm) open choke flowed oil at rates of 15 257BOPD and 16 872BOPD respectively, from perforated intervals 8152–8192ft and 8020–8115ft (2485–2497m and 2444–2473m).

15/17-3, an exploratory well, was drilled about 1.4 miles (2.3km) south of 15/17-1a on the downthrown side of a prominent fault where the reservoir was expected to be considerably lower than in the first two wells. This third well also encountered a full oil column in the same reservoir between 8210–8325ft (2494–2537m). A production test of perforated intervals 8210–8290ft and 8307–8322ft (2494–2519m and 2524–2528m) flowed at a rate of 15 509BOPD through a 2-in (51mm) open choke.

At this point, an oil/water contact had not yet been established, although the proven oil column was in excess of 800ft (243m). The Ocean Victory was then moved into block 15/12 where the Burmah group drilled at a down-dip location on the Piper structure. 15/12-1 encountered thick Piper reservoir sands but found them wet.

15/17-4, located on the north flank of the structure between 15/12-1 and 15/17-2 encountered Piper reservoir sands between 8441 and 8710ft (2564–2646m). Log analysis and nearly continuous coring of the reservoir established an oil/water contact at 8588ft (2609m), i.e. 8510ft (2585m) sub-sea. The Ocean Victory was then moved from the field area to drill an exploratory well elsewhere.

A tighter seismic grid was shot after the initial discovery and the refined seismic data plus velocity and geologic information from the appraisal wells defined the field limits such that a single production platform was envisaged.

However, the south, east and west flanks of the Piper structure are complicated by faulting and it was thought necessary to determine if the oil/water contact was the same over the entire field and if the faults were sealing or non-sealing. The determination of oil/water contacts was important

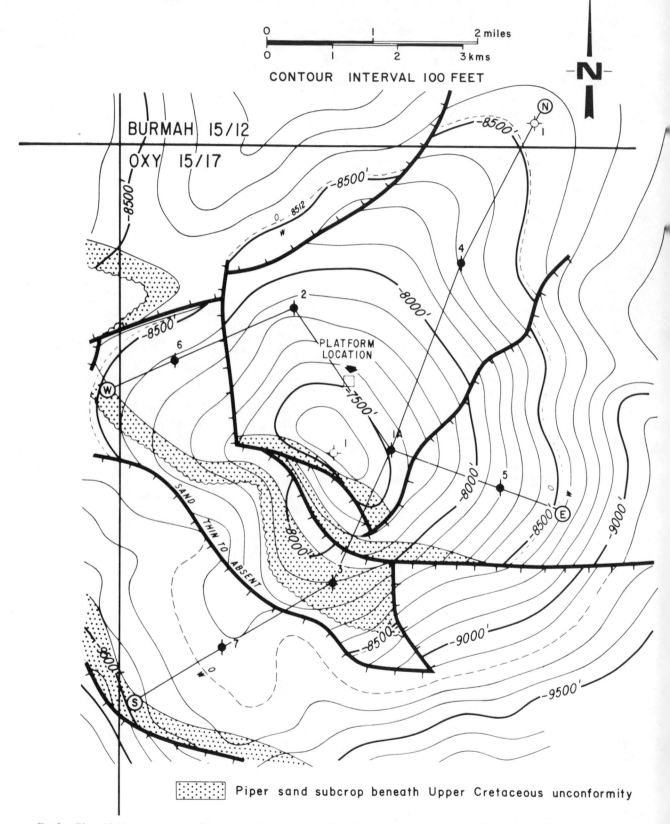

FIG 2.   Piper Field structure map with contours drawn on top of the Piper sandstone, showing well locations and lines of cross-sections in Figs 9 and 10.

in order to define the field boundaries for placement of the production platform being designed and to prove whether the field could be developed from one platform. Wells 15/17-5 and 6 both encountered oil sands and confirmed the oil/water contact found in 15/17-4.

The main Piper Field, delineated by the above mentioned appraisal wells, covers an area of about 8000 acres (3.24km²) with horizontal oil/water contact at approximately 8510ft (2585m) subsea and a vertical oil column of about 1150ft (349m).

The 15/17-7, an exploratory well, drilled south of the main Piper field tested the downfaulted nose where Jurassic sands were expected structurally lower than the oil/water contact of the main Piper field. This well encountered oil bearing Piper sandstone at 9225ft (2812m) and a new pool with oil/water contact at 9272ft (2826m) (9199ft (2804m) subsea) was established. The new pool will be delineated by drilling from the production platform (Fig 2).

PERMIAN AND MESOZOIC GEOLOGICAL HISTORY

The Mesozoic geological history of the Piper field area is best visualized by following a series of block diagrams (Figs 4, 5, 6, 7 and 8). Data available for deciphering the Mesozoic geological history is fairly plentiful as all but two wells penetrated the entire Mesozoic section. Additionally, 15/17-1a on the Piper structure as well as 15/11-1 and 14/19-1 on the other two nearby structures penetrated through the Permian evaporites and into several hundred feet of Carboniferous sediments. The Mesozoic section typical of the Piper field (Fig 3) is compiled from several wells to best illustrate stratigraphy and is an aid when reviewing the block diagrams.

Permian–Triassic

A restricted Permian Sea covered most of the North Sea Basin area resulting in widespread deposits of Zechstein evaporites. Thick Permian salts are present in the southern North Sea Basin. In the Piper area, approximately 400ft (122m) of anhydrite with thin-bedded dolomite and shales, from which a rich assemblage of Upper Permian palynomorphs was recovered, are considered to be a facies of the Zechstein formation. The evaporites in the Piper area were deposited, at least partially, in a sabkha environment, and lie unconformably on middle Carboniferous sandstones and shales.

Extensive non-marine red shales, claystones and sandstones were deposited in north-west Europe throughout the Triassic. Although no diagnostic fossils are found in the red shales underlying the Piper area, they are considered to be of Triassic age based on regional correlation. The isopachs shown on Fig 4 illustrate the red shale thickness prior to deposition of the Middle Jurassic sequence. Thinning of the unit may be either depositional or due to erosion over the crest of an incipient north-east trending structure which began forming during the Triassic or Lower Jurassic period.

The maximum thickness of Triassic red shales deposited in the Piper area cannot be determined from seismic data and will remain unknown until basinward wells are drilled.

Middle Jurassic

The Triassic red shale is overlain by non-marine siltstone, shales, lignite layers, thin-bedded fresh water limestone, tuffs, agglomerates and basalt flows. A Middle Jurassic Bathonian age was determined from beds bearing ostracodes and from lithologic similarities with onshore sections. No discordance between Triassic and Middle Jurassic sediments is observed in dipmeters or seismic records.

A key marker-bed is the basalt flow (shown on Fig 5) up to 100ft (30m) in thickness and present in several wells. The basalt thins northwards and is absent, apparently by non-deposition, along the north-west side of block 15/17. Several hundred feet of basalt in a similar stratigraphic position is present a few miles south of the Piper area. The pre-basalt section also thickens southwards into the basin. Several hundred feet of grey-brown shales, siltstones and lignite layers as well as tuffs and agglomerates were deposited following the basalt flow. Erosion has removed most of the the post-basalt section over the Piper structure except for several hundred feet preserved on the south-east flank.

Upper Jurassic

Detailed stratigraphy of the reservoir sandstones will be discussed in a later section. In general, the sandstones which were deposited during the first major marine transgression in this area, are Upper Jurassic (Upper Oxfordian/Kimmeridgian) in age and unconformably overlie the non-marine Middle Jurassic rocks. The reservoir sequence is a series of high energy sands deposited mostly above wave base in a beach/bar environment. They are an alternating sequence of thin transgressive and thicker regressive sands, with occasional shales indicating the most extensive transgressive pulses. The top of the sand sequence grades upwards into dark grey to black, highly organic Kimmeridgian-age siltstone and shale. These were deposited in a deeper water, lower energy environment than the underlying sands, and represent the final Upper Jurassic transgression. Log-correlations show southwards thinning of the Kimmeridgian shale as well as the reservoir sandstone with the thickest section located on the north-east side of the field. The sands thin south-westwards toward 15/17-7 well (Figs 6, 11), where the basalt "subcrops" immediately beneath the sand, indicating the pre-sand topographic high and the area of maximum erosion.

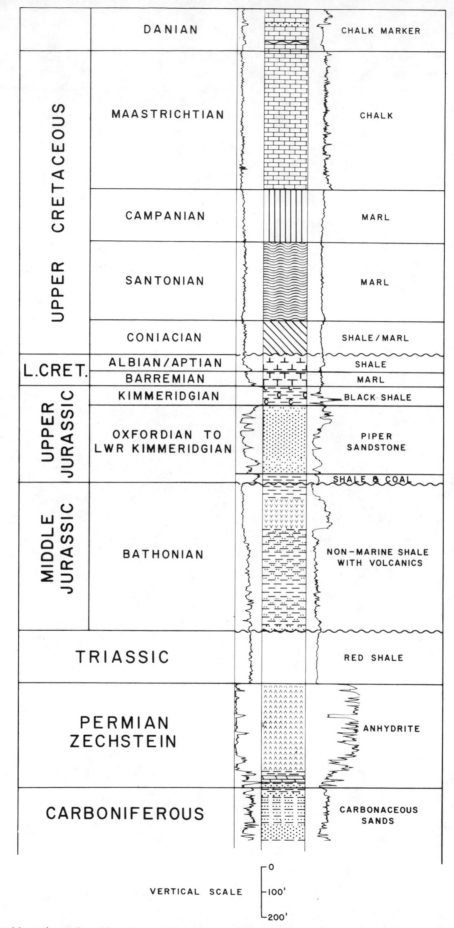

Fɪɢ 3.   Composite Mesozoic stratigraphic column and wireline log of Piper Field. Gamma-ray curve on the left and resistivity on the
right. The lithological symbols on the column correspond with those of the block diagrams in Figs 4—8.

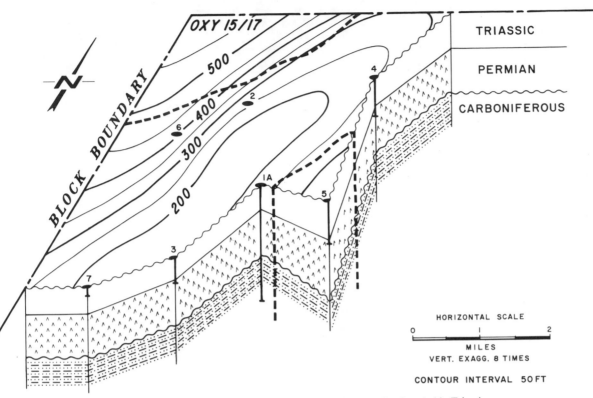

FIG 4. Palaeogeological surface and isopach map of red shale of probable Triassic age.

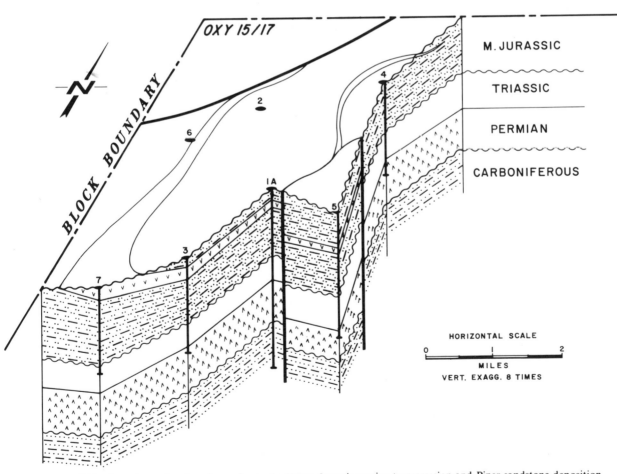

FIG 5. Middle Jurassic erosional surface prior to the Upper Jurassic marine transgression and Piper sandstone deposition.

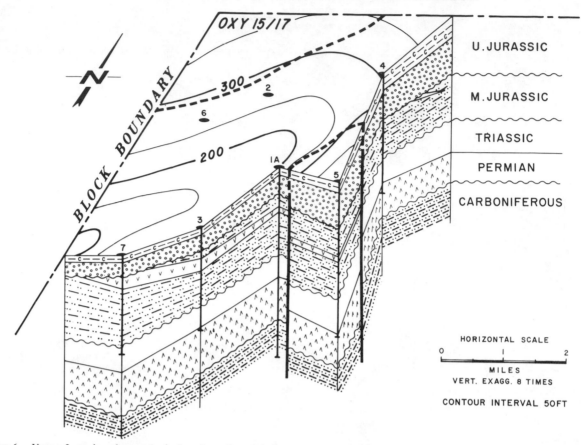

FIG 6. Upper Jurassic palaeogeological surface, illustrating the Kimmeridgian highly organic shale and siltstone surface prior to deposition of Lower Cretaceous marl and shale. The isopachs are of the gross Piper sandstone thickness, which underlies the Kimmeridgian shale and siltstones.

Lower Cretaceous

Barremian and Albian/Aptian shales and marl were deposited over Kimmeridgian shale. The latest Jurassic and earliest Cretaceous rocks are absent over the Piper structure. There is no obvious angularity between Jurassic and Lower Cretaceous strata on dipmeter or seismic records but well-data indicates the Lower Cretaceous onlaps the Jurassic. As seen in Fig 7, Albian/Aptian overlaps Barremian in the north-east and Upper Jurassic sediments in the south-west indicating that the south-west area remained structurally high throughout Lower Cretaceous time. It is possible that a complete Lower Cretaceous–Upper Jurassic section is present basinwards of the Piper structure, but is not yet confirmed by drilling.

Post-Albian faults trending west to north-west and thrown down to the south formed north-dipping tilted blocks. Subsequent post-Albian to pre-Turonian/Coniacian erosion resulted in the pre-Upper Cretaceous palaeogeology as shown in Fig 7.

Upper Cretaceous

The faulting and tilting mentioned above continued and was contemporaneous with pre-Maastrichtian, Upper Cretaceous deposition. Onlapping of Turonian, Coniacian, Santonian and Lower Campanian sediments took place over the Piper structure and progressively thicker sediments were deposited on flanks and on the downthrown side of faults (Fig 8). By late Campanian times sedimentation was fairly uniform over the structure and subsequent Maastrichtian chalk varies in thickness only from about 550–600ft (170–185m).

Tertiary

Structural cross-sections in Figs 9 and 10 illustrate the present day structural configuration over the field. A major sedimentation change from chalk to sand and shale occurred at the close of the Cretaceous and during Danian times. A commonly held theory as to the cessation of carbonate deposition is that the water temperature changed from warm to cold due to the influx of arctic waters following rifting in the area between the North Sea and Greenland. At the same time uplift and erosion of the Scottish Highlands, Orkneys and the Shetlands resulted in the deposition of massive sand, silt and shale units. Subsidence and deposition of deltaic sands and shales in the Northern North Sea Basin continued throughout Palaeocene and Eocene times. The pre-Tertiary domal Piper structure shows continued growth in the Lower Tertiary, probably as the result of

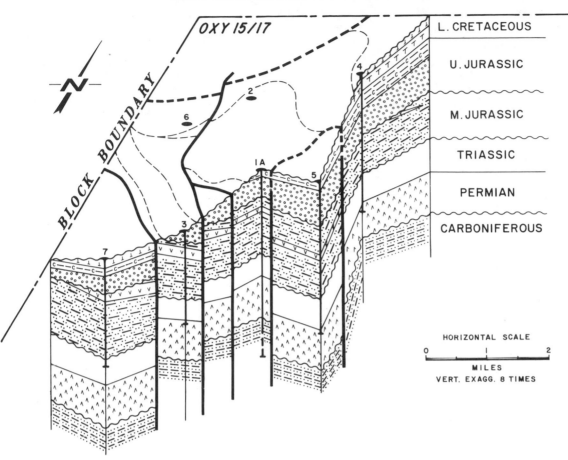

FIG 7.   Palaeogeology prior to Upper Cretaceous deposition. Post-Albian faults trending west to north-west and thrown down to the south formed north-dipping tilted blocks.

sedimentary drape, differential compaction and minor faulting. The structure was completely masked in middle Tertiary times by sedimentation during regional subsidence and topographic expression of the structure is absent on the ocean bottom.

STRATIGRAPHY OF PIPER SANDSTONE

Detailed studies of the reservoir sandstone are continuing at the time of writing this paper. Hopefully, the basic conclusions presented here will not be greatly changed—only expanded. The Piper reservoir is a series of stacked and possibly imbricated barrier-bar and other littoral and shallow shelf marine sand bodies, separated only by thin and largely discontinuous siltstones and silty shales. The gross sandstone thickness varies from 115 to 360ft (40–110m) across the Piper field (Fig 11). Porosities range from 22 to 27 percent (average 23.3 percent) and permeabilities range from 150md to 10d (arithmetic average 1495md). The quality of the reservoir sands is partly related to depositional energy. Sedimentary features and modes of bioturbation observed in cores and

relative deflection of gamma-ray logs (a good grain-size indicator), suggest most sands are high energy deposits laid down above wave base. These sands were deposited as upper offshore sheet sands and in the upper and middle shoreface zone of barrier bars. Wave action was dominant and tidal influence is interpreted as minimal. Therefore, most sands are clean, extremely well sorted, loosely packed and virtually uncemented. Sands in cored intervals are devoid of body fossils, although generally extensively bioturbated. Mega- and micro-fossils, however, are present in the interbedded silty shales.

It is important to view these bars, stacked one on top of the other, in perspective. Individual barrier bars, catalogued from recent and ancient sediments typically vary from 30 to 50ft (9–15m) in thickness, extend tens to hundreds of miles in length and from one to a hundred miles in breadth. We see good continuity of individual sand bodies by correlation between the Piper wells over an area of about 15 square miles (39km^2).

The gross reservoir-sequence can be subdivided into three main sandstone intervals separated by two silty shale intervals which are mapped nearly continuously over this area. The subdivisions are referred to as Lower Sandstone, Lower Shale,

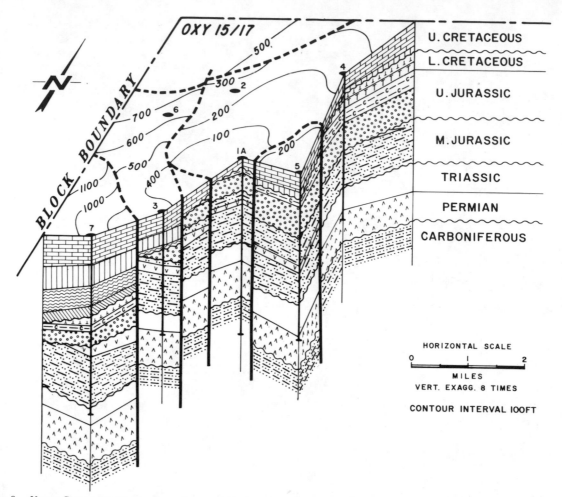

FIG 8. Upper Cretaceous Campanian palaeogeological surface and isopach map, prior to Maastrichtian Chalk deposition.

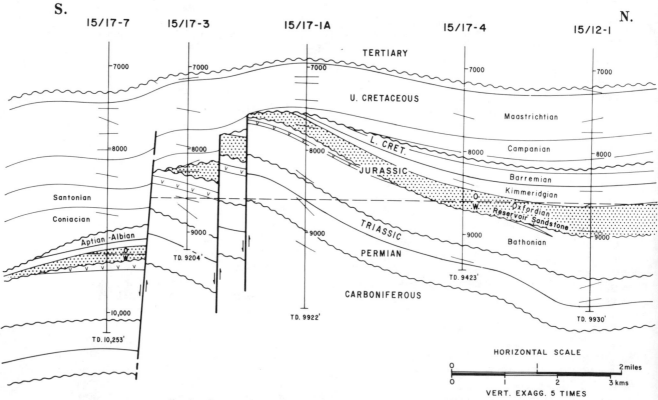

FIG 9. Structural cross-section drawn north—south across Piper Field.

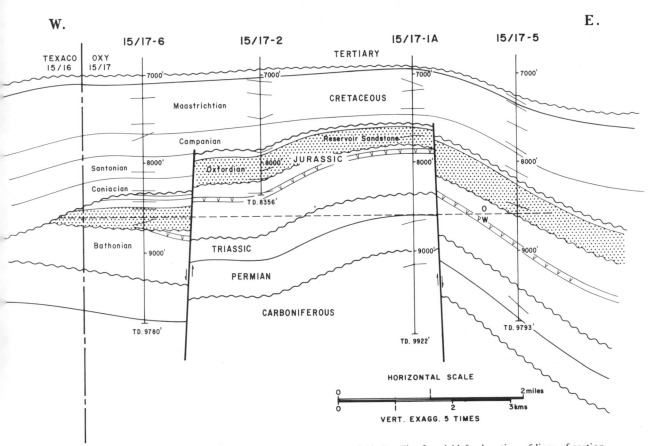

FIG 10.   Structural cross-section drawn east—west across Piper Field. See Figs 2 and 11 for location of lines of section.

Middle Sandstone, Middle Shale and Upper Sandstone. The Upper Sandstone grades upwards into the Kimmeridgian Shale (Fig 12).

Lower Sandstone

The Lower Sandstone, 20–100ft (6–30m) thick comprises a basal transgressive sand and shale unit, overlain by a regressive sand unit followed by a transgressive sand unit. The depositional phase is inferred generally from grain-size gradation, *i.e.* coarsening upward for a regressive and downward for a transgressive phase.

The basal transgressive unit, 15–40ft (4.5–8m) thick is very fine- to fine-grained sandstone, silty and clay rich in part, with a thin basal quartzitic conglomerate, grading upward into silty micaceous organic shale. The fauna is dominated by marine lucinoid bivalves, thin-shelled pectinid bivalves and cerithid gastropods and ostracodes. Both mega- and micro-fauna are indicative of an Upper Jurassic Oxfordian age. The basal sand/shale unit represents the initial Upper Jurassic marine transgression across the erosion surface of Middle Jurassic sediments. The faunal assemblage in the organic shales in the upper part of the basal transgressive unit suggests open marine, lower energy deposition with brackish or fresh water influence.

The low energy open marine shales which mark the end of the initial marine transgression grade

upwards into fine- through course-grained sandstone. This upward increase in grain-size from siltstone to coarse sand indicates a shallowing regressive environment. Coarse sands of the regressive unit are extremely friable and bioturbated, highly porous and average 20–30ft (6–9m) increasing to 70ft (21m) in thickness in 15/17-5, and are interpreted as a barrier bar.

Log correlations indicate that a sand unit up to 20ft (6m) in thickness overlies the regressive unit over most of the structure. This sand grades upward from coarse- to fine-grained and into the overlying shale, indicating a continuous deepening of water and maximum transgression.

Lower Shale

The Lower Shale subdivision is mapped continuously over the Piper area as seen on the fence diagram (Fig 12) and is 10–40ft thick (3–8m). It is highly organic, micaceous and silty, and contains fragmented pectinid bivalves and cerithid gastropods. The ammonite *Amoeboceras prionodes* (S. Buckmann) is present and confirms an Upper Jurassic age.

This organic shale, based on faunal assemblage, is considered to have been deposited in a low energy open marine environment with brackish or fresh water influences. The Lower Shale grades upwards into Middle Sandstone, indicating

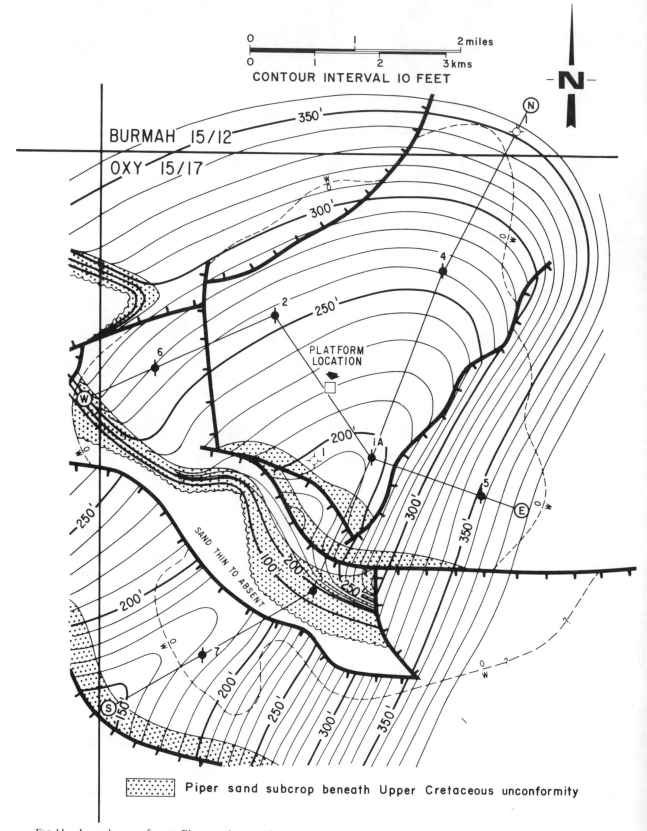

FIG 11. Isopach map of gross Piper sandstone thickness, showing well locations and lines of cross-sections in Figs 9 and 10.

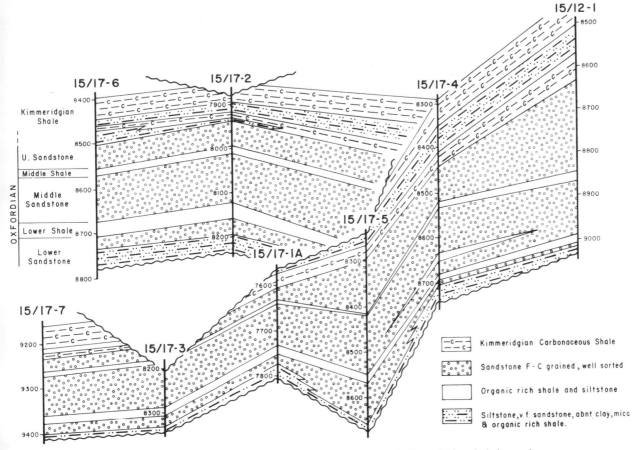

FIG 12. Fence diagram of Upper Jurassic Piper Sandstone and Kimmeridgian shale interval.

regression and a return to clean winnowed sand deposition in a high energy wave zone.

Middle Sandstone

The Middle Sandstone is comprised of sand bodies 20–60ft in thickness (6–18m) and the total thickness of the member is 100–160ft (30–47m). The sands are extremely porous, permeable, well sorted, bioturbated and generally grade upward from fine- to coarse-grained size. This unit has the best reservoir characteristics throughout the field. The trace fossils and bioturbation of the sandstone are the results of reworking by burrowing organisms in medium to high energy zones. Bioturbation has destroyed most sedimentary structures but non-bioturbated sands generally exhibit cross-bedding. The absence of animal life in the non-bioturbated cross-bedded sands was probably due to increased sedimentation rates in upper foreshore or beach zones. Very shallow water with intermittent subaerial exposure would not have allowed the burrowing organisms to survive.

The lowermost Middle Sandstone grades upwards from the Lower Shale through fine to coarse grains, indicating a regressive cycle. The succeeding sand body which also grades from fine to coarse is marked by a sharp contact and indicates the end of one cycle and beginning of

another, due to a change in current direction, a lateral shift in shoreline wave action, or a very rapid minor transgression followed by regression. Thin flaser-bedded lenticular laminated siltstones, sandstones and shales interbedded with sand bodies are of possible intertidal and shallow subtidal origin.

Middle Shale

The Middle Shale which is 10–20ft (3–6m) in thickness, can be mapped over the field area, and, as in the case of the Lower Shale, is an open marine deposit in a deeper water transgressive phase. The shale is moderately bioturbated, organic, micaceous and silty, with ripple laminated lenticles of fine sand and lignite fragments.

Upper Sandstone

The Upper Sandstone grades upward from the Middle Shale and is comprised of several sandstone bodies, 6–20ft (2–6m) thick and averages 70ft (21m) in total thickness over the main Piper area while thickening to 200ft (61m) northeastward. Individual sand bodies in the Upper Sandstone are thinner bedded and finer grained than in the Middle Sandstone. Generally the sands are fine to very fine grained, poorly cemented, friable, often ripple laminated or highly bioturbated and numerous layers of large bivalve shells (*Ostrea*) occur throughout. Sandstones are interbedded with numerous thin silty shales. The depositional nature of the Upper Sandstone is one of fluctuating current direction and depositional environment. The Upper Sandstone grades upward into the silts and carbonaceous shales of the overlying Kimmeridgian shale.

Kimmeridgian Shale

Highly organic shales and silts 70–230ft (21–70m) in thickness in the Piper field area were deposited following the deposition of the high energy sandstone. The shale deposits represent low energy sedimentation of the final Upper Jurassic transgression and they are overlain by low energy open marine sediments which were deposited throughout Cretaceous time. Faunal assemblage and regional correlation indicate that the shale overlying the Piper reservoir sandstone is of Kimmeridgian age.

RESERVOIR DATA AND PRODUCTION POTENTIAL

The following are some of the basic reservoir data obtained from the appraisal drilling. Gross sand thickness as seen on the isopach map (Fig 11) ranges from 115–360ft (35–110m) and the average net sand thickness above the oil/water contact is 146ft (45m). Average porosity is 23.3 percent based on log calculations verified by core analysis. Average permeability from core analysis is 1495md (arithmetic) and 761md (geometric). The oil gravity is 36° API and GOR is 446. The oil/water contact in main Piper Field is at 8518ft (2595m) subsea. The area of main Piper Field is about 8000 acres (32.2km²) and the new pool discovery on the south side of the field is a minimum of 600 acres (2.4km²) with an oil/water contact at 9199ft (2805m) subsea. Therefore, to date, approximately 8600 acres, 13.5 square miles (34.6km²) are proved to be productive.

As previously discussed, several discreet sand bodies comprise each of the members which, collectively, comprise the gross Piper reservoir interval. Production testing and model studies show little difference in production capabilities of the different units, no doubt because all the sands exhibit good porosity and permeability and one sand body is overlain by another with virtually no barriers to fluid communication.

Detailed engineering analysis and simulation studies have been made of the main Piper reservoir to determine the oil in place within the field and to make predictions of its performance and ultimate oil recovery. Oil in place, as determined by Occidental engineers, is about 1.55 billion barrels and ultimate recovery efficiency will be about 50 percent resulting in recoverable reserves of about 650–900 million barrels utilizing either a natural or artifical water drive. The excellent quality of the

reservoir, and indicated effectiveness of the water drive, either natural or artificial, will allow the wells to produce at high rates. The field should be capable of producing at a peak rate of more than 200 000BOPD.

These studies do not include the south Piper new pool discovery which will add additional recoverable oil when fully delineated by drilling from the production platform.

We are constructing a production platform with eight legs which will be launched by barge and which will accommodate 36 wells, and be installed at the field location shown on Fig 2. With 36 well-slots the capability is provided for drilling an adequate number of water injection wells, adjustment in the number and location of the producing wells based on results from early producing wells and development of south Piper pool extension. It is expected that, within the limits of safe practice, wells will be turned to production as they are completed and that producing and drilling operations will be carried out simultaneously.

The strength of the natural water drive and the degree to which the faults will act as impediments to fluid flow are not known at this time. However, the development and production schemes which are planned will permit early assessment of the natural water drive and timely determination of the degree to which the field faults will act as production barriers. We will, therefore, be able to adjust the relative number and placement of injection and production wells to ensure maximum economic recovery from the field.

Oil and gas will be separated on the platform at approximately 100psig and the oil will be transferred to a terminal onshore through an undersea pipeline 30 inches (760mm) in diameter and approximately 130 miles (209km) in length. The capacity of the line will be 500 000BOPD. The terminal and marine facilities are being built on the island of Flotta, in the Orkneys, and will provide crude and gas processing facilities, oil storage and tanker loading facilities.

CONCLUSIONS

The Piper Field, the first commercial oil accumulation found in the East Moray Firth sub-basin of the northern North Sea, was discovered on December 22nd, 1972. The reservoir is an Upper Jurassic Oxfordian/Kimmeridgian age series of stacked and imbricated sand bodies, deposited in a high energy shallow marine environment. The sandstone series averages 250ft (76m) in thickness over the Piper area. The sands were deposited as a beach/barrier bar complex above wave base and are friable, clean, poorly cemented, very porous and permeable. Thicker individual sand bodies are regressive and normally show a gradual increase in mean sand grain-size from base sand to top.

Thinner sands whose grain-size increases downwards to a coarse to pebbly base are interpreted as transgressive. Piper reservoir sands were deposited during a major Upper Jurassic marine transgression on a north-east dipping surface underlain by Middle Jurassic non-marine sediments. The Piper sandstone grades upward into siltstones and organic black shales of Kimmeridgian age. These low energy open marine siltstones and shales were deposited in the final Jurassic transgressive phase.

In the Piper Field area, dominant structural growth occurred during the post-Albian pre-Upper Campanian stages of the Cretaceous and again in the Lower Tertiary. A combination of faulting, erosion, and contemporaneous faulting and deposition during Cretaceous times resulted in the Piper structure being a partially masked series of slightly tilted fault blocks, downthrown to the south. Uplift and differential compaction in Lower Tertiary time and complete masking of the structure by middle Tertiary times during North Sea basinal subsidence resulted in the structural configuration of the Piper area as we see it today (Figs 9 and 10).

ACKNOWLEDGEMENTS

This paper has previously been presented before the American Association of Petroleum Geologists at San Antonio, Texas, on the 2nd April, 1974. It is published by permission of the management of the North Sea group operated by Occidental of Britain, inc. and including Getty Oil International (England) Limited, Allied Chemical (North Sea) Limited and Thomson Scottish Petroleum Limited.

We wish to thank our co-workers in Occidental of Britain, Inc., particularly Harold Lee, Chief Geophysicist, and also the individuals within the two research companies conducting studies on Piper Field rocks, *i.e.* Robertson Research International and Exploration Research International.

BIBLIOGRAPHY

Conner, D. C. and Kelland, D. G. 1974. Piper Field, U.K. North Sea, Interpretive log analysis and geologic factors Jurassic reservoir sands, presented at *3rd European Formation Evaluation Symposium, 14–15th October, 1974.*

Rigby, H. K. and Hamblin, W. K. (Eds). 1972. Recognition of ancient sedimentary environments. *Soc. Econ. Paleo. and Min. Spec., Pub. No. 16.*

Shelton, John W. 1973. Models of sand and sandstone deposits: A methodology for determining sand genesis and trend, *Bull. Oklahoma Geol. Surv.,* **118,** 122 pp.

Thomas, A. N., Walmsley, P. J. and Jenkins, D. A. L. 1974. Forties Field, North Sea. *Bull. Am. Ass. Petrol. Geol.,* **58,** 396–406.

28

The Occurrence of Jurassic Volcanics in the North Sea

By FRANK HOWITT

(BP Exploration Company Limited, London)

ELIZABETH R. ASTON

(Independent Consultant)

and MAURICE JACQUÉ

(Total Oil Marine Limited, Paris)

SUMMARY

Lavas and tuffs have been encountered below Cretaceous rocks during exploration of the area near the Forties and Piper Oil-fields in the UK sector of the North Sea. Some of the volcanic horizons lie within sediments of undoubted Jurassic age.

These volcanic beds appear to form one major province which is associated with a positive magnetic anomaly, and overlies a highly rifted floor. The lavas are interbedded with grey deltaic sediments in the north and with barren red beds in the south. A Bathonian/Bajocian age can be assigned to several of the northern sections, but in the south only a post-Permo/Triassic age can be inferred.

The lavas are undersaturated, porphyritic, vesicular, olivine basalts and have alkaline affinities. There is one occurrence of saturated lavas. The central and southern part of the province is dominated by lavas, but these pass outwards and particularly northwards to sequences composed mainly of clastic volcanic rocks.

There are probable Jurassic volcanic occurrences in other North Sea and related areas, and these are noted.

INTRODUCTION

The first Forties Field wildcat well, 21/10-1, unexpectedly encountered a thick series of volcanic rocks beneath the Cretaceous, and several other deep wells in the area proved their extension to the north and west. Apart from being pre-Cretaceous their age was unknown, but radioactive dating consistently suggested Mesozoic. Despite this, as the volcanic rocks were interbedded with barren red beds and petrographically strongly resembled lavas from the Midland Valley of Scotland, early opinion as to their age varied between Permo-Carboniferous and Triassic.

Later exploration further north in Degree Quadrangle 15 (see Figs 1 and 3 for location) also encountered pre-Cretaceous volcanics but many of these were interbedded with sediments of undoubted Jurassic age and confirmed the likelihood that the radioactive dating of the southerly occurrences was correct.

Generally the volcanic rocks consist of basaltic lava flows (from 1m to 9m thick where cored), together with agglomerates, tuffs and tuffaceous sediments. Few true pyroclastics have been identified in the available samples and explosive activity was probably low. The epiclastics vary from pebble-sized agglomerates to sand-grade tuffs. The tuffaceous sediments are difficult to distinguish from heavily weathered lavas and boles. The well sections (Fig 2) have been subdivided into three major rock types only—(a) lavas, (b) tuffs/epiclastics and (c) weathered lavas and tuffs and/or tuffaceous sediments. It is impossible to distinguish between pyroclastics and epiclastics from log-data alone.

Occasionally one lava cross-cuts an earlier one, but this is a localized feature common in a volcanic province, and not comparable with major intrusive environments. In fact no good evidence of large scale or discrete intrusive bodies has been found. Sediments have been baked locally and have been described as hard, grey or red-brown, with a metallic lustre or a spotted appearance. However this metamorphism is restricted, occurring only beneath lava flows, and where cored the zone was less than 1m thick.

DISTRIBUTION OF THE VOLCANICS

Pre-Cretaceous volcanics were encountered in the Forties Field area in wells 21/3-1A, 21/9-1, 21/10-1, 21/10-4 and 22/6-1. However they are absent in wells to the south, east and west.

Farther north pre-Cretaceous and post-Triassic volcanics are present in wells 14/15-1, 15/12-1, 15/13-1, 15/17-1A and other Piper Field wells, 15/21-1 and 15/26-1, but they are absent in most wells of the Degree Quadrangle 16 and in some others to the west and north in the Degree Quadrangles 14 and 15.

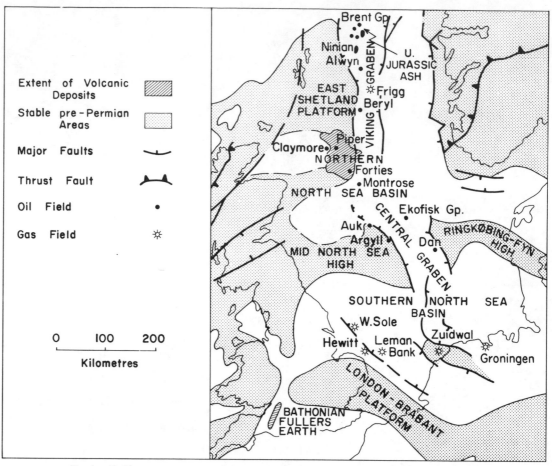

FIG 1. Outline structure map of the North Sea, indicating location of Jurassic volcanics.

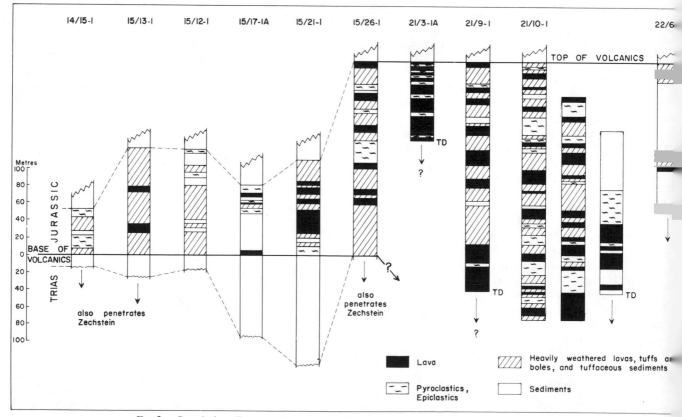

FIG 2. Correlation diagram of volcanic sections in wells near the Forties and Piper oil-fields.

The base of the volcanic section has not been reached by the drill in the Forties area (more than 743m were penetrated in well 21/10-1) where the section profiles are dominated by lava. This drilling supports the aeromagnetic evidence which indicates a magnetic high anomaly centred to the north of Forties at about the Total 21/3-1A well location. (See map Fig 3, where the higher intensity contours from a reconnaissance grid of 35km flown by SAPA for Total Oil Marine are outlined.) A more detailed line-spacing of 3km was flown by the Aero Service survey of 1963, but this does not go farther north than the 58°N parallel. (See map Fig 4 where total magnetic intensity contours are reproduced from the 1963 Aero Service survey with the kind permission of the Aero Service Division of Western Geophysical Company.)

Dr. F. Rimbert has studied these surveys for Total Oil Marine and has computed depths and thicknesses of the volcanic rocks. She first had measurements of magnetic susceptibility in the terrestrial field made on core samples from 21/10-1 and 21/3-1A, by the Laboratoire de Geomagnetisme de St. Maur. The highest measurement recorded from 3 samples in 21/10-1 was 2776×10^{-6} u.e.m., and from 21/3-1A was 377×10^{-6} u.e.m. Since in other localities the ratio of basaltic volume to sedimentary volume in the section is not known, an arbitrary choice of 2000×10^{-6} u.e.m. mean value was decided upon for the province. Strong anomalies were isolated by considering the theoretical shape in a terrestrial magnetic field of inclination 70°N, which is a couple formed by a negative anomaly to the north and a somewhat stronger positive anomaly to the south. Figure 4 illustrates nine such anomalies in

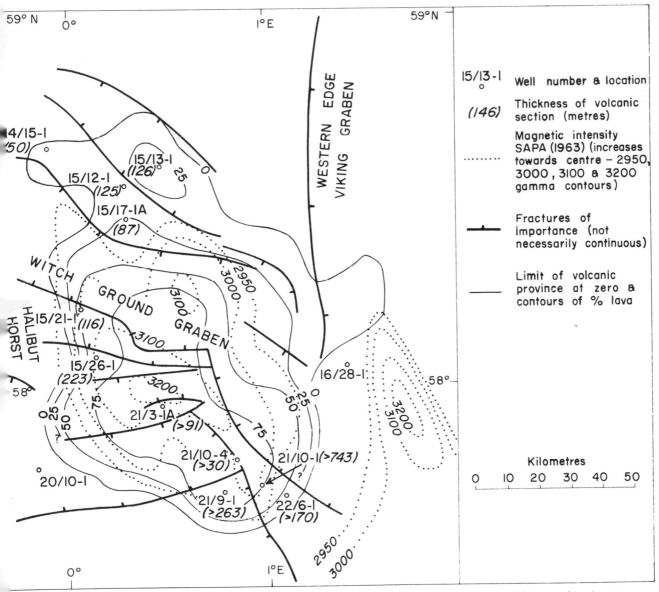

FIG 3.   Map of volcanic province near the Forties and Piper oil-fields, showing magnetic intensity on reconnaissance grid and contours of lava percentage from well evidence.

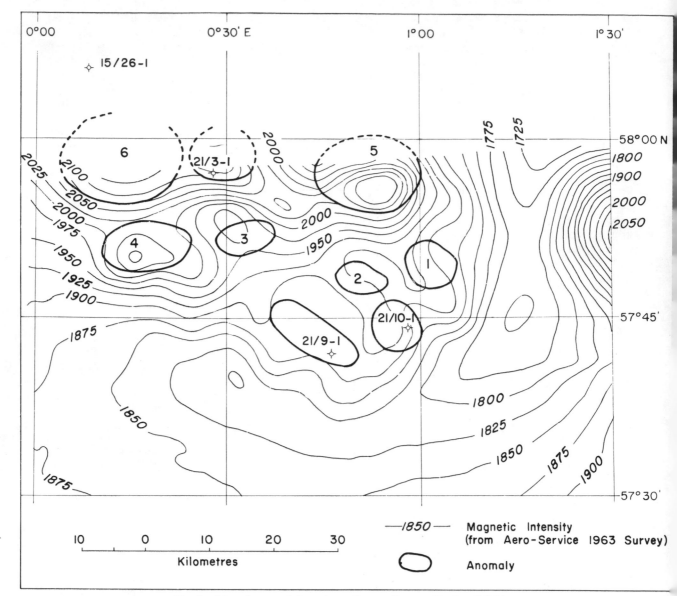

FIG 4. Map of magnetic intensity and anomalies near Forties oil-field.

the area, three associated with known well locations of basalt—21/31-A, 21/9-1 and 21/10-1. The thickness of basalt at the anomalies was deduced from magnetization per unit volume at a computed depth. Dr. Rimbert constructed from these calculations a simplified isopach map of basalt in the province as shown in Fig 5. Calculated thicknesses at three of the anomalies were greater than 1500m and more than likely represent extrusive centres or vents.

There is no associated magnetic anomaly on the reconnaissance map, Fig 3, in the area of volcanic rocks in the vicinity of wells 14/15-1, 15/13-1, 15/12-1 and 15/17-1, but this is perhaps not surprising as the known percentage of basalt is less than 30 percent in a total thickness of volcanics of less than 100m.

The location of the possible lava vents could have been controlled by the junction of a fault system at the entrance to the Moray Firth Basin,

where the major north–south rift system of the Viking Graben and the Central Graben is broken by a more or less east–west series of faults forming the Witch Ground Graben which could be reactivated splays of the Great Glen Fault system (see Figs 1 and 3). This intersection of systems would naturally create a weak crustal point, and this noticeably coincides with the biggest magnetic anomalies, and hence suggests the upwelling of the basalt at the time of rifting, particularly of the Witch Ground Graben.

From the well-evidence the region displays a relatively simple zoning, from a lava-dominant centre to an outer rim of clastic volcanics (see Fig 3). The outermost zone has a weighting to the north, which could be due to a prevailing wind of the time, blowing the ashes in that direction, although evidence to the south is to some extent lacking because of pre-Cretaceous erosion and poor well-control.

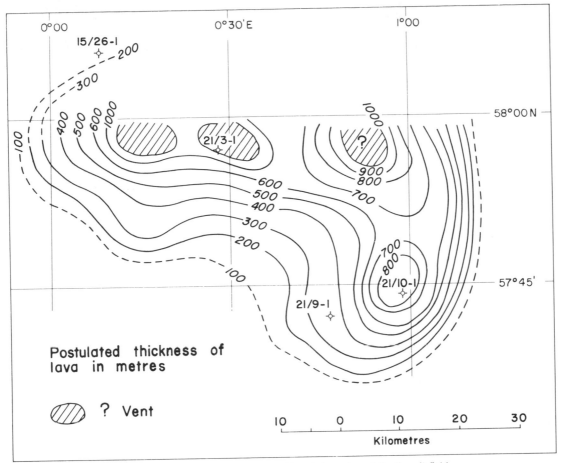

FIG 5. Map of calculated isopachs of basalt in region of Forties oil-field.

AGE OF VOLCANICS

Where interbedded with sediments the latter are either barren red beds (southern sector) or sparsely fossiliferous grey beds (northern sector). Core samples have been dated radiometrically, but this has not provided reliable ages. The dating has been attempted on whole rock samples which are sheared, altered and show secondary mineralization, and therefore all dates are minimum ages. On behalf of Total Oil Marine Ltd, Robertson Research International obtained a radiometric age of 165 ± 4my by the argon method on the core of 21/3-1A. On behalf of Monsanto they recorded an age of 152 ± 6my on a core sample from well 15/21-1. For BP, Geochron Inc. similarly recorded ages of 144 ± 5 and 160 ± 12my on two samples from well 21/10-1, and the Scottish Universities Research Reactor Centre recorded an age of 109 ± 2my on a sample from well 21/9-1. The minimum age spread is thus Aptian to Bathonian.

Taking the province in general the volcanic beds are undoubtedly pre-Cretaceous and in Degree Quadrangle 15 the sections overlie barren red beds of inferred Triassic age, which themselves overlie strata dated as Zechstein (see Fig 2). In Degree Quadrangles 21 and 22 however the base of the sequence has not been penetrated. Here some of the lavas are interbedded with barren red beds, and, together with nearby well-evidence, seismic sections strongly suggest that the sequence is post-Zechstein.

In the northern sector where grey sediment interbeds occur, floral/faunal dating has been attempted. In well 15/13-1 in a sidewall core from beds above the volcanics D. J. Batten and G. L. Eaton of BP found a rich microplankton assemblage of dinoflagellate cysts, including *Systematophora orbifera, Tenua spp., Gonyaulacysta spp.* (including *G. jurassica*) and a single specimen of *Endoscrinium luridum*, together suggesting a late Oxfordian/early Kimmeridgian age. From beds beneath the volcanics they found a Middle Jurassic (Bathonian/Bajocian) miospore flora of *Araucariacites australis, Inaperturopollenites turbatus, Perinopollenites elatoides,* rare *Cerebropollenites mesozoicus, Tsugaepollenites dampieri, Classoppolis torosus* s.l., *Densoisporites velatus, Duplexisporites, Heliosporites reissingeri, Ischyosporites* and other taxa. In well 14/15-1 D. J. Batten and G. L. Eaton recorded a spore assemblage from a sidewall core just above the volcanic sequence, which although comprised of many long ranging forms, had a Bathonian/Bajocian aspect. Spores found in a core immediately beneath the volcanics were earliest Bajocian to late Toarcian. We know, via well-trade information and are authorized to report, that Robertson Research International for Burmah *et al.* recorded from beds above the volcanics of well 15/12-1 dinoflagellates including *Gonyaulacysta longicornis* and *Scriniodinium dictyotum*, dating these beds as Kimmeridgian or possibly Oxfordian. From samples just above and within the volcanics of this well Robertson's found foraminifera *Lenticulina muensteri*, ostracodes *Macrodentina bathonica*, *Darwinula incurva*, *Belekocytheridea punctata*, miospores *Cerebropollenites mesozoicus*, *Dictyophyllidites harrisii*, *Gleicheniidites senonicus*, *Sestroporites pseudoalveolatus*, *Tsugaepollenites dampieri*, *T. trilobatus*, which together suggest a range from Oxfordian to Bajocian. Towards the base of these beds in 15/12-1 numerous *Klukisporites variegatus* were found, an addition which, with the ostracode evidence, strongly favours a Bathonian/Bajocian age. Similarly in well 15/17-1A for Occidental, Robertson's recorded one Upper Jurassic species of palynomorph—*Classopollis echinatus*—above the volcanics and Middle Jurassic ostracodes associated with miospores ranging down into Bajocian in the interval containing volcanics. Monsanto *et al.* well 15/21-1 also contained a Middle Jurassic flora above and within the volcanics. Implied Trias and certain Zechstein occur deeper in this well and also in wells 14/15-1, 15/12-1, 15/13-1, 15/17-1A and 15/26-1. The latter well was largely barren beneath the Cretaceous, except for one sidewall core at the top of the volcanic section, where J. E. Williams of BP recorded a few simple bisaccate pollen, one *Cerebropollenites mesozoicus* and one *Systematophora sp.*, indicating at least post-Bathonian age.

The occurrence of interbedded grey sediments in the northern wells and of a sequence of barren red beds in the southern sections could be significant. A simple lithological comparison suggests a Jurassic age in the north and a Triassic age in the south. However, the presence of red mudstones with contained flora of at least post-Bathonian age, which overlie red anhydritic sediments of implied Trias age in well 15/26-1, suggests that the sediment change could also be due to differences in facies (deltaic versus continental). In any case, because of the very close proximity, a Middle Jurassic age is assigned to all the volcanic beds mentioned in Degree Quadrangles 14, 15, 21 and 22.

PETROGRAPHIC CHARACTERS

The province is dominated by undersaturated basalts containing olivine (now altered to serpentines, etc.) and pyroxene phenocrysts. They tend towards an alkaline affinity, with sodic or calcic plagioclase feldspars, diopsidic or titaniferous augites and occasional analcime. In one locality, well 21/3-1A, more saturated rocks occur, containing several late-stage or secondary minerals, notably quartz, biotite and hornblende.

Explosive activity was of somewhat lesser importance. Tuffs (vitric, pumice, crystal and lithic varieties) occur, often interbedded with lavas. Weathered lavas, agglomerates and tuffaceous sediments also occur and are often difficult to distinguish from each other and from boles, laterites and sometimes true sediments. Detailed petrographic descriptions were carried out for BP by Dr. R. Walls, to whom the authors are indebted, as the following is a collation of his work. Wells showing affinities have been grouped together.

Forties Field wells

In the wells BP 21/9-1, 21/10-1, 21/10-4 and Shell/Esso 22/6-1 a thick series of lavas occurs together with some tuffs and agglomeratic/tuffaceous horizons. The lavas are grey or purplish, vesicular and often partially altered, or in some cases completely lateritized. Large, fresh, early pyroxene phenocrysts (augite or diopsidic or titaniferous varieties), and smaller rounded olivine phenocrysts (now serpentine), are set in a groundmass of second generation pyroxene and feldspar. The latter are often zoned with An_{40-70} centres and An_{5-30} rims. Some alkaline lavas (oligoclase or andesine basalts) also occur. Analcime, apatite, magnetite and ilmenite are commoner accessories.

Autobrecciation (due to flow) and late hydrothermal activity have altered the lavas. Often the large pyroxene phenocrysts alone remain unaltered, and new minerals—zeolites, chlorite, carbonate and hematite—occur as replacements of earlier minerals, and by infilling vesicles and lining shear planes.

The associated tuffs are sand- to gravel-grade. They comprise lava and phenocryst clasts, or primary pumice and ?lapilli fragments. Some of the associated red sediments (weathered lava tops) also contain volcanic grains, and flows must have been intermittent to allow such deep weathering profiles.

Well Total 21/3-1A

Lavas are dominant also at this locality, although some tuffs have been recorded. Unlike Forties, these lavas are saturated trachybasalts with alkaline plagioclase feldspars (oligoclase or andesine) and small pyroxene phenocrysts. Olivine and large pyroxene phenocrysts as well as vesicles are rare, whilst quartz, biotite and hornblende are present (although as with Forties some of the latter are partially secondary). Chlorite, ?analcime and epidote occur from late-stage alteration.

As the well 21/3-1A has only penetrated the top of a probably very thick sequence of lavas, undersaturated types may be present below.

Small intrusions have been noted in the cores of both wells 21/3-1A and 21/10-1. They are of comparable petrography to the lavas they intrude, and are considered to be small localised events typical of any volcanic area.

Wells Monsanto *et al.* 15/21-1 and BP/Deminex 15/26-1

These two wells show a thinner sequence of volcanics, mainly lavas, but with a higher proportion of tuffaceous material than in the Forties and 21/3-1A areas.

The lavas are basalts ranging from olivine-bearing types (as in 15/21-1), small pyroxene phenocryst types (both wells), and trachytic types (15/26-1). The presence of ?oligoclase and other zoned feldspars in several samples, suggests alkaline tendencies. However these rocks are severely altered (olivine replaced by bowlingite, pyroxene replaced by chlorite/serpentine/carbonate mixtures and feldspars which have been sericitised, and hence classification is difficult. Magnetite (10 percent of one sample in 15/21-1), apatite and ilmenite are common accessories with secondary hematite.

The lavas are interbedded with red lateritic or tuffaceous mudstone horizons and with tuffs. Volcanic clasts include trachytic, basaltic, vitric and flow-banded lava types, glass shards and possible ash lapilli. The volcanic material is usually green and chloritic, whilst the weathered debris is usually red-brown.

The undersaturated nature of these lavas also separates them from those in well 21/3-1A. The higher percentage of clastic volcanic material and sediment interbeds, indicates they are also further from a volcanic source. Lava and volcanic debris in general tends to increase upwards through the section and the sediment interbeds are usually barren red beds, although lignite, quartz and shell fragments, grey shales and coal seams do occur.

Wells Burmah *et al.* 15/12-1, Occidental *et al.* 15/17-1A and BP 14/15-1

A small horst is associated with this area and the sequences are dominated by clastic volcanic material. Lavas are rare—possibly one or two thin flows of olivine basalt may occur in 15/12-1 (or indeed these may be large boulders).

True sediments in the sequences appear to be grey in colour with traces of carbonaceous material or coal seams. Red "shale" interbeds, often spotted or speckled, are believed to be weathered tuffs and tuffaceous mudstones. Tuffs do occur, and these contain pumice, lithic and vitric grains and flow structures have been noted. These tuffs are typically poorly sorted, very fine sand- to pebble-grade and varicoloured (purple-green, black or grey, greens and yellow-brown). Calcareous cement is common and in places forms hard veined rocks.

The dominance of clastic volcanic material

(probably mainly due to weathering but including some explosive fragments) indicates that this area was farthest from the volcanic centres. The occurrence of red shales containing volcanic material, and of ferruginized or weathered tuffs also indicates that accumulation was fairly slow.

Well BP/NIOC 15/13-1

This is the most northerly well of the province, in our knowledge, to penetrate volcanic material ascribed to the Jurassic. It has a lava-dominated sequence with sediment interbeds. Weathered tuffs and tuffaceous mudstones overlie a sequence of interbedded tuffs and lavas.

The lavas are olivine rich (now altered to serpentine or iddingsite) with very large pyroxene phenocrysts (comparable to those of the Forties Field lavas) set in a ground mass of feldspar laths (An_{50-70}), pyroxene (?titaniferous) and magnetite. Vesicular varieties are infilled with serpentine. The associated tuffs and agglomerates comprise weathered grains of trachytic, basaltic and glassy lava together with pumice and shard fragments.

This basaltic province is hence dominated by undersaturated lavas, with the exception of the saturated, rather sodic type of well 21/3-1A. Pyroxene phenocrysts are ubiquitous and large ones (5–20mm) occur in Forties Field and 15/13-1 wells. Alteration is common, particularly of the olivine (usually to serpentine) and sometimes of the groundmass pyroxene (to chlorite mixtures), and feldspars (to sericite). Common accessory minerals are analcime, apatite, magnetite and some ilmenite, together with secondary minerals chlorite, calcite and hematite. The lavas are commonly porphyritic with sub-ophitic and granulitic groundmass textures, and traces of flow-banding and auto-brecciation. Vesicles are common, usually infilled with one of the secondary minerals.

The closest petrographic similarity of this volcanic province is with that of the Midland Valley of Scotland, where the Permo-Carboniferous lava suites are also porphyritic, alkali olivine basalts. This similarity, particularly with the lavas of the Forties Field, initially suggested a comparable age, despite the radiometric dating obtained. The earlier Devonian episodes of vulcanicity in Scotland produced rocks dominantly of calc-alkaline basalts and andesites. The Permian spilites of north German wells are also quite distinct petrographically and do not resemble those of the Forties area.

OTHER PROBABLE JURASSIC VOLCANIC OCCURRENCES

Aeromagnetic data adjacent to the presently-known province suggest at least one small area which could be an extension of the Forties-Piper development. This area is indicated on the map, Fig 3, seems to be restricted to blocks 22/3,

22/4, 22/8 and 22/9, and appears to be likely to contain thick lavas beneath the Cretaceous.

Volcanic ash has been recovered in a core containing sediments of Upper Jurassic age, probably Kimmeridgian, from a well in the Brent area (M. J. Bowen, personal communication).

It may also be significant that certain rocks found within Permian sediments near the Auk Field, although of completely different aspect to the Forties occurrence, have apparently been "overprinted" with a Jurassic/Lower Cretaceous event. Radiometric dating of these demonstrably older rocks consistently indicates an age of not younger than Lower Cretaceous (M. J. Bowen, personal communication). This is confirmed in Total Oil Marine well 39/2-1 where volcanics underlying Zechstein carbonates and overlying Namurian have been given a radiometric age of 182 ± 5my by Robertson Research International—i.e. about Upper Lias.

Ashes also seem to have been deposited over quite a large area of southern England, near Bath, to provide the bentonites in the Fullers Earth Formation. These are more precisely dated as Upper Bathonian (*Retrocostatum* Zone). Grim (1933) even recorded the presence of glass shards in these beds, but this fact has not been confirmed by the more recent examinations of Hallam and Sellwood (1968). Nevertheless they adduced and concluded that the Fullers Earth was a montmorillonite rock produced from the alteration *in situ* of volcanic ash. Our recordings of probable Bathonian volcanics in the North Sea certainly adds weight to their arguments. It seems unlikely, however, that the Forties–Piper area vulcanism provided the source for the Bath occurrence.

Nearer to the south of England, however, is yet another probable Jurassic volcanic occurrence. In Zuidwal, north-west Netherlands, a post-Triassic, pre-Lower Cretaceous volcano has provided the structural foundations for the gas field in the overlying Lower Cretaceous rocks. (See Cottençon *et al.*, 1975.)

Thus it now seems certain that volcanicity, no doubt connected with rifting, occurred during Jurassic times, probably commencing in the Bathonian and petering out in the Kimmeridgian. Precise centres of activity are known only in northwest Holland and the Forties–Piper area of the North Sea. Others perhaps have been, or will be, discovered.

ACKNOWLEDGEMENTS

This paper would not have been possible without the cooperation of all the owners of well-information mentioned, and permission to publish was granted by them without hesitation. The main operators were Burmah Oil (North Sea) Ltd., Occidental of Britain, Inc., Monsanto Oil Co. of the UK Inc., Shell UK Exploration and Production Ltd., Total Oil Marine Ltd. and BP Petroleum Development Ltd., but in all cases many partners were also consulted and gave their approval. To all of them the authors express thanks and appreciation. We also thank contractors and consultants, already mentioned by name for agreeing to the publication of rock ages by geochemical or fossiliferous means and the magnetic data upon which our story depended. Many colleagues in Total Oil Marine and BP assisted in collating the information and reading the manuscript. We thank them also.

REFERENCES

Cottençon, A., Parant, B. and Flacelière, B. 1975. Lower Cretaceous gas-fields in Holland. *Paper 30, this volume.*

Grim, R. E. 1933. Petrography of the Fuller's Earth deposit, Olmstead, Illinois, with a brief study of some non-Illinois earths. *Econ. Geol.*, **28**, 344–63.

Hallam, A. and Sellwood, B. W. 1968. Origins of Fuller's Earth in the Mesozoic of South England. *Nature, Lond.*, **220**, 1193–5.

DISCUSSION

Dr B. W. Sellwood (Reading University): In a recent paper Dr A. Hallam and I expressed our satisfaction that the Bathonian volcanics predicted by ourselves and Grim many years earlier had at last been vindicated (*Nature, Lond.*, **220**, pp. 1193–5). We also took the story a stage further. The existence of the volcanic suite close to the triple trough-junction of Forties–Central Graben–Viking suggested the presence of a plume. If volcanism was associated with rifting, then the precursory phases of rifting should be marked by extensive regional updoming over its site. We suggested that spreads of clastic rocks in the Middle Jurassic of Yorkshire and eastern Scotland resulted from this early phase.

Is there supporting evidence to indicate what the provenance of the coarse sediments are in the Jurassic of the troughs? From petrological studies onshore, I suspect that an Upper Palaeozoic (Carboniferous) source is the most likely and that very rapid shedding of sediment occurred during updoming.

Dr F. Howitt: We are happy to confirm Jurassic volcanism near the UK, as Drs Hallam and Sellwood predicted from their examination of the Fuller's Earth (Bathonian) of south-west England. We are not personally aware of the provenance of the coarser Jurassic sediments within the North Sea graben system, but the likelihood is the Scottish Caledonian massif contemporaneously uplifted during Middle Jurassic times.

D. B. Morris (Hunting Geology and Geophysics Ltd): First, I would like to comment that in addition to the aeromagnetic surveys by SAPA and Aero Service, the UK North Sea is covered by a non-exclusive aeromagnetic survey flown in 1970 by Huntings.

Secondly, I would like to know if there is any relation between the Jurassic faulting and the deeper basement structures seen on the aeromagnetic survey.

Dr F. Howitt: The Hunting 1970 aeromagnetic survey does not conflict with the surveys commented on in our paper but was not thought to be in sufficient detail to work out depths of discrete anomalies. In our view there does not appear to be any clear relationship between known faulting affecting Jurassic beds and the deeper basement structures in the vicinity of Forties, although this could be masked by the volcanic effect.

D. G. Roberts (Institute of Oceanographic Sciences): In view of the evidence for Jurassic–Lower Cretaceous volcanism in the North Sea, English Channel and Newfoundland, could you say if Mesozoic volcanics have been encountered in wells drilled to the west of Shetland?

Dr F. Howitt: I am not aware at this time that any statement has been made that Jurassic beds exist in wells west of Shetland, nor indeed of any volcanic rocks except in Tertiary deposits.

29

The Base of the Cretaceous: A Discussion

By R. J. JOHNSON

(Burmah Oil (North Sea) Limited)

SUMMARY

The sediments representing the basal intervals of the Cretaceous in the northern areas of the North Sea possess a distinctive contact with the underlying deposits of the Upper Jurassic. The contact is marked in most areas by a major unconformity and represented by a seismic reflector traceable over a wide region.

This paper attempts to look closely at the nature of the boundary between the Cretaceous and the Jurassic and utilizes the results of clay mineral analysis, organic and non-clay mineral studies and micropalaeontology to establish a relationship with the lithological variations observed over a selected interval in the northern North Sea.

The characteristics of the Lower Cretaceous/Upper Jurassic sequence are discussed in the light of recent work on oil occurrence and plate tectonics and a correlation is suggested between the Lower Cretaceous of southern and eastern England, the northern North Sea and the Jameson Land area of East Greenland.

INTRODUCTION

One of the most characteristic features of the northern part of the North Sea is the distinctive contact between the Jurassic and the Cretaceous. This contact has been shown by the results of recent drilling to embody a major unconformity which separates strata of markedly different lithological type and which gives rise to a well-defined seismic reflector.

To obtain a closer view of such phenomena and to study the wider implications of this gap in the geological succession, a series of samples were obtained from a representative drilled section in the northern North Sea and subjected to a number of detailed analyses. These samples extended over an interval 45m below the Jurassic/Cretaceous unconformity and 120m above it. In addition to well-cuttings obtained every 3m a total of twenty-six sidewall cores was used to supplement the lithological description and to enable assessment of clay-mineral content, organic and trace-element composition and palaeontological variation.

NATURE OF SAMPLED INTERVAL

Lithological Content

Figure 1 shows the lithological content. A cuttings-log is inserted next to the interpreted section and indicates the broad tonal contrasts within the samples. Sidewall cores were taken from the positions indicated to the right and provide the basis for identification of rock type. A sonic/gamma-ray log curve clearly illustrates the position of the Jurassic/Cretaceous boundary which is used as a datum. The thickness of the section studied is shown in relation to the overall stratigraphic succession on the far left.

Both Jurassic and Cretaceous samples contain a predominance of argillaceous material. The Jurassic is non-calcareous and consists of a dark grey claystone typical of the Kimmeridgian whilst, in striking contrast, the Cretaceous is represented by light grey or white calcilutites containing thin bands of almost pure limestone. Above this zone, highly calcareous claystones take on a bright orange or reddish hue which persists for about 100m before medium grey, fine-grained claystones become dominant.

Clay Mineral Content

Figure 2 shows a qualitative expression of the clay mineral content. Kaolinite and illite are particularly apparent, both in the Jurassic and in the Cretaceous, whilst vermiculite is only significant within the first 250m of the Cretaceous. The occurrence of montmorillonite and illite/montmorillonite is to some degree interdependent. Unlike kaolinite and illite, montmorillonite is not stable at depth and commonly alters to illite/montmorillonite within a certain depth range. The parameters governing such a diagenetic function have been fully described by Powers (1967) and it is interesting to observe that this process of conversion appears to become complete at a level just below the top of the sampled zone. Montmorillonite is conspicuously absent in samples deeper than this level whilst the content of illite/montmorillonite increases.

Organic and Mineral Content

Figure 3 depicts the organic and mineral content. Presence of calcite closely correlates with sonic curve fluctuation and illustrates the contrast between the basal Cretaceous carbonates and the non-calcareous Jurassic. Trace-metal determinations on organic extracts show zinc, copper and

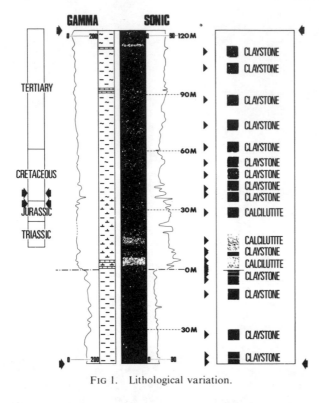

FIG 1.   Lithological variation.

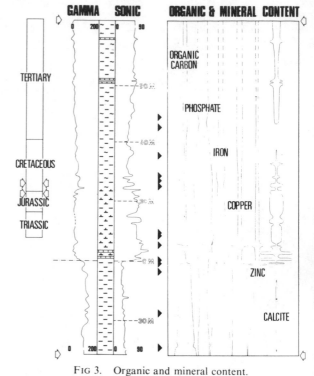

FIG 3.   Organic and mineral content.

iron to diminish sharply at the contact although these minerals primarily reflect the degree of organic carbon which is abundant in the Jurassic and scarce in the Cretaceous. Phosphate content is relatively high, reaching a maximum in the Lower Cretaceous.

Palaeontological Content

Figure 4 shows a qualitative assessment of the palaeontological content. The nannoplankton are noticeably absent in the Jurassic but appear at intervals in the Cretaceous. Conversely, the microplankton (mostly of algal type) and the

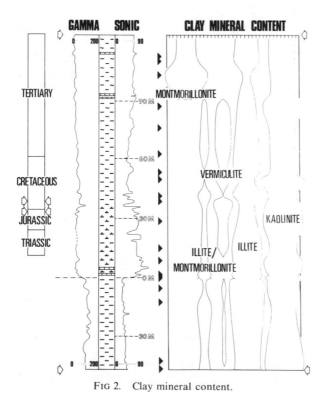

FIG 2.   Clay mineral content.

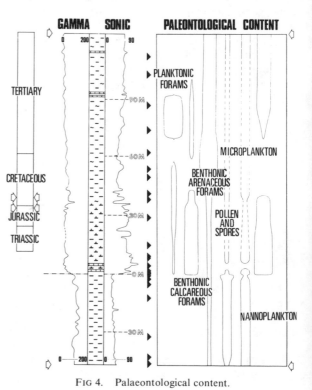

FIG 4.   Palaeontological content.

spores and pollen typify the beds below the contact but became less abundant above it, where incomplete sampling is represented by dashed lines.

STRATIGRAPHY OF SAMPLE INTERVAL

On palaeontological evidence, the sequence studied was placed in the Upper Jurassic and Cretaceous of the northern North Sea. The unconformable contact between Kimmeridgian and Lower Cretaceous Barremian is signified by the absence of Volgian, Ryazanian, Valanginian and Hauterivian strata. In the upper part of the interval, the presence of Cenomanian and Turonian beds indicates the Upper Cretaceous.

Figure 5 shows the effect of superimposing these subdivisions on the results of the detailed analysis.

The depositional break between Kimmeridgian and Barremian is indicated by wide fluctuations in nearly all parameters whilst each stage demonstrates a distinctive group of characteristics. The Kimmeridgian is immediately apparent by its consistent variance with the strata of the Lower Cretaceous. As described earlier, this unit contains virtually no calcareous benthonic or planktonic fauna but is characterized by a high percentage of spores and pollen. Carbonate presence is low but

organic carbon and trace-metal content of iron, copper and zinc is high. Kaolinite and illite form a dominant part of the clay minerals, although this is not peculiar to the Kimmeridgian. Vermiculite is poorly represented.

The association of these factors in an extremely dark-toned claystone indicate restricted anaerobic conditions where poorly oxygenated waters inhibited the decay of organic matter and prevented the spread of bottom-living organisms. Marine conditions are suggested by the microplankton but the abundance of pollen and spores reveals the proximity of widespread landmass regions enjoying a temperate climate. The low carbonate content suggests such land areas were responsible for a steady detrital influx but the slight preponderance of illite over kaolinite and the fine-grained nature of the sediment indicate that the provenance was low-lying and the rate of deposition relatively slow.

Immediately above the unconformity, this pattern is completely altered. The analysis of the Barremian shows it to be distinguished by a high calcareous content with abundant nannoplankton and calcareous benthonic foraminifera. Trace metals diminish but since these were derived from organic residues they also reflect the sharp decline in organic carbon. A brief resumption of Kimmeridgian conditions 9m above the base of the

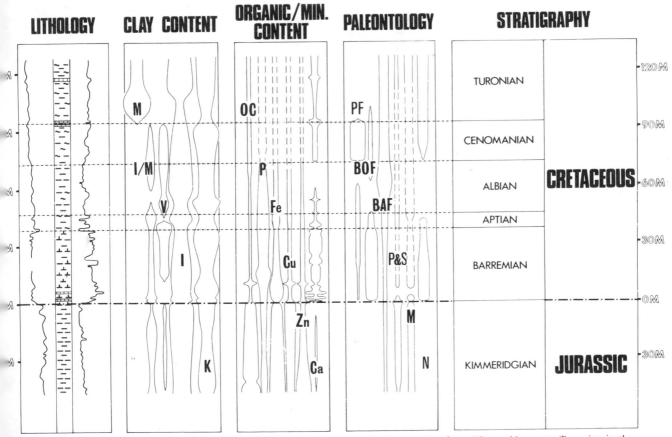

FIG 5. Variations in lithology, clay content, organic/mineral content and palaeontology from Kimmeridge up to Turonian in the northern areas of the North Sea.

Barremian is suggested by a sudden increase in carbon content again, but the lack of any effect on the microfauna suggests the narrow interval of dark sediments may merely consist of reworked Upper Jurassic material.

The relatively low content of planktonic foraminifera in the Barremian suggests fairly shallow water, these forms tending to become more abundant in a seawards direction. The low level of spores and pollen indicates little terrestrial influence whilst the position of the Barremian immediately above the much older Kimmeridgian points to the unit being of a transgressive nature.

It is this aspect which possibly explains why the white or light grey basal sequence takes on the red colouration about 20m above the contact. The increase in kaolinite content, and particularly the relative sudden increase in vermiculite as the redness becomes more dominant, suggest that the Lower Cretaceous seas were progressively inundating an extensively weathered land surface of low relief. The clay minerals point to this land surface being composed of a high percentage of crystalline rocks rich (in the case of kaolinite) in alkaline material undergoing an acid-leaching process (Weaver and Pollard, 1973) whilst the presence of vermiculite suggests metamorphic rocks containing a high proportion of biotite mica and chlorite. The red colouration may be a derivative of extensive laterite formation and can be compared with the red and pink chalks described from the southern North Sea and from onshore outcrops in eastern England. Since weathered products would not be expected to travel too far, the red staining most likely emanated from the exposed surface of the old Caledonides massif areas, either to the west where a low-lying East Shetland Platform/Scottish Highland region may have been contributing, or to the east where a Bergen Ridge/Norwegian Highland area probably lay above sea level.

The contact between the Barremian and the Aptian is marked by a sudden decrease in both lime content and calcareous benthonic foraminifera. The red colour also lessens in the vicinity of the change from calcilutite to claystone and suggests a minor gap in the sedimentary sequence. This reduction in the lime-favouring elements of the fauna, which also affects the nannoplankton, heralded the onset of the Albian.

The Albian appears to be a period in the Lower Cretaceous when the normally high calcareous content of the sediments shows a marked drop. Not only does the recorded content of calcite diminish in the samples but it is also shown on the sonic log. In the Albian, the calcareous nannoplankton and calcareous benthonic foraminifera are almost entirely absent but, conversely, the numbers of arenaceous foraminifera show a marked expansion. This points to an increased water depth associated with a high content of suspended sediment. In the clay minerals, the relative proportion of illite also increases and, since this tends to be carried further basinwards than the other members of the suite, its variation suggests a heightening of the transgressive nature of the unit, thus placing the sampled interval further away from the basin margin than in earlier times. The widespread incursion of the Albian over the continental shelf of north-west Europe is well documented elsewhere (Casey and Rawson, 1974).

The Cenomanian interval forms the base of the Upper Cretaceous and is again well documented as a major transgressive phase. It is distinguished in the sample interval by a sudden abundance of planktonic foraminifera suggesting fairly deep open marine waters which possess a good connection to oceanic areas allowing free circulation. The reappearance of the nannoplankton points to a high oxygen content and these forms significantly parallel a sudden increase in the carbonate content. Kaolinite declines at the top of the Cenomanian and is accompanied by the disappearance of vermiculite. This suggests that the landmass areas which had so influenced the earlier sediments had dwindled in significance and now lay far away—an observation supported by the fine-grained nature of the grey claystones which characterize both the Cenomanian and the overlying Turonian.

Planktonic foraminifera disappear at the onset of the Turonian. This and the similar decline in benthonic calcareous foraminifera suggest a further deepening of water. This conclusion is supported by the continuation of the benthonic arenaceous foraminifera.

REGIONAL CORRELATION OF SAMPLED INTERVAL

Whilst palaeoenvironmental interpretation of the sampled section sheds an interesting light on the sedimentation in the northern North Sea immediately above and below the Jurassic/Cretaceous boundary, the presence of chronological gaps may be placed in regional perspective by a comparison with other areas. To facilitate this, a series of vertical sections spanning the unconformity were selected from published data and illustrated in Fig 6. They pass south to north in a line extending from the Dorset area of southern England to East Greenland, with the sampled interval from the northern North Sea inserted midway in the series.

Precise correlation of stratigraphic intervals within the late Jurassic and early Cretaceous is the subject of extensive discussion (Hallam, 1971; Jeletsky, 1974; Casey, 1974). It is hampered on a regional scale by the appearance of a strong provinciality expressed in nearly all shallow marine invertebrate faunas of the time and has resulted in the employment of a dual classification based on the Ammonoidea. Differentiation is particularly

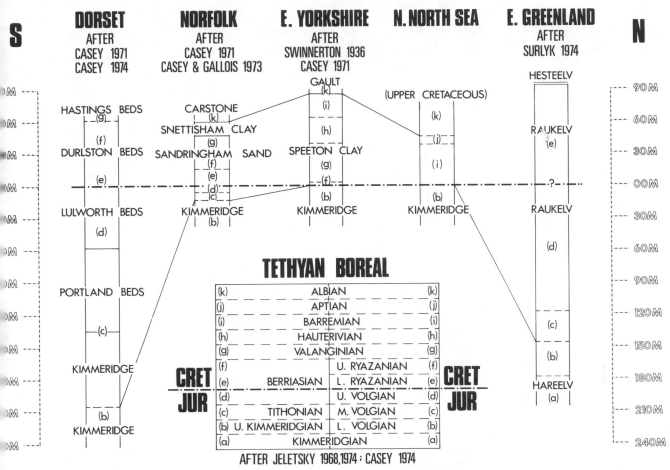

FIG 6.   Regional correlation: Jurassic—Cretaceous boundary.

well documented for this group and enables a Tethyan faunal realm to be distinguished to the south of the British Isles and a less diverse Boreal realm to be distinguished to the north. The larger Tethyan realm extends in a wide circum-global belt aligned along the present Mediterranean area whilst the Boreal realm includes all of onshore Britain, the entire North Sea, East Greenland, north-western Canada and parts of the Arctic. It is also represented in the Moscow region, the northern Urals and in central Siberia (Zakharov, 1970).

At the close of the Jurassic, when the isolation of these two regions was at its highest, the Tethyan ammonoid faunas are marked by a predominance of the Beriasellidae, whilst those of the Boreal region are distinguished by the Dorsoplanitidae and Craspiditidae. Difficulties in correlation of these and other forms has given rise to the Tithonian–Berriasian subdivision based on the former, and the Volgian–Ryazanian subdivision based on the latter.

Much work has been done in an attempt to reconcile these differences (e.g. Jeletsky, 1974) and some authors would prefer to see a uniform Berriasian stage applied to both Tethyan and Boreal realms (Saks and Shulgina, 1974). However, as Casey (1974) has stated: "work

continues on what has remained one of the outstanding problems of Mesozoic stratigraphy."

The correlation table displayed in Fig 6 maintains the separate Tethyan/Boreal sub-division and sets out to express the most recent views on the overall correlation of the two faunal realms.

The sections displayed can all be placed within the Boreal realm. Although each succession contains faunal assemblages affiliated to the Volgian–Ryazanian classification, a wide degree of local variation makes this extremely difficult in some areas. Facies developments are primarily responsible for this, particularly during the period of extensive regression and uplift that occurred in the closing stages of the Jurassic. These movements resulted in the formation of a rapidly subsiding area of deposition in southern England containing both brackish and freshwater Portlandian, Purbeckian and Wealden sequences largely isolated from the more stable shallow marine deposits of Norfolk, Lincolnshire and Yorkshire (Casey, 1971). Casey and Bristow (1964) have described the nature of the land barrier—the Bedford Isthmus—which separated these two regions.

In Dorset, the Lulworth and Durlston Beds represent the Purbeckian sequence straddling the

Jurassic/Cretaceous boundary. Continuous sedimentation, interrupted only briefly by an occasional hiatus, is markedly in contrast to the thin and condensed sequences described from eastern England. Here, in the Lower Cretaceous of Norfolk, Casey and Gallois (in press) have shown that the Sandringham Sands represent a long phase of marine sedimentation occupying an interval extending from the Middle Volgian to the Valanginian. Although of lesser thickness, it represents the equivalent of the top Portlandian, the Purbeckian and the Hastings Beds.

In the same fashion, Fletcher (1974) has demonstrated that the Speeton Clay of Yorkshire extends from the Upper Ryazanian to the Hauterivian. Both these successions, and the nearby Spilsby Sandstone of Lincolnshire, compare closely with the type Volgian section of the Moscow region.

Set against vertical sections from southern and eastern England, the sampled interval from the northern North Sea represents an even more reduced and condensed sequence. This characteristic compares with certain contemporary deposits in East Greenland. In particular, the Wollaston Foreland exposures described by Maync (1949) and Haller (1971) suggest a common tectonic influence affecting both areas more or less simultaneously.

At Wollaston Foreland, the Jurassic/ Cretaceous boundary is marked by an interval of non-deposition. Within it, structural discordancy induced by step-faulting and rotational slippage caused transgressive Lower Cretaceous deposits to accommodate to a complex surface of tilted blocks. The continuation of gentle adjustments on this topography resulted in a form of syntectonic sedimentation with the crestal portions of the blocks receiving only a thin and incomplete section whilst the adjacent troughs contain much thicker and more complete sequences showing fairly continuous passage from Jurassic to Cretaceous. Differential subsidence of this type had evidently ceased by Valanginian times as these deposits transgress uniformly over the entire region, although in central East Greenland this transgression is of Albian–Cenomanian age.

A different pattern of sediments is described from the Jameson Land area by Surlyk (1974). Seemingly more marginal than Wollaston Foreland (450km to the north), the succession appears to have been subjected to much milder tectonism with the result that a fairly continuous sequence of marine shales and sandstones reflect the Dorset sequence on the southern edge of the Boreal realm.

THE JURASSIC/CRETACEOUS BOUNDARY

It is evident from the sections displayed in Fig 6 that the contact between Jurassic and Cretaceous is more often than not indicated by tectonic activity and structural displacement. This is certainly very marked in the northern North Sea and at Wollaston Foreland. The resultant stratigraphic break is commonly termed the Cimmerian unconformity and as Fig 5 illustrates, usually represents the absence of Volgian, Ryazanian and Valanginian strata. Unfortunately, the term "Cimmerian" is also applied (cf. early, mid or late) to other Mesozoic epeirogenic phases and much confusion exists as to its correct usage.

The influence of the "late Cimmerian" tectonic phase becomes appreciably less marked towards the southern margin of the Boreal realm. Casey (1967, 1971) has directed attention to the significance of depositional breaks in the Portland, Purbeckian and Wealden succession and suggests that one of these may be related to the major depositional gap further north. In particular, a band of phosphatic nodules and chert pebbles in the Middle Volgian has been closely studied and traced across the palaeogeographic barrier of the London–Brabant Platform to similar nodular phosphorites near the Jurassic/Cretaceous boundary in Buckinghamshire, Norfolk and Lincolnshire. It is suggested that this hiatus may indicate the period of maximum tectonic disturbance in the "late Cimmerian" break observed in the northern North Sea. To support such a view, earlier work by Neaverson (1925) on the provenance of the chert pebbles seen in southern and eastern England suggests that these appear to have come from Wales (in the case of Dorset) and from the Carboniferous Limestone of Derbyshire (in the case of Norfolk and Lincolnshire). In addition, analysis of the heavy-mineral content in beds at the same Middle Volgian horizon suggests certain of these could only have been derived from a metamorphic provenance such as the Scottish Highlands.

Since such minerals are peculiar to this zone, it would suggest that much of the Caledonides area of the western and northern British Isles was subjected to a phase of uplift and rejuvenated drainage that was responsible for sending a wave of sediment far to the south. With this in mind, it is tempting, therefore, directly to relate such uplift to simultaneous activity in the northern North Sea and ascribe the major phase of "late Cimmerian" tectonism to the Middle Volgian.

Undoubtedly one of the most striking features of the Middle Volgian or late Cimmerian unconformity is the noticeably large shift in gamma response. Inasmuch as the calcareous Lower Cretaceous sediments possess a fairly low and uniform level of natural radioactivity (approximately 40 API units), which is reasonably consistent over the entire North Sea, the additional deflection of the curve as it enters the Kimmeridgian suggests a much greater influence than that normally seen between carbonates and shales. Whilst this may be less than 50 API units in

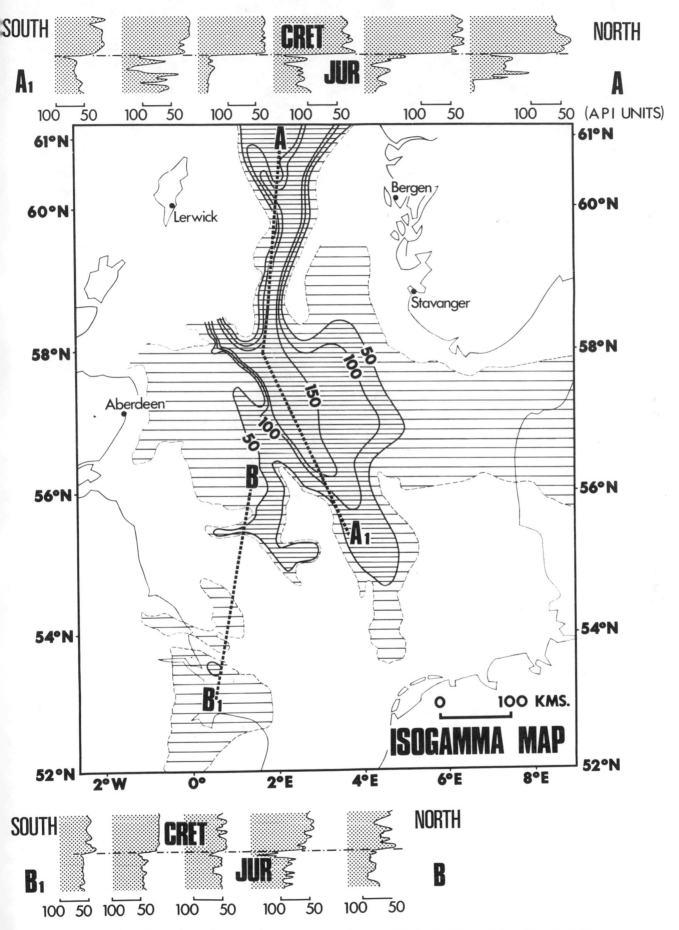

FIG 7. Regional isogamma map showing variation in natural radioactivity for the Kimmeridgian of the North Sea.

the southern area, natural radioactivity shows an increase of over 150 API units in the extreme northern area. Figure 7 displays the variation in deflection along two south–north sections which form a longitudinal line extending across the North Sea. These are simply to demonstrate the contrast in log character. They cover only a very short distance above and below the unconformity and are shown reverse. Such a fluctuation in value recorded from a number of control points has been superimposed on the general distribution of the Kimmeridgian to produce what has been termed a regional isogamma map. This map, showing zones of roughly equal gamma deflection, may be seen to bear a marked relationship with the pattern of sedimentation at the close of the Mesozoic. The region of highest value shows an interesting co-incidence with the line of the Viking Graben and increases to the north.

Earlier discussion has shown that the Kimmeridgian, apart from its influence on the gamma log, has other unusual features. Figure 5, for instance, reveals the dark claystone to possess an unusual content of trace metals which Dunn (1974) has shown to relate largely to the high organic fraction. Iron, copper and zinc and phosphorus are evidently associated in some way with the abundance of marine life and it is likely that similar inorganic trace metals such as actinium are primarily responsible for the radioactivity which so affects the log curve. Casey (1971) has noted in the onshore sections of southern and eastern England how the general presence of phosphorus in the stratigraphic succession suddenly materializes in the Late Jurassic and is strongly evident in the Cretaceous. Figure 5 shows this mineral to be present throughout the sampled interval and to increase slightly in the Lower Cretaceous.

In all its aspects, the Kimmeridgian represents a palaeoenvironment where a superabundance of marine life flourished in conditions which allowed the normal process of oxygenation after death to be greatly inhibited and it is of little surprise that the interval represents one of the most important sources of hydrocarbons currently being discovered in the northern North Sea.

ORGANIC SEDIMENTS AND THE OPENING OF THE NORTH ATLANTIC

Recent studies by Irvine et al. (1974) have related the development of organically-rich strata not only to climate but especially to the presence of mineral nutrients. The planktonic base of the food chain would appear to be encouraged to exceptional productivity by the occurrence of a high content of life elements like phosphorus, sulphur and nitrogen, and other workers, speculating on the ultimate cause of Boreal and Tethyan provinciality, have added a further factor by suggesting that seasonal illumination changes may

also play an important role (Reid, 1973; Hallam, 1972). During the "winter" period, reduction of light in high latitudes (similar to that which occurs in the much less equable climate of the present-day) would cause a sufficiently severe inhibition of the photosynthetic activity of planktonic forms to result in unnaturally short life spans. The high annual death rate thus produced would cause a constant charging of the sediments with organic matter and, in an anaerobic environment, produce the type of deposits seen in the Kimmeridgian.

Since global availability of nutrients ultimately depends on their release from the mantle by volcanic processes, both Irvine et al. (1974) and Valentine (1971) suggest optimum conditions can be directly related to zones of mantle upwelling associated with plate tectonics.

This is directly attributable to certain tectonic developments in the North Sea at the time of the late Jurassic and early Cretaceous. Burke and Dewey (1973) and Naylor et al. (1974) have described plate-tectonic activity associated with the Viking Graben which, as the isogamma map of Fig 7 demonstrates, reflects the maximum degree of organic activity occurring in the Kimmeridgian. Naylor described this zone of subsidence as the "failed arm" or non-spreading member of a triple junction possessing two other arms which progressed into divergent plate margins and became the line along which the northern Atlantic Ocean opened up. During the stresses prevalent along the rifted arms of this rrr junction, prior to the selective development of these two as divergent plate margins, it is suggested that the mineral nutrients necessary to produce organically rich sediments were released from the mantle in the Viking Graben area of the North Sea and exerted a particular influence on the sediments of the late Jurassic. This situation seems to have radically changed by early Cretaceous times, however. The presence of organic carbon appears to be drastically reduced as a totally different palaeoenvironment prevails which implies that such changes in sedimentation were directly related to the failure of the Viking Graben arm during the period of regression marked by the Jurassic/Cretaceous boundary. Such a failure had a far-reaching impact and may be correlated not only with contrast in lithological type but to the regional uplift of the Scottish Highland and Welsh areas deduced by Neaverson from heavy mineral analysis. Since Casey (1971) has effectively dated this zone to be of Middle Volgian age in the Dorset sections, it is suggested that the failure of the Viking arm and the concurrent subsidence of the Viking Graben (with all its associated step and block-fault developments) took place largely in Middle Volgian times. Furthermore, since the failure of one arm meant the development of the remaining two, it is additionally suggested that divergence and hence the opening of the northern Atlantic Ocean also commenced in Middle Volgian

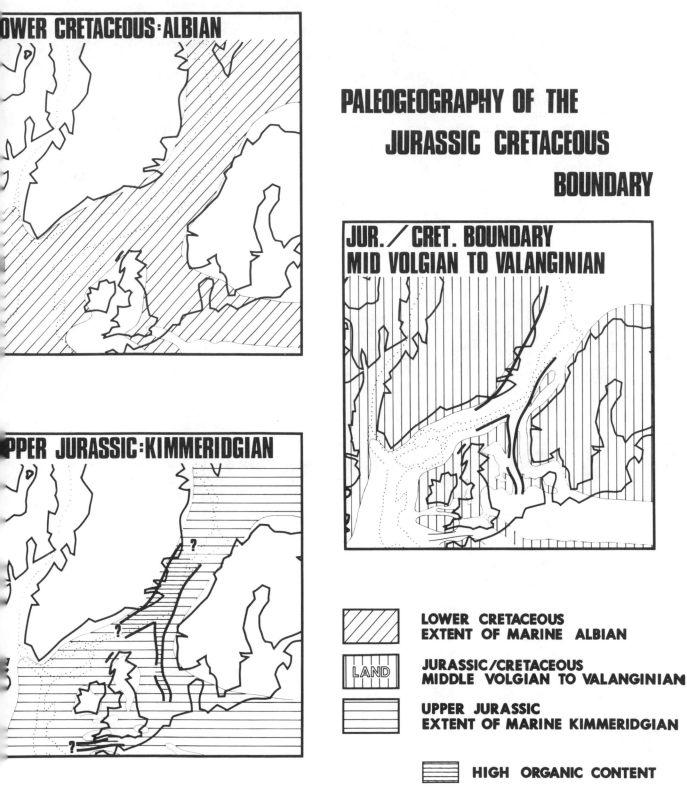

FIG 8.   Palaeogeography of the Jurassic—Cretaceous boundary.

times although its impact on sedimentation was probably not felt until the early Ryazanian. The view concurs with palaeobiogeographical studies by Basov *et al.* (1972) and with the conclusions of Phillips and Forsyth (1972) utilizing magnetic JOIDES data.

A series of three maps, incorporating these assumptions, is shown in Fig 8. They illustrate the position of the triple junction rift pattern and the impact of this feature on the palaeogeography of the Kimmeridgian; the Middle Volgian to Ryazanian interval, and the Albian.

Based on a reconstruction of the continental masses in the Boreal Realm and adjacent areas described by Owen (1974), they further illustrate the probable degree of separation that had occurred by the latter part of the Lower Cretaceous to allow the resricted depositional pattern of the Kimmeridgian to be completely replaced by open marine sedimentation.

Map 1 shows the palaeogeography towards the end of Kimmeridgian times with maximum organic content concurring within the failed arm of the Viking Graben. The trilete form of the suggested triple junction prior to divergence is also shown with the junction lying to the north of the Shetlands. Kimmeridgian deposits are thick along the narrow strait separating Greenland and Europe and spread into the area of the North Sea as shown on the map. The fact that the other two arms of the junction lay to the west and north of the British Isles suggests that the organic shales of Dorset reached this point from a westerly direction and relate to a series of broadly trilete troughs extending along the axis of the Western Approaches–English Channel area (Naylor *et al.*, 1974). Marginal sediment in the vicinity of the London–Brabant Platform and the Pennine–Scottish area to the north suggest these persistent highs were regions of non-deposition. South and east of the British Isles, the Kimmeridgian enters a calcareous or non-marine facies.

The impact of the Middle Volgian uplift on sedimentation caused a regressive palaeo-geographic setting to develop and this is shown in Map 2 of Fig 8. In the North Sea, it reflects the distribution of Kimmeridge shale shown on the isogamma map of Fig 7, and substantiates the concept of relatively isolated basins of sedimentation suggested by Casey (1971). Although this map indicates the broad outline of these basins at the height of Boreal faunal provinciality, post-depositional erosion may have considerably modified the true extent of late Jurassic sedimentation. The non-marine, fluviatile deposits of the Dorset area are divided from those of Norfolk, Lincolnshire and Yorkshire by the Bedford Isthmus forming part of the London–Brabant massif. In eastern England, Casey (1974) suggests that such deposits be termed the Spilsby Province. These are regarded as being temporarily isolated not only from the Dorset basin to the south but from the larger Boreal area to the north by a short-lived resurrection of the Mid North Sea High as a faunal barrier. However, in the Upper Ryazanian this pattern of local basins appeared to have largely broken down. Palaeontological evidence suggests the Spilsby Province once more became part of the much larger Boreal realm and established links with Eastern Greenland and Siberia via the northern route and Denmark and Poland via an eastern seaway.

Map 3 of Fig 8 shows the marked influence that the widening North Atlantic had on the sedimentary pattern. The narrow strait to the north had opened up (Harland, 1969) as a result of plate-divergence and normal marine circulation had ensued. Carbonate sedimentation had become dominant and many areas within the Boreal realm show faunal assemblages containing appreciable numbers of migrants from the Tethyan realm to the south. The Barremian, the Aptian/Albian, the Cenomanian and finally the Maastrichtian were the result of a progressive flooding of the continental margin areas as the belt of new oceanic crust continued to widen down the centre of the newly formed North Atlantic Ocean.

CONCLUSIONS

The detailed study of the sampled section in the northern North Sea has suggested that the following sequence of events may have occurred at the Jurassic/Cretaceous boundary.

1. The climax of the "late Cimmerian" tectonic phase may be traced to a sedimentary hiatus of Middle Volgian date in southern England. Contributory evidence suggests a period of uplift caused renewed drainage in western and northern areas of the British Isles at this time.

2. The association of organically-rich sedimentation with the presence of mineral nutrients derived from zones of mantle upwelling suggests the Kimmeridgian of the central part of the northern North Sea was laid down at a time when the triple-junction system centred north of the Shetlands was approaching a crucial stage.

3. The failure of the Viking Graben arm and the development of the two Atlantic arms into divergent plate margins were associated with step-faulting and rotational block movement which may be contemporaneous with the uplift observed to the west. In this case, failure and subsequent opening of the North Atlantic Ocean can be dated as of Middle Volgian age.

4. The condensed Lower Cretaceous sequence of the sampled section of the northern North Sea may be closely compared with contemporaneous strata outcropping in the Wollaston Foreland area of East Greenland and represents a form of syntectonic deposition.

5. The progressive widening of the strait between Greenland and Norway after the Middle Volgian break resulted in a change in the pattern of sedimentation. The restricted Kimmeridgian facies was replaced by open marine carbonate deposits. The red colouration seen in parts of this interval relate to progressive transgression of the Lower Cretaceous over a weathered surface of alkaline and metamorphic rocks probably of Caledonide date. Gradual flooding of the continental margin areas during the Cretaceous was the result of continued Atlantic divergence.

REFERENCES

Basov, V. A., Krimgolts, G. Ya, Mesezhnikov, M. S., Saks, V. N., Shulgina, N. I. and Vakhrameev, V. A. 1972. The problem of continental drift during the Jurassic and Cretaceous in the light of paleobiogeographical data. *24th Int. geol. Congr. Canada 1972, Sec. 7*, 257–64.

Burke, K. and Dewey, J. F. 1973. Plume generated triple junctions. *Jl. geol.*, **81**, 406–33.

Casey, R. 1967. The position of the Middle Volgian in the English Jurassic. *Proc. geol. Soc. Lond.*, **1640**, 246–7.

Casey, R. 1971. Facies, faunas and tectonics in late Jurassic–early Cretaceous Britain. *In* F. A. Middlemiss, P. F. Rawson and G. Newall (Eds.) Faunal Provinces in Space and Time. *Geol. Jl. Spec. Iss.*, **4**, 153–68.

Casey, R. 1974. The ammonite succession at the Jurassic–Cretaceous boundary in Eastern England. *In* R. Casey and P. F. Rawson (Eds.) The Boreal Lower Cretaceous. *Geol. Jl. Spec. Iss.*, **5**, 193–266.

Casey, R. and Bristow, C. R. 1964. Notes on some ferruginous strata in Buckinghamshire and Wiltshire. *Geol. Mag.*, **101**, 116–28.

Casey, R. and Gallois, R. W. (*In press*) The Sandringham Sands of Norfolk. *Proc. Yorks. Geol. Soc.*

Casey, R. and Rawson, P. F. 1974. A review of the boreal Lower Cretaceous. In: R. Casey and P. F. Rawson (Eds.) The Boreal Lower Cretaceous. *Geol. Jl. Spec. Iss.*, **5**, 415–30.

Dunn, C. E. 1974. Identification of sedimentary cycles through Fourier analysis of geochemical data. *Chemical Geology*, **13**, 217–32.

Fletcher, B. N. 1974. The distribution of Lower Cretaceous (Berriasian–Barremian) foraminifera in the Speeton Clay of Yorkshire, England. *In* R. Casey and P. F. Rawson (Eds.) The Boreal Lower Cretaceous. *Geol. Jl. Spec. Iss.* **5**, 161–8.

Hallam, A. 1971. Provinciality in Jurassic faunas in relation to facies and paleogeography. *In* F. A. Middlemiss, P. F. Rawson and G. Newall (Eds.) Faunal Provinces in Space and Time. *Geol. Jl. Spec. Iss.*, **4**, 129–52.

Hallam, A. 1972. Diversity and density characteristics of Pliensbachian–Toarcian molluscan and brachiopod faunas of the North Atlantic margins. *Leithaia*, **5**, 389–412.

Haller, J. 1971. *Geology of the East Greenland Caledonides.* Wiley, Interscience, London, 413 pp.

Harland, W. B. 1969. Contribution of Spitzbergen to understanding the tectonic evolution of North Atlantic Region. *Mem. Am. Ass. Petrol. Geol.*, **12**, 817–51.

Irvine, E., North, F. K. and Couillard, R. 1974. Oil, climate and tectonics. *Can. Jl. Earth Sci.*, **11**, 1–17.

Jeletzky, J. A. 1974. Biochronology of the marine boreal latest Jurassic, Berriasian and Valanginian in Canada. *In* R. Casey and P. F. Rawson (Eds.) The Boreal Lower Cretaceous. *Geol. Jl. Spec. Iss.*, **5**, 48–80.

Maync, W. 1949. The Cretaceous beds between Kuhr Island and Cape Franklin (Gauss Peninsula), northern East Greenland. *Meddr. Grønland*, **133**, 1–291.

Naylor, D., Pegrum, R., Rees, G. and Whiteman, A. 1974. The North Sea trough system. *Noroil*, **2**, 17–22.

Neaverson, E. 1925. The petrography of the Upper Kimmeridge Clay and Portland Sand in Dorset, Wiltshire, Oxfordshire and Buckinghamshire. *Proc. Geol. Assoc.*, **36**, 240.

Owen, H. G. 1974. Ammonite faunal provinces in the Middle and Upper Albian and their paleogeographical significance. *In* R. Casey and P. F. Rawson (Eds.) The Boreal Lower Cretaceous. *Geol. Jl. Spec. Iss.*, **5**, 145–54.

Phillips, J. D. and Forsyth, D. 1972. Plate tectonics, paleomagnetism, and the opening of the Atlantic. *Bull. Geol. Soc. Am.*, **83**, 1579–1600.

Powers, M. C. 1967. Fluid-release mechanisms in compacting marine mudrocks and their importance in oil exploration. *Bull. Am. Ass. Petrol. Geol.*, **51**, 1240–54.

Reid, R. E. H. 1973. Origin of the Mesozoic "Boreal" realm. *Geol. Mag.*, **110**, 67–9.

Saks, V. N. and Shulgina, N. I. 1974. Correlation of the Jurassic–Cretaceous boundary beds in the Boreal Realm. *In* R. Casey and P. F. Rawson (Eds.) The Boreal Lower Cretaceous. *Geol. Jl. Spec. Iss.*, **5**, 387–92.

Surlyk, F. 1974. The Jurassic–Cretaceous boundary in Jameson Land, East Greenland. *In* R. Casey and P. F. Rawson (Eds.) The Boreal Lower Cretaceous. *Geol. Jl. Spec. Iss.*, **5**, 81–100.

Valentine, J. W. 1971. Plate tectonics and shallow marine diversity and endemism, an actualistic model. *System. Zool.*, **20**, 253–64.

Weaver, C. E. and Pollard, L. D. 1973. *The Chemistry of clay minerals. Developments in Sedimentology*, **15**, Elsevier, Amsterdam, 213 pp.

Zakharov, V. A. 1970. *Late Jurassic and Early Cretaceous bivalve molluscs in the north of Siberia and their habitat.* Academy of Sciences of the USSR. Siberian Branch. Inst. of Geol. and Geophys. Pt 2 Fam. Astaskidal. "Nauka" Moscow. 136 pp.

DISCUSSION

Dr D. H. Matthews (Cambridge University): We have heard three examples of Jurassic volcanism (Forties Field, one from Dr W. H. Ziegler, one from Dr P. Heybroek) within the graben system, and Mr R. J. Johnson has suggested that the graben may be an incipient mid-ocean ridge type of plate-margin. While reserving my judgement on the latter point, the coincidence during Middle Jurassic time of volcanism, graben structure and plenty of (Permian) salt suggests an analogy with the Red Sea, where metalliferous suspensions are concentrated in hot brine-filled depressions within the graben floor. It seems possible that sooner or later the drill may encounter iron-rich sedimentary ore bodies formed in such an environment.

R. J. Johnson: A comparison of the plate-tectonic aspects of the Viking Graben with that of the Red Sea is a little outside the scope of this paper. The occurrence of mineralization in failed arms is noted by Burke and Dewey (1973) and reference can be made to a description of a Pre-Cambrian rift by Kanasewich *et al.* (Kanasewich, E. R., Klowes, R. M. and McCloughan, C. M., 1968, A buried Pre-Cambrian rift in western Canada; *Tectonophysics*, **8**, 513-27) where such a phenomenon is thought to occur.

It is possible that at certain times during the development of the North Sea *rrr* junction system, the formation of sedimentary ore-bodies may have taken place on the floor of the graben. However, the thickness of the Tertiary and Mesozoic sedimentary succession has deterred extensive exploration in this particular zone and little is known of the deeper horizons within the Viking Graben.

Professor Arthur Whiteman (Aberdeen University): As one of the group proposing lithospheric plate and thermal uplift failed-arm hypotheses to account for the evolution of the Viking Graben System and the central and southern North Sea grabens, perhaps I should speak up in defence of these hypotheses.

We have heard a great deal about the palimpsest hypothesis from Dr Walter Ziegler and Donald Blair of Esso, and I do not think that the wrench-fault derived hypothesis they propose is tenable.

The proposition that a stress-field, derived by identifying the Great Glen Fault and the Tornquist Line as part of transcurrent movements resulting from north-south compression of the lithosphere, and by identifying the sub-Upper Cretaceous graben system of the North Sea as part of the associated tension pattern, is unacceptable on the grounds that a system of this type could not have existed from early Palaeozoic to Cenozoic times.

Certainly the pre-Permian basement has exerted controls on late structural development, but in a number of places where these controls have operated, they can be recognized, *e.g.* Moray Firth-Great Glen Fault area and Celtic Sea. A complete palimpsest explanation as proposed by Dr Walter Ziegler is unacceptable.

Dr Peter Ziegler has shown how some kind of crustal thinning-thermal uplift-collapse process probably operated under the central North Sea and under the Rhine Graben System and David Naylor, Dick Pegrum, Graham Rees and myself have propounded a lithospheric plate-thermal uplift-failed-arm hypothesis (*Noroil*, 1974, **2**, pp. 17-22; *Tectonophysics*, in press), to account for the origin of the graben system.

This is not the place to go into a discussion of these hypotheses but I think that an explanation of the evolution of the North Sea trough-system should take into account the predominant structural tensional style proven and that thin and anomalous crust exists under the Vøring-Stadt Basin arm; under the Viking Graben (in process of study by Bergen University Seismological Observatory; Dr Hinz of the Bundesanstalt für Bodenforschung; Durham University and myself at Aberdeen University); and under the West Shetland-Faeroe Channel. Indeed, Professor Martin Bott and D. G. Roberts at this meeting have told us that continental crust is absent along the whole of this last-named zone. The crust is thin under the Rhine Graben System and the Oslo Graben, as it is under many "failed arms" throughout the world.

Many people now recognize that troughing and crustal thinning are closely inter-related and I think the Viking Graben and the fault-controlled troughs to the south may well be underlain by thinned crust. West Sole Trough is associated with thinned crust by B. J. Collette (1968, On the subsidence of the North Sea area. *In D. T. Donovan (Ed.) Geology of Shelf Seas*, Oliver and Boyd, London), although he saw his results in a different way than we do. The hypothesis needs further testing beneath the graben system.

The question I should like to ask Drs Peter Heybroek and Myles Bowen is—can they tell us whether they have seen any Ziegler-style transcurrent faults among the many tensional faults they have shown on their maps of the Viking and Dutch grabens and on the many miles of seismic sections they have examined?

Dr W. A. Read (Institute of Geological Sciences): Could Mr Johnson comment on the possibility that the Kimmeridge black shales may represent algal blooms?

R. J. Johnson: The closest modern analogy to the black shale conditions represented by the Kimmeridgian seems to be the products of "red-tides" noted off the west coasts of South Africa, South America, north-west Africa and California. These have been described by Bogorov (1967) and, particularly, by De Buisonje (1971).

According to the observations of these authors, red-tides are basically caused by an enormous expansion of phytoplankton (usually one particular species of either the Diatomea or the Dinoflagellata) which increases to such an extent that not only are most other phytoplankton outnumbered but the resulting overdose of metabolic substances causes conditions to rapidly become lethal for virtually all forms of marine life. An example quoted by De Buisonje: during a red-tide occurrence in 1948 near the coast of Florida, not only did the death of vast numbers of such phytoplankton occur but millions of fish, crabs, shrimps and other animals were killed—the water acquiring an oily yellow appearance.

Red-tides require special conditions: good illumination and a relatively warm surface-water temperature (at least for part of the year) and—most significantly—the presence of upwelling sea-water rich in nutrients such as phosphates and nitrates.

Reference is made in my paper to the work of Irvine *et al.* (1974), invoking similar relationships between abundance of phytoplankton, climate (especially temperature) and availability of nutrients. Valentine (1971) suggests, furthermore, that optimum nutrient conditions can be directly related to mantle upwelling associated with plate-tectonic activity of the type indicated in the axial part of the North Sea.

The nature of the Kimmeridgian black shales thus corresponds very closely with the sort of deposit produced by modern red-tides. Although the actual phenomena may itself persist for only a few days, or exceptionally a few weeks, the effect is to concentrate abnormally large amounts of dead organic material on the sea-floor and create toxic conditions inimical to life. This material, and frequently the sea water above it, contains a high percentage of H_2S and is almost or completely devoid of oxygen—conditions which almost perfectly mirror the anaerobic type of environment envisaged for the Kimmeridgian. Decomposition under the absence of oxygen means that not only fatty substances of dead organisms are preserved but the musculature also—the whole being converted into a rather rigid soapy mass which suggests itself as an ideal basis for the later generation of hydrocarbons.

Although mass mortality of organisms is obviously a built-in characteristic of modern red-tides, the widespread nature of the Kimmeridgian deposits suggests additional factors may have amplified the pattern of repeated proliferation and then fatality of marine organisms. One of these may be the kind of seasonal illumination changes proposed by Reid (1973) and Hallam (1972).

These workers were primarily concerned with understanding the cause of differentiation between the Boreal and Tethyan realms as expressed in forms such as the Ammonoidea. They concluded that in extra-tropical latitudes such as the late Jurassic North Sea region, the effects of seasonal fluctuation in the degree of illumination and an associated variation in the temperature of surface waters could well have caused a considerable influence on the ability of the simple organisms at the base of the marine food chain to carry out photosynthesis. This fluctuation may have provided a seasonal impetus to the recurring pattern of widespread red-tide occurrence in "summer" followed by mass mortality in "winter" and encouraged the gradual build up of considerable thicknesses of the organic shales seen in the North Sea.

Malcolm Butler (Transworld Petroleum): (1) Has Mr Johnson seen any indication of a major angular unconformity between the Upper Jurassic and Lower Cretaceous in the northern North Sea?

(2) Would he agree that the Lower Cretaceous limestones and red shales sitting on these fault-blocks in the northern North Sea represent condensed sequences possibly similar to those which were developed on Schwelles in the Devonian of Germany?

R. J. Johnson: (1) There is no evidence of a major angular discordance between the Upper Jurassic and the Lower Cretaceous, although some discordance certainly exists. Differential removal of the Kimmeridgian on the crestal regions, for instance, suggests rotational tilt-movements of the fault-blocks did take place at this time, although studies of the development of such features indicates that movement—followed by subsequent erosion and development of a hiatus or unconformity—was not restricted to any one particular point in time. Rather, a series of progressive adjustments appear to have taken place with the Middle Volgian movement certainly seeming to be the one associated with the most widespread changes—ushering in a general regression and the onset of the completely new palaeoenvironmental conditions of the Lower Cretaceous.

(2) The Lower Cretaceous strata encountered on the crestal regions of the blocks certainly suggests a condensed sequence—particularly when (as in Fig 6) the succession is placed in juxtaposition with the sequences spanning the Jurassic/Cretaceous boundary in Dorset and certain parts of East Greenland. A comparison with the Schwellenfacies of the Devonian in Germany could be made and a relationship developed which suggests a similar mode of deposition over a series of submarine ridges and troughs. However, a clear interpretation of the various factors involved in the development of the Schwellenfacies is still awaited. Perhaps a better example may be found in sediments of Lower Cretaceous age described by Maync (1949) and Haller (1972) from the Wollaston Foreland area of East Greenland.

REFERENCES

Bororov, V. G. 1967. Biological transformation and the exchange of energy and matter in the ocean. *Okeanologiya*, **7**, 649–65.

De Buisonje, P. H. 1972. Recurrent red-tides, a possible origin of the Solnohofen Limestone. *Proc. Kun. Ned. Akad. Wetenschap, Ser B.*, **75**, 152–77.

30

Lower Cretaceous Gas-Fields in Holland

By A. COTTENÇON and B. PARANT

(ELF-RE)

and G. FLACELIÈRE

(Petroland NV)

SUMMARY

Petroland in The Netherlands has discovered three gas-fields in the sandstones of the lower Cretaceous: Leeuwarden and Harlingen on shore and Zuidwal in the Waddenzee. The production of Leeuwarden averages 115 000m³ per day. The fields of Zuidwal and Harlingen have not yet produced.

The gas-bearing sandstones, of Valanginian age, form a part of a basin located in the north-west part of the country, the sedimentation of which commenced locally during the Wealden. They accumulated on a Zechstein or Triassic palaeo-relief. At Zuidwal the sandstones cover an old volcanic formation upon which the structure is moulded.

A sedimentary study of the sandstones has enabled the identification of a barrier inland deposit which represents a succession of transgressive episodes during the early Cretaceous. At Zuidwal, in the centre of the basin, the bars are piled up vertically, while in the more coastal zone of Leeuwarden they are deposited separately along the shore.

INTRODUCTION

The gas fields discovered by Petroland in the Lower Cretaceous are located in the northern part of the Netherlands. Three fields have been discovered: the Harlingen and Leeuwarden fields onshore in the province of Friesland; and the Zuidwal field, the largest of the three, discovered offshore in the shallow water of the Waddenzee, between the mainland and the Frisian Islands.

These three fields are located in the same Lower Cretaceous sub-basin, the "Zuidwal basin", which is of local extent in the north-west of the Netherlands. Other small contemporaneous basins are known in the eastern part of the country and in the Hague region, where both oil and gas are produced.

The oil exploration work conducted by Petroland in this area was initiated in 1964. Often hindered by access and environmental problems, which are frequent in the Netherlands, they were carried out in three stages. The discovery of the two onshore fields took place in 1964 and 1965 with the first wells Harlingen 1 and Warga 1 drilled in accordance with the existing seismic data. During this first stage, the work done on free lands culminated in the drilling of 6 wells: three in the Harlingen area (Har 1, 2, 3) and three others in the Leeuwarden area, where the gas-wells of Eernewoude 1 and Nijega 1 confirmed the discovery of Warga 1.

In 1966–1967, drilling activities were delayed by the Dutch authorities until new mining laws were issued. Work was resumed in 1968 with delineation during 1968–1969, then the development during

1969–1970 of the Leeuwarden field, on which eleven wells have been drilled: LW 1 through 11. In March 1969 Petroland obtained a concession covering the Harlingen and Leeuwarden fields. Production had been delayed one year until October 1971 when Gasunie completed the necessary extension of its main gas pipeline system to carry the Leeuwarden gas.

In the Waddenzee, where a major magnetic anomaly was known to exist, the work was hampered by the nature of the territory, which remained more or less flooded at low tide. Since 1969, the seismic studies have been improved by the use of air-cushion vehicles. These made the delimitation of the Zuidwal structure possible and proved the existence of a local thick Lower Cretaceous section.

Drilling, in accordance with the drilling permit obtained in March 1969, has been made possible by the use of self-elevating barges (Transocean 1 and 2), brought in by navigation channel. The Zuidwal 1 well led to the discovery of the third gas-field in the Lower Cretaceous at the end of 1970. Beneath the sandstone reservoir, the well entered volcanic breccia and was stopped after having penetrated more than 1000m of this formation.

The ZU 2 and 3 wells, as well as OI 1, drilled in 1971–1972, permitted a better understanding of the extent of the gas accumulation and of the reservoir characteristics.

Furthermore, the exploration of the pre-Cretaceous series showed that the Zuidwal structure follows the shape of a former volcano which underlies the location of ZU 1 (Fig 1).

This paper is based on the geological, structural

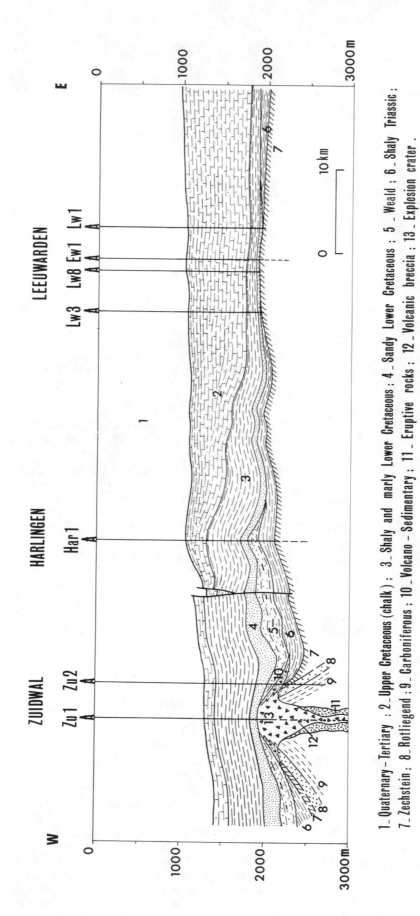

Fig 1. Structural cross-section of The Netherlands Lower Cretaceous gas-fields.

1 – Quaternary – Tertiary ; 2 – Upper Cretaceous (chalk) ; 3 – Shaly and marly Lower Cretaceous ; 4 – Sandy Lower Cretaceous ; 5 – Weald ; 6 – Shaly Triassic ; 7 – Zechstein ; 8 – Rotliegend ; 9 – Carboniferous ; 10 – Volcano – Sedimentary ; 11 – Eruptive rocks ; 12 – Volcanic breccia ; 13 – Explosion crater .

and reservoir data resulting from the 21 wells already mentioned as well as from five others drilled by Petroland on the borders of the Neocomian basin. It is also based on regional studies concerning the Lower Cretaceous, conducted by the French group since the beginning of its activities in the Netherlands.

DESCRIPTION OF THE SECTION

Many wells have been drilled into the Lower Cretaceous in the northern part of the Netherlands. Some of them have encountered a sandy sequence at the base of the Lower Cretaceous and their stratigraphic position has been defined by Petroland through exploration wells in the fields of Zuidwal, Harlingen and Leeuwarden.

Age of the Lower Cretaceous Sands

The microfaunal assemblage of the sands is characterized by the occurrence of the arenaceous foraminifera *Ammovertella*, *Triplasia*, *Citharina*, *Lenticulina saxonica* and the ostracodes *Schuleridea* cf. *thoerenensis*, an association which characterizes the Valanginian stage. Their palynological assemblage includes (in the upper part) *Pseudoceratum*, *Muderongia crucis*, and *Odontochitina*; (in the middle and lower part) *Muderongia neocomica*, *M. simplex*, *Hystrichosphaxidium pulchrum*, *Hystriochosphaeridium*, *Scuncodinum*

apatelum. This association also gives a Valanginian age. A fauna of ammonites has been found in a core from Harlinge I at 1695.4m: *Polyptychites quadrificus* v. Koenen and *P.* cf. *keyserlingi* Neum. and Uhl., both of Middle Valanginian age, *i.e.* the *Polyptychites keyserlingi* Zone.

Age of the Underlying Sequence

The Valanginian sandstones rest on different formations: they overlie Zechstein anhydrite on the crest of the Leeuwarden anticline and on Lower Triassic red clays on its flanks. In the Harlingen area they rest on sands and clays which show an ostracode-assemblage of the German Wealdian 1-2: *Cypridea granulosa granulosa*, *Kliena alata*, *Scabriculocypris trapezoides*. In the Zuidwal field, the first well drilled on the crest of the structure encountered Valanginian sandstones resting on a basaltic breccia, probably of Purbeckian age. On the flanks of this anticline, wells have again encountered sands and clays of the Wealdian.

A geological map of the formations lying beneath the base of the Cretaceous transgression has been made (Fig 2). The south-west part of this area shows the Texel High with a south-east to north-west trend: here the Lower Cretaceous rests on the Carboniferous while to either side appears the Rotliegendes, in turn flanked by the Zechstein series. This Zechstein subcrop shows a north–south high axis. The Lower Trias sediments are widely

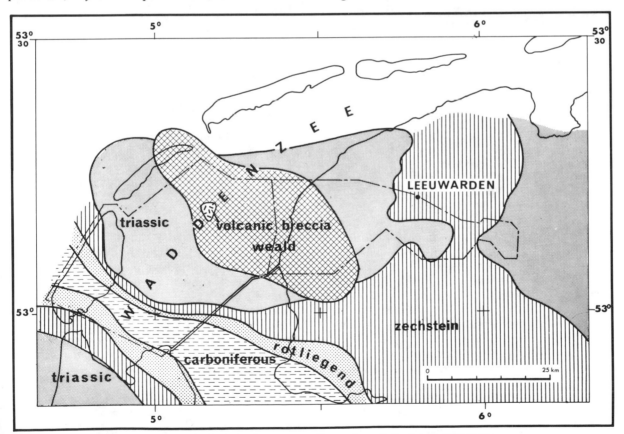

FIG 2. Geological sub-crop map of strata beneath the Lower Cretaceous.

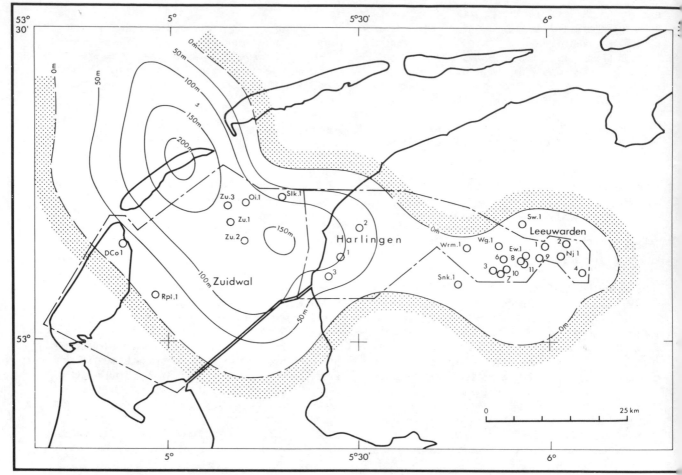

FIG 3. Isopach map of Lower Cretaceous Sandstones.

represented in the northern part of the area. Jurassic sediments are missing, but a small Wealdian basin present between Harlingen and Zuidwal shows that lagoonal conditions prevailed in the area just before the Lower Cretaceous (Valanginian) marine transgression.

Age of the Overlying Sequence

The Valanginian sands are covered by clays which show the following microfaunal assemblage of Hauterivian age: *Haplophragmium aequale, Nodosaria sceptrum, Citharina harpa, C. sparsicostata, Lenticulina crepidularia, L. guttata, Hechtina antiqua* and the ostracodes *Schuleridea* cf. *thoerenensis, Protocythere triplicata*. The Hauterivian age of these clays is equally confirmed by their palynological assemblage.

Lithology and Sedimentology of the Valanginian Sands

The Valanginian series consists of fine to very fine, grey to buff sandstones with argillaceous cement of irregular distribution. Individual clay layers are absent, the sediment being more or less a clayey sandstone. The quartz grains are angular to subangular. The sandstone is often bioturbated with horizontal and vertical worm-burrowings.

The thickness of the sandstones varies (Fig 3). At Leeuwarden it ranges from 9 to 32m, at Harlingen from 60 to 90m, and at Zuidwal from 90 to 140m. The thicker section at Zuidwal is characterized by a succession of five individual palynozones. Three of them are represented in the Harlingen section and only one in the sandstone of the Leeuwarden 10 well (Fig 4). The Leeuwarden 10 well, like the Leeuwarden 8 (Fig 5), has probably encountered an episode of the Valanginian, which might constitute a single cycle.

Analysis of the sandy facies leads to the conclusion they have resulted from a "barrier island" type of sedimentation. The lower part of the sequence has indeed a negative evolution, *i.e.* the clayey proportion is higher at the base than at the top (Fig 7). Also bioturbated sediments are more prevalent near the top of the sequence.

The upper part of the section is represented by a positive sequence, the clay proportion becoming higher near the top. This sequence is also characterized by the appearance of a new facies—an oolithic ironstone and a highly ferruginous black clay. Dynoflagellates are more frequent in the lower negative sequence, while pollens are more abundant in the upper positive sequence.

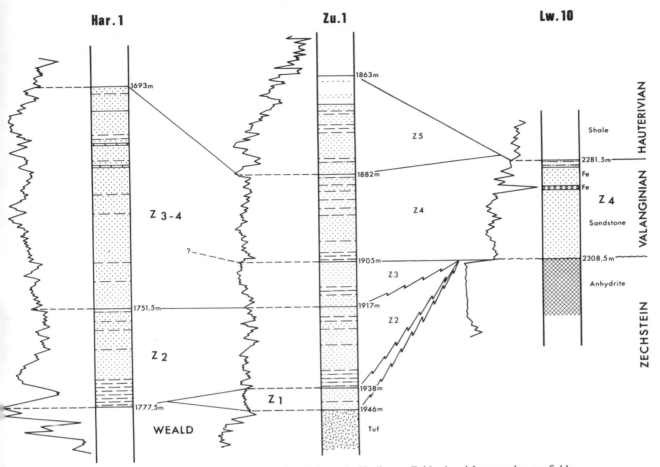

FIG 4.   Stratigraphic and gamma logs of wells from the Harlingen, Zuidwal and Leeuwarden gas-fields.

In this sequence the more marine episode occurs at the base. Following this the sea became more and more shallow and the upper part of the section was deposited in a restricted circulation environment with considerable fresh water influence.

In the Leeuwarden area, several barrier island deposits have been recognized. They may sometimes be without inter-communication and the related gas-pools have individual water-levels: —1840m at Leeuwarden 6 and —1940m at Leeuwarden 8.

At Zuidwal and at Harlingen, because of greater subsidence several sedimentary episodes of the same type are superimposed. Some wells, in the southern part of the area, like Riepel 1, show a sandy facies, with only a positive sequence, in which pollen of lacustrine environment is present.

On the other hand, in the north some wells show a Valanginian deposit which includes only a negative sequence, in which the basal argillaceous facies is more developed, due to more marine conditions.

During the Valanginian stage, the Texel axis constituted a high, probably an emergent zone. The open sea was located north of this high and the sandy barrier islands were deposited along the shore, and more or less parallel to it.

STRUCTURAL AND RESERVOIR DATA

The Neocomian basin sediments of Friesland–Waddenzee were deposited on a former platform area where the thin basal Triassic is regionally preserved above the Permian and beneath the transgressive Lower Cretaceous. The lack of Jurassic sediments probably implies non-deposition during a period of pre-Cretaceous erosion. The present orientation of the Lower Cretaceous sandy deposits and their detailed structure is the result of a conjunction of two tectonic trends: (a) a north-east–south-west system, corresponding to the reactivation of the Hercynian direction, so well marked in the Netherlands at the end of the Carboniferous, particularly along the Groningen–Friesland anticline axis; (b) north-west–south-east system, following the direction of the great regional faults along the edges of the Hague Graben and parallel to the Texel–IJsselmeer High, which characterizes the maximum pre-Cretaceous erosion.

The Zuidwal field (Figs 6 and 7)

The Zuidwal structure, as defined by the top of the Valanginian reservoir, includes a main culmination of 8×8km and a secondary one of 5×2.5km. Both culminations trend north-west to

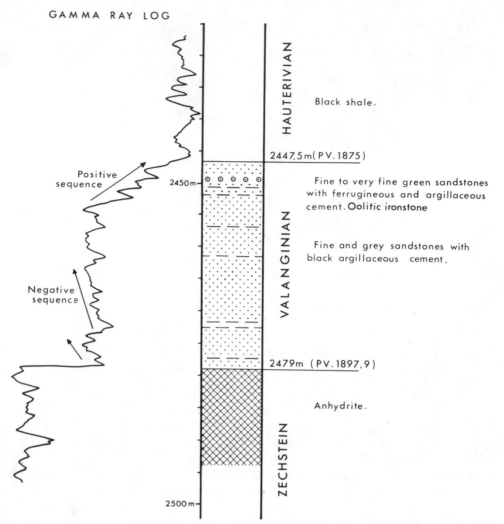

FIG 5.   Log of Leeuwarden 8 (deviated well).

south-east, but the detail of the closed area includes several finger-like north-west to south-east trending parts. The vertical closure is 90m. The entire structure is localized on a dome created by a former volcano. The cross-section of Fig 1, which gives the configuration of this dome, is interpreted from geophysical and subsurface data.

The Carboniferous—the oldest of the known formations—was encountered in ZU 2 and OI 1 at about 2500m beneath a condensed, though normal, pre-Cretaceous section. This formation has been intruded by a volcano, which ZU 1 has penetrated immediately beneath the Neocomian sandstones at about 1950m, and has been drilled to a depth of 3000m.

Above the Carboniferous, the Permian strata seem to have been raised around the crater by the volcanic intrusion and, due to the pre-Cretaceous erosion, were displayed in concentric aureoles (Rotliegendes followed by Zechstein). Beneath the Wealdian layers small patches of Triassic are present.

Finally, the transgressive sandstones of the Valanginian, which form the gas-bearing reservoir,

overlap the former volcanic dome. It should be noted that the influence of the resistant dome has persisted throughout the Wealdian, the deposits of which are absent in the vicinity of the volcanic neck. The dome evidently remained high during the deposition of the sandstones and the Neocomian clays, since these formations are thinner at ZU 1 than in the peripheral wells. It seems certain that the present Zuidwal structure at the Lower Cretaceous level is the result at least in part, of the differential compaction of these deposits around the resistant dome.

The thickness of the Neocomian sandstones of the reservoir ranges from 102 to 160m. Their mean porosity is of the order of 15 percent. The gas reserves in place are evaluated at 41.6×10^9 std m^3 for the whole structure.

The exploitation of the Zuidwal gas faces severe constraints in so far as problems of access and environment are concerned. The development plans of the field have now been established. They provide for a jacket for the production wells on the field itself. The treatment centre will be established on the mainland in Friesland. In this manner

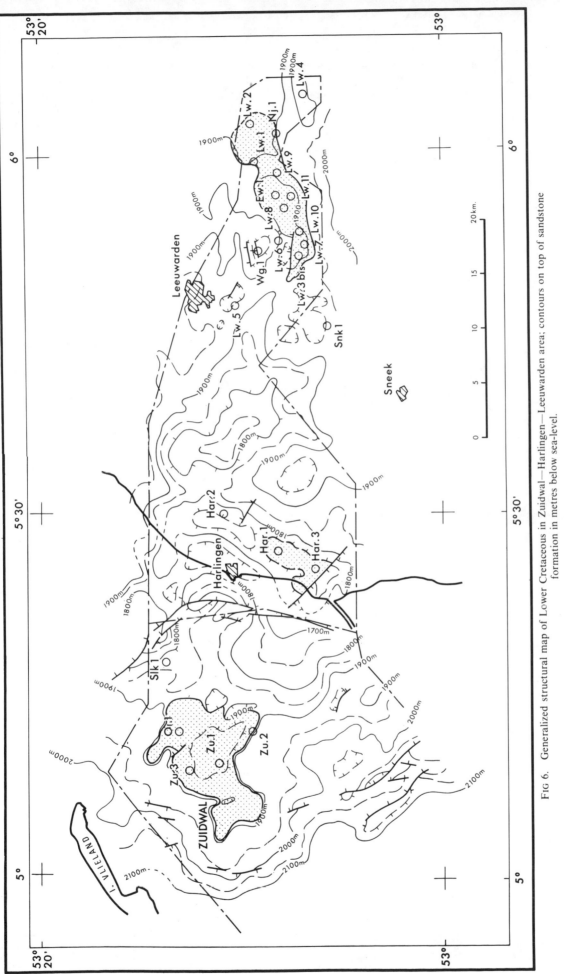

Fɪɢ 6. Generalized structural map of Lower Cretaceous in Zuidwal—Harlingen—Leeuwarden area; contours on top of sandstone formation in metres below sea-level.

production will be possible on time and with respect for the ecological desiderata. Development will be executed after the granting of a production licence by the Dutch authorities.

The Leeuwarden Field (Figs 6 and 9)

The Leeuwarden field gas comes from a complex trap resulting from several factors: (a) structural position of the reservoir sandstones at the base of the Neocomian; (b) stratigraphic trapping, with sandstone pinch-out; (c) permeability trapping, where the sandstones become highly argillaceous. In short, the producing area forms a sinuous belt about 17km long and 3 to 4km wide with the main orientation varying from east-north-east–west-south-west to east–west, and a secondary one from north-west–south-east to west-north-west–east-south-east. The vertical closure reaches some 60m. The structure of the discovery well Warga 1 represents a small independent accumulation which is not in production. The thickness of the sandstone reservoir reaches a maximum of 25 to

30m with a mean porosity of 15 percent and a fluctuating permeability of 0.8 to 3.5md. The original gas reserves in place have been evaluated at 11.7×10^9 std m³. The exploitation, which was begun in 1971, continues at a mean rate of 115 000 std m³ per day. The annual production of the Leeuwarden field was 414×10^6 std m³ in 1973.

The Harlingen Field (Figs 6 and 8)

The Harlingen structure forms a part of a north-east–south-west high axis parallel to the shore. The gas is trapped in the southern culmination of the structure and forms a small closed area against a transverse fault. It is 2.5km long and its mean width is 1.5km. The vertical closure is in the order of 30m. This culmination is incompletely filled. Only the upper part of the sandstone is gas-bearing at Harlingen I over a thickness of 19m. The reservoir has rather favourable characteristics: mean porosity 22 percent and permeability 30md. The reserves in place are evaluated at 2.5×10^9 std m³.

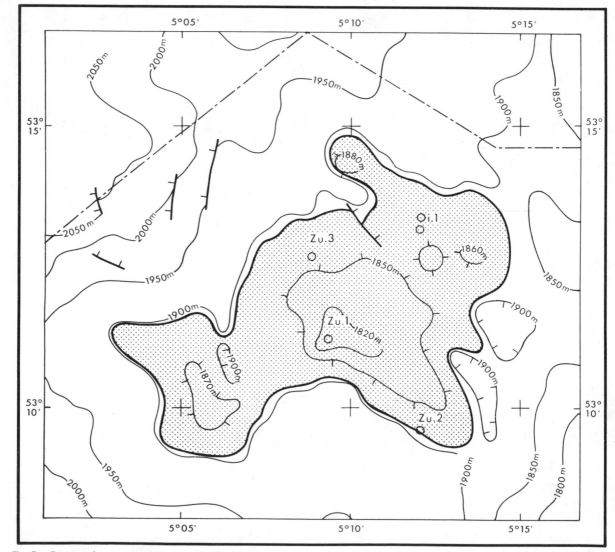

FIG 7.   Structural contour map on top of Lower Cretaceous in Zuidwal gas-field. Contours in metres below sea-level. Scale as for Fig 9.

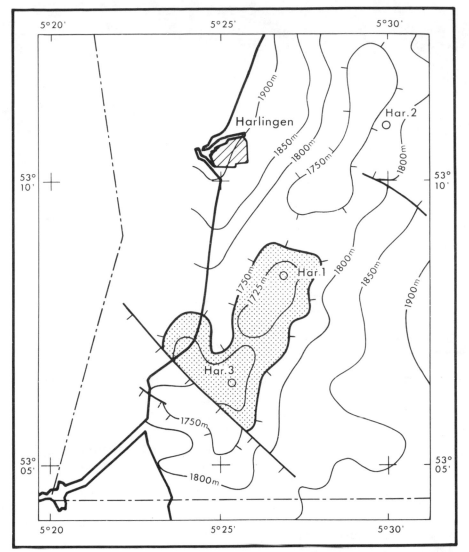

FIG 8.   Structural contour map on top of Lower Cretaceous in Harlingen gas-field. Contours in metres below sea-level. Scale as for Fig 9.

CONCLUSION

The final result of the exploration of the Netherlands Lower Cretaceous sub-basin was the discovery of three individual gas-fields, whose cumulative reserves represent some 56×10^9 std m³ of dry gas.

The table below shows this gas has minor variations of composition and the fields of Harlingen and Leeuwarden have the higher N_2 content (figures are in percentages).

	N_2	CO_2	CH_4	C_2H_6	C_3H_8
Zuidwal	3.05	1.26	88.72	5.22	1.17
Harlingen	19.70	0.02	77.10	2.38	0.49
Leeuwarden	15.00	0.27	80.96	2.72	0.41

The occurrence of these gas reserves results from both the local geological conditions described above and the regional palaeogeography. We assume that the Zuidwal basin is the southernmost extension of the Netherlands Graben, which extends farther north through the "L" and "F" offshore blocks. However, the Netherlands Graben resulted from a subsidence which originated in Upper Triassic times and terminated at the end of the Jurassic, coupled with great halokinetic activity. The Zuidwal basin resulted from a later period of subsidence and contains only a Lower Cretaceous sequence. When the subsidence ended, the entire former subsiding area was progressively uplifted during the deposition of the Upper Cretaceous Chalk which reached its maximum thickness towards the west, i.e. on the former Texel High, and toward the east of the basin (Fig 2).

This Upper Cretaceous uplift and the additional Tertiary section which has a wide regional extension, provided good conditions of burial and trapping in the Zuidwal basin. In addition a

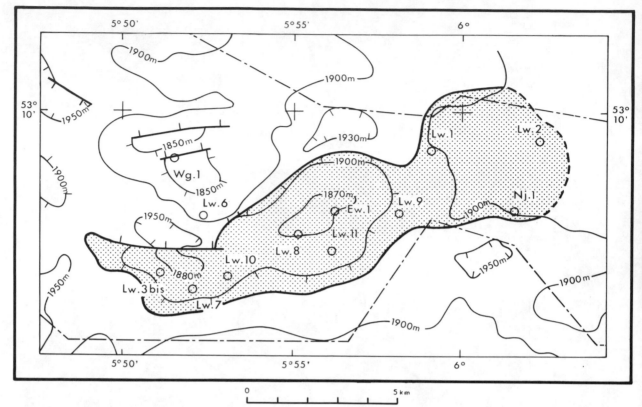

FIG 9.   Structural contour map on top of Lower Cretaceous in Leeuwarden gas-field. Contours in metres below sea-level.

relatively high degree of coalification, at a mean depth of 2000m, has been observed in the Lower Cretaceous series, mainly in the Wealdian sequence. The reflectance values range from 0.4 to 0.8 in the wells. These conditions allow us to conclude that the gas originates from the Wealdian coals and/or from the Neocomian clays though it may also come from the Carboniferous section *via* the Rotliegendes reservoir.

Regionally the Permian structural pattern is of a broad and low synclinal area, bordered by a higher area eastwards and also by pre-Cretaceous eroded areas to the south and west (Zuidwal anticline).

This structural pattern implies a high probability of complete migration of all pre-Cretaceous gas accumulations. However, we may assume that new gas production might have occurred from the Carboniferous source rocks in a later period. Part of this gas may have reached the Lower Cretaceous reservoir of the Zuidwal basin, particularly in the Zuidwal anticline where the Rotliegendes reservoir underlies the Cretaceous transgression.

31

Chalk of the North Sea

By J. M. HANCOCK

(Department of Geology, King's College, London)

and P. A. SCHOLLE

(US Geological Survey)

SUMMARY

Upper Cretaceous sediments are the most extensive Mesozoic deposits in the North Sea, being present almost everywhere more than 60km offshore except over inversion axes in the Southern North Sea Basin. The usual thickness is not more than 500m, but in grabens and marginal troughs it reaches 1000-1600m. North of the London-Brabant High almost the whole succession is in basinal facies. In the southern North Sea the Upper Cretaceous is mostly pure chalk, but northwards in the Central Graben the chalk becomes slightly argillaceous. In the Viking Graben chalk almost disappears and the succession is of clay, which itself becomes silty northwards. All these northward changes are more marked in the Santonian-Campanian. The clastics in the Viking Graben are probably derived from Greenland which may have had a more seasonal climate than north-west Europe.

Chalk is a distinctive limestone chiefly because it was deposited as low-magnesian calcite which is stable at surface pressures and temperatures. However, chalks undergo a consistent sequence of diagenetic changes as a result of pressure-solution and reprecipitation with increasing depth of burial. From isotopic analyses of the chalks it is possible to determine maximum depth of burial, palaeogeothermal gradients, and proximity to zones of deformation. In chalk reservoirs, oil migration into the rock can be dated because its entry stopped further diagenesis.

1. Distribution of facies and general geology

By J. M. HANCOCK

INTRODUCTION

The Upper Cretaceous of Europe north of the Alpine-Carpathian belt shows a broadly regular pattern of facies. Over upstanding regions of Precambrian and Palaeozoic rocks, variously called massifs, cratons, swells, platforms or highs, there is a set of standard facies which ranges from sediments deposited in fresh water (variegated facies) to those formed in seas 100m or a little more in depth (chalk marl facies) (Bushinskii, 1954; Cieśliński and Tröger, 1964; Diener, 1967; Hancock, 1974; Kahrs, 1927; Klein and Soukup, 1966; Marcinowski, 1974; Pożaryski, 1969; Voigt, 1929). During the Upper Campanian, the time of highest sea-level, most of these massifs were submerged. It is notable that of the few left unsubmerged, two of them, the Scottish Highlands and Norway, adjoin the North Sea. The thickness of clastic facies associated with massifs does not usually exceed 200m and may be much less, but in northern Czechoslovakia reaches 700m.

Between these old massifs were basins which subsided relative to the massifs, and independently of world changes of sea-level. In these basinal regions the dominant facies is a range of chalks. The usual basinal thicknesses go up to 400-500m.

Within the basins were major trenches in which the Chalk was exceptionally thick, reaching up to nearly 2000m in north Denmark (Sorgenfrei, 1969). Voigt (1963), who first realized the significance of these trenches, called them "randtröge" (marginal troughs) because they usually lie alongside massifs. Such marginal troughs (e.g. Sole Pit Basin and North Saxony Basin) developed during the Jurassic, but with time migrated sideways, so that the Upper Cretaceous trenches were developed alongside, not over, thick Jurassic sediments. The thick Jurassic is subsequently uplifted to form an inversion axis (e.g. Sole Pit Inversion, Pompeckj's Swell) in relation to which the new marginal trough is compensatory.

GENERAL DISTRIBUTION OF THE UPPER CRETACEOUS UNDER THE NORTH SEA

A glance at the isopach map (Fig 1) shows that under the North Sea there are a number of concealed massifs and highs.

(1) At the southern limits of the North Sea, between East Anglia and Belgium, is the concealed

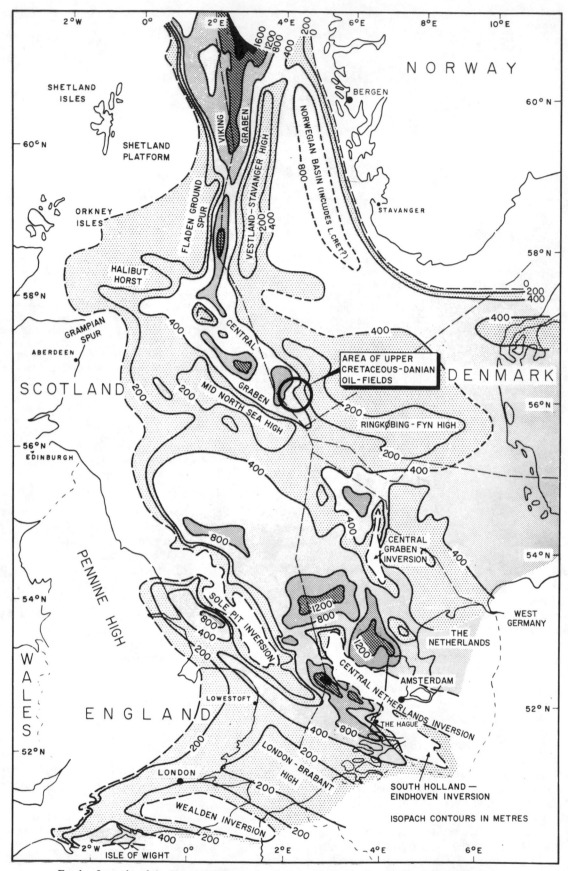

FIG 1.   Isopachs of the Upper Cretaceous combined with the Danian Chalk of the North Sea.

London–Brabant High, but this was already submerged in large part by the beginning of the Upper Cretaceous, all of which is represented by chalk.

(2) The *Mid North Sea High* was largely submerged in the Albian, and here also the Upper Cretaceous is developed as chalk. At a few places, *e.g.* the Auk field, there is Upper Cretaceous resting directly on Trias, but still the basal facies is chalk.

(3) I have no evidence from the *Ringkøbing–Fyn High*, but by analogy with the extension in Denmark the Upper Cretaceous will all be chalk or marly chalk.

(4) *Shetland Platform*, including Fladen Ground Spur and Halibut Horst: where Upper Cretaceous is known to be present—and then it is sometimes only Upper Campanian–Maastrichtian—it is almost wholly chalk.

(5) *Vestland–Stavanger High*: there is no certain evidence on facies available.

These massifs and highs occupy a minority of the North Sea leaving two major basinal regions, one in the north and one in the south, connected by a col along the line of the Central Graben between the Mid North Sea High and the Ringkøbing–Fyn High.

(1) Southern North Sea Basin

This really includes the East Midlands Shelf and the Anglo-Dutch Basin. With an area of about 180 000km², it is a similar area to the Danish–Polish Furrow, and about twice the size of the Paris Basin. With an area of about 100 000km² in which the thickness is 400m or more, it is also one of the bulkiest Upper Cretaceous basins in Europe, and certainly the bulkiest in the continent to be composed almost entirely of chalk.

The greatest thicknesses in the Southern North Sea Basin are more than 1600m which is reached in both the basin on the south-west flank of the Central Netherlands Inversion, and to the south-west of the Central Graben Inversion (Fig 2).

There are three inversion axes where the Upper Cretaceous is now absent, although the earlier Upper Cretaceous, at least, was deposited across the whole of these inversions. Each inversion involved two main phases of uplift, and each of these would have caused thinning of stages close to the inversion region.

Sole Pit Inversion. This has a compensatory trough and a faulted margin on its south-west flank. Even on the unfaulted north-eastern margin most of the effects of the inversion disappear very rapidly: as little as 25km from the region of total Upper Cretaceous removal there may be no obvious breaks in the succession. The gradual thinning of isopachs against the inversion implies that the earlier phase of movement (Turonian to Santonian) was probably the greater, but there was a second phase of uplift during the Maastrichtian.

Central Netherlands Inversion. This also has a compensatory trough and a faulted margin on its south-west flank. The first uplift here is usually recorded as Turonian (Heybroek, 1974), but the seas were shallower here than over the Sole Pit Inversion, and it is not clear how much of the Turonian break is due to local uplift, and how much is due to the general fall in sea-level during the Turonian. The first definite phase of inversion movement probably occurred in the Santonian, but some of the evidence of this was destroyed by the erosion which accompanied the stronger second phase of uplift during the Maastrichtian–lower Palaeocene.

Central Graben Inversion. (Dutch offshore region.) This shows a different pattern from the other two in the North Sea. The chief uplift is reported by Ziegler (1974) as being Maastrichtian–Upper Palaeocene, but, as he points out, this uplift was synchronous with "the last phase of downfaulting of the central and northern parts of the North Sea Central Graben", and thus presents a different pattern from most other inversion axes in northern Europe. I suspect, moreover, that movements were uneven along the length of the inversion.

Movements of the first phase of uplift are commonly grouped under the name "Sub-Hercynian Phase", and those of the second phase as "Laramide". As Heybroek (1974) has very rightly pointed out, implications of correlation between different inversions, such as Stille theory would imply, should be avoided. On land, where there is a rigorous stratigraphic control, movements that could be called "Sub-Hercynian", not only extended over several stages, *e.g.* in Poland from Upper Turonian to Santonian (Pozaryski, 1960), but within one region could occur at different times in different places. Nor was the pattern uniform within the area of a single inversion axis: we know that the West Netherlands Inversion was less strongly uplifted during the Upper Cretaceous than its extension in the Roer Valley Graben (Heybroek, 1974).

(2) Northern North Sea Basin

There are three major subdivisions of this large basinal region: the *Viking Graben*, the *Central Graben*, and the *Norwegian Basin*. Very little information is, as yet, available about the last. The Viking and Central Grabens are true fault-bounded troughs.

The rift system of the Central Graben had started to develop during the Triassic, and movements continued intermittently into the early Palaeocene (Ziegler, 1974), although there was little rifting actually during the Cretaceous (Parker, 1975). The 400-m isopach roughly marks the boundary of the graben, and the maximum thickness is slightly in excess of 1400m. Similar remarks could be made about the Viking Graben, although this is more elongated, and the maximum thickness barely exceeds 1200m except in the far north.

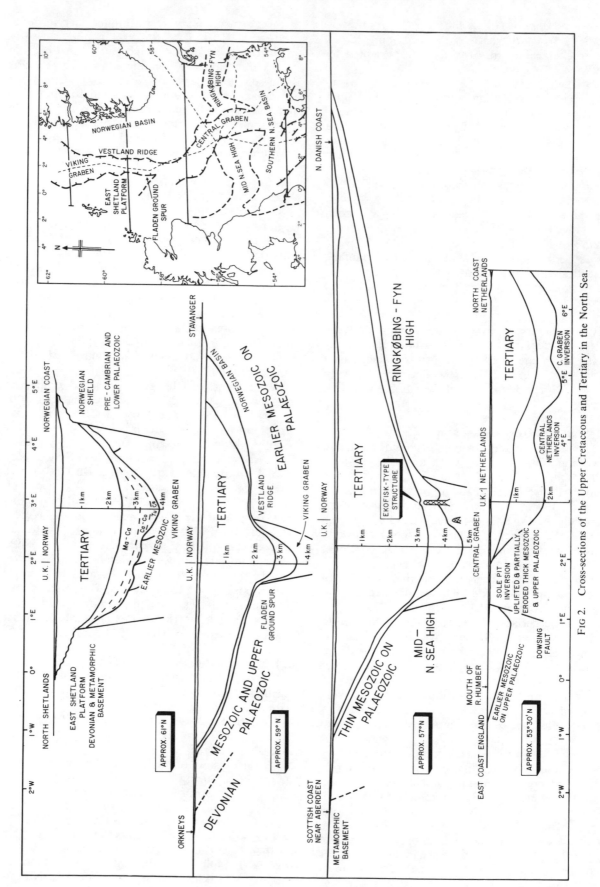

FIG 2.   Cross-sections of the Upper Cretaceous and Tertiary in the North Sea.

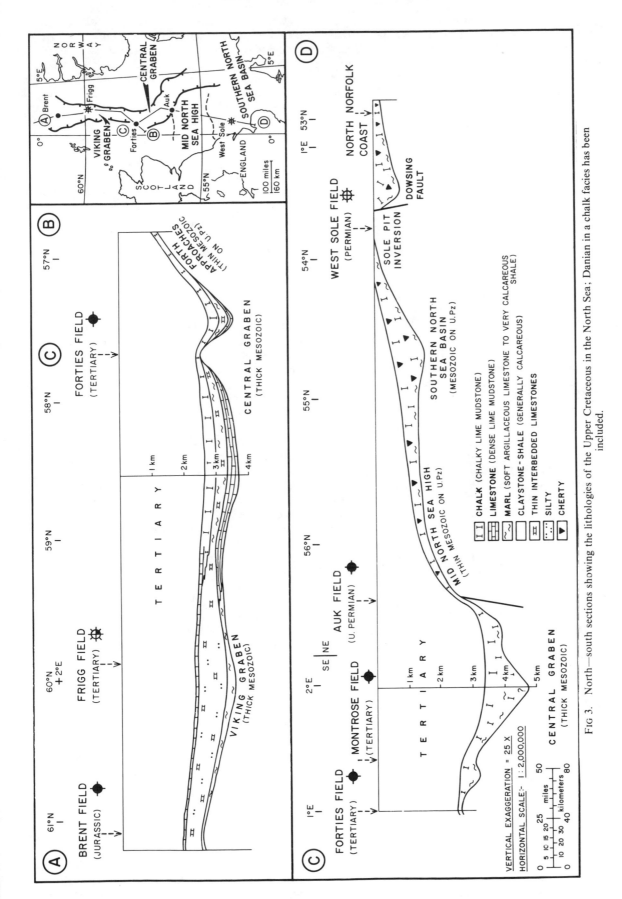

Fig 3.   North—south sections showing the lithologies of the Upper Cretaceous in the North Sea; Danian in a chalk facies has been included.

In most of the Central Graben the Upper Cretaceous is represented by chalks, but northwards of about latitude 57° lateral facies changes begin (Fig 3). These involve a decrease in the chalk component and the incoming of a clastic component. As one goes further north along the Viking Graben, more and more of the succession is clay (Howitt, 1974). In the northern part of the Viking Graben there is hardly any chalk at all.

CHALK OF THE SOUTHERN NORTH SEA

The bulk of the Upper Cretaceous in the southern North Sea is chalk. This is a sediment dominantly composed of calcitic crystals secreted by planktonic marine algae of the class Haptophyceae (Christensen, 1962). Most of these calcite crystals are tablet-shaped (called plates or laths), and in the living algae are arranged in rings called coccoliths, in which the calcite plates often overlap one another (see McIntyre and Bé, 1967). In some chalks a high proportion of the coccoliths are still preserved (*e.g.* Frey, 1972, Pl 15, Fig 1), whilst in others most of the coccoliths have broken into their component plates (*e.g.* Hancock and Kennedy, 1967, Pl 3, Fig a).

The most distinctive feature of chalk is that the calcite of the coccolith plates is low magnesian (Thompson and Bowen, 1969), which means that it is stable at ordinary temperatures and pressures (discussion in Bathurst, 1971), and is effectively protected from pressure-solution at loads up to 1000m overburden by the Mg content of sea-water (Neugebauer, 1973; 1974). Although coccolith-rich sediments are known today in which there are also large amounts of aragonite and high magnesian calcite, *e.g.* in British Honduras (Scholle and Kling, 1972), the bulk of the Upper Cretaceous chalk of Europe lacked significant quantities of both aragonite and high magnesian calcite at the time of deposition. The result is that chalk is free from the type of early lithification which affects most shelf-sea carbonate sediments. Compared with the much studied carbonate sediments of the Caribbean and the Persian Gulf, the Chalk was deposited in deeper and colder water. For pure chalks the depth was seldom less than 80m, more commonly 200–300m, and possibly reached 1000m or more in places.

For fuller discussion of conditions of chalk deposition reference should be made to Black (1953), Carter (1972), Hancock (1974), Hancock (*in press*), Kennedy and Garrison (*in press*), Reid (1968), Scholle (1974) and Wolfe (1968).

There are a number of sub-facies of chalk, most of which were formed in the shallower range of chalk deposition (Hancock, 1974; Kennedy and Garrison, *in press*). Because the southern North Sea was already submerged when chalk deposition started in the Cenomanian, the water depths in this region were generally considerable, and deeper-water chalk facies predominate, although it has been impossible to date to make a proper survey of the evidence available. Nevertheless, the earlier phase of uplift of all the inversion axes would have been accompanied not only by submarine erosion, but probably by the formation of such facies as hardground chalk. Shallower water chalk facies are known to occur in the Santonian–Campanian near the Central Netherlands Inversion. Unfortunately, much of this would have been removed by erosion during the second phase of uplift.

Those familiar with the Chalk of southern England would find that of the southern North Sea different, chiefly in that much of it, but not all, is a hard limestone like the Chalk of the Yorkshire coast, in spite of it having been deposited as low magnesian calcite. This is a diagenetic effect, and it is variation in forms of diagenesis that has provided most of the lateral and vertical variation in the lithology of the North Sea Chalk. This is discussed in the second half of the paper.

In addition to chalks there are a few other facies in the southern North Sea. Thin marls at the top of the Cenomanian, which are presumably the Plenus Marls, are widespread. Near the Central Netherlands Inversion there are mid-Senonian shallow marine clastic facies (Ziegler, 1974). The Maastrichtian in the vicinity of inversions, especially near the unfaulted margins, probably includes tuffeau facies similar to the Maastricht Tuffeau.

Over most of the Southern North Sea Basin all the stages of the Upper Cretaceous are represented. The proportional thicknesses are essentially the same as are found in Yorkshire and north Norfolk, but the absolute thicknesses reached are sometimes greater than in Yorkshire, and much greater than those in Norfolk which are reduced by the underlying London–Brabant High. On the other hand the total thicknesses of the Upper Cretaceous are much greater than that found in Yorkshire, largely because on the land the younger parts of the Cretaceous have been removed by erosion. In the southern North Sea the Maastrichtian stage alone provides around 20 to 50 percent of the total thickness. Moreover, chalks continue into the Danian (Lower Palaeocene).

CHALK OF THE CENTRAL GRABEN

Most of the Upper Cretaceous in the Central Graben is chalk, and it is only north of about latitude 57° that marly chalks become prominent in the middle of the succession (Fig 3). However, even south of 57°, and perhaps more particularly to the east, the Santonian–Campanian Chalk contains a small but appreciable amount of clay. Harper and Shaw (1974) report that in the Norwegian Chalk reservoirs there is up to 20 percent or more of clay and clay-sized quartz particles in the Senonian, which would be about four times the quantity of clay in the Santonian–Lower Campanian of Yorkshire. The clay is sometimes red-brown such as would be provided by erosion of an Old Red Sandstone facies. Purer chalks occur everywhere in the Maastrichtian, with clay contents probably below 4 percent.

There is some indication that part of the Santonian–Campanian occurs in 10–30cm rhythms, related to varying proportions of the clay and chalk components. Such rhythms become accentuated by early carbonate-solution-diagenesis so that the top and bottom of each cycle is marked by concentrations of anastomosing wispy marl seams (sometimes called "horsetails") (Plate I, d). Such wispy marl seams show a pattern similar to those in the Upper Cenomanian of Sussex rather than those in the Santonian–Lower Campanian of Yorkshire where most of the concentration of clays is accompanied by, or follows, stylolite development; stylolites are not prominent in the Santonian–Campanian of the Central Graben.

The clay content also shows up a variety of burrows, much of it indeterminate mottling, but *Chondrites* and *Zoophycos* can be recognized. The occasional flints are usually blue-grey and often zoned. Patches of pyrite are sometimes common; it occurs as clusters of discrete crystals rather than the familiar pyrite nodules of the Cenomanian in southern England.

The thicknesses of the individual stages of the Upper Cretaceous show some important differences from those in the Southern North Sea Basin. In the northern parts of the Central Graben the Cenomanian and Turonian are absent, or so thin as to be unrecognized palaeontologically in wells. Only in the central region of the Graben does one find combined Cenomanian–Turonian thicknesses that run up to several hundred metres. In contrast, the Coniacian, where it can be distinguished, is unusually thick, and may reach 300m. The Santonian–Campanian–Maastrichtian thicknesses show similar proportions between themselves as are found in the Southern North Sea Basin, the Maastrichtian being the thickest of the three.

Although there is a tendency to have data from structural highs where representations are thin, such as is illustrated by Harper and Shaw (1974, Fig 4), the common absence of Cenomanian and/or Turonian is too widespread to be explained in this way. It seems a genuine oddity when one considers that we are discussing a basinal region not dependent on clastics for deposition. It should be remembered also that the Coniacian is often decidedly thick.

R. Eckert and H. Soediono (personal communications) assume that the Cenomanian–Turonian strata were deposited in deeper water which might suggest the sea-bed was below a carbonate compensation depth in parts of the northern North Sea region. There is insufficient space here to discuss this properly, but it is a hypothesis that should be considered. The wispy marls of the Santonian–Campanian show similarities to the subsolution phenomena in believed deep-water carbonates described by Garrison and Fischer (1969). There are clear signs of corrosion on coccolith plates in the Central Graben Chalk, which is not seen in southern England.

These features are not sufficient to support a compensation depth control on the thickness of the Cenomanian–Turonian. The Central Graben is not peculiar in this pattern of thicknesses. The Danish trough has a maximum Cenomanian of only 45m, combined with a Senonian–Maastrichtian which reaches more than 1700m of carbonates (Sorgenfrei and Buch, 1964). In fact, unusual thicknesses of the Campanian–Maastrichtian stages compared with earlier stages, and particularly when compared with Cenomanian and Turonian, are a feature of pelagic sediments of several regions in the Pacific (Douglas, 1973, Schlanger et al., 1973) It seems that we are seeing the effects of varying rates of plankton production.

Carbonate sedimentation, including chalk, continues into the Palaeocene in the Torfelt–Ekofisk area (Harper and Shaw, 1974). Elsewhere the Lower Palaeocene is more marly or even a shale, but it includes a formation with considerable quantities of carbonate in it. This is the "mélange" which is a slump deposit along the borders of the south-eastern part of the Central Graben, and up to 20km from the faults where the material has slid from. The lumps are of Maastrichtian and Danian age in an upper Danian matrix. Much of the rock is an unsorted conglomerate of pebbles of chalk, blue-grey to white flint, and occasional glauconitic or chamositic marl in an unbedded but often irregularly streaked chalky argillaceous matrix. Some of the chalk pebbles are composite which would indicate that there was more than one period of slumping. The style of slumping suggests that the matrix was still near-liquid in properties. It differs in many respects from known slump deposits in the Upper Cretaceous of western Europe as have been described by Giers (1958), Voigt (1962) and Kennedy and Juignet (1974), notably in the slurry state of the matrix and the lack of consolidation of the pebbles at the time of slumping.

THE VIKING GRABEN

The Viking Graben contains the largest body of Upper Cretaceous clastic material in Europe north of the Alpine–Carpathian belt. As will be seen from Fig 3, it contains typically about 1000m of shales and clay sandwiched between marly sediment at the base and marly limestones (which include chalk) at the top, but the vertical changes are gradational. Even more commonly than in the Central Graben, the Cenomanian and Turonian are thin or missing, although their combined thickness can run up to nearly 300m. The data tend to be biassed towards areas where the Lower Cretaceous is thin or missing because of early Cretaceous tectonic uplift (so-called late Cimmerian movements) but this does not seem to be the explanation for the thinness of the Cenomanian–Turonian. The base of the shales or clay is already Coniacian or Santonian, and the bulk of the succession is Campanian–Maastrichtian.

As one goes north in the Viking Graben the chalk component gradually disappears, although even at the northern end there can still be a few metres logged as chalk high in the Maastrichtian. Not only does the clastic component increase northwards, but the ratio of silt to clay increases; the coarsest detritus is quartz-sand which occurs as rare streaks in the Santonian–Campanian. A few aragonitic bivalves, with the aragonite apparently preserved, have been seen in the Maastrichtian. There is little sign of burrowing.

The closest described facies to this argillaceous development is probably lithology 8 of Kauffman (1969) as represented by the Hartland Shale in central Colorado. In that region the Hartland Shale grades upwards into Blue Hill Shale (Kauffman's lithology 6) which is a slipper-clay facies (Hancock, 1974), like the Gault clay of southern England and northern France. Compared with the average Gault, the Viking Graben clays tend to be less silty, less glauconitic, and more micaceous. The slipper-clay facies is believed to have been deposited in around 120–200m of water.

Such a large volume of clastic material is unusual in the Upper Cretaceous of northern Europe, and is unique in the Campanian–Maastrichtian. High sea-levels during the Campanian meant that there was little land available for erosion, and even with falls of sea-level in the Maastrichtian, the usual near-shore sediments were carbonates. The few examples of upstanding massifs in the Campanian failed to provide any noticeable amounts of detritus, probably because the climate was non-seasonal (Wilson, 1973). Such conditions existed in the Hebridean region of Scotland, in Northern Ireland, around the Ardennes Massif in Belgium, and in southern Sweden. The sediments in the Viking Graben are not merely there because of the relative proximity of land masses in the Scottish Highlands and Norway, but are a reflection of a seasonal climate that allowed erosion of the land. The source of the detritus may well have been Greenland since it was as close as Norway at the time (Briden et al., 1974; Smith et al., 1973). The Scottish Highlands is the least likely of these three: because of the known thinness of the Campanian in the Hebrides; because it is difficult to visualize transport of silt and clay across the Shetlands Platform which lacks an appropriate clastic succession; and because it was farther away than Norway or Greenland. To some degree, derivation from Norway would involve the problem of transport across the Vestland–Stavanger High. It will be seen from the isopachs (Fig 1) that the Viking Graben not only extends north of the region of recent North Sea exploration, but is actually thickening northwards. This supports a Greenland provenance for the sediment in the Viking Graben.

ACKNOWLEDGEMENTS

I am much indebted to Shell UK Exploration and Production Ltd for the generous provision of data and samples from their wells, and for draughting Figs 1–3. I am grateful to Mr Peter Woods and Dr H. J. Milledge of the Department of Crystallography at University College, London, for the use of their scanning electron microscope, and to Mr Ray Parish for photographic work.

2. Chalk diagenesis
By PETER A. SCHOLLE

Because chalks are essentially micro-coquinas of coccoliths and planktonic foraminifera, and because, in many cases, they have only small amounts of terrigenous material, they often act as simple and stable chemical systems. With the bulk of the material in chalks having originated as low-Mg calcite and with widespread uniformity of chalk facies, these sediments react to changes in environmental conditions (temperature, pressure, water chemistry) quite differently from the more frequently studied shallow-water carbonates which are generally composed of aragonite and high-Mg calcite. These latter sediments are unstable in fresh water: thus, even the small fluctuations of sea-level which commonly occur during deposition will suffice to bring about wholesale dissolution, cementation and other alteration of these sediments. Complex depositional patterns coupled with irregular patterns of water input and flow yield extremely varied and complex patterns of alteration and porosity occlusion.

Chalks, on the other hand, are generally deposited in deeper waters and thus probably see less fluctuation in pore-water chemistry. Furthermore, even when meteoric waters enter chalks the effects are relatively less important than in the Bahamian-type sediments. Chalks in onshore areas often act as aquifers for meteoric water flow and yet have suffered little alteration by those waters. Because chalks have uniform facies with wide distribution and normally see changes in water chemistry only during later diagenetic, broad regional uplift, they show patterns of diagenesis which are far simpler and more regional than those of most other carbonate rocks.

The two main processes which are most important in chalk diagenesis are sea-floor cementation and burial cementation. Sea-floor cementation is the formation of "hardgrounds" or syn-sedimentary lithification surfaces at the sea-floor during hiatuses in sedimentation. Considerable work has been done on this type of alteration (Bromley, 1968; Kennedy and Garrison, *in press*). Although it is a very widespread process affecting chalks, it achieves real significance in the cementation of chalks only in certain zones which are recognizable on the basis of chemical as well as physical and biological criteria (Scholle and Kennedy, 1974). In most chalks, the processes of burial diagenesis appear to be more important than any near-surface processes.

Burial diagenesis involves progressively increasing overburden pressure which brings about dissolution of slightly more unstable components (certain species of foraminifera or coccoliths), or smaller crystals, or those portions of crystals which are directly at grain-to-grain contacts. This carbonate goes into solution and is reprecipitated very locally in a suitable micro-environment (within foraminiferal chambers or as overgrowths on coccolith plates). These processes have been observed in nannofossil oozes and chalks of the Pacific Ocean (Schlanger *et al.*, 1973) and the same processes can be shown to have affected both onshore and offshore European chalks.

In the present study, samples have been collected from fresh outcrops in England and Northern Ireland and from 20 wells in the North Sea. Both cores and cuttings were used from the offshore areas. Samples were examined by means of scanning electron microscopy, thin-section microscopy, porosity-permeability analysis and oxygen isotopic determination. The chalks from both onshore and offshore areas yielded patterns of alteration which appear to be internally consistent and which seemingly are a response to regional patterns of tectonism and fluid migration.

The most obvious relationship found was a correlation between increased depth of burial and decreased porosity, decreased permeability and increased sample hardness. Chalks from onshore areas which had never seen significant overburden or tectonic stresses and samples from offshore areas where the chalk was near the surface showed porosities in the range of 38 to 48 percent (with corresponding permeability of 4 to 13md). SEM photos (Plate I, a) show a porous, non-interlocking mass of irregularly-shaped crystals derived from

the breakdown of skeletal organisms (mainly coccoliths).

Burial to progressively greater depths apparently yields a continuous loss of porosity so that by the 1500–2000m (5000 to 7000ft) level, most samples have porosity values in the 15 to 30 percent range (with permeabilities of 0.1 to 1md). SEM photos (Plate I, b) show a rock with considerable overgrowth of original crystals yielding rather euhedral crystal shapes with considerable interlocking of adjacent crystals, although there is still extensive open pore space.

Depths of burial in the 2700–3300m (9000 to 11 000ft) range have led to a very great loss of porosity. Samples in that depth range normally show porosity values of 2 to 25 percent and permeabilities of 0 to 0.5md unless porosity and permeability have been enhanced or retained by one of a number of factors such as fracturing or early oil entry into the rock. SEM photos (Plate I, c) generally show complete welding together of crystals with virtually total loss of porosity, although the outline of original coccolith or foraminiferal fragments can still be seen locally.

The measurement of oxygen isotopic values of bulk chalk samples leads to a better quantitative understanding of the processes involved in the previously observed qualitative sequence of chalk burial–alteration. Figure 4 shows the relationship between porosity and oxygen isotopic values (δO^{18}) for onshore chalks from England and Northern Ireland and for cores from DSDP Leg 12, site 116 in the Hatton–Rockall Basin of the North Atlantic. The line fitted to the data is a least squares fit based only on the onshore samples of Upper Cretaceous chalks. The DSDP samples (which are Tertiary) have been added for reference only and the oxygen values of these samples have been plotted as 2 per mil more negative than they really are to correct for initial differences between Cretaceous and Tertiary palaeotemperatures recorded by the coccoliths and foraminifera. This correction is clearly only an approximation and it does not affect the slope of the line.

The clear relation (correlation coefficient 0.72) between loss of porosity and increasingly negative oxygen isotopic values is an effect of progressive burial diagenesis. As a chalk ooze with an initial porosity of 80 percent is buried and compacted the original biological material, which records a depositional palaeotemperature, is partly dissolved and reprecipitated in equilibrium with a sequence of higher temperatures at depth. Thus, the bulk isotopic value of a chalk reflects the amount of unaltered original sediment still present plus an average of the isotopic values of all the small increments of cement which were deposited in that rock throughout its burial history.

Figure 4 amplifies this interpretation. With increased burial (and loss of porosity) the isotopic values of chalks take on progressively more negative values. The data in Fig 4 are based on

PLATE I.   See text for explanation.

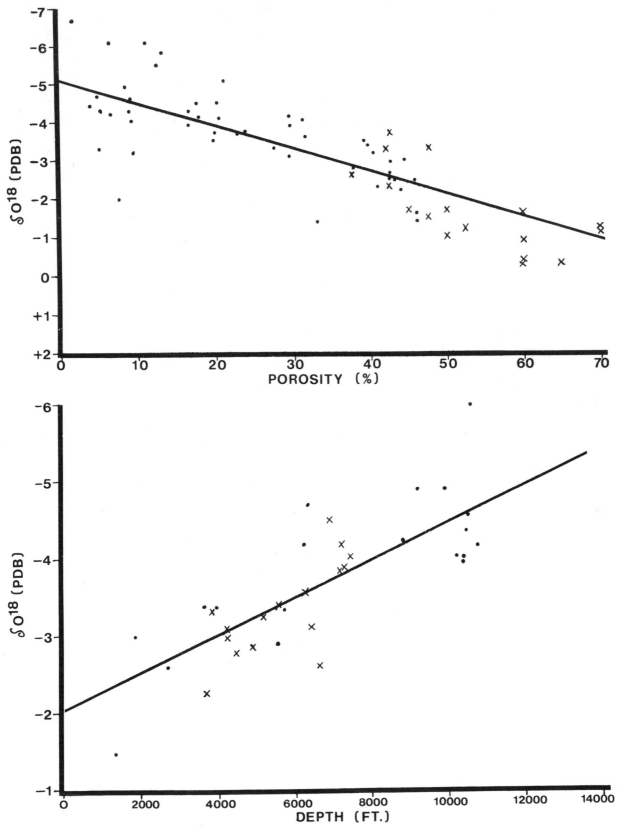

FIG 4. See text for explanation.

averages of data for 20 North Sea wells. Also, individual analyses for a single well are plotted on the same diagram. The data are based largely on cuttings. The scatter in the data may be due to a number of factors such as caving of samples, uplift of the section, early (sea-floor) cements, retention of porosity due to the presence of oil, and abnormal geothermal gradients, in addition to analytical error. In fact, the data are remarkably consistent (correlation coefficient of 0.76) despite all the potential errors.

This may offer several interesting applications. First, these isotopic values record the maximum depth of burial (pressure and temperature) seen by the section and they can be used for regional determinations of basinal subsidence and uplift. Also, if enough data are available (even from cuttings), one can generally determine a palaeogeothermal gradient. Areas of high palaeogeothermal gradients have a much steeper depth isotope ratio relation than that shown on Fig 4. This work can be done on material which has had no special sample handling, an advantage over many geochemical or palynological techniques.

Another application is the timing of oil entry into chalk reservoirs. While thermal maturation of organic material continues, chalk diagenesis is largely terminated by oil entry into the rock. The reasons are complex but reduction of differential pressure within the rock and restriction of water flow clearly are important factors. In some North Sea wells where oil is produced from chalk, the chalks have anomalously high porosities and anomalously positive isotopic values—both indicate oil may have entered the rock much earlier in its burial history. In the Ekofisk area isotopic data indicate a burial depth of 1800–2400m (6000 to 8000ft) at the time of oil entry.

Finally, the isotopic values of chalks also respond to the differential stresses encountered in tectonic zones. Preliminary data indicate that in both onshore and offshore areas one can detect major zones of deformation by isotopic gradients in chalks from considerable distances from the deformed area.

Thus, an understanding of chalk diagenesis in an area such as the North Sea can provide a number of advantages. It allows the prediction of petrophysical characteristics of chalks at depth, predicts patterns of tectonism, determines maximum depth of burial and palaeogeothermal gradients, and, in special cases allows determination of the time of oil migration.

ACKNOWLEDGEMENTS

I would like to express my thanks to numerous oil companies, especially Shell, Cities Service, Esso, Phillips and Conoco for their contributions of samples and, in some cases, funds for the completion of this study. The bulk of the analytical work was done through Shell Development Co. in Houston, Texas with the very great help of Dr R. M. Lloyd. Support for fieldwork and other analyses came from National Science Foundation Grant GA-36696.

EXPLANATION OF FIG 4 AND PLATE I

Fig 4

Upper graph: Plot of porosity *versus* oxygen isotopic ratio (relative to PDB standard). Dots represent Upper Cretaceous outcrop samples from various regions in England and Northern Ireland, and the solid line is a least squares fit to that data. The crosses represent samples from the subsurface Tertiary section in DSDP Leg 12, site 116 core. These samples have been corrected for an average temperature difference between the Upper Cretaceous and middle Tertiary of about 10°C. Thus, their oxygen isotopic values are plotted as 2 per mil more negative than the actual analytical results. This does not affect the slope of the data points.

Lower graph: Plot of depth *versus* oxygen isotopic ratio (relative to PDB standard). Dots represent the average of numerous data points for a single North Sea well; each dot stands for a different well. The crosses represent individual analyses on cuttings from a single well, Conoco/Petroland L/16-1 in the Netherlands North Sea. The solid line is a least squares fit to the dots only.

Plate I. Chalks of the North Sea

a. Scanning Electron Photomicrograph of a chalk from the Conoco/NCB well 49/21-3. Sample depth is 378m. Scale bar is about 5 microns. Note high porosity, irregular, anhedral grain shapes dominated by coccolith-derived material.

b. Scanning Electron Photomicrograph of a chalk from the Shell–Esso well 30/16-3. Sample is from 2231m depth. Scale bar is about 5 microns. Matrix grains mostly overgrown and now show rather euhedral outlines. Considerable porosity loss has occurred but some intercrystalline porosity remains.

c. Scanning Electron Micrograph of a chalk from Shell–Esso well 30/19-1. Sample is from 3100m depth. Crystals are almost completely interlocked and porosity has been nearly entirely obliterated yet coccolith origin of most grains is still recognizable.

d. Chalk rhythm of varying proportions of calcareous plankton and clay; the top and bottom of the rhythm has been accentuated by post-depositional solution of the carbonate. Probably Campanian. Natural size. Shell–Esso well 30/16-1; depth 2211m.

REFERENCES

Bathurst, R. G. C. 1971. *Carbonate Sediments and their Diagenesis, Developments in Sedimentology 12.* Elsevier, Amsterdam. 620 pp.

Black, M. 1953. The constitution of the Chalk. *Proc. geol. Soc.,* **1499,** lxxxi–lxxxvi.

Briden, J. C., Drewry, G. E. and Smith, A. G. 1974. Phanerozoic equal-area world maps. *Jl. Geol.,* **82,** 555–74.

Bromley, R. G. 1968. Burrows and borings in hardgrounds. *Meddr. dansk. geol. Foren.,* **18,** 247–50.

Bushinskii, G. I. 1954. Lithology of Cretaceous deposits of the Dnieper–Donets basin. *Akad. nauk SSSR, Trudy Inst. geol. nauk.,* **156,** 1–307 [in Russian].

Carter, R. M. 1972. Adaptations of British Chalk Bivalvia, *Jl. Paleont.,* **46,** 325–40.

Christensen, T. 1962. Alger. *Botanik,* **2** (2), 178 pp.

Cieśliński, S. and Tröger, K.-A. 1964. Epikontynentalna kreda górna Europy środkowej. *Kwart. geol. (Warszawa),* **8,** 797–809.

Diener, I. 1967. Zur Paläogeographie der Oberkreide Nordost Deutschlands. *Ber. deutsch. Ges. geol Wiss., ser. A (Geol. Paläont.),* **12,** 493–509.

Douglas, R. G. 1973. Planktonic foraminiferal biostratigraphy in the central north Pacific ocean. *Initial Rep. Deep-sea Drill. Proj.,* **17,** 673–94.

Frey, R. W. 1972. Paleoecology and depositional environment of Fort Hays Limestone Member, Niobrara Chalk (Upper Cretaceous), west-central Kansas. *Paleont. Contr. Univ. Kans.,* Art. 58 (Cretaceous III), 72 pp.

Garrison, R. E. and Fischer, A. G. 1969. Deep-water limestones and radiolarites of the Alpine Jurassic, *Spec. Publs. Soc. econ. Paleont. Miner. Tulsa,* **14,** 20–56.

Giers, R. 1958. Die Mukronatenkreide in östlichen Münsterland. *Beih. geol. Jb.,* **34,** 148 pp.

Hancock, J. M. 1974. The sequence of facies in the Upper Cretaceous of northern Europe compared with that in the Western Interior, *in* Caldwell, W. G. E. (ed.), Cretaceous System in the Western Interior of North America—Selected Aspects, *Spec. Pap. geol. Ass. Can.,* **13.**

Hancock, J. M. *(in press).* Chalk, *in* Fairbridge, R. W. (ed.), *Encyclopedia of Earth Sciences,* 6, Sedimentology.

Hancock, J. M. and Kennedy, W. J. 1967. Photographs of hard and soft chalks taken with a scanning electron microscope, *Proc. geol. Soc.,* **1643,** 249–252.

Harper, M. L. and Shaw, B. E. 1974. Cretaceous–Tertiary carbonate reservoirs in the North Sea. *Offshore North Sea Technology Conference, Stavanger, Norway,* G-IV/4.

Heybroek, P. 1974. Explanation to tectonic maps of the Netherlands. *Geologie Mijnb.,* **53,** 43–50.

Howitt, F. 1974. North Sea oil in a world context, *Nature, Lond.,* **249,** 700–3.

Kahrs, E. 1927. Zur Paläogeographie der Oberkreide in Rheinland-Westfalen. *Neues Jb. Miner. Geol. Paläont. BeilBd B.,* **58,** 627–87.

Kauffman, E. G. 1969. Cretaceous marine cycles of the western interior. *Mount. Geologist,* **6,** 227–45.

Kennedy, W. J. and Garrison, R. E. *(in press).* Morphology and genesis of nodular chalks and hardgrounds in the Upper Cretaceous of southern England, *Sedimentology.*

Kennedy, W. J. and Juignet, P. 1974. Carbonate banks and slump beds in the Upper Cretaceous (Upper Turonian–Santonian) of Haute Normandie, France. *Sedimentology,* **21,** 1–42.

Klein, V. and Soukup, J. 1966. The Bohemian Cretaceous basin, *in* Svoboda, J. *et al.* (Eds.), *Regional Geology of Czechoslovakia, part 1, the Bohemian Massif,* Ústred. ústav geolog., 487–512.

Marcinowski, R. 1974. The transgressive Cretaceous (Upper Albian through Turonian) deposits of the Polish Jura Chain. *Acta geol. Pol.,* **24,** 117–217.

McIntyre, A. and Bé, A. W. H. 1967. Modern Coccolithophoridae of the Atlantic Ocean. 1. Placoliths and Cyrtoliths. *Deep-Sea Res.,* **14,** 561–97.

Neugebauer, J. 1973. The diagenetic problem of chalk—the role of pressure solution and pore fluid. *Neues Jb. Geol. Paläont. Abh.,* **143,** 223–45.

Neugebauer, J. 1974. Some aspects of cementation in chalk. *Spec. Publs int. Ass. Sediment.,* **1,** 149–76.

Parker, J. 1975. Lower Tertiary sand development in the central North Sea. *Paper 35, this volume.*

Pożaryski, W. 1960. An outline of stratigraphy and palaeogeography of the Cretaceous in the Polish lowland. *Pr. Inst. geol. (Warszawa),* **30,** 377–440.

Reid, R. E. H. 1968. Bathymetric distributions of Calcarea and Hexactinellida in the present and the past. *Geol. Mag.,* **105,** 546–59.

Schlanger, S. O., Douglas, R. G., Lancelot, Y., Moore, T. C. and Roth, P. H. 1973. Fossil preservation and diagenesis of pelagic carbonates from the Magellan Rise, central north Pacific Ocean. *Initial Rep. Deep Sea Drill. Proj.,* **17,** 407–27.

Scholle, P. A. 1974. Diagenesis of Upper Cretaceous chalks from England, Northern Ireland, and the North Sea, *Spec. Publs int. Ass. Sediment.,* **1.**

Scholle, P. A. and Kennedy, W. J. 1974. Chalk diagenesis. *Abstr. Progr. geol. Soc. Amer.,* **6** (7), 943.

Scholle, P. A. and Kling, S. A. 1972. Southern British Honduras: lagoonal coccolith ooze. *Jl. sedim. Petrol.,* **42,** 195–204.

Smith, A. G., Briden, J. C. and Drewry, G. E. 1973. Phanerozoic world maps, *in* Hughes, N. F. (ed.), Organisms and continents through time. *Spec. Pap. Palaeontology,* **12,** 1–42.

Sorgenfrei, T. 1969. Geological perspectives in the North Sea area. *Meddr dansk geol. Foren.,* **19,** 160–96.

Sorgenfrei, T. and Buch, A. 1964. Deep Tests in Denmark 1935–1959. *Danm. geol. Unders.,* **36** (III), 1–146.

Thompson, G. and Bowen, V. T. 1969. Analyses of coccolith ooze from the deep tropical Atlantic. *Jl. mar. Res.,* **27,** 32–38.

Voigt, E. 1929. Die Lithogenese der Flach- und Tiefwassersedimente des jüngeren Oberkreidemeeres. *Jb. halle. Verb. Erforsch. mitteldt. Bodenschätze,* **8,** 1–165.

Voigt, E. 1962. Frühdiagenetische Deformation der turonen Plänerkalke bei Halle/Westf. *Mitt. geol. StInst. Hamb.,* **31,** 146–275.

Voigt, E. 1963. Über Randtröge vor Schollenrändern und ihre Bedeutung im Gebiet der Mitteleuropäischen Senke und angrenzender Gebiete. *Z. dt. geol. Ges.,* **114** (for 1962), 378–418.

Wilson, L. 1973. Variations in mean annual sediment yield as a function of mean annual precipitation. *Am. Jl. Sci.,* **273,** 335–49.

Wolfe, M. J. 1968. Lithification of a carbonate mud: Senonian Chalk in Northern Ireland. *Sedim. Geol.,* **2,** 263–90.

Ziegler, P. A. 1974. The geological evolution of the North Sea area in the tectonic framework of northwestern Europe. *Proc. Bergen North Sea Conference, Norg. geol. Unders.*

DISCUSSION

John Church (Robertson Research International Ltd): Could Dr Hancock comment upon the origin of the pink staining that occurs at several horizons in the Chalk of the North Sea. In addition can he suggest what environment prevailed during the deposition of the Lower Campanian and Santonian in the Viking Graben.

Dr J. M. Hancock: Pink staining in Chalk under the North Sea is irregular in distribution (as it is in the Chalk on land). Its association with the clay content of the Chalk suggests a derivation from older pink rocks exposed to erosion. A superficial examination of the colour in chalk of the Central Graben gives a better match with Old Red Sandstone than New Red Sandstone, for what that is worth!

The little that I can say about the environment of the Viking Graben is mostly included in the paper. In so far as the Santonian and Lower Campanian may have differed from the Upper Campanian and Maastrichtian, there is some indication from the fauna that the earlier sediments were deposited in deeper water, more than 200m, but I would not hazard a figure of maximum possible depth.

Dr A. Hallam (Oxford University): Could Dr Scholle explain why there is a correlation between δO_{18} values and depth of burial in the North Sea Chalk, which appear also to correlate with progressive increase in interstitial cementation as a result of pressure solution of the coccolith constituents. The best-documented fractionation processes are those involving a change from marine to continental environments, but the system in question has remained beneath the sea since its formation.

Dr P. A. Scholle: While fresh-water alteration could produce the observed isotopic changes, the geological setting and observed depth-related patterns of alteration make this interpretation untenable. The isotopic shift, more reasonably, is a result of dissolution of original sedimentary particles and their reprecipitation as cement. The original grains are isotopically "heavy" and reflect Cretaceous and Tertiary sedimentary palaeotemperatures. The reprecipitated cements, however, are isotopically "lighter" as they are precipitated at higher temperatures encountered in the subsurface. Thus with progressive burial and cement formation, the average oxygen isotopic value of the chalk gets progressively lighter as the pore-waters get isotopically heavier. Exact values depend on whether the pore-fluid system is open, half-open, or closed.

Dr N. V. Dunnington (Cluff Oil Co.): Can any information be provided on the relative timing of fracturing, entry of oil into reservoir and stylolite development, in any of the fields concerned?

Dr P. A. Scholle: The Ekofisk samples that I have examined show preserved primary porosity and abnormally heavy oxygen isotopic values, which indicate that oil entered the rock rather early in its diagenetic history or that the reservoir zone became over-pressured early, presumably before 200m burial depth. I have no data on fracture and stylolite timing.

R. A. James (BP Development Ltd): Does Dr Scholle think that the Upper Cretaceous could have acted as an effective cap-rock anywhere in the northern North Sea?

Dr P. A. Scholle: In those areas of the North Sea, especially the northern part of the Central Graben and the Viking Graben, where the Upper Cretaceous is argillaceous, the answer clearly is yes. In other areas it depends on the degree of diagenesis of the chalk and the time of hydrocarbon movement. If the chalk was deeply buried and thus extensively cemented before migration of oil, the unit could act as a cap-rock. At a porosity of about 10 percent (a reasonable value for chalk at 3000m depth) chalk would have a permeability of 0.1 to 0.01md.

32

Geology of the Dan Field and the Danish North Sea

By F. B. CHILDS and P. E. C. REED

(Gulf Oil Company–Eastern Hemisphere)

SUMMARY

The Dan field was discovered in 1971 by the Dansk Undergrunds Consortium's fifteenth offshore wildcat, which encountered oil and gas in Maastrichtian and Danian Chalk at the subsea depth of 5790–6565ft (1765–2001m). Production of some 800 barrels per day from each of five wells began in July 1972.

The field lies on the eastern flank of the North Sea Tertiary basin and near the axis of the Central Graben, a deep trough filled with a thick sequence of Permian to Cretaceous sediments. Upper Cretaceous–Danian Chalk at the top of the sequence provides the reservoir for several further hydrocarbon accumulations in offshore Denmark. Geochemical studies indicate that deeper Upper Jurassic marine shales are the probable source beds for these accumulations.

The Dan field is a halokinetically induced domal anticline. The Chalk reservoir has an average porosity and permeability of 28 percent and 0.5md respectively. The solution GOR is 600ft^3 (17m^3)/bbl and the crude oil is 30° API with low sulphur content (0.29 percent).

INTRODUCTION

The Dan field, named after the first king of Denmark, lies in the south-western portion of the Danish North Sea, in a water depth of 140ft (43m), approximately 205km west of the port of Esberg and 170km south-east of the Norwegian Ekofisk oil field (Fig 1). Discovered in May, 1971, by the Dansk Nordsø M-1x well, which encountered a total hydrocarbon column of approximately 750ft (229m) in Lower Tertiary (Danian)–Upper Cretaceous Chalk, Dan was brought on stream in July 1972, and currently produces approximately 4500BOPD.

HISTORY OF EXPLORATION IN OFFSHORE DENMARK

Concession Activity

In 1962 the concession for exploration and recovery of oil and gas in the onshore and inland waters of Denmark was granted to the Danish company, A. P. Møller. To carry out the work, the Dansk Undergrunds Consortium (DUC) was formed, with A. P. Møller as concessionaire, Gulf as operator, and Shell as assisting party.

In 1963 the concession area was extended to include the Danish continental shelf. A few years later, Chevron and Texaco joined the joint venture group in the continental shelf only. In 1970 Gulf withdrew from a portion of the shelf, retaining only the area referred to as "Area A-South-west" and Chevron became operator within the portion that Gulf relinquished, referred to as "Area A-North-east".

As a result of decisions reached within the International Court of Justice in 1969, a border treaty between Denmark and Germany was signed in 1971 whereby Denmark released its claim to a certain portion of the continental shelf. Prior to the border treaty, only one well, Dansk Nordsø B-1x, had been drilled in the area now belonging to Germany.

Seismic Surveys

The first seismic survey in the Danish North Sea was conducted in 1963. Using 50-pound dynamite charges, 8700 line miles (14 000km) of seismic reflection data had been acquired by the end of 1967. CDP multiplicity varied from 100 to 600 percent. Mapping of good reflections representing the top and bottom of the Lower Tertiary–Upper Cretaceous Chalk and the top of Permian (Zechstein) salt outlined several large structural closures.

From 1968 to the present, an additional 4500 line miles (7245km) of seismic reflection data have been acquired. By using air guns, aquaflex, and aquapulse as energy sources, and increasing the CDP coverage to 1200 to 4800 percent, significant improvements in data quality have been achieved. These improvements have increased the reliability of correlation of deep events and have led to delineation of more subtle structures.

Positioning throughout all seismic surveys was carried out with the Decca system and/or satellite navigation.

Drilling Activity

The first offshore well in Denmark, the Dansk Nordsø A-1x, was drilled in 1966 on a large Zechstein salt dome in the south-western part of the Danish shelf and found oil and gas in non-commercial quantities in Lower Palaeocene

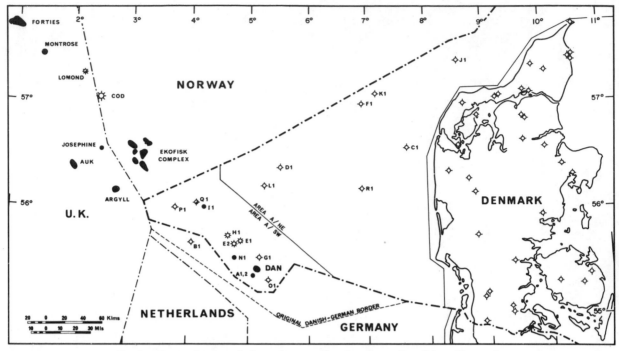

FIG 1.   Concession boundaries and status of exploration wells, offshore Denmark.

(Danian) Chalk. This well was the first to establish the presence of oil in the North Sea.

Up to the present time, 21 additional exploratory wells have been drilled in the Danish North Sea. Beyond the Dan field discovery, the occurrence of oil and gas has been established in four and three wells respectively. The proven hydrocarbon accumulations are all in the south-western area and, with the exception of the Q-1x and M-8x wells, in the Danian–Upper Cretaceous Chalk. Oil shows in the Q-1x and M-8x were from a formation stratigraphically below the Chalk.

GEOLOGY OF THE DANISH NORTH SEA

Tectonic Setting

The Danish continental shelf lies on the eastern flank of the North Sea Tertiary basin, an elongate epeirogenic downwarp filled with over 10 000ft (3048m) of sediments along its depocentre (Fig 2). Below the Tertiary Basin, the tectonic framework of offshore Denmark is dominated by a major horst complex, the Ringkøbing–Fyn High, which extends westwards from onshore Denmark toward the Mid North Sea High. These two structural highs are separated by a deep trough, termed the North Sea Central Graben, trending north-west–south-east through the western portion of offshore Denmark and filled with a thick section of Permian to Cretaceous sediments. The Dogger High is an uplifted fault block within the Central Graben (Fig 3). Other major Permo-Mesozoic sedimentary basins: the Danish Embayment and the North German Basin (Sorgenfrei and Buch, 1964) flank the Ringkøbing–Fyn High to the north

and south, respectively, and are connected by the Horn Graben.

These major tectonic elements were clearly established by late Permian time and continued to control sediment distribution into the Cretaceous (Fig 4). Normal faulting along the graben margins was particularly pronounced during the Triassic and late Jurassic. By late Cretaceous time, however, differential subsidence had ceased, and both basins and highs were blanketed by chalky limestones. The central North Sea subsided as a single basin during the Tertiary.

Dan field and other hydrocarbon discoveries in offshore Denmark are confined to the Central Graben area.

Geological History (Fig 5)

Pre-Permian. During the Devonian and Carboniferous, offshore Denmark was part of the emergent, low-lying foreland flanking the elevated Caledonian mountain belt, which extended from Scotland through the northern North Sea to Norway (Ziegler, 1974). Local incursions of the sea onto this foreland during the Carboniferous, however, are indicated by the presence of a few hundred feet of Viséan–Namurian shales, sands, and limestones unconformably overlying slightly metamorphosed Lower Palaeozoic shales in the Central Graben area.

Permian. Differential subsidence across the Ringkøbing–Fyn–Dogger–Mid North Sea High trend first occurred during early Permian, separating the North Sea into two intracratonic basins.

As in the southern North Sea (Glennie, 1972),

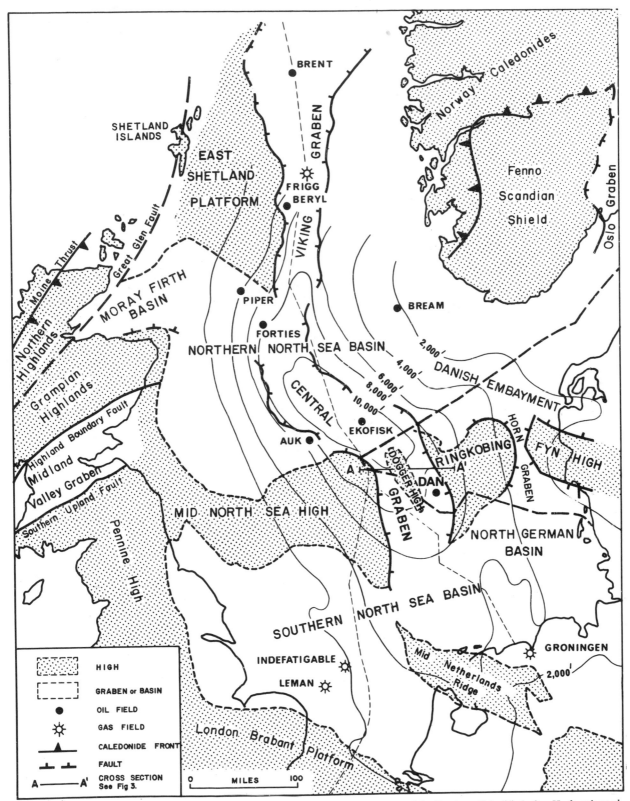

FIG 2. Geotectonic map of the North Sea, showing structural contours on the base of the Tertiary. (Modified after Heybroek *et al.*, 1967 and Rhys, 1974, Fig 8.)

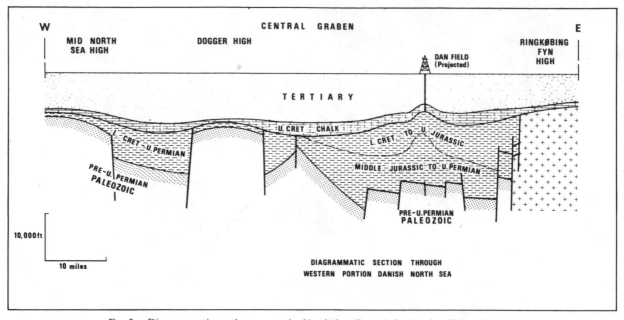

FIG 3.   Diagrammatic section across the North Sea Central Graben in offshore Denmark.

red-bed sequences of shale and sand (Rotliegendes) were deposited under arid desert conditions in the Central Graben and Danish Embayment. Weathered volcanic basalt flows are characteristic of the lower Rotliegendes encountered in offshore Denmark wells.

During Late Permian, continued subsidence resulted in marine transgression. While thin sequences of anhydrite, carbonates, and minor shales accumulated on the upper flanks of the Ringkøbing–Fyn and Dogger Highs, thick salt was

deposited in the Northern and Southern North Sea Basins and in the Central Graben. Later halokinesis of this salt, beginning in the Late Triassic, has caused local thickness variations of Jurassic, Cretaceous and Tertiary sediments.

Triassic. With retreat of the Zechstein seas at the end of the Permian, an arid continental depositional regime again prevailed in the central North Sea, and thick sequences of Triassic red-beds were deposited in the Danish Embayment and

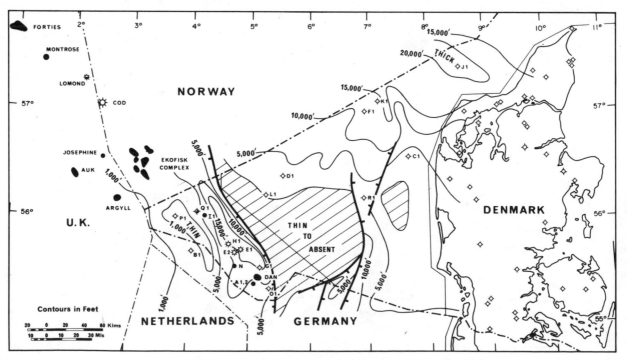

FIG 4.   Isopach map of Triassic to Lower Cretaceous, offshore Denmark.

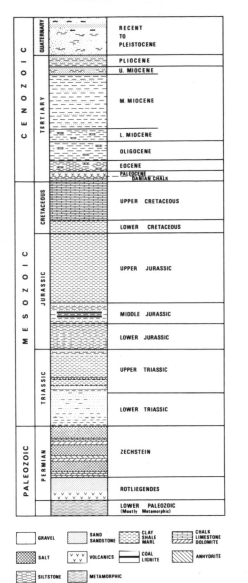

FIG 5. Composite stratigraphic column, offshore Denmark.

the Central Graben. Seismic reflection data suggest that the Horn Graben was developed mainly during the Triassic.

Predominantly sandy near the base, the composite Triassic section known from drilling in offshore Denmark becomes increasingly shaly upwards, indicating a gradual reduction in relief of the sediment-source areas. During the late Triassic, saline lakes were periodically developed, as shown by the presence of beds of evaporitic carbonates, anhydrite, and, in the Central Graben, salt. Locally, thicker Triassic salt beds show evidence of halokinetic movement.

Latest Triassic (Rhaetian) time was marked by deposition of grey sands and shales in lagoonal environments, heralding the return of marine conditions which have prevailed in the Central Graben area to the present time.

Jurassic. By early Jurassic time, fully marine conditions were established throughout the basinal

areas, where Lower Jurassic sediments are represented by dark grey to greyish brown shales with minor quantities of very fine-grained sandstone and siltstone. These shallow-marine deposits are succeeded by Middle Jurassic shales with interbedded sandstone and coal, indicative of marine regression and deltaic to paludal deposition.

A marked change in the rate of subsidence and in the depositional environment occurred in the late Jurassic in the Central Graben. In this region, Middle Jurassic deltaic deposits are abruptly overlain by several thousand feet of dark grey to black carbonaceous shales, occasionally with very thin limestone interbeds. Lithology and the presence of radiolaria indicate a deep-water, restricted-marine environment of deposition.

The rapid subsidence and sediment accumulation, accompanied by faulting and minor folding along the eastern margin of the graben and on the flanks of the Dogger High, may have occurred in response to the main opening of the Atlantic ocean basin (Hallam, 1971).

The Danish Embayment contains a much thinner section of Upper Jurassic shale, predominantly medium to light grey and slightly silty and sandy but becoming black to dark grey near the top.

Cretaceous. Deposition of dark grey shales, commonly pyritic, continued into the early Cretaceous in areas of thick Upper Jurassic sediments. However, sub-littoral microfaunas indicate shallowing conditions and gradual cessation of differential uplift across the Ringkøbing–Fyn and Dogger Highs. During the Aptian and Albian, these highs were gradually onlapped by reddish brown to brownish grey, marly shales grading upwards into dolomite and limestone.

By late Cretaceous time, most of the Dogger and Ringkøbing–Fyn Highs were submerged below sea-level and had ceased to shed clastic sediments into the adjacent lows. As a result, thick accumulations of clean, white to light grey, chalky limestone were deposited throughout offshore Denmark under quiet, clear-water, open-marine conditions. Over 3000ft (900m) thick in some parts of the Central Graben area, the Chalk thins to less than 1000ft (300m) where it onlaps and overlaps the Ringkøbing–Fyn Highs (Fig 6). Pronounced thickness changes across Zechstein salt domes attest to active halokinetic movement during the Late Cretaceous.

Tertiary. Chalk deposition continued in offshore Denmark in the early Palaeocene (Danian), although reworked Maastrichtian faunas in the Danian Chalk attest to local uplift and erosion, probably related to salt doming. Gradually deepening waters in Late Danian time terminated

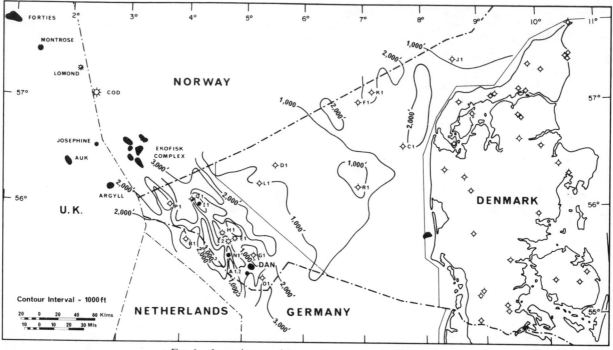

FIG 6.   Isopach map of Upper Cretaceous—Danian chalk.

chalk deposition, and led to a resumption of clastic sedimentation.

The post-Danian history of the central North Sea is one of gradual epeirogenic subsidence and basin-filling (Fig 3). The Palaeocene to Oligocene sequence is almost entirely shale, with a thin interval of volcanic tuff in uppermost Palaeocene basal Eocene and thin limestone and dolomite stringers in the overlying Eocene and Oligocene sections. Microfaunal assemblages indicate bathyal to outer shelf conditions throughout early Tertiary time.

Gradually shallowing shelf conditions are indicated for the Miocene to Pliocene interval, characterized by shales and clays in the Lower Miocene, grading upwards into interbedded clay and shale in the Middle Miocene, which in turn grade up into fine- to coarse-grained sand, with subordinate clay and shell interbeds, in the Upper Miocene to Pliocene section.

Quaternary. Quaternary deposits in offshore Denmark consist of fine- to coarse-grained sands and shell beds with occasional lenses of lignite and soft grey to grey-brown clays.

DAN FIELD

The stratigraphic column penetrated by the Dan field discovery well consists of 1900ft (579m) of Upper Miocene to Recent sand and clay, 3890ft (1186m) of Lower Palaeocene to Middle Miocene clay and shale, 960ft (293m) of massive Upper Cretaceous to Danian chalk, 210ft (64m) of Lower

Cretaceous carbonates and shales, and over 350ft (107m) of Upper Jurassic shales. The hydrocarbon column lies in the Danian–Upper Cretaceous Chalk unit, immediately overlain by early Palaeocene shales which provide the reservoir seal. Geochemical characterization of the organic matter in formations above and below the Chalk indicates that the underlying Upper Jurassic shales were the source beds of the oil.

Reservoir. The stratigraphy, lithology, and petrophysical characteristics of the Dan field reservoir are best known from the M-1x well, which was continuously cored through most of the hydrocarbon-bearing interval (Fig 7). On the basis of nannofossils, the uppermost 130ft (40m) of Chalk are dated as Danian and the underlying 800ft (244m) as Maastrichtian. The lowest Danian nannofossil zone identified in offshore Denmark is missing in this well, indicating an unconformity at the Danian–Maastrichtian contact. The thick Maastrichtian Chalk conformably overlies 30ft (9m) of Campanian and possibly older late Cretaceous Chalk which rests, possibly unconformably, on Lower Cretaceous (Albian to Valanginian) dolomite, limestone, and arenaceous shales.

Megascopically, the Chalk is firm to hard, white at the base of the section, becoming cream to buff and tan in the middle and upper Maastrichtian, and white to greyish white in the uppermost Maastrichtian and the Danian. These colour variations are attributable, in part, to variable oil-staining. Highly inclined to vertical fractures, generally intersecting and sometimes crossed by low-angle fractures, are common in the

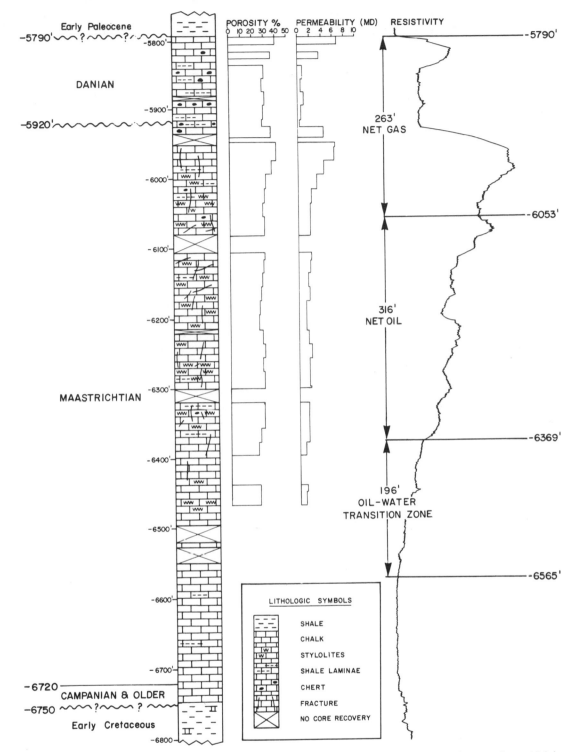

FIG 7.   The Dan field reservoir at M-1x well. Porosity and permeability based on 486 plug analyses and averaged over 10-ft intervals.

Maastrichtian interval but rare and of very restricted extent in the Danian Chalk. Similarly, stylolites are abundant in the Maastrichtian below 6000ft (1829m), but virtually absent from the uppermost Maastrichtian and Danian sections. Thin shale laminae, occasionally pyritic, occur at several horizons throughout the Chalk. Chert is common in the Danian and the upper part of the Maastrichtian.

In thin-section, the Chalk is classified mostly as fine-grained, mud-supported biomicrites, grading through fossiliferous micrites to micrites containing virtually no skeletal debris. Recognizable skeletal grains, generally constituting 5–10 percent of the rock, but occasionally up to 25 per cent, include "calcispheres" (*Oligostegina*), planktonic foraminifera, sponge spicules, radiolaria, echinoid plates and spines, ostracodes, benthonic foraminifera, and rare molluscan and bryozoan fragments. Whereas calcispheres are the dominant skeletal grains throughout the Maastrichtian and older Chalk, planktonic foraminifera are the dominant skeletal material in the Danian Chalk.

Although porosity is not evident in thin-sections, scanning electron microscopy shows moderate to good, primary interparticle porosity within the micritic matrix, the greater part of which appears to consist of coccolith debris. Individual coccolith plates, generally about 5μ in size, are subordinate to plate fragments, which appear as minute euhedral to subhedral crystals of rhombohedral calcite with an average size of less than 1μ. Minor welding of the crystals is common, probably a result of incipient pressure-solution at particle contacts. Secondary cement is only rarely observed.

Core analysis measurements, averaged over 10ft (3m) intervals, show that most of the Chalk section in the M-1x well has porosities of 25–30 percent and permeabilities less than 2md. However, significantly higher porosities (30–40 percent) and permeabilities (3–7md) occur at two levels, one at the top of the Danian Chalk and one immediately below the Danian–Maastrichtian contact. In both cases, these higher porosity–permeability zones have abrupt upper boundaries and gradational lower boundaries. Pore-size distribution analyses by the mercury injection method tend to suggest that the dominant pore size in these zones may be somewhat higher than that in the intervening and underlying sections, i.e., 1μ versus less than 0.85μ. The origin of these higher porosity–permeability zones is still not clear, but it is tentatively suggested that they represent secondary leaching at early and late Danian unconformities.

As shown in Fig 7, the hydrocarbon column in the Dan field can be subdivided into a gas cap 262ft (80m) thick, an oil zone 316ft (96m) thick, and an oil–water transition zone 196ft (58m) thick. Although measured core permeabilities through the oil zone in the M-1x well show an average permeability of about 2md, effective permeability determined from production tests in all wells is only 0.5md. This apparently indicates that the observed primary fractures do not contribute to reservoir permeability.

Reservoir characteristics for the oil zone, averaged over the entire field, are tabulated below.

Structure

Figure 8 shows the structural configuration of the Dan field reservoir as known from well information and seismic data. The field is a symmetrical domal anticline approximately 7km in diameter. Formation dip on the flanks varies from three to five degrees.

The structure is bisected by a major normal fault trending north-west–south-east with approximately 200ft (61m) of throw down to the north-west. This fault is clearly evident in seismic profiles. Subsidiary branch faulting shown in Fig 8 represents one of several possible interpretations to account for the different levels of fluid contacts encountered in the wells. However, it is recognized that changes in capillary pressure, resulting from variations in pore size and porosity, may account equally well for the different fluid levels.

As evident from seismic reflection data, the Dan field structure is a result of halokinetic uplift which began in the late Jurassic and continued into the early Tertiary. Associated faulting is believed to have helped provide the paths for migration of the hydrocarbons from their Upper Jurassic source beds into the Chalk reservoir.

Development

Installation of the production system was begun in mid-1971, soon after the M-2x well had confirmed the field discovery. Feasibility studies determined that the most economical production plan, designed with a maximum of safety, called for installation of a four-pile, six-well drilling platform and a four-pile production platform with associated three-pile flaring platform. It was decided to utilize an oil tanker for storage and transport of the produced oil, and therefore a single-buoy-mooring loading system was installed.

The drilling platform and flare structure were fabricated in the United States and transported by barge to the Dan field. The drilling platform was installed in September 1971, and the drilling rig "Britannia" was moved over the platform to drill and complete four development wells. The M-1x was also re-entered and completed. Installation of the flare structure proceeded in October 1971, but was interrupted by bad weather and was resumed in early 1972.

The production platform and single-buoy-mooring system were fabricated in Holland and

Net thickness	160ft (49m)
Porosity	28 percent
Effective permeability	0.5md
Water saturation	30 percent
FVF	1.30
Reservoir pressure	3800psi at 6200ft (1890m) BSL.
Reservoir temperature	163°F (73°C) at 6200ft (1890m) BSL.
Solution gas–oil ratio	600:1 SCF/BBL
Producing mechanism	Solution-gas drive
Oil gravity	30° API
Sulphur content	0.29 percent

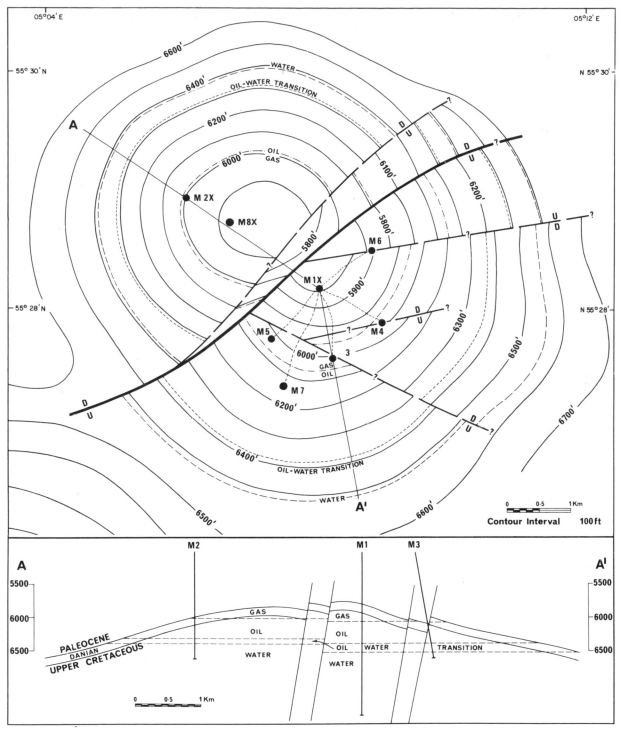

FIG 8. Structural contours on top of Danian Chalk in the Dan field and structural cross-section across field.

installed after installation of the flare structure. The five wells were hooked up and production commenced on July 4, 1972. As a result of studies undertaken to determine better completion techniques, a sixth well was drilled in May 1974.

The wells are stimulated by acid-fracture treatments which have been successful in improving initial productivity. However, with the intrinsic low permeability of the Chalk, each well has exhibited a decline in flow rate and tubing pressure during production. Initial production rates of each well were from 2000BOPD to 3000 BOPD, but within a period of several weeks the rates had stabilized at about 800BOPD. This decline in productivity is typical of low-permeability reservoirs.

Future Plans

Since commencing production, tests have been conducted to analyse the reservoir characteristics of the Chalk. The data from these tests have been used in simulation studies to ascertain the optimum completion and well-spacing procedures for development of the field. As a result of these studies, together with more favourable economic conditions, an expansion programme for the development of the field is now planned.

A second drilling platform is planned for completion during the early part of 1975. The platform will be designed to accommodate six wells and will be situated adjacent to the existing production platform. This should increase the production of the field to some 800BOPD, approximately twice the present production rate.

ACKNOWLEDGEMENTS

The authors express their appreciation to the companies A. P. Møller, Shell, Chevron, Texaco and Gulf of the Dansk Undergrunds Consortium for their permission to publish this paper. Special acknowledgement is due to Gulf Oil Company of Denmark and Gulf's research and technical centres for assistance in preparation of the paper.

REFERENCES

Glennie, K. W. 1972. Permian Rotliegendes of Northwest Europe interpreted in light of modern desert sedimentation studies. *Bull. Am. Ass. Petrol. Geol.*, **56**, 1048–71.

Hallam, A. 1971. Mesozoic geology and opening of the North Atlantic. *Jl. Geol.*, **79**, 129–57.

Heybroek, P., Haanstra, U. and Erdman, D. A. 1967. Observations on the geology of the North Sea Area. *Wld. Petrol. Congr. 7, Mexico, Proc.*, **2**, 905–16.

Rhys, G. H. (Compiler, 1974). A proposed Standard Lithostratigraphic Nomenclature for the southern North Sea and an outline structural nomenclature for the whole of the (UK) North Sea. A Report of the Joint Oil Industry–Institute of Geological Sciences Committee on North Sea Nomenclature. *Rep. Inst. Geol. Sci.*, **74/8**, 14 pp.

Sorgenfrei, T. and Buch, A. 1964. Deep tests in Denmark. *Danm. geol. Unders. Raeke III*, **36**, 1–146.

Ziegler, P. A., 1974. The North Sea in a paleogeographic framework. *Proc. Bergen North Sea Conference. Norg. geol. Unders.*

33

Geology of the Ekofisk Field, Offshore Norway

By W. D. BYRD

(Phillips Petroleum Company)

SUMMARY

The Ekofisk Field is located in Block 2/4 of the Norwegian sector of the North Sea. A complex history of pre-Chalk deposition left a deep depression which was filled with progressively overlapping Chalk deposits. The environment of deposition of the Chalk must have been one of very quiet water, as evidenced by lack of high energy faunal types, and by the occasional presence of intact, fragile coccospheres.

The Danian–Upper Cretaceous interval is a chalky limestone consisting of a very fine-grained calcite silt matrix, in which are set varying amounts of coarser-grained constituents.

The matrix porosity of the reservoir consists of the original intergranular spaces between the loosely-packed coccolith grains, but the tectonic fractures are considered to be the single most important characteristic which allows the Chalk to produce hydrocarbons at prolific rates. Hydrocarbon migration from the Tertiary shales into the reservoir is due to differential fluid pressure.

LOCATION

The Ekofisk Field is located near the geographic centre of the North Sea, 200 miles (322km) south-west of Stavanger, Norway. Figure 1 shows the location of Ekofisk in the Norwegian Block 2/4 and other fields discovered in the greater Ekofisk area.

GEOLOGICAL HISTORY

A complex history of pre-Chalk deposition left a deep depression which was filled with progressively over-lapping Chalk deposits. In general, the bathymetric lows were filled first. Complete inundation of the pre-Chalk strata occurred during Maastrichtian time. Salt movement occurred after the deposition of up to 3000ft (914m) of Upper Cretaceous carbonate. Carbonate deposition continued into Danian time and the old salt structures continued to grow as evidenced by thin or absent units over the structural highs. Intense fracturing in the Danian and Upper Cretaceous Chalk along structural crests suggests additional salt movement some time after deposition and lithification.

DEPOSITIONAL ENVIRONMENT

The environment of deposition of the Chalk must have been one of very quiet water, as evidenced by lack of high energy faunal types, and by the occasional presence of intact, fragile coccospheres. Burrows indicate that the sediment was not significantly disturbed by waves or currents after deposition, thus supporting the theory of a low-energy depositional environment. Foraminiferal assemblages, sediment composition, and primary sedimentary structures all suggest that the Chalk was deposited in deep water, probably of the order of hundreds of metres. The lack of coarse terrigenous clastics and paucity of terrigenous clays indicate that the area was isolated from any major source of terrigenous influx during late Cretaceous and most of Danian time.

CHARACTERISTICS OF THE DANIAN–UPPER CRETACEOUS ROCKS

Although the Upper Cretaceous carbonate has not been penetrated completely on the Ekofisk structure, seismic data and nearby well-information indicate a total Upper Cretaceous thickness of approximately 2500 to 3000ft (762 to 914m). At Ekofisk, only the upper 250ft (76m) of the Upper Cretaceous is of interest. The overlying 450ft (137m) of Danian Chalk is the main objective. Reservoir pressure data indicate that at Ekofisk, the Danian–Upper Cretaceous reservoir is a single accumulation.

This interval is a chalky limestone consisting of a very fine-grained calcite silt matrix, in which are set varying amounts of coarser-grained constituents. These coarser grains are skeletal in origin, comprising mainly planktonic foraminifera with minor amounts of sponge spicules, echinoderm plates and the remains of other benthonic organisms.

Scanning electron microscope (SEM) studies have given insight to the fine-grained texture and composition of the Chalk. The deposit is a chalk in the strictest sense of the word, composed mainly of calcareous unicellular algal remains, called

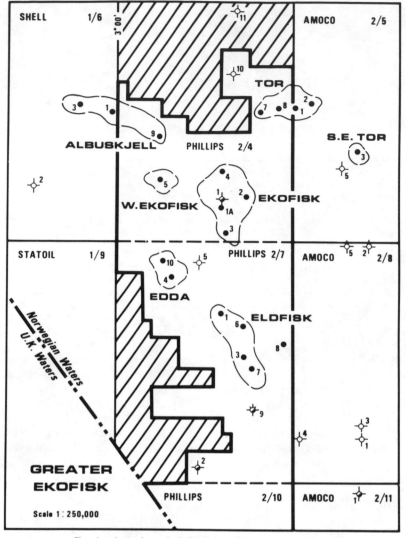

FIG 1.   Location of oil-fields in the Greater Ekofisk area.

coccospheres (Plate I, A), and their disarticulated gear-shaped skeletal plates called coccoliths (Plate II, A). These algae were present in extreme abundance in the upper layers of the open ocean. When they died their skeletal remains fell to the ocean floor. Most coccospheres break up into their component coccoliths before they finally come to rest, but under extremely quiet water conditions, an occasional intact coccosphere was incorporated in the resulting sediment. The coccoliths themselves may be further disaggregated into tiny calcite plates. Most coccospheres noted in the Ekofisk sediments range from 15 to 30 microns. Coccoliths average approximately 5 microns and the individual coccolith plates are usually somewhat less than 1 micron.

Non-calcite components of the Danian–Upper Cretaceous interval include varying minor amounts of clay, dolomite, gypsum, pyrite, siderite, barite, and silica. Of these, silica appears to be the most abundant. Some zones within the Chalk are highly silicified. Plate II, B is an SEM photomicrograph of a partially silicified chalk that

was etched with dilute hydrochloric acid before photographing. Note the insoluble grains standing in relief above the dissolved calcite matrix. These grains include silicified sponge spicules and radiolarians, as well as siliceous infillings of foraminiferal chambers. In some other samples, the Chalk matrix itself has been silicified. Where it occurs, this silicification obviously reduces porosity.

In the upper portion of the Danian, an increase in terrigenous clay content reflects a transitional change in depositional regimes. At the close of Danian time, terrigenous clay influx increased to the point where the resulting sediment is a pure clay shale.

POROSITY AND PERMEABILITY

The matrix porosity of the reservoir rock consists of the original intergranular spaces between the loosely-packed coccolith grains. The average measured porosity of core samples through the

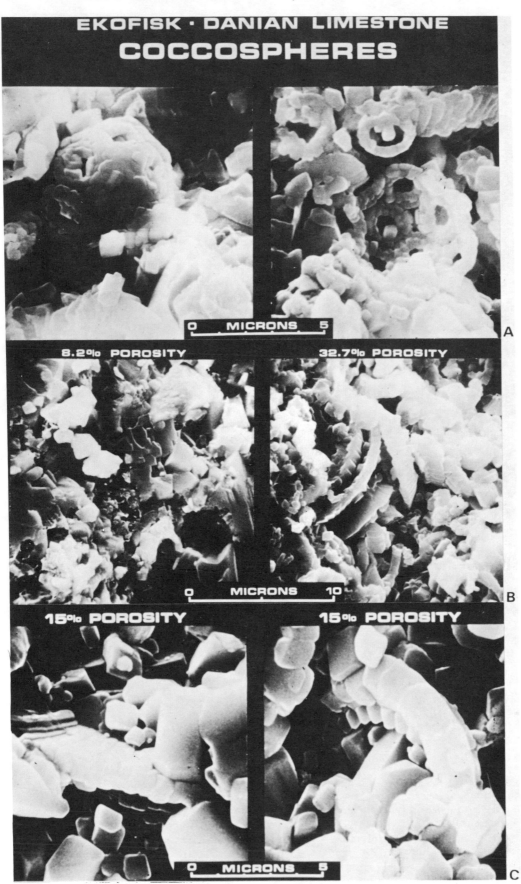

PLATE I. See text for explanation.

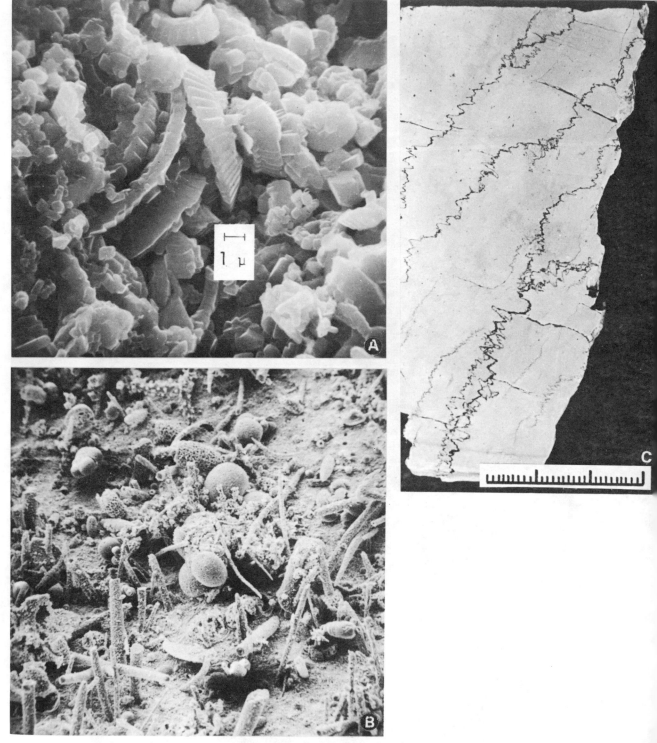

PLATE II.   See text for explanation.

Danian–Upper Cretaceous interval at Ekofisk is 28 percent. In Plate I, B a tight sample (porosity less than 15 percent) is contrasted with a representative sample from an interval containing measured porosities up to and slightly exceeding 40 percent. The sample on the right illustrates the loose packing of coccoliths and the mere "spot-welding" at grain point contacts. In other samples, an increase in secondary calcite cement and possible tighter packing of the coccolith grains reduces the intergranular porosity. Plate I, C illustrates how secondary calcite cementation has reduced matrix porosity to 15 percent.

Very thin detrital clay laminae were noted within the Chalk interval. Where present, this clay obviously reduces porosity and permeability.

The source of the calcite that cements the reservoir rock and reduces the porosity is problematical but may be related to pressure-solution and reprecipitation associated with the formation of stylolites. Numerous stylolites were observed in the tighter portions of the reservoir sections. Plate 2, C shows a core slice containing several stylolites.

The Ekofisk reservoir rock is highly fractured. This fracturing is a result of tectonic stresses induced by the growth of the salt structures on which Ekofisk Field is situated. Indeed, this intense tectonic fracturing is considered to be the single most important characteristic of the Ekofisk reservoir rock that allows the chalk to produce hydrocarbons at the prolific rates that have been abundantly demonstrated.

HYDROCARBON MIGRATION AND ACCUMULATION

Comparison of chemical analyses of oils from the Danian–Upper Cretaceous reservoir at Ekofisk and oils extracted from sidewall cores taken in the overlying Tertiary shales suggest that the Tertiary shales are the source rock for the oil at Ekofisk. From the Δt shale plot of a typical Ekofisk well in Fig 2, an initial deviation from the normal North Sea gradient can be seen at 4200ft (1280m), and then more significantly at 5300ft (1615m). This

abnormal pressure continues down to just above the Danian, where there is a slight pressure regression before passing into the over-pressured Danian reservoir. Such a situation suggests that the over-pressured shales are preventing vertical migration of hydrocarbons out of the Danian reservoir and providing an effective trapping mechanism. Since these overlying shales are believed to be the source of oil, and since they have a higher fluid pressure than the accumulation, fluid flow should be from the shales into the underlying Chalk.

Ekofisk was the first major oil-field discovered in the North Sea and has been producing from four wells at a rate of 40 000 barrels of oil per day since July 1971. Oil is expected to be flowing through a pipeline to the Teesside loading facility in June 1975. Gas will be re-injected until the gas pipeline to Emden, Germany is completed in early 1976.

ACKNOWLEDGEMENTS

I thank Mr Don Dalrymple, a colleague, who made the original scanning electron microscope interpretation of the Ekofisk cores, for his valuable guidance in the preparation of this paper. Also, I thank Messrs Jim Davis and Colin Wilkinson for their constructive criticism of the manuscript, and Phillips Petroleum Company for permission to publish this paper.

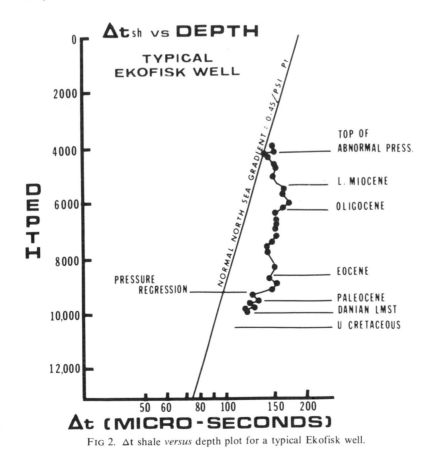

FIG 2. Δt shale *versus* depth plot for a typical Ekofisk well.

EXPLANATION OF
PLATES

Plate I.
A.   Scanning electron microscope (SEM) photographs of coccospheres.
B.   SEM photographs of contrasting porosities within the Chalk.

C.   SEM photographs of porosity reduction by secondary calcite cementation.

Plate II.
A.   SEM photograph of coccoliths.
B.   SEM photograph of partially silicified chalk etched with dilute hydrochloric acid, showing sponge spicules and foraminiferal casts.
C.   Core slice containing several stylolites.

DISCUSSION

Dr J. M. Bowen (Shell UK Exploration and Production Ltd): Dr Ziegler and Mr Childs have indicated subaerial erosion taking place around the top-Chalk levels. Particularly at the Dan Field, leaching was suggested at top-Maastrichtian and top-Danian levels. Could Mr Byrd say whether such erosion and leaching occurred at Ekofisk?

W. D. Byrd: Several horizons within the Chalk suggest erosion or non-deposition at Ekofisk. Whether this is due to erosion, leaching or non-deposition is the question. SEM studies do not support the leaching theory in the intervals near the top of the Maastrichtian and Danian chalks.

Another possibility is non-deposition on positive structural crests as the structures were growing during Chalk deposition.

There is no difference in the Chalk faunal constituents suggesting shallowing of the water to facilitate subaerial erosion. Further, the overlying Palaeocene clastics are deep-water shales, suggesting a continuity of deep-water deposition from chalk to clastics.

Dr N. V. Dunnington (Cluff Oil): I would like to ask if any information can be provided on the relative timing of fracturing, entry of oil into reservoirs and stylolite development in any of the fields concerned.

W. D. Byrd: Fracturing and oil migration into the reservoir must have taken place during late salt-movement, probably during Miocene time, as shown by seismic evidence. By late Miocene time sufficient overburden (5000–6000ft) was present to cause pressure-solution phenomena such as stylolites.

Dr W. G. Townson (Shell UK Exploration and Production Ltd): I would like to ask Mr Byrd, in view of Dr Scholle's remark that in Ekofisk Chalk diagenesis ceased when a burial depth of about 6000ft had been attained, does he think that the Palaeocene was already over-pressured at this depth and was generating hydrocarbons, apparently in Miocene times. The Jurassic would presumably be deep enough at that time, however.

W. D. Byrd: The over-pressured system may have started in Miocene time. The gradient in the Palaeocene–Eocene rocks appears to have a parallel but offset trend to the present day gradient. This suggests that compaction ceased in Miocene time.

The accepted depth of initial hydrocarbon generation is in the 6000-ft range. It is reasonable to believe that migration may have started in Miocene time.

A. Ford (Shell International Petroleum Maatschappij, BV): Dr Scholle mentioned anomalously high porosities at Ekofisk in reservoir chalk below 10 000ft. In an earlier discussion on structural style Mr Blair said that solution of chalk was an important reservoir parameter at Ekofisk. Mr Byrd made no mention of solution in his talk on Ekofisk. In spite of Mr Byrd being unable to reply to Mr Bowen's question on unconformities at end Maastrichtian and end Danian, I ask Mr Byrd if he would acknowledge the role of solution at Ekofisk and comment on how it came about.

W. D. Byrd: The dominant solution factor or process at Ekofisk is stylolitization and the redistribution of calcite as overgrowths.

Some minor etching of coccolith plates has been demonstrated in the SEM studies, but it is not considered a major factor in porosity enhancement.

Kjell G. Finstad (Saga Petroleum): Can the Jurassic shales be considered as a possible source for the oil found in the Chalk?

W. D. Byrd: We consider both the Jurassic and the Palaeocene shales as potential source rocks. We favour the Palaeocene because of the overlying over-pressured system and the lack of shows in the Chalk below the reservoir. Any shows in the lowermost Chalk are probably from the Jurassic source rocks.

34

Lower Tertiary Sand Development in the Central North Sea

By J. R. PARKER

(Shell UK Exploration and Production Limited)

SUMMARY

The distribution of Lower Tertiary sands in the UK sector of the central North Sea is closely related to the regional tectonic setting. Within the Palaeocene interval large scale topset, foreset and bottomset units have been recognized on seismic sections. These units form a sedimentary complex which has built out eastwards and south-eastwards into the basin. Wells drilled in the topset units have penetrated thick composite sand intervals with interbedded lignites, interpreted as having been deposited under shallow marine to coastal plain conditions. The foreset beds, consisting of clay with thin sand beds, separate the topset sediments from their lateral equivalents which are a thinner sequence of shales and turbidite sands deposited in a deep marine environment. These turbidite sands form the reservoirs of the Forties and Montrose fields. Similar sedimentary units can be recognized in the Eocene but to date few sand intervals have been found in the bottomset (turbidite) interval.

INTRODUCTION

The North Sea Tertiary basin extends 600 miles (960km) southwards from the present day continental margin north of the Shetland Islands to the Netherlands and is 300 miles (480km) wide between Denmark and the UK. The outline of the basin coincides with the surrounding present day coastline and the deepest part of the basin, over 10 000ft (3050m) below sea-level, lies in the geographical centre of the North Sea (Fig 1). This deep central portion of the basin overlies a Mesozoic graben system bounded by north to north-west trending tensional faults.

Following major faulting and rifting at the end of the Jurassic (Late Cimmerian movements), only minor rifting movements occurred during the Cretaceous, although the graben system continued to subside. Sedimentation of the Chalk continued without significant interruption from the Upper Cretaceous into the Danian. The last phase of rifting occurred during the Lower Palaeocene (Laramide movements) when reactivation of the fault systems resulted in the formation of fault-scarps, local erosion of the Chalk and deposition of the "mélange" (slump-deposits containing chalk fragments in a siliciclastic matrix of Lower Palaeocene age) on the down-thrown side of these fault-scarps. At the same time there was rapid subsidence of the graben system. Coinciding with these Laramide movements, chalk deposition ceased and the area was thereafter dominated by clastic deposition.

The combination of lithological, sediment-ological, geophysical and palaeontological data suggests that a deep-water basin existed in the centre of the North Sea in early Tertiary times. Turbidites and mass-flow deposits were laid down in the deep-water basin while much thicker shallow-water sediments were deposited at the basin margin.

PALAEOCENE CLASTIC INTERVAL

Seismic Expression

The Palaeocene clastic interval can be defined on seismic sections by a lower event corresponding to the top of the Chalk (or top of the "mélange" if

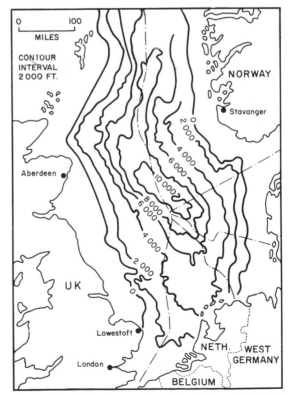

FIG 1.   Top chalk structure map (from Dunn *et al.*, 1973).

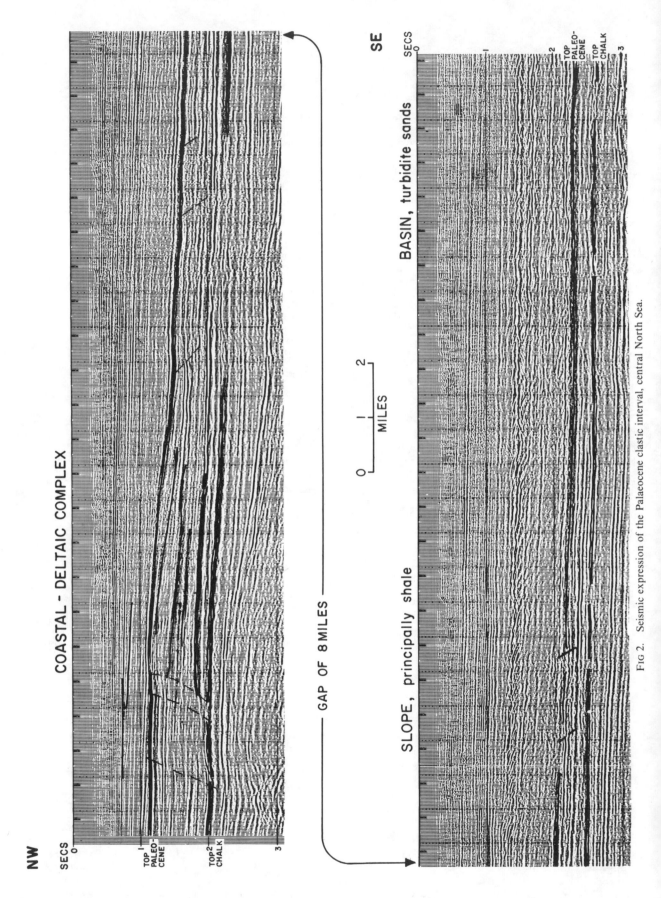

Fig 2.   Seismic expression of the Palaeocene clastic interval, central North Sea.

present) and by an upper event corresponding to a regional volcanic tuff layer which can also be distinguished on logs. Palaeontologically the top of the Palaeocene is close to the base of this tuff horizon.

Within the Palaeocene interval a series of sedimentary units with topset, foreset and bottomset geometries can also be recognized on seismic sections (Fig 2). These units form a huge eastwards and southwards outbuilding sedimentary complex, produced, at least in the upper part, by a process of deltaic progradation.

Wells drilled in the top-set zone have penetrated thick composite sand intervals with interbedded lignite beds, interpreted as a sequence of outbuilding coastal, deltaic and interdeltaic sediments. The fore-set zone corresponds to the basin-slope and here wells have penetrated predominantly clayey sections. This zone separates the thick coastal-deltaic sequence from its deep water equivalent which is a thinnner sequence of basinal shales and turbidite-sands. These turbidite-sands constitute the reservoirs of the Forties and Montrose Fields. An idealized relationship between these seismic geometries and their environments of deposition is shown in Fig 3.

Coastal-deltaic Sequence

The coastal-deltaic sequence is very sandy. Massive friable sands are interbedded with thin grey clays and lignite beds. The lignites are very persistent and can be mapped seismically. The sands are very variable in grain-size, sorting and rounding, and frequently pyritic and micaceous. Marine intervals, interbedded with the non-marine sediments, contain a foraminiferal assemblage which suggests an inner sublittoral environment of deposition, that is, water depths less than 600ft (183m).

Slope Sequence

This is principally an argillaceous sequence. There is a lack of detailed lithological control (*e.g.* cores), but logs indicate some sand units. The foraminiferal assemblage suggests a very variable environment of deposition ranging from inner sublittoral to bathyal. However, the shallower water fauna is interpreted as having been displaced, as might be expected in a slope-facies.

Basinal Sequence

The basinal sequence consists of alternations of green to dark grey shales and fine- to very coarse-grained quartz and feldspathic sands. The latter vary in occurrence from very thin beds to units over 200ft (61m) thick.

The sands are subangular to rounded, moderately to poorly sorted, friable, with porosities ranging from 20 to 30 percent. Grading is fairly common in the thinner beds whereas the thick beds are often massive with very few internal structures. Load-casts, contorted bedding and groove-marks are common in the thinner beds. Large scale cross-bedding is conspicuously absent, even in the thick-bedded coarse-grained intervals; where present cross-bedding is restricted to the centimetre scale. Dish structures (water-escape structures (Wentworth, 1967; Lowe and LoPiccolo, 1974)) are present in some of the thicker sand beds.

The alternation of sand beds of varying thickness with finely laminated dark shales, the association of sedimentary structures, the presence of graded bedding and the absence of large scale cross beds strongly suggest deposition from turbidity currents, mass-flows and slumps.

Palaeontological evidence from the intercalated shale units suggests water of bathyal depth, that is, probably in the order of 600–3000ft (180–910m). The bathymetric relief thus indicated

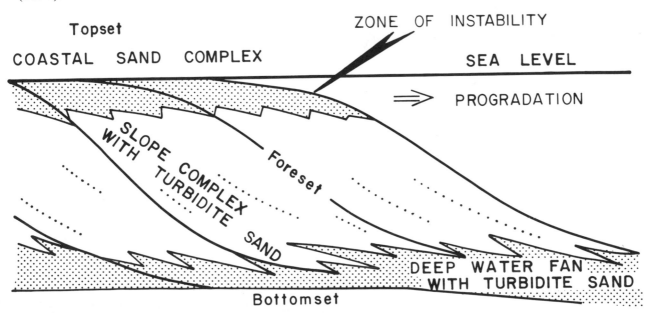

FIG 3.   Idealized cross-section through prograding deltaic slope.

palaeontologically between the shelf area—600ft (180m) or less—and the basinal area agrees with the relief seen between the topset and bottomset sedimentary packages on the seismic profiles.

DISTRIBUTION OF SAND WITHIN DEEP WATER TURBIDITE BODIES

Studies of both recent and ancient deep-water turbidite sand bodies (e.g. Kruit et al., 1972; Normark, 1970) show that the major deposition commonly occurs at the base of the slope, producing a submarine fan. The fan deposits are sandstones with intercalated shales which pinch out updip into slope sediments and downdip into basin floor sediments. Such fans can be divided into three main areas (Mutti and Ricci Lucchi, 1972; Walker and Mutti, 1975). An upper part shows bulk transportation and deposition restricted to elongate channel fill ("feeder") sand bodies, often with persistent levees. These elongate sand bodies appear to occur as fairly straight deposits varying in width from a few hundred feet to several miles. A channel on the upper fan may be entrenched below the general level of the fan surface or it may be aggrading and its floor may be built up above the normal fan surface. The relief of such a channel and the height of its levees generally

decreases down the fan. In the middle part of the fan numerous meandering and upbuilding channels occur. These channels lack persistent levees and pass downcurrent into a braiding and rapidly migrating system that disappears downfan (Haner, 1971). Rapid deposition within these channels and just below their ends accelerates upbuilding in the middle part of the fan, forming depositional lobes ("supra fans") on the fan surface. The lower part of the fan is characterized by broad sheet-like flows and the absence of channels. It grades into the basin-plain where slowly accumulating pelagic sediments alternate with thin-bedded, fine-grained, laterally persistent turbidite flow deposits.

Sediment transport from the shelf into the deep basin can either be channelled in submarine canyons cut in the slope (Fig 4a) or by movement across a wide front from a depositionally active prograding slope (Fig 4b). Both types of transport appear to have occurred during the deposition of the Palaeocene sediments of the North Sea Basin.

Using the model outlined above a generalized sand-distribution map for the deep-water Palaeocene sands can be constructed (Fig 5). Although the basin continued to subside throughout the Tertiary, the top Chalk structure map (Fig 1) gives a rough approximation of the sea-floor topography on which the turbidites were

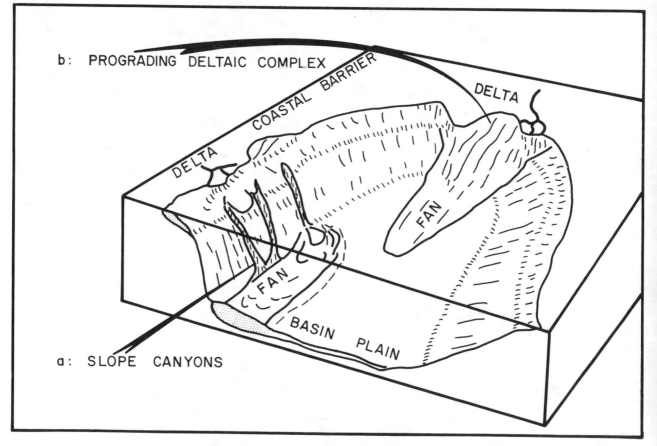

FIG 4. Block diagram showing models of sediment transport.

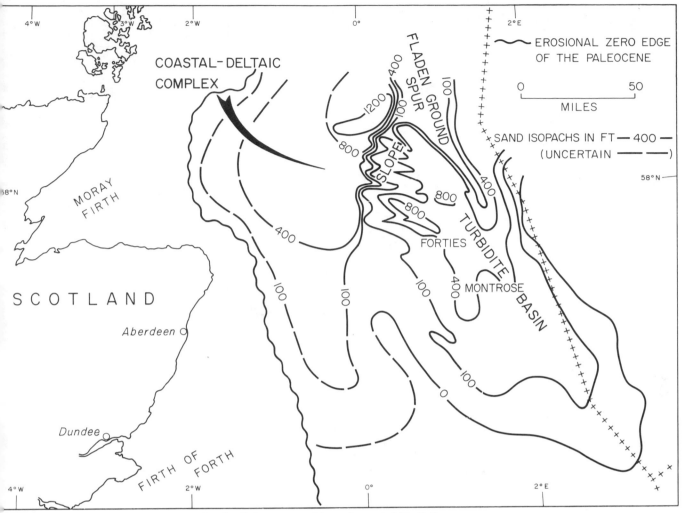

FIG 5.   Palaeocene sand isopachs, central North Sea.

deposited. From the updip point of turbidite sand deposition on the fore-set slope to the base of the slope an increase in sand percentage is assumed. Beyond the base of the slope a decrease in sand percentage, bed thickness, grain size and total thickness of the Palaeocene is expected. A cross plot of net sand thickness against total Palaeocene thickness based on well data confirms this picture of decreasing sand percentage in the assumed downcurrent direction and also laterally away from the fan.

Reservoir quality is good in the sands of the upper and middle parts of the fan, the Forties and Montrose fields being located in this area (Thomas *et al.* 1974; Walmsley, 1975; Fowler, 1975). The reservoir quality deteriorates to the south and east, away from the sediment source.

DEPOSITIONAL HISTORY OF THE PALAEOCENE

In early Palaeocene times the outer Moray Firth and Northern North Sea Basins were deep-water areas in which turbidite-sands were deposited, as far south as 56°N. The sands deteriorate in quality towards the south and east. The coast from which these sands were derived appears to have been close to the present day Scottish coast and there is evidence from the mapped sand distribution that the sediment supply was from a series of widely separated point sources, perhaps canyons in the basin slope.

Later during the Palaeocene, there was a change in depositional style. A major coastal-deltaic complex prograded eastwards over the Moray Firth area, overlying the earlier turbidites. This coastal-deltaic complex produced a large influx of sediment into the deeper turbidite basin from across the whole width of the prograding slope (Figs 5 and 6).

These younger turbidite sands only extended as far south as 56°40′N. Levees are characteristically developed on the upper surface of the fan at this time.

The end of this cycle of deposition is marked by the widespread volcanic tuff marker.

No detailed work has been carried out in the provenance of the Palaeocene sands but the lithologies observed and the basin model are

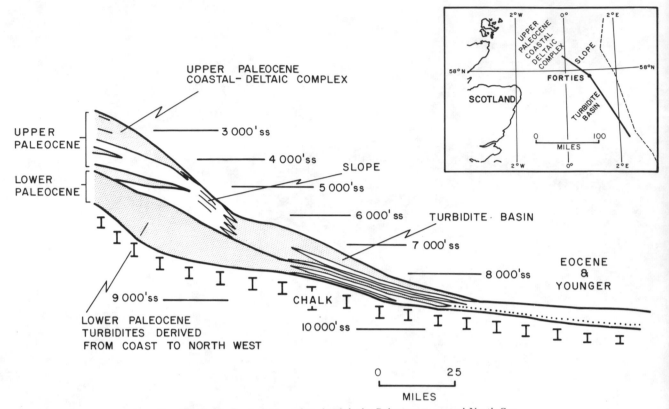

FIG 6.   Generalized cross-section through the Palaeocene, central North Sea.

consistent with derivation from the Northern Highlands–Orkney area.

EOCENE SAND DEPOSITION AND DISTRIBUTION

The major Tertiary sand deposition occurred during the Palaeocene. However, during the Eocene, sand deposition continued in the coastal-deltaic environments but on the slope and in the deep water basin deposition was principally of shale. Some channelling is seen on the basin slope and the few sand bodies in the deep water basin appear to originate from these channels.

ACKNOWLEDGEMENTS

This paper owes much to concepts developed by Mr E. Oomkens. The help of both Shell and Esso staff is gratefully acknowledged. Permission for publication was given by Shell UK Exploration and Production Limited and Esso Exploration and Production (UK) Inc.

REFERENCES

Dunn, W. W., Eha, F. and Heikkila, H. H. 1973. North Sea is a tough theater for the oil-hungry industry to explore. *Oil and Gas Jl.*, **71**, 122–8.

Fowler, C. 1975. The geology of the Montrose Field. *Paper 36, this volume.*

Haner, B. E. 1971. Morphology and sediments of Redondo submarine fan, Southern California. *Bull. geol. Soc. Am.*, **82**, 2413–32.

Kruit, C., Brouwer, J. and Ealey, P. 1972. A deep water sand fan in the Eocene Bay of Biscay. *Nature phys. Sci.*, **240**, 59–61.

Lowe, D. R. and LoPiccolo, R. D. 1974. The characteristics and origins of dish and pillar structures. *Jl. Sedim. Petrol.*, **44**, 484–501.

Mutti, E. and Ricci Lucchi, F. 1972. Le Torbidit dell'Appennino settentrionale: introduzione all'analisi di facies. *Mem. Soc. Geol. Italiana*, **11**, 161–99.

Normark, W. R. 1970. Growth patterns of deep sea fans. *Bull. Am. Ass. Petrol. Geol.*, **54**, 2170–95.

Thomas, A. M., Walmsley, P. J. and Jenkins, D. A. L. 1974. Forties Field, North Sea. *Bull. Am. Ass. Petrol. Geol.*, **58**, 396–406.

Walker, R. G. and Mutti, E. 1973. Turbidite facies and facies associations. In *Turbidites and deep water sedimentation. Soc. Econ. Paleont. Mineral, Short Course notes*, 119–57.

Walmsley, P. J. 1975. The Forties Field. *Paper 37, this volume.*

Wentworth, C. M. 1967. Dish structure, a primary sedimentary structure in coarse turbidites. *Bull. Am. Ass. Petrol. Geol.*, **51**, 485.

DISCUSSION

Dr J. P. B. Lovell (Edinburgh University): What is the rate of change in the thickness of the turbidite sandstones away from supposed source, especially on the lower part of the fan?

Dr J. R. Parker: The sand-percentage of the total Palaeocene interval is around 60 in the Forties area, in the upper part of the fan, decreasing over a distance of some 100 miles to 10 percent and less in the distal part of the fan. A similar decrease in sand-percentage is seen laterally across the fan. The sand isopach-map (Fig 5) and the cross-plot of net sand-thickness against total Palaeocene-thickness, referred to in the text, are based on data from 77 wells over an area of 25 000 square miles, plus extensive seismic coverage. The North Sea Palaeocene is thus one of the better controlled examples of an ancient deep-water fan.

G. Wind (Amoco Europe): Which species indicate the deep-water environment of deposition?

Dr J. R. Parker: The shales intercalated with the turbidite sands generally contain a pure arenaceous fauna with scattered occurrences of radiolaria and planktonic foraminifera. The arenaceous assemblage consists predominantly of the following genera: *Rhabdammina* (*Bathysiphon*), *Bolivinopsis*, *Glomospira*, *Ammodiscus*, *Cyclammina* and *Haplophragmoides*.

D. G. Roberts (Institute of Oceanographic Sciences): Could Dr Parker comment on the end-Palaeocene relief of the delta and in particular on the immediate post-Danian relief of this basin?

Dr J. R. Parker: The immediate post-Danian relief of the basin was probably in the order of 3000ft. A deep-water basin still existed at the end of the Palaeocene with somewhat shallower water depths, say, 2000-2500ft.

Dr J. D. Collinson (Keele University): It is suggested that fans occur in two situations: at the foot of a fixed slope fed by canyons and at the foot of a prograding delta-slope. Is there any apparent difference between the fans attributed to these different situations? Are the delta-front fans seen to be fed by major channels in the supposed delta slope?

Dr J. R. Parker: The two situations are postulated on the basis of mapped sand-distributions: a fan fed by a canyon has a triangular shape, the apex corresponding to the proximal part of the fan, while a fan fed from a delta slope has a much wider proximal part. There is no seismic evidence for any major channels in the North Sea Palaeocene delta-slope.

Dr D. H. Matthews (Cambridge University): Do you have the information to say whether the axis of Neogene sedimentation follows the axis that one would deduce from structural contours on the base of the Palaeocene?

It is clear that we have in the North Sea basin the best-studied example of a sinking epeirogenic basin on Earth. The reason for such sinking is unknown and it is a major objective of the Geodynamics Project to seek for it. One thing that one could do is to shoot a long refraction line along the axis of the presently sinking basin in order to find the velocity structure of the upper mantle below the Moho. (There are difficulties—the basin has a kink in it, which limits the range.) Such lines have been shot in Europe by French and German seismologists and, last summer, in the UK by 80 teams controlled from the Universities of Karlsruhe and Birmingham. We plan to shoot such a line, perhaps in 1977, which is why I ask a question so little related to oil resources.

Dr J. R. Parker: Although there has been very little oil-company interest in the Neogene because of its lack of prospects and also the absence of any regional seismic reflectors in the Upper Tertiary, the information available suggests that the Neogene axis of sedimentation does follow the axis that we would deduce from the base-Palaeocene structural contours, this in turn over-lying the earlier Mesozoic graben-system. The Cenozoic subsidence of the North Sea is discussed by Clarke (1973, *Earth and Planetary Sci. Lett*, **18**, pp 329-32) who suggests an increase in subsidence rate throughout the Cenozoic. Our well data would substantiate this suggestion.

35

Lower Tertiary Tuffs and Volcanic Activity in the North Sea

By M. JACQUÉ

(Total Oil Marine Limited)

and J. THOUVENIN

(Compagnie Française des Pétroles)

SUMMARY

Traces of basic explosive volcanic activity are widespread in the lower Tertiary of North Sea wells, stretching 1000km north to south and 300km west to east. Volcanic ashes and minute glass debris are included in the sedimentary section, often associated with diatoms; their preservation varies with the conditions of deposition.

Several phases can be distinguished. The acme of the major one occurs during the most regressive period of the Lower Tertiary, immediately preceding the uppermost Palaeocene–Lower Eocene transgression. It forms an excellent time marker, associated with a seismic horizon plotted over most of the North Sea. The Danish and north German outcrops seem contemporaneous with this phase.

An earlier episode, probably in lowermost Thanetian, is now known in the Halibut Horst and Forties areas and in a few wells of the Viking Graben. Minor recurrent episodes are known from dispersed locations in the Lower Eocene. The Harwich outcrop probably correlates with one of these.

INTRODUCTION

Over about 400 000km², covering the whole North Sea, tuffaceous debris is present, interbedded with the shales and sands of the Lower Tertiary section. In more than 200 wells from a total 260 examined by us (Fig 1), their presence has been recognized either from direct examination of cores, sidewall cores and cuttings, or from the characteristic shape of the logs. The great abundance of a particular diatom is also diagnostic of the tuffaceous horizons. In the few wells where they have not been identified, the time equivalent formations are either shallow and drilled in large diameter (southern North Sea) or have been deposited in conditions unfavourable to the preservation of these fragile components. For the time being, four periods of activity, of very uneven importance, have been recognized.

GENERAL DESCRIPTION

In the absence of pelagic foraminifera, our stratigraphy is derived from a zonation based on the palynoplankton. The nature and relative importance of the miospores and the micro-plankton on the one hand and unconformities observed in certain wells on the other, suggest a correlation with the formations observed in the Paris and Anglo-Belgian basins (Fig 2) and the main regressions and transgressions of the Lower Tertiary. Occasional arenaceous and very rare pelagic foraminifera have also been used in establishing the correlations. A tentative parallel has been drawn between the absolute dating, foraminiferal zones, and main tectonic events relating to the opening of the North Atlantic. This may be subject to revision with further progress of our knowledge.

One main zone, characterized both by the abundance of material and areal extent, occurs during the most regressive phase of the Lower Tertiary, during our middle nt 2a palynozone, dominated by the abundance of taxiodaceae. During that episode, from about 55 to 53my, the 'Sparnacian' lignitic shales of the Paris Basin, the Woolwich Beds and upper part of the Reading Beds in the London–Hampshire Basin, were deposited, and the Artois Sill acted as a barrier for the first time, probably more as a consequence of the general regression rather than its uplift, as is often stated in the textbooks. Early manifestations of this period of main activity are present in the uppermost part of the preceding palynozone, our lower nt 2a, dominated by the dinoflagellate *Wetzeliella hyperacantha*, in sidewall cores from the TOM 3/14-1 and 3/15-1 wells. Final minor manifestations of the same period have been recognized in the basal part of the overlying transgressive upper nt 2a palynozone, correlated with the Sinceny Sands of the Paris Basin and the Blackheath and Oldhaven Beds of the London Basin (*Cyclonephelium ordinatum* and *Deflandrea oebisfeldensis* zone), for example in our TOM 3/25-1 well.

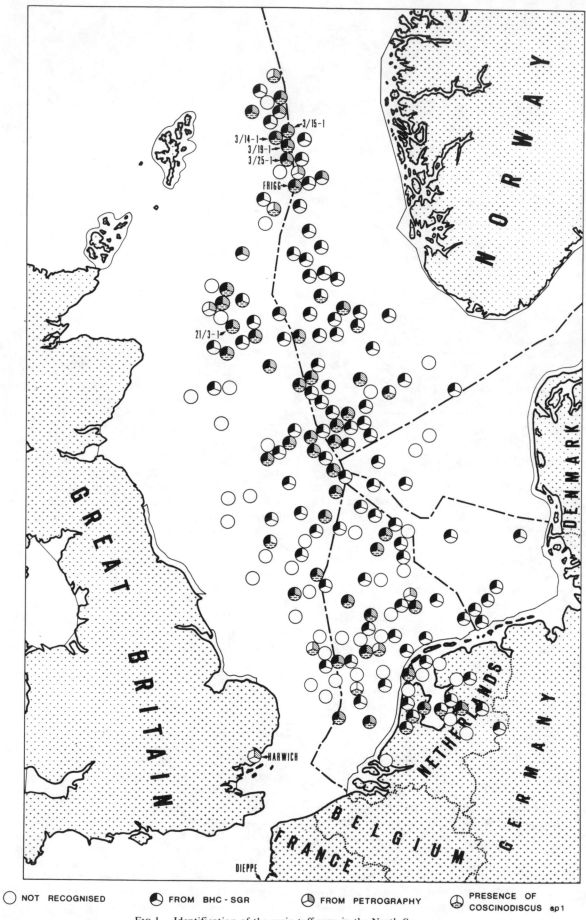

FIG 1.   Identification of the main tuff zone in the North Sea.

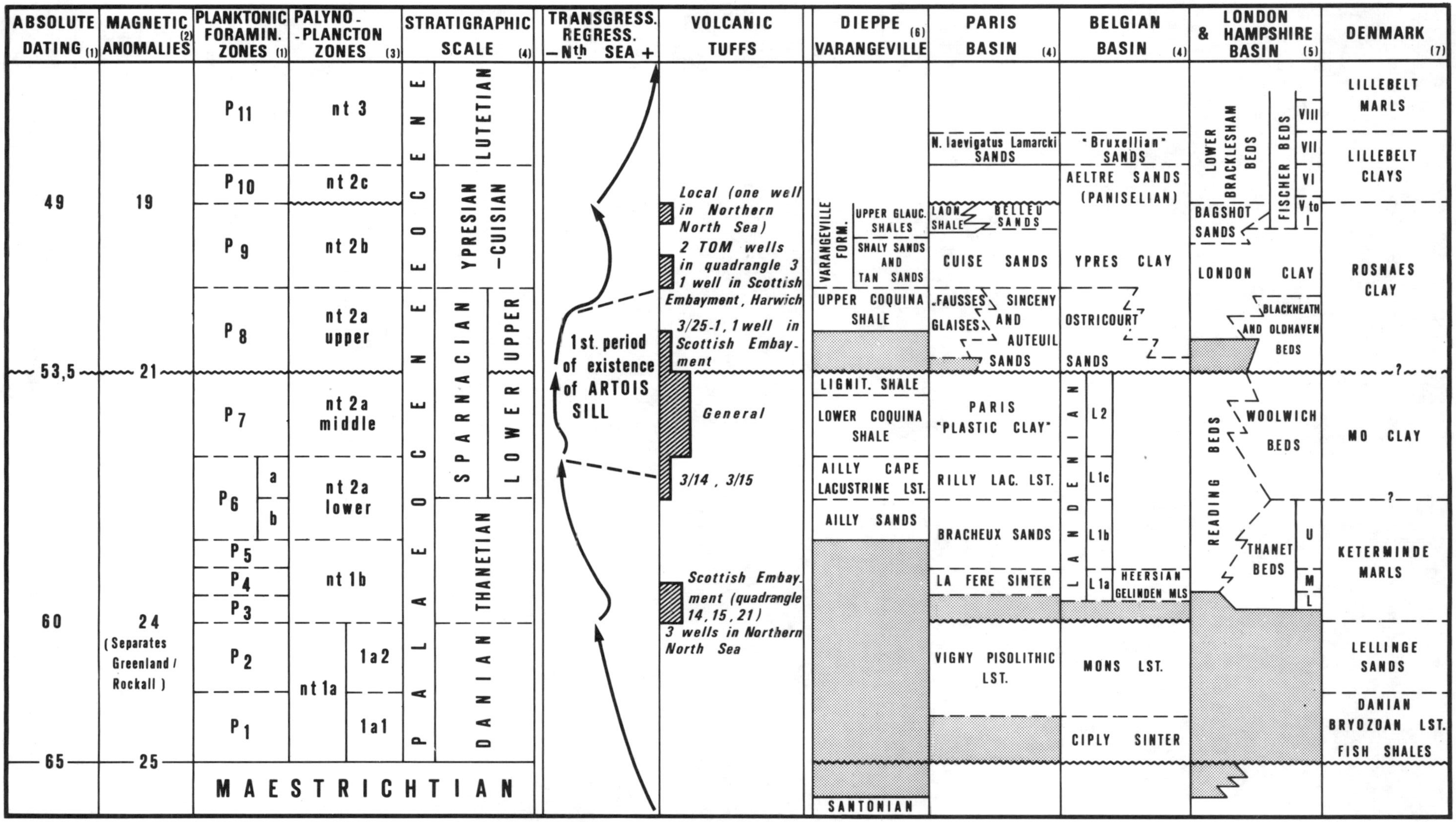

FIG 2. Tentative correlation of stratigraphic, tectonic and volcanic events in Lower Tertiary.

(1) From BERGGREN, 1972
(2) From VOGT and AVERY, 1974
(3) Joint TOTAL/ELF zonation

(4) Adapted from FEUGUEUR, 1963 and CURRY, 1967
(5) From CURRY, 1965
(6) From BIGNOT, 1973

(7) Palynologically unchecked

An earlier volcanic episode, separated from the main zone, occurs towards the base of our nt 1b or *Eisenackia crassitabulata* palynozone, which corresponds to the base of the Thanetian, about 60my (magnetic anomaly 24). First observed in our TOM 21/3-1A well, it has since been found again in many wells of quadrangles 14, 15 and 21 and more recently in TOM 3/19-1 and two other wells in the Viking Graben.

Later minor episodes occur in the basal Eocene nt 2b or *Wetzeliella edwardsi* zone (Ypresian–Cuisian), during the deposition of the Ypres and London clays in the Anglo-Belgian basin and the Cuise Sands in the Paris basin. Near the base of this zone, tuffaceous elements have been observed in the TOM wells 3/14-1 and 3/19-1 and in one well from the Halibut Horst area. The Harwich occurrence (Elliott, 1971) appears contemporaneous with this. A final minor episode, at the end of the same period, is so far limited to one well of the northern North Sea. Lower Tertiary volcanism seems to cease about 49my (magnetic anomaly 19). At least, no younger tuffs have been observed so far in the Palaeogene of the North Sea wells in our possession.

LITHOLOGY AND MINERALOGY

A thorough examination of the lithological, mineralogical and chemical characters is possible from a Schlumberger MCT core, cut in the main zone of our TOM 3/15-1 well (Fig 3). The variations in lithology reveal, in geochronological order, a decimetric succession of generally clear vertically sorted coarse-grained levels, followed by fine to very fine darker levels, either volcanic or sedimentary, often thinly bedded. Under transmitted light, the coarse levels contain weathered volcanic glass shards, plagioclase phenocrysts and a pyritic and very calcareous cement (Plate I, B). The shards, brown, angular, with vesicular clasts, are characteristic. An epitaxic calcite film often covers the glass or the detrital material, giving it a particular sheen. Under reflected light (Plate I, E) the vitreous elements, coated by clay material derived from weathering, are characteristic in shape and colour, with their brownish heart. The periphery varies from greyish to different shades of green, the colour variation not being a criterion of any particular horizon. The grains of the coarse volcanic beds are sorted, ranging in size from 200μ at the base to 1μ at the top, an arrangement typical of an aeolian fall of volcanic ashes, from the early coarse elements to the later tropospheric fine ones (Plate I, A). The core is, then, a succession of rhythmic volcanic beds of explosive origin (Plate I, D) alternating with either a second volcanic sequence or, if there is a period of rest, with the sedimentary deposition of shales and silts with minor quartz, feldspars, micas and occasional lignitic debris (Plate I, C). Such a

rhythmicity is evidently impossible to observe from sidewall cores and cuttings.

Under the microscope, the volcanic shards, generally aphyric and vitreous, do not afford any clue to the original petrographic nature of the lavas (Plate I, F). However some strongly weathered dark ash granules contain some small microlithic plagioclases. Occasionally, some vesicles in the vitreous elements are filled with a certain amount of tabular crystals of zeolites. The nature of these cannot be determined optically. In addition to these primary minerals, others derive from the alteration of the glass (chlorites, shales, iron and titanium oxides) or are secondary infilling of the voids (mainly carbonates).

A total mineralogical X-ray analysis has been conducted. The tuffaceous horizons are characterized by the low proportion, or absence, of quartz and the presence of minerals directly linked to the volcanic debris or their alteration products: plagioclases, anatase, zeolites (analcime and clinoptinolite), montmorillonite, and swelling chlorite. Calcite, siderite and pyrite are also present, derived mainly from the environmental deposition. The interstratified detrital sedimentary levels have a very constant composition with about 50 percent quartz, 6 percent orthoclase and 5 to 8 percent pyrite, the rest being shale or very poorly crystallized glauconite.

Some coarse levels of the core have also been submitted to fluorescence X-ray analysis. The total rock samples show elements derived at the same time from the volcanic glass, the microlithic granules and the sedimentary environment. Most of the calcium, iron, and probably the basic ions indispensable to zeolite formation apparently came from the latter. This is confirmed in some analyses by abnormally high CaO content (17 to 22 percent) for a volcanic rock, due to secondary calcite. The other analyses give, for the major elements, a normal distribution for a basic volcanic rock, and the results are considered significant. The mineral composition corresponds to a basalt, or better to a basanite with analcime (the analcime detected from the X-ray analysis being primary). The volcanism is consequently basaltic, basic and continental. Its explosive character is not linked to its nature, but to the presence of water.

THE INTERSTRATIFIED SEDIMENTS

The sediments interbedded with the volcanics vary in lithology according to the environment. In addition, the source of the detrital material influences the preservation of the volcanic material. In a shaly sequence of low energy, as in 3/15-1, the tuffs are preserved with their bedding, texture and sorting. In such cases, very characteristic shapes of the sonic and gamma-ray logs are observed (Fig 4). The sonic peaks may be due to rapid successions of volcanic sequences,

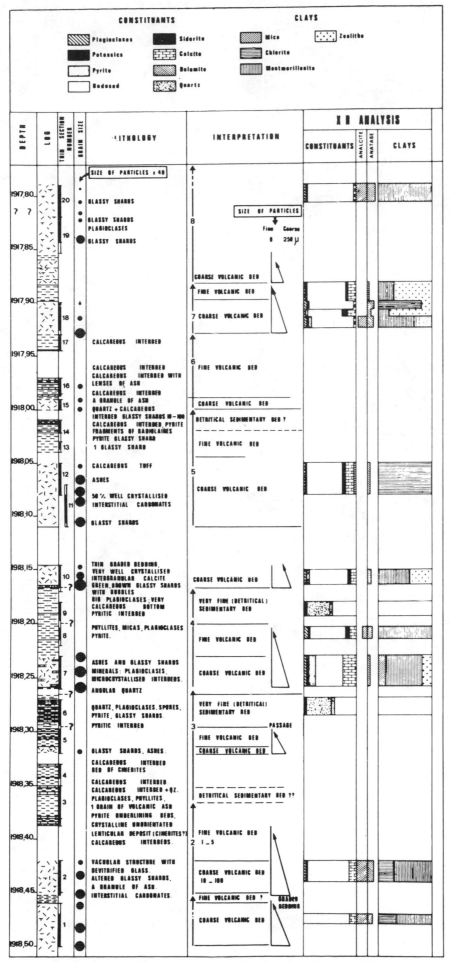

FIG 3.   Petrographical and sedimentological log of core from Well 3/15-1.

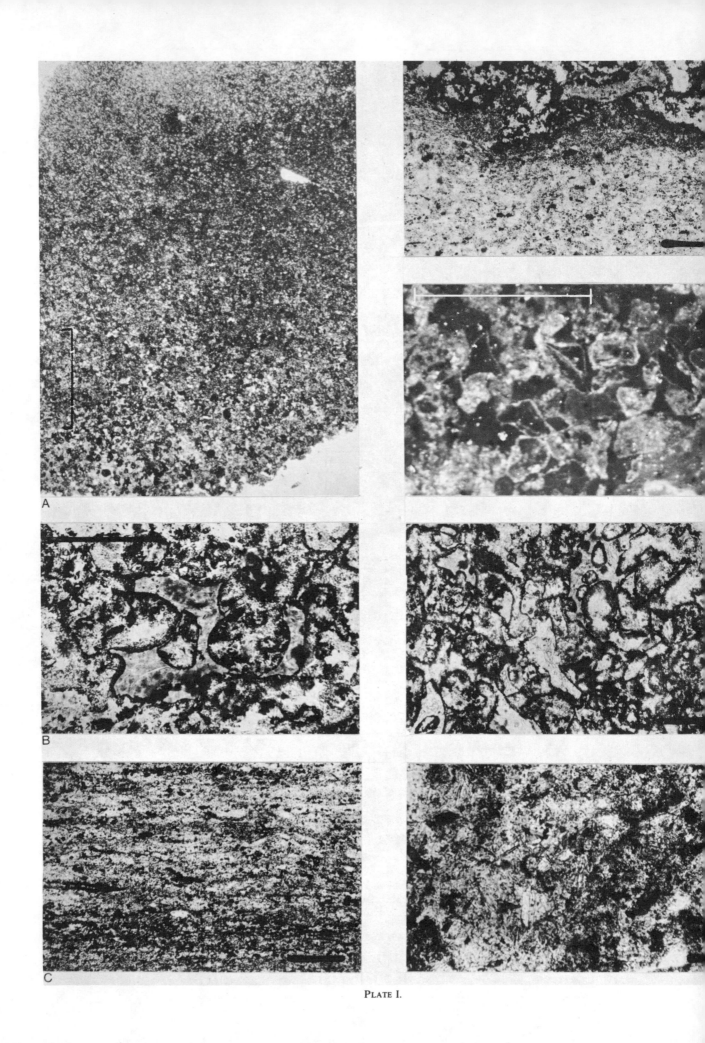

PLATE I.

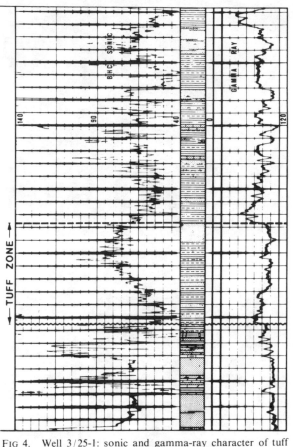

FIG 4. Well 3/25-1: sonic and gamma-ray character of tuff zone in a shaly environment.

perhaps with calcareous cementation; the low radioactivity is linked with the relatively small amount of argillaceous material. In a more sandy deposition, as for the Thanetian volcanism of 21/3-1A or the Frigg wells, the preservation is not as good, the volcanic elements being stirred up with the sand grains in this medium energy environment. In addition, their texture is disarranged or obliterated. The log shape varies with the lithology.

In a high energy environment—beach or continental deposits—reworking and weathering would rapidly destroy the tuffs and this probably accounts for the fact that they have not been recognized in some wells close to or on the East Shetland Platform, or in Varangeville or many Dutch wells.

Environment and rate of deposition have, then, strongly influenced the thickness, individualization and preservation of each volcanic episode. It is thus impossible to use the tuff-zone isopachytes to trace the sources of volcanic material, as the "net ash" cannot be properly established.

The faunal content of the main tuff zone is dominated by the great abundance of *Coscinodiscus sp.1*, the often pyritized internal mould of a large diatom. This typical strongly biconvex microorganism was first described from the tuff zone of northern Germany. Its existence is not limited to this particular horizon, and it may occasionally be found as high as the Miocene (Plate II, A-E). Another species, *Coscinodiscus sp. 2* is biconcave and found throughout the Lower Tertiary (Plate II, F). The extraordinary blooming of the species during the volcanic episode may be due to the enrichment of the sea water in silica and/or CaO, linked with the volcanic phenomenon itself.

THE OTHER TUFFACEOUS HORIZONS

These have been briefly described before, none of them having at present the large areal extension of the main zone. The mineralogy, petrography and morphology of the tuffaceous horizons of the lower zone (nt lb) are similar to those of the main zone, the only difference being in the more clastic and higher energy depositional environment, which prevents the vertical sorting of the volcanic material, which is mixed with sand grains. In one well however, in quadrangle 15, the presence of strongly weathered brownish microlithic ashes has been observed. These differ from the microlithic debris of the MCT core of 3/15-1 in the main tuffaceous zone and have not so far been noted at the same time-level in other wells. In these circumstances, it is difficult to decide if this is a particular bed limited to a small area, or if it is influenced by the detrital supply, or even is a different alteration of the common microlithic glass.

The same absence of individual character continues into the lower Ypresian horizon (lower nt 2b). The Upper Ypresian (upper nt 2b) so called red tuffs (Plate I, G) are known from only one well in the northern North Sea and consist of vitreous angular vacuolar shards, brownish to colourless, dispersed in a pale reddish matrix made of organic fragments (radiolaria, sponges). Their morphology, very similar to the better preserved fragments of our MCT core are mixed with atypical sediments, which may suggest a reworking of the lower zone. An isolated late volcanic episode of small extent is also possible.

SURFACE OCCURRENCES

As early as 1918, surface outcrops were sampled and measured in ash tuff levels in the Lower Tertiary of north and north-east Limfjord in Denmark, associated with diatomite horizons (Andersen, 1938; Madirazza and Fregerslev, 1969). The sequence, analysed in detail, has shown as many as 179 successive grain-sorted millimetric to decimetric volcanic cycles, similar to those observed in our core. Correlation and thickness variations between several outcrops have shown a northern to north-western localization for the source. From chemical analyses, the petrography

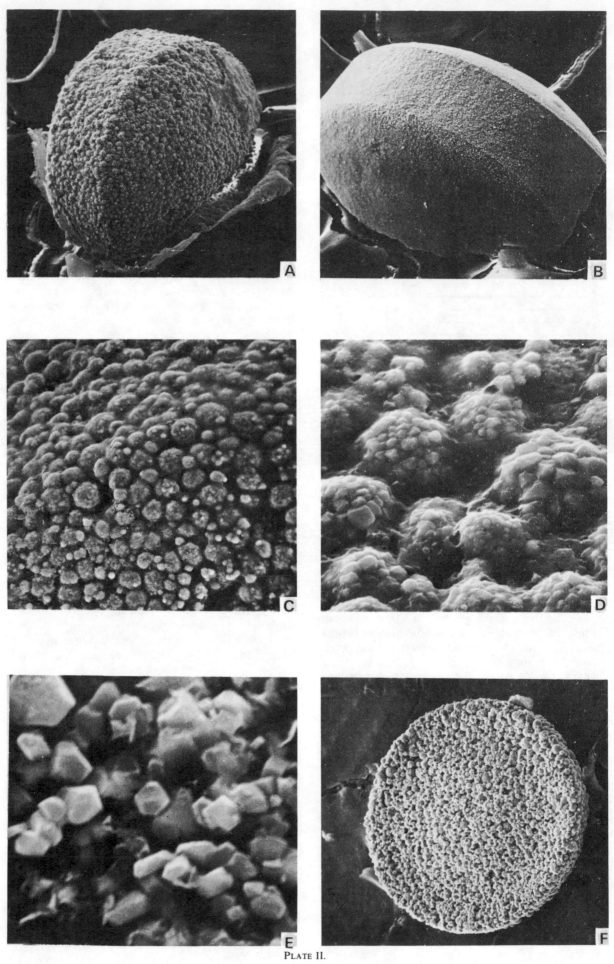

PLATE II.

ranges from basalts to dacites and from andesite to liparite.

In the North Jutland province of northern Germany a similar section of alternating tuffaceous levels and sediments has been measured (Andersen, 1938; Wirtz, 1939). The volcanic material is dark green, grain-sorted and magnetic and passes into clearer finely bedded sedimentary levels. The volcanic shards are angular, acicular or sub-rounded and of basaltic composition. An X-ray analysis shows the presence of the zeolite heulandite and of montmorillonite. Correlations are possible with the Danish outcrops and confirm the north-western origin. Unfortunately, we did not have the opportunity to examine Danish and German samples and to analyse their palynoplankton content for comparison.

The calcareous bands of the London clay were first described by Elliott (1971) as containing some volcanic material, in an outcrop occurring on the foreshore at Harwich. We have re-examined these and their petrography is similar to that of the North Sea wells. They include plagioclase phenocrysts, fine pyrite crystals, microlithic ash granules and some lignitic debris. The palynoplankton examination places them in our basal nt 2b (Ypresian) horizon, as in TOM wells 3/14a-1 and 3/19-1.

Despite our thorough sampling and examination, we have been unable to find any trace of the time equivalent volcanism in the section of Cap d'Ailly-Varangeville, near Dieppe, in France. The formations of the regressive phase of the late Palaeocene were, however, deposited in a very restricted, lagoonal to lacustrine environment, unfavourable to the preservation of the volcanic material.

INTERPRETATION OF ERUPTIVE DYNAMISM

The morphological, petrographical and mineralogical uniformity of the vitreous fragments of all our tuff zones show the repetitive character and uniqueness of the volcanic phenomenon producing the tuffs. The cyclicity finds expression in several paroxysmal phases of unequal magnitude, separated in time, but with a single magma type. The same type of material has been dispersed over a very large area, of the order of 400 000km². Each period of activity can be subdivided into a certain number of volcanic episodes, alternating with periods of rest or very reduced activity, as shown in our MCT core.

An estimate of the amount of volcanic material deposited during a single phase, assuming its continuity over the whole North Sea, can give us some points of comparison with similar phenomena, historic or at the present time.

For a cumulated 1cm thickness of net volcanic material over the whole North Sea, the volume of pyroclastic material would be 4km³. Single beds of 8cm have been identified, which would give a total volume of 32km³ for a single episode. As a comparison, the Santorini eruption has disseminated over 200 000km² of volcanic products, with thicknesses up to several tenths of metres. Other volcanic explosions have spread volumes equal or larger than 10km³, as for example Bezimianni with 25km³.

Each tuffaceous horizon can be considered as representing, in a stratigraphical succession, violent paroxysmal explosions, separated by periods of rest of several hundreds to thousands of years. The importance of the ash volume for each episode—several cubic kilometres—equates such volcanic eruptions with Santorini (Mediterranean Sea), Mount Katmai (Alaska) or Hekla (Iceland).

Generally, this type of explosive dynamism characterizes an acid, not a basaltic volcanism. However, a high water content may allow a basaltic magma to become as explosive as an acid one. Such examples are known, the most recent being the 1947 Hekla eruption, which dispersed ½km³ of pyroclastic material of basaltic composition. A basaltic magma acquires such an explosive potential whilst passing through water-saturated formations during its ascent to the surface. Such conditions might have produced the North Sea tuffs, the eruptions taking place in areas where large quantities of water were trapped underground, providing the water for such an explosive phenomenon. In early phases of shallow submarine eruptions, similar conditions may occur, but they would cease rapidly with the growth of the volcano above sea-level. On the other hand, deep submarine eruptions are unable to provide large scale pyroclastic dispersion, producing, rather, localized palagonite type tuffs.

ORIGIN AND AGE OF THE VOLCANISM AND ITS RELATIONSHIP TO NORTH ATLANTIC OPENING

The hypothesis of a single source for all the tuffs of the North Sea is plausible but is neither proved nor necessary. Several synchronous volcanoes, belonging to the same volcanic province and thence having the same type of dynamism and chemical composition, are more probable. The spreading over Denmark and northern Germany of a very large volume of pyroclastics (evaluated as 4200km³ by Elliott, 1971) implies the vicinity of the source; this has been suggested, from magnetic and gravimetric anomalies, to lie in the Skagerrak (Sharma, 1970). Tuffs in the wells of the Dutch and German sectors of the North Sea may have the same origin.

For the central and northern North Sea other sources are to be found, inasmuch as the wind transportation of ashes must have been from north and north-west to south and south-east. This was earlier demonstrated by the study of the German and Danish outcrops. In addition, an ecological study of the faunal content of the Lower Tertiary of Denmark (Bonde, 1974) suggests an upwelling along the eastern coasts of the North Sea which implies the same predominant winds. The main source of the pyroclastic material is consequently to be found in the volcanism of the "British Province" of western Scotland and its possible north-western extensions, now disappeared.

The period of volcanic activity shown in our correlation chart extends from 60 to 49my, compared with absolute datings ranging from 65 to 49my in the British Province (Evans *et al.*, 1973), beginning at the Cretaceous–Palaeocene boundary.

Traces of later Danian activity may yet be identified in North Sea wells, but we have so far found only one unquestionable glass shard, in a well from quadrangle 21, not taken into account in this discussion, and this possibly came from polluted cuttings. We must also acknowledge, however, that there is a discrepancy between our period of acme (53.5 to 55my) and that given by other authors (about 60my) for which we do not at present have a satisfactory explanation. On the other hand, the extinction of volcanism is synchronous to both interpretations.

In any case, the volcanism and tuff horizons of North Sea wells can be correlated and correspond to an essential volcanic phase during the geological history of the Lower Tertiary, and associated with a major regressive phase of worldwide implications (Rona, 1973). Such a volcanism must be connected with important tectonic events, such as vertical or tangential movements on at least a north European scale. The North Atlantic opening at this very moment went through a very dynamic phase, with a maximum activity around 55my which fits with our whole figure (Pitman and Talwani, 1972; Vogt and Avery, 1974). More detailed correlations will be possible in the future with a better knowledge of these movements and of the related magnetic anomalies, but at present such detail is still out of our reach.

ACKNOWLEDGEMENTS

This paper is the result of a work group involving several specialists from the Central Laboratory of the Compagnie Française des Pétroles in Bordeaux and MM. Jourdan and Gauthier of the Laboratory of Petrography of Professor Brousse, at Orsay University (Paris XI), who have also made an important contribution to its preparation. The authors thank the management of the Compagnie Française des Pétroles for permission to publish.

EXPLANATION OF PLATES

Plate I
A.   3/15-1. 1918.14–1918.17m
Thin section; direct light.

Scale, in lower left corner, is 5mm.
B.   3/15-1. 1918m.
Transmitted natural light.
Scale, in top left corner, is 500μ.
C.   3/15-1. 1918.27–1918.38m
Interstratified sediments; transmitted natural light.
Scale, in lower right corner, is 500μ.
D.   3/15-1. 1918.23–1918.26m
Contact between sequences; transmitted natural light.
Scale, in lower right corner, is 500μ.
E.   Main tuff zone; reflected light.
Scale, in upper left corner, is 500μ.
F.   3/15-1. 1918.06–1918.08m
Transmitted natural light.
Scale, in lower right corner, is 500μ.
G.   Upper Ypresian "red tuff" horizon; transmitted natural light.
Scale in lower right corner, is 500μ.

Plate II
A.   Coscinodiscus Sp. 1 × 280.
B.   Coscinodiscus Sp. 1 × 250.
C.   Coscinodiscus Sp. 1 × 1000.
D.   Coscinodiscus Sp. 1 × 3500.
E.   Coscinodiscus Sp. 1 × 10 000.
F.   Coscinodiscus Sp. 2 × 280.

REFERENCES

Andersen, S. A. 1938. Die Verbreitung der eozänen vulkanischen Ascheschichten in Danemark und Nordwestdeutschland. *Z. Geschiebeforsch.*, **14**, 179–207.

Berggren, W. A. 1972. A Cenozoic time scale—some implications for regional geology and paleobiogeography. *Lethaia*, **5**, 195–215.

Bignot, G. 1965. Le gisement éocène du Cap d'Ailly (près de Dieppe, Seine Maritime). *Bull. Soc. Geol. Fr.*, (7) **VII**, 273–83.

Bignot, G. 1973. Esquisse stratigraphique et paléogéographique du Tertiaire de la Haute-Normandie. *Bull. Soc. Geol. Norm.*, **16**, 23–47.

Blondeau, A., Cavelier, C., Feugueur, L. and Pomerol, C. 1965. Stratigraphie du Paléogène du bassin de Paris en relation avec les bassins avoisinants. *Bull. Soc. Geol. Fr.*, (7) **VIII**, 200–21.

Bonde, N. 1974. Palaeoenvironment as indicated by the Mo Clay Formation (Lowermost Eocene of Denmark). *Tertiary Times*, **2**, 29–36.

Curry, D. 1965. The Palaeogene beds of South East England. *Proc. Geol. Ass.*, **76**, 151–73.

Curry, D. 1966. Problems of correlation in the Anglo–Paris–Belgian Basin. *Proc. Geol. Ass.*, **77**, 437–67.

Elliott, G. F. 1971. Eocene volcanics in South East England. *Nature phys. Sci., Lond.*, **230**, 9.

Evans, A. L., Fitch, F. J. and Miller, J. A. 1973. Potassium argon age determination on some British Tertiary igneous rocks. *Jl. geol. Soc. Lond.*, **129**, 419–43.

Feugueur, L. 1963. L'Yprésien du bassin de Paris. Essai de monographie stratigraphique. *Mem. Serv. Cart. Géol. det. Fr.*, 568 pp.

Fregerslev, S. 1969. An X-ray powder diffraction study of the Lower Eocene tuff sequence from Mønsted, Nth. Jutland. *Meddr. dansk. geol. Foren.*, **19**, 311–18.

Madirazza, I. and Fregerslev, S. 1969. Lower Eocene tuffs at Mønsted, North Jutland. *Meddr. dansk. geol. Foren.*, **19**, 283–310.

Norin, R. 1940. Problems concerning the volcanic ash layers of the Lower Tertiary of Denmark. *Geol. Foren. Stockh. Forh.*, **62**, 31–44.

Pitman, W. C. and Talwani, M. 1972. Sea floor spreading in the North Atlantic. *Bull. Geol. Soc. Am.*, **83**, 619–46.

Rona, P. A. 1973. Worldwide unconformities in marine sediments related to eustatic changes of sea level. *Nature phys. Sci. Lond.*, **244**, 25–7.

Ross, C. S. and Smith, R. L. 1961. Ash flow tuffs: their origin, geologic relation and identification. *Prof. Pap. U.S. geol. Surv.*, **366**, 81 pp.

Sharma, P. V. 1970. Geophysical evidence for a buried volcanic mount in the Skagerrak. *Meddr. dansk. geol. Foren.*, **19**, 368–77.

Vogt, P. R. and Avery, O. E. 1974. Detailed magnetic surveys in the North East Atlantic and Labrador sea. *Jl. geophys. Res.*, **79**, 363–89.

Wilcox, R. E. 1959. Some effects of recent volcanic ash falls with especial reference to Alaska. *Bull. U.S. geol. Surv.*, **1023-N.**, 476 pp.

Wirtz, D. 1939. *Das Altertertiär in Schleswig Holstein.* Gesellsch. für Geschiebforschung.

36

The Geology of the Montrose Field

By CLIVE FOWLER

(Amoco (UK) Exploration Co., London)

SUMMARY

The Montrose Field was discovered in December 1969 and was the first oil-field to be found in the United Kingdom sector of the North Sea. The field is located 130 miles (209km) east of Aberdeen in 300ft (91m) of water. The reservoir is a sandstone of Palaeocene age at a depth of approximately 8000ft (2438m) below mean sea level. This sandstone lies at the base of a thick section of Tertiary sediments comprised mainly of claystones and shales and is underlain by carbonates of Danian age. The Montrose structure is a large low-relief domal feature with an apparent area of closure of some 70 square miles (181km^2) at the top of the Palaeocene. The structure is only partially filled and has a maximum gross oil column of 190ft (58m). Recoverable reserves in the field range from 107 to 166×10^6 barrels from 329 to 450×10^6 barrels of oil in place.

INTRODUCTION

The Montrose Field was discovered in December 1969 with the completion of the successful oil well 22/18-1. It was the first oil-field to be found in the United Kingdom sector of the North Sea. The blocks containing the Montrose Field, 22/17 and 22/18, are parts of 1964 licences P 019 and P 020 which are operated by Amoco on behalf of the group composed of Amoco (UK) Exploration Company, The Gas Council (Exploration) Limited, Amerada Petroleum Corporation of the United Kingdom Limited and North Sea Incorporated. The field is shown on the geographical map of the North Sea (Fig 1), and is located some 130 miles (209km) due east of Aberdeen, Scotland, in depths of water varying between 290 and 310ft (88–94m). Named after the celebrated Scottish general, the Marquis of Montrose, the field lies 32 miles (51km) south-east of BP's Forties Field and 28 miles (45km) west-north-west of the Amoco group's Lomond Field.

HISTORY OF DISCOVERY AND DELINEATION

Early seismic reconnaissance work carried out in the northern North Sea prior to 1964 had indicated the presence of a large north–south trending, low-relief, anticlinal structure with 30 square miles (78km^2) of closure at the base of the Tertiary in United Kingdom blocks 22/17 and 22/18. The two blocks were awarded to the Amoco Group in 1964, as part of the first round of British Government licence allocations. At the time of the award, the geology of that part of the North Sea was largely unknown. Significant gas discoveries had been made onshore in the Netherlands, and there were high hopes of finding gas in commercial quantities in the southern North Sea. In 1964, however, the hydrocarbon potential of the northern North Sea was regarded as somewhat uncertain and speculative, although it was realized that the area could be oil-prospective. Exploration through the mid and late 1960s was largely centred on the southern North Sea gas province, with considerable success. The majority of the exploration activity in the northern North Sea was concentrated in the Norwegian sector, and despite some small discoveries of oil in the Danish sector and gas/condensate in the Norwegian sector, the results until late 1969 were disappointing. Few northern wells were drilled in United Kingdom waters, and by May 1969, when the Amoco 22/18-1 wildcat well was spudded, only 5 wells had been drilled north of Latitude 57°N.

22/18-1 was spudded on May 1, 1969 using the semi-submersible rig "Sea Quest". It was located to test the low relief domal structure at the base Tertiary which had been defined by the original reconnaissance seismic and detailed by some 450 miles (724km) of additional programme carried out in 1967 and 1968. The detailed seismic had indicated that the prospect had two culminations on its longer north–south axis, and the 22/18-1 well was situated primarily to test sandstones of the basal Tertiary on the southern feature. The well discovered oil in Palaeocene sandstones which were encountered below 8180ft (2493m) subsea. 100 gross ft (30m) of oil-bearing sandstone were found above the oil–water contact at 8256ft (2516m) subsea, and on a short term test the well flowed 40° API gravity, low sulphur oil at a maximum rate equivalent to 2160 barrels per day on two 1-in chokes at a flowing tubing pressure of 200psi. The testing programme was curtailed by bad weather, and the well was completed on December 28, 1969. In view of the low-rate test and the relatively small oil-column (92 net ft (28m)), the discovery was considered marginal and it was decided not to pursue development plans at that time.

A further 160 miles (257km) of seismic programme was shot over the prospect between late 1969 and early 1971 and the interpretation of these data indicated the likelihood of additional closure on the structure together with a third culmination on its north-west flank. The discovery was confirmed by delineation well 22/18-2 in November 1971. This well, located approximately 5 miles (8km) north of the discovery well, again

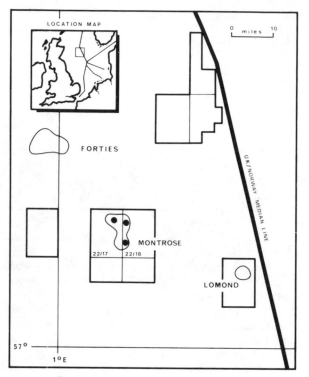

FIG 1.   Location map of Montrose Field.

found oil in the Palaeocene sandstones below 8050ft (2453m) subsea. 185 gross ft (56m) of oil-bearing sandstone were encountered, and on test, a maximum flow-rate equivalent to 4190 barrels of oil per day was recorded at a flowing tubing pressure of 725psi. This increased flow rate together with the confirmation of the seismic and geological interpretation gave more encouragement to delineate the structure, and the field was further defined in June 1972 when the 22/17-1 well was successfully completed some 2.4 miles (3.9km) west of 22/18-2. 170 gross ft (52m) of oil-bearing sandstones were encountered below 8125ft (2477m) subsea, and the well tested oil at a rate of 2200 barrels per day with 700psi tubing pressure. 22/17-1 thus completed the appraisal of the structure, which was proved to be only partially filled with oil. The three wells therefore confirmed the Montrose Field with a maximum oil column of 190 gross ft (58m) and a closed area above the oil–water contact of some 24 square miles (62km²).

STRATIGRAPHY

The Montrose Field is situated in the central portion of the large northern North Sea Tertiary Basin some 26 miles (42km) west-south-west of the median line dividing the United Kingdom and Norwegian sectors of the Continental Shelf. The axis of deposition of this Tertiary basin is in a general north-west to south-east direction and Montrose is situated on trend with other proved hydrocarbon accumulations in the area such as the Forties, Lomond and Cod Fields. At Montrose, a thick sequence of Quaternary and Tertiary sediments consisting primarily of grey claystones and shales is present. Sandstones occur in the Plio–Pleistocene interval and become the dominant lithotype in the Palaeocene, while white Danian limestones are present at the base of the Tertiary. The Tertiary is underlain by Upper Cretaceous Chalk.

Post-Palaeocene

The Montrose delineation well 22/18-2 has been selected as the type-well for the field, and its stratigraphic column is presented in Fig 2. Holocene to Pleistocene sediments occupy the first 1500ft (457m) of section, and the interval is characterized by medium to dark grey, soft and sticky clays which are variably calcareous and carbonaceous. Also present are numerous bands of poorly sorted, unconsolidated sands in which shell debris is very common. These loose sands are particularly important since, as possible shallow gas or lost circulation zones, they represent potentially serious drilling hazards.

The post-Palaeocene Tertiary sequence in the Montrose Field area is characterized by a thick sequence of argillaceous sediments which range in age from Pliocene to Lower Eocene. The sequence has been dated primarily by micropalaeontological assemblages assisted where possible by log correlation. The section is essentially an extremely monotonous succession of dominantly grey claystones which are occasionally brownish and greenish grey. The claystones give way to shales which are variably calcareous, rarely glauconitic and pyritic. In the Oligocene and Eocene, minor stringers of argillaceous limestones and dolomites occur and these are generally brownish or beige in colour. Despite the uniform nature of the succession, three distinctive and useful lithological markers occur. The first of these horizons is found at approximately 5000ft (1524m) where the incidence of a dark brown shale together with a distinct change in character of the Gamma Ray log marks the top of the Oligocene. The second occurs at a depth of some 7500ft (2286m) where a varicoloured grey, blue and green shale unit together with a Gamma Ray log character change enables the top of the Middle Eocene to be identified. The third prominent lithological marker is found towards the base of the Eocene at an

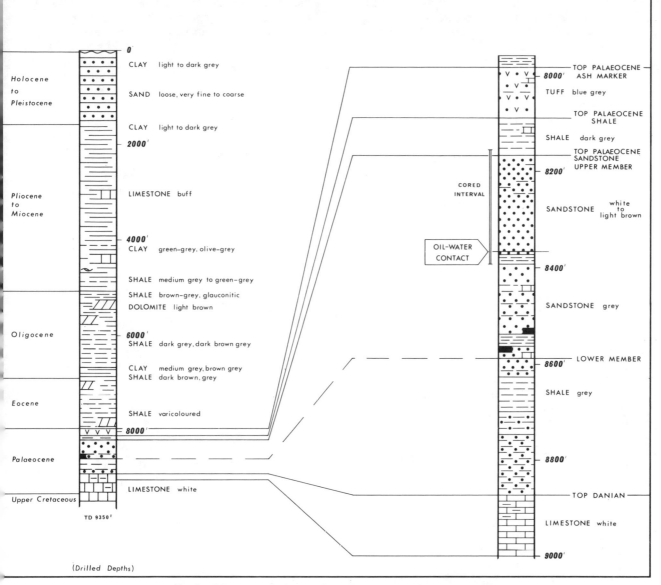

FIG 2.   Montrose Field. Graphic section of Amoco Well 22/18-2.

approximate depth of 7800ft (2377m), and is represented by the appearance of a distinctive red-brown shale.

Palaeocene

The Palaeocene succession in the Montrose Field has been divided into four basic lithostratigraphic units. The youngest of these has a distinctive lithological and log character and is referred to as the Palaeocene Ash Marker. It is a good reflector upon which seismic maps can be constructed and it usually occurs in conjunction with the appearance of the foraminifera *Globigerina triloculinoides*, which is generally believed to be indicative of a Palaeocene age. The top of the Palaeocene in Montrose is therefore conveniently picked at the appearance of the Ash Marker, although Palaeocene microfossils can occur for a limited distance above this horizon. The

Ash Marker is an interbedded sequence of light grey shales, blue-grey, mauve and black-flecked tuffs, minor thin carbonates and rare fine-grained sandstones. The unit is typically some 100ft (30m) in thickness. The origin of the volcanic tuffs is somewhat uncertain. Similar tuffs lying beneath Lower Eocene sediments are present in Denmark and it has been suggested that their source might be from volcanic centres beneath the Skagerrak. North Sea tuffs, however, are very widespread and their farthest known westward occurrence is closer to the Palaeocene Hebridean igneous province of Scotland rather than to a possible volcanic centre in Denmark. The Montrose tuffs may therefore have had their origin to the west of the field in the Hebridean province of the British Isles.

The second lithostratigraphic unit forms the cap-rock to the oil accumulation in the underlying reservoir sandstones. It is approximately 80ft

(24m) thick and is termed the Palaeocene Shale. The unit is typically composed of a dark grey to black micaceous and pyritic shale which is variably carbonaceous, often containing plant fragments. Thin stringers of very fine-grained sandstone grading to siltstone may occur, and rare stringers of muddy limestone may also be present. The Palaeocene Shale is generally barren of microfossils which may suggest its deposition in a restricted fresh-water to brackish lagoonal environment.

The Palaeocene Sandstone, which forms the third unit, constitutes the reservoir horizon in the Montrose Field and may itself be divided into two members. These are an upper massive sandstone sequence and a lower sequence comprising predominantly of interbedded sandstones and shales. The upper massive sands reach 400 gross ft (122m) in thickness, are white to light grey in colour, becoming light brown or buff when oil-stained, and are frequently very friable. Sorting is usually poor to moderate, with grain sizes ranging from fine to very coarse, but in general, the sequence tends to fine in an upward direction. The sand grains are subangular to well rounded, and are set in an interstitial illitic clay matrix which can reach a proportion of some 20 percent of the lithotype. The illitic matrix is frequently mixed with carbonaceous material and is rarely replaced by fine-grained calcite. The sand grains comprise 60–70 percent quartz, up to 10 percent feld-spar (principally fresh plagioclase), 5 percent mica of which the dominant form is muscovite, and up to 10 percent of rock fragments and accessory minerals. The rock fragments include metaquartzite, altered basalt and gneiss, while common accessory minerals are glauconite, zircon, sphene and tourmaline. Porosities and per-meabilities in the massive sands are generally moderate to high. The massive sands contain rare stringers of very calcareous tight sandstones which are generally grey and medium-grained, and which contain up to 50 percent of calcite cement. The massive Palaeocene sand member contains frequent intercalations of medium to dark grey, silty, micaceous shale, while plant fragments and thin, impure coals can also be present. The principal sedimentary structures in the sequence are gentle cross-bedding and laminations, and rare slump structures are also evident. Slight graded-bedding is sometimes seen within laminated sections which exhibit cross-bedding.

The high proportion of metamorphic rock fragments in the sandstones suggests a metamorphic source area, while the many well-rounded quartz grains are indicative of a high degree of transport and may represent reworked aeolian sedimentary source rocks of possible Devonian or Permo-Triassic age. The most likely source area for the Palaeocene sands is considered to be the Scottish Massif situated to the west of the Montrose Field. The environment of deposition of the reservoir sands is a subject which is still under study and which will probably remain unresolved until further geological data become available from the development wells. During the Palaeocene, sand-size material was carried into the northern North Sea Tertiary Basin by rivers flowing off the highland source areas, and it has been suggested that the Palaeocene producing horizons are formed of turbidites which were laid down at the base of the original Continental Slope. However, petrographic studies of cores taken through the massive sands in the Montrose Field suggest an alternative hypothesis—that they may be a product of a deltaic environment with occasional shallow marine incursions into the delta indicated by the presence of the calcareous sandstone stringers.

The lower member of the Palaeocene Sandstone is a sequence of interbedded sandstones and shales which in the 22/18-2 delineation well is approximately 300 gross ft (91m) in thickness. Shales are the major lithotype and are usually medium to dark grey or greenish grey, silty, slightly micaceous and pyritic. The sandstones are typically similar to those of the upper member though generally finer in grain size. Other lithologies present include brown siltstones and rare white limestones.

The oldest unit in the Palaeocene sequence of the Montrose Field is termed the Danian and this may also be divided into an upper and lower member. The thickness of the upper member was 280 gross ft (85m) in the 22/18-2 well, and creamy white, soft chalk with interbedded grey or greenish grey shales forms the major rock types, although brownish argillaceous, sandy limestones and light grey marls may also be present. Although assigned a Danian age on micropalaeontological grounds, it is possible that the upper part of this section is stratigraphically younger, since the Danian fauna may have been reworked. The upper member of the Danian may represent a transition between the underlying clean carbonate depositional phase and the predominantly clastic deposition of the overlying succession. This change may reflect an increase in the supply of terrigenous material to the basin, caused by uplift and erosion of the upland source areas. It could be related to the opening of the North Atlantic between Greenland and North West Europe dated at some 60my.

The lower member of the Danian, which reached 170 gross ft (52m) in 22/18-2, is almost entirely composed of greyish white, slightly argillaceous chalk. Rare marl beds may also be present, and these are typically light grey in colour.

The boundary between the Danian and the underlying Upper Cretaceous is not well defined lithologically or palaeontologically. In Montrose, this boundary is conveniently picked at a prominent character change on the Gamma Ray-Sonic log where the distinction between the slightly argillaceous Danian chalk and the clean chalk of the Upper Cretaceous can be readily seen.

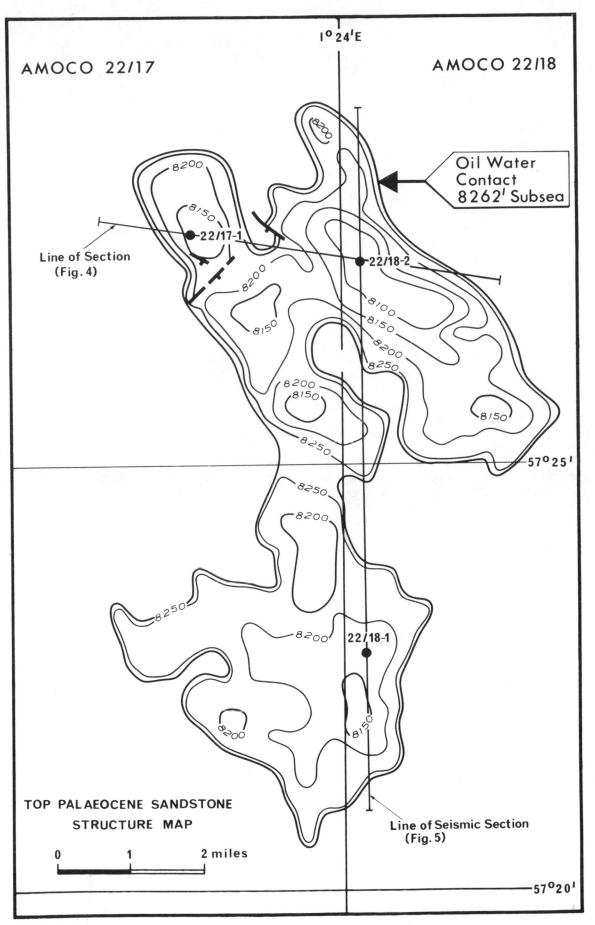

AMOCO 22/17

AMOCO 22/18

1° 24' E

8200

8200

8150

22/17-1

Oil Water
Contact
8262' Subsea

Line of Section
(Fig. 4)

8200

8150

8100

8150

8200

8250

22/18-2

8200

8150

8200

8150

8250

8150

57° 25'

8250

8200

8250

8200

22/18-1

8150

8200

8150

Line of Seismic Section
(Fig. 5)

TOP PALAEOCENE SANDSTONE
STRUCTURE MAP

0   1   2 miles

57° 20'

FIG 3.   Contours on the top of the Palaeocene Sandstone.

STRUCTURE

The Montrose structure is essentially a gentle low-relief domal feature. At the top of the Palaeocene it has areal closure of approximately 70 square miles (181km²), and a maximum vertical closure of some 400ft (122m). The structure, however, is only partially filled and has a maximum oil column of 190 gross ft (58m) within an area of hydrocarbon limits of 24 square miles (62km²). The contour-map at the top of the Palaeocene Sandstone (Fig 3) shows the Montrose Field to consist primarily of three main structural culminations. The two culminations which make up the northern portion of the field trend in a north-west to south-east direction, while the southern culmination appears to have a north–south direction superimposed on this underlying trend. The field is largely unaffected by faulting, although some minor dislocation is mapped on the north-west flank.

The exact position of the oil–water contact in Montrose has not yet been finally resolved. The three wells drilled so far have been located on the three structural culminations and each has had a slightly different oil–water contact. Although the range of difference is only of the order of 60ft (18m), the oil–water contacts observed in each of the wells indicate the possibility that the field may consist of three discrete oil accumulations. However, the difference in the water contacts could be the result of undefined structural, lithological or hydrodynamic phenomena, and this question is another which will probably remain unanswered until further data from the development wells becomes available. For the purposes of reserve calculations, however, it has been assumed that the field is single oil accumulation, and the oil–water contact shown in Fig 3 is therefore the average of the three contacts observed.

A structural cross-section of the North Montrose area is presented in Fig 4. The structure was probably initiated by block-faulting in the late Palaeozoic as a result of the Hercynian phase of earth movements, and was subsequently modified during the late Jurassic Cimmerian phase. The area remained positive throughout the Lower Cretaceous, and was then subject to the widespread Upper Cretaceous marine transgression which resulted in the deposition of a complete Chalk sequence. The Montrose area remained relatively stable during the late Mesozoic and early Cenozoic, although slight thickness variations within the Chalk suggest that it was weakly affected by the late Cretaceous Laramide phase. A full Tertiary succession was then deposited and the Palaeocene structure developed to its full extent by the Miocene Alpine mountain building episode. Structural expression diminishes progressively above the Palaeocene and disappears in the Pliocene, as can be seen in the seismic section presented in Fig 5. Growth on the structure had therefore probably ceased by the late Miocene

although subsequent differential compaction could have occurred to allow structural expression to persist into the Pliocene.

Much research has been carried out since the discovery of the field in an attempt to determine the origin of the Montrose oil. It seems likely that organic shales of Jurassic age are the primary source although some oil and gas may have been generated by Palaeocene shales.

RESERVOIR PARAMETERS AND RESERVES

The Montrose Field has a maximum oil column of 190 gross ft (58m) while the average net is only of the order of 60ft (18m). Available geological evidence indicates that the field is completely underlain by water, and no gas–oil contact was observed in any of the wells so far drilled.

Core analysis indicates an average porosity within the oil-bearing massive Palaeocene sand member of 22.8 percent. In this determination, calcareous sandstones were deleted from the data set and porosity values reduced by 0.9 percent porosity units to allow for the effect of confining pressure. The average porosity of 22.3 percent determined from the 22/18-2 well logs for this interval compares favourably with the core data results. Because of this favourable comparison between core and log porosity within the hydrocarbon-bearing section and also the greater availability of log data, the log-derived porosity has been used in the determination of original oil in place.

The average horizontal permeability of the combined core sets is 39.3md, and again allowances have been made for the effect of confining pressure and the presence of calcareous sandstones. This average figure obtained from the core compares closely with the horizontal permeability obtained from analysis of the well test data. Such values obtained from oil tests showed permeabilities in the 35–40md range. Vertical permeability was determined from test data and in 22/17-1 was 1.21md. The corresponding horizontal figure was 39.0md, and the resulting ratio of vertical to horizontal values is very low. The significance of this low ratio is to suggest that water coning should not be detrimental to ultimate oil recovery.

Initial reservoir pressure at a subsea datum of 8150ft (2484m) is 3750psia and this represents a normal gradient of approximately 0.46psia per ft. Gas–oil ratios for the field are in the approximate range of 425–575ft³ (12.0–16.3m³) per barrel, and the Montrose crude is undersaturated by roughly 1000psia. The density of the oil at bubble point pressure is 0.642gm/cc.

Original oil in place has been determined using the above parameters. As has previously been

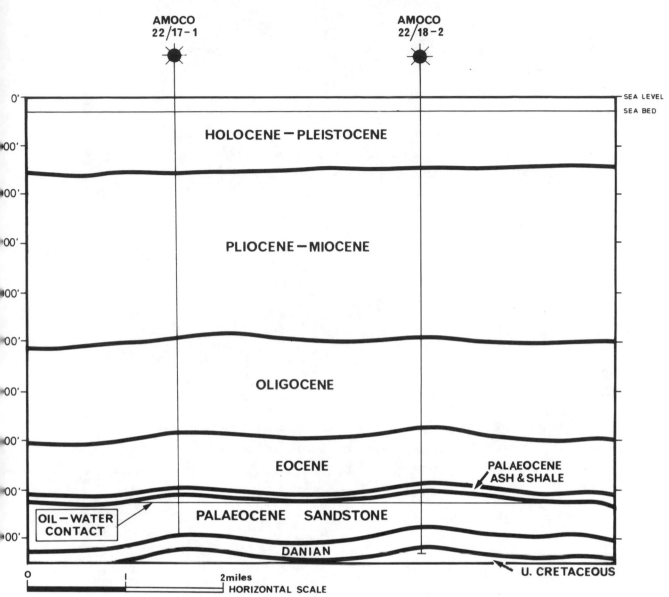

FIG 4.   Montrose Field: west-north-west/east-south-east structural cross section.

discussed, the field appears to consist of three structural culminations with North Montrose made up by two of these and South Montrose by the third. Initially it is planned to develop the northern area, and development of the southern portion will be decided following the results of a study of initial reservoir performance. Since there is a possibility that the field may consist of two sections, original oil in place has been calculated as ranging from 329 to 450MMstb.

Recoverable reserves for the field are limited by the drainage area, which in this context has been assumed as the area occupied by production wells plus the area encompassed within 1500ft (457m) of the outer wells. Recoverable reserves in the field-wide drainage area have been estimated as ranging from 107 to 166MMstb. In order to estimate recoverable reserves, an overall recovery factor of 49 percent has been applied to the original oil in place contained in the drainage area of the field.

FIELD DEVELOPMENT

The present plan calls for an initial phase of one fixed platform to develop the northern structural culminations with the location as shown on Fig 6. Following a study of initial reservoir performance, a decision will be made for the development of the southern portion of the field.

It is planned to drill 24 production wells from the Montrose "A" platform, and, as the claystones and shales of the Miocene and Pliocene tend to hydrate and swell during drilling to become very sticky, the maximum deviation of production wells will be limited to 45°. The average oil column thickness for the 24 wells in Montrose North is 120ft (37m), while in the South the average thickness is 70ft (21m). In both North and South Montrose, a well spacing of 200 acres (81 hectares) has therefore been selected as the optimum distance to ensure full recovery.

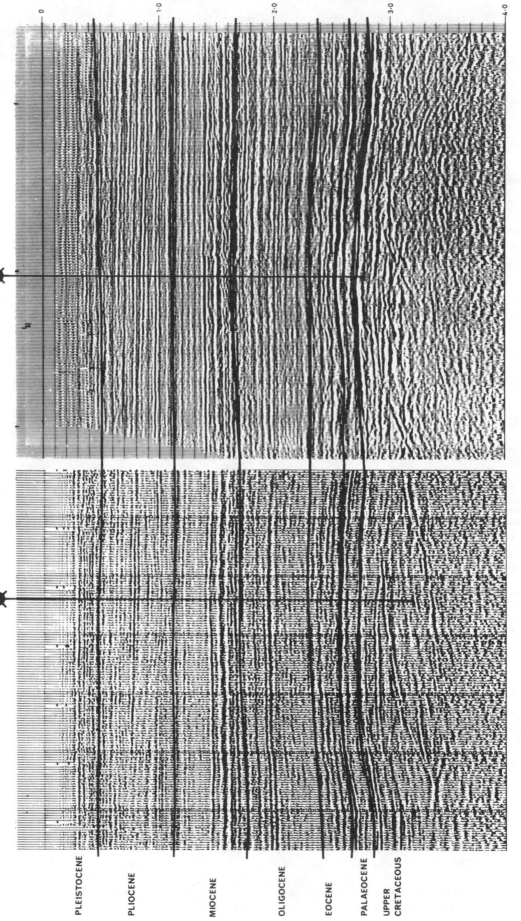

N

AMOCO
22/18-2

AMOCO
22/18-1

S

0   1   2 miles

PLEISTOCENE

PLIOCENE

MIOCENE

OLIGOCENE

EOCENE

PALAEOCENE

UPPER
CRETACEOUS

Fig 5. Montrose Field : south/north seismic section.

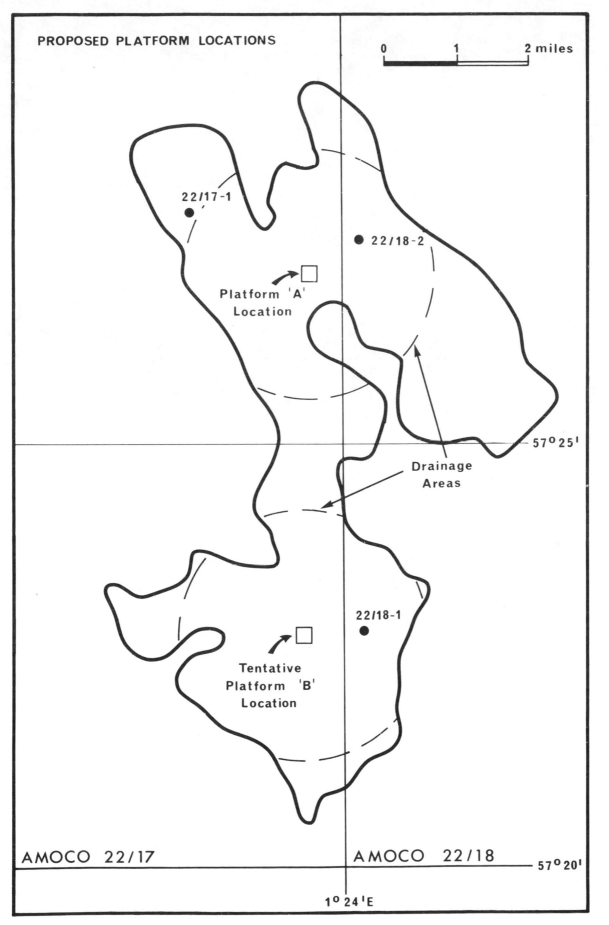

FIG 6.   Proposed development of Montrose Field.

Productivity indices based on well-test data show that, in North Montrose, producing rates of between 1500 and 5000BOPD with 1000psia drawdown are feasible, depending on pay-thickness. Applying a rate of 2500BOPD to the 24 wells, a maximum capability of 60 000BOPD is estimated for the North Montrose area. Oil production should remain at this plateau rate for a period of three years and with a start-up planned for 1976, production should peak in 1979, and decline will probably begin in the early 1980s. Montrose South has a thinner net pay and a range of well-productivities from 1000 to 2000BOPD. If the southern platform is installed then its maximum yield is estimated at some 3500BOPD. Plans for well-completions are based on the premise that the reservoir will produce by either natural or induced water drive. As the oil column is thin, completion intervals will be selected structurally high in each well bore.

The Montrose "A" fixed platform will be of a conventional template design and will be an 8-legged structure with pile foundations. The complete platform will be 544ft (166m) high from the seabed to the top of the communications tower.

The jacket will support an 8-legged superstructure accommodating drilling, production and service modules. The jacket will be constructed on land, skidded on to a barge, towed to the field and then launched on to location. The equipment modules are designed to be fabricated onshore and lifted onto the already installed jacket. The platform has been designed to process 60 000BOPD, and this production system includes facilities for oil-processing, gas- and water-handling, artificial lift, water-injection, electrical power generation, hydraulic safety mechanisms and corrosion controls.

The Montrose crude will be transported using a direct loading scheme. Facilities will exist on the platform for the loading of oil directly into tankers via two single-buoy moorings. Loading lines to the moorings will be 10in (0.25m) in diameter and they will extend one mile (1.6km) from the platform.

The Montrose "A" platform jacket is now under construction at the manufacturing yard of UIE in Cherbourg and the production modules are currently being fabricated in Scotland, Holland and France. Development drilling is planned to start in the Montrose Field by early 1976 and oil production is scheduled to begin later that year.

37

The Forties Field

By P. J. WALMSLEY

(BP Petroleum Development Limited)

SUMMARY

The Forties Field was discovered in 1970. It is located 175km east of Peterhead in 90–130m of water. Oil occurs in a sandstone reservoir of Palaeocene age at a depth of 2135m below a thick Cenozoic section consisting primarily of mudstone. The Palaeocene sandstone and mudstone section lies above Danian and Maastrichtian micritic limestones which in turn overlie presumed Jurassic volcanics. The structure is a broad low-relief anticlinal feature with a closed area of 90km^2 and a gross oil column of 155m. Recoverable oil reserves are estimated to be 1.8×10^9 bbls.

INTRODUCTION

The Forties Field lies in the British Sector of the North Sea about 175km east of Peterhead (Fig 1). Approximate co-ordinates are latitude 57° 45′ N, longitude 0° 45′ E. About 95 percent of the field falls within BP Licence Block 21/10, the eastern end extending into Shell/Esso Block 22/6. The field is named "Forties" after the meteorological area in which it was discovered. In one regard this is a misnomer because the meteorological area owes its name to a large sector of the North Sea that is 40 fathoms deep, whereas water depths across the Forties Field vary from 50 to 70 fathoms (91–131m) (Fig 8).

HISTORY OF DISCOVERY

The discovery of the Forties Field was made in October 1970. Block 21/10 formed part of a 2nd Round Licence awarded in November 1965. At that time only five marine wells had been drilled in British waters. Little was known of the northern North Sea other than that it covered a large Tertiary sedimentary basin possibly overlying in part a thick development of older sediments, and oil prospects in the area were considered at best to be highly speculative. By the end of 1969 over 50 exploration wells had been drilled in the northern North Sea and apart from minor shows (including the Cod gas discovery) little real encouragement had been obtained. However in December 1969 Phillips made their important Ekofisk oil discovery in south-west Norwegian waters, and simultaneously Amoco/Gas Council discovered the smaller Montrose oilfield in UK Block 22/18 (Fig 1). These successes quickly accelerated the tempo of exploration activity and it was in this climate that BP spudded its first well in 21/10 using Sea Quest in August 1970.

21/10-1 was located near the crest of a low amplitude structure with a closed area at the base of the Tertiary of 40km^2, centred on a large structural nose which plunged south-eastwards across the block defined from an early 5×5km seismic reconnaissance survey. At 2098m subsea the well entered oil-bearing Palaeocene sands. An oil–water contact was established at 2217m subsea and subsequently tested 37° API, low-sulphur oil at a rate of 4730bbl/day on a 54/64-in surface choke. A field of major proportions had been discovered.

A detailed 1.5×1.5km seismic survey was shot immediately to supplement the earlier work and the combined data interpreted in conjunction with that obtained from the discovery well. The new interpretation indicated a larger closed area than originally had been envisaged.

The first appraisal well 21/10-2 was spudded in June 1971 some 5.5km to the north-west of the discovery well and found an oil column of 32.5m with the same oil–water contact. The second appraisal well 21/10-3A was drilled 7km west of 21/10-1 and this found an oil column of 126m, again with the same oil–water contact. At the same time Shell/Esso drilled a successful outstep in Block 22/6, 3.8km south-east of 21/10-1. Here the sand development of the oil-bearing section in the upper part of the Palaeocene showed considerable deterioration compared with other Forties wells. The same oil–water contact was present. By this time plans for the construction of four platforms to develop Forties were well advanced. To select finally the sites for the platforms an additional well was drilled 5.5km south-west of 21/10-1. This was completed in September 1972 and was again successful.

The first five wells have confirmed a major oil field with a maximum oil column of 155m in Palaeocene sandstones and occupying an area of about 90km^2 (Fig 6).

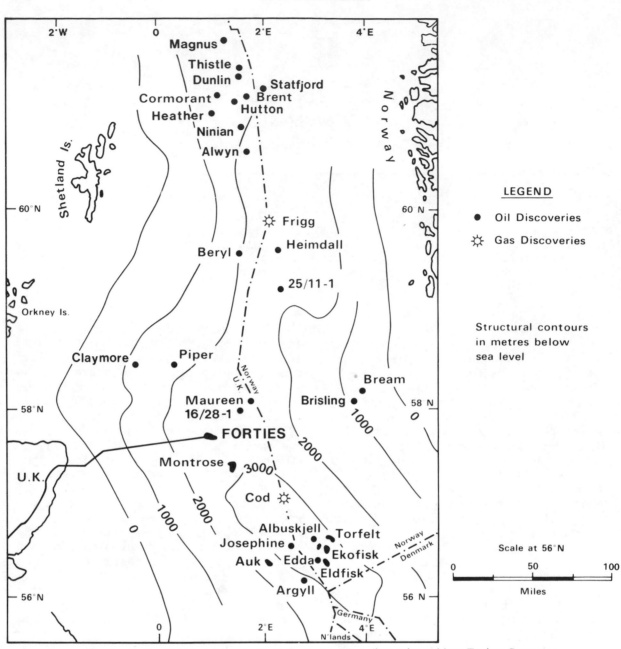

FIG 1.   Forties Field in relation to North Sea hydrocarbon discoveries and base Tertiary Contours.

GENERAL STRATIGRAPHY

The stratigraphic column of the Forties Field is typified by the discovery well 21/10-1 (Fig 2). Sandstones of Palaeocene age provide the oil reservoir. These underlie a thick monotonous section of grey-brown, variably calcareous and carbonaceous mudstones, ranging from Upper Palaeocene to Holocene. Sandstones are present in the Plio-Pleistocene and thin beds of limestone in the Eocene, but the post-Palaeocene succession is primarily argillaceous. The stratigraphic subdivisions shown for the Tertiary in Fig 2 are based on micropalaeontology.

In 21/10-1 the Palaeocene is 506m thick. The Eocene/Palaeocene boundary is placed at 2015m

subsea on palaeontological evidence. The Palaeocene is predominantly terrigenous, but the basal 31m consists of white-grey, slightly argillaceous and sandy micritic limestone of Danian age. This overlies cleaner and more compact micrite of the Maastrichtian at 2521m subsea, the boundary being poorly controlled palaeontologically and therefore chosen on the basis of gamma ray–sonic log character (Fig 4).

The Maastrichtian micrites are 114m thick and overlie the pre-Cretaceous section unconformably. 753m of pre-Cretaceous strata were penetrated in 21/10-1, the section being largely comprised of volcanic material. These volcanics are mainly composed of alkaline olivine-basalts with large pyroxene phenocrysts and altered porphyritic

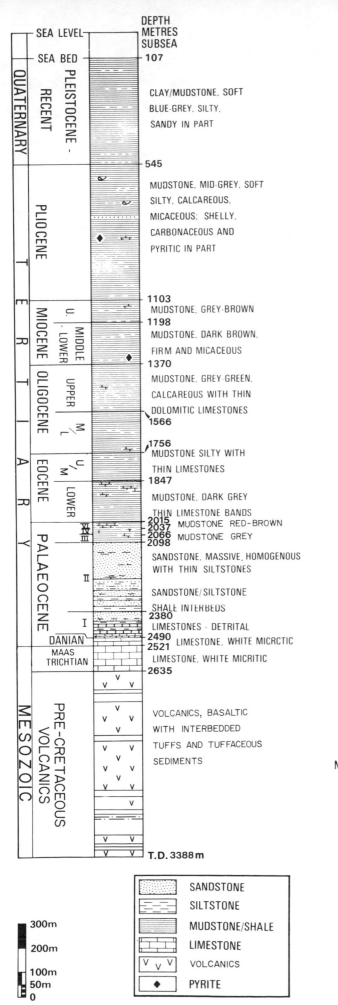

FIG 2. Forties Field stratigraphy: Well BP 21/10-1.

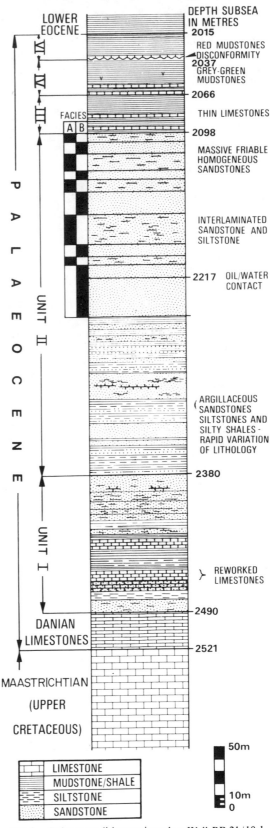

FIG 3. Palaeocene lithostratigraphy: Well BP 21/10-1.

rocks with agglomerates, pyroclastics and tuffs interbedded with red-brown siltstones and mudstones. The volcanics are considered to be extrusive and generally non-explosive and the tops of the flows are weathered. It has been concluded that the volcanic section is Jurassic in age, probably Bathonian. Further details are given in a companion paper (Howitt *et al.*, 1975) and will not be repeated here. 21/10-1 is the deepest well in the Forties Field and terminated at 3388m subsea.

PALAEOCENE STRATIGRAPHY

The post-Danian section of the Palaeocene has been described in detail in an earlier paper (Thomas *et al.*, 1974). Since then regional studies of the Palaeocene have caused BP geologists to amend the four-fold subdivision previously

described, and the new subdivisions are now shown in Fig 3.

There are two principle changes. One is that Unit IV of L. Aston and the late M. J. Wolfe has been redefined as three units of which two, Unit VI and Unit IV are present at Forties. Unit V is absent and is marked by a minor disconformity at 2037m subsea in well BP 21/10-1. The other is that Unit II has been expanded to encompass the major part of the upper member of Unit I. Our newly defined lithostratigraphic units are now as follows (no formal terminology is proposed at this stage).

Unit I: 2380–2490m subsea. This lowest unit contains a basal sandstone 4.6m thick overlain by a sequence of calcareous sandstone, limestone and mudstone and is characterized palaeontologically by reworked Danian and Cretaceous faunas indicative of contemporaneous erosion at the end of the Mesozoic.

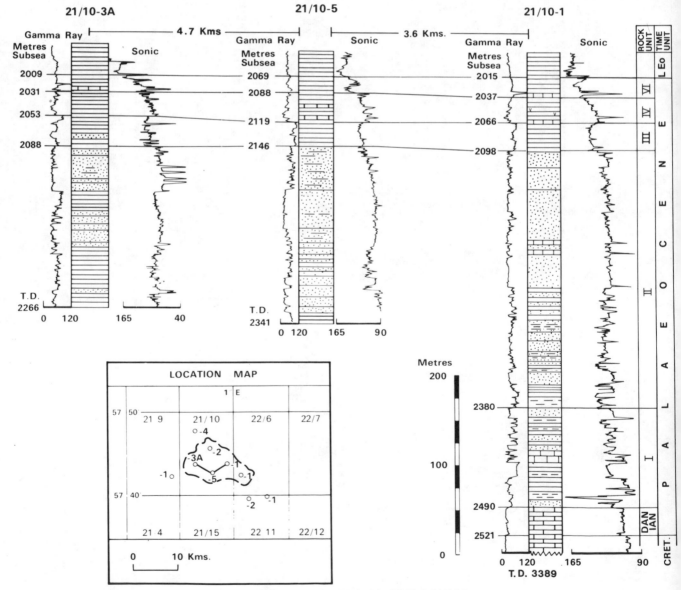

FIG 4.   Well correlation 21/10-3A, 21/10-5, 21/10-1.

Unit II: 2098–2380m subsea. Unit II contains all the massive sandstones within the Palaeocene of the Forties area and is the producing "formation" of the Forties Field. The dominant lithologies are sandstone and mudstone, the sandstone predominating in the upper part of the unit over most of the field. Clean homogeneous sandstones, individually up to 80m provide the major reservoir rock (Fig 4). These vary in colour from brown to almost white, in grain size from fine to coarse, and are typically clean and friable. Sorting is poor to moderate but reservoir properties are excellent.

The clean sandstones are separated by intervals containing fine-grained, locally silty sandstone, commonly with detrital mica and lignite, interbedded with laminated siltstone and shale as upward-fining units commonly less than 1.5m thick. The sandstones have moderate to good reservoir characteristics and make up to 20–55 percent of this facies. Load casts and ball-and-pillow structures are common.

The facies of Unit II becomes more shaly towards the southern and particularly the eastern part of the field where grey kaolinitic shales, graded siltstone-shale couplets and burrowed green waxy shales become increasingly predominant. The central part of the field has the highest percentage of clean sand and has the best production potential.

The depositional environment of the Palaeocene sandstones is the subject of a companion paper (Parker, 1975).

Unit III: 2066–2098m subsea. This unit forms the cap rock to the reservoir sands. It consists of dark grey silty lignitic shaly mudstones rich in montmorillonite. Its thickness and lithology are constant across the field and its contact with the underlying Unit II is transitional.

Unit IV: 2037–2066m subsea. This is essentially a mudstone unit. The mudstones are green-grey, slightly calcareous and contain several thin pale grey clays with a distinctive mineralogy indicative of degraded volcanic ash. It has a characteristic

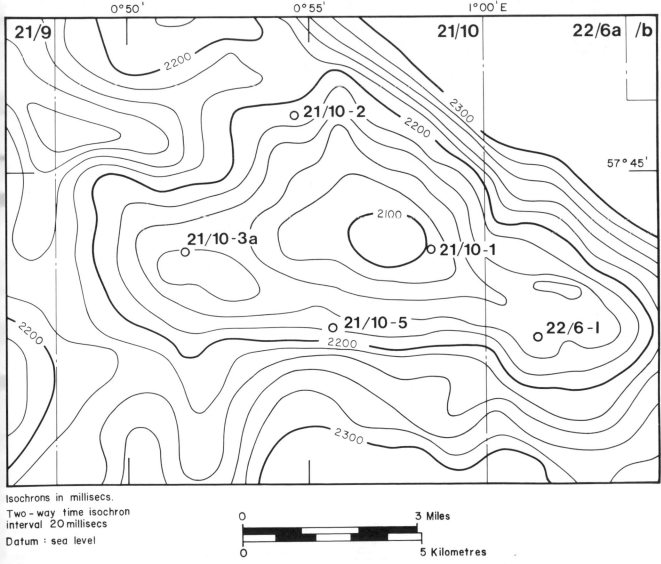

FIG 5.   Isochrons on a seismic reflector overlying the Palaeocene reservoir.

sonic-log pattern increasing in velocity with depth. This acts as a good log correlation feature (Fig 4), and also gives rise to the prominent seismic horizon used to map the configuration of the top of the reservoir.

Unit VI: 2015–2037m subsea. At Forties this unit overlies Unit IV unconformably, Unit V which is recognized to the north and west being absent. Unit VI is comprised of red mudstones and contains a characteristic red-stained foraminiferal assemblage in which *Globigerina* cf. *triloculinoides* predominates.

STRUCTURE

The regional tilt at the base of the Tertiary in the Forties area is down to the east. In Block 21/10 this regional tilt is interrupted by a large east–south-east trending nose. Reversal of dip on this nose provides closure for hydrocarbon accumulation in

the overlying Palaeocene sands. Isochrons on a seismic horizon 35m above the top of the sands are given in Fig 5 and the depth contours on the top of the reservoir in Fig 6. A structural cross-section east–west across the field is shown in Fig 7. There is an approximate coincidence between the structural spill point and the oil-water contact at 2217m subsea.

The depth contours on the top of the reservoir indicate a broad dome 16km east–west by 8km north–south having a closed area of about 90km² and a vertical closure of 155m.

The Palaeocene structure overlies a faulted high at base-Cretaceous unconformity level, in turn overlying complex pre-Cretaceous block-faulting with associated thick volcanics. Structural expression remained relatively positive during late Mesozoic and early Tertiary time until the late Miocene. Uplift probably ceased earlier with differential compaction allowing structural expression to persist through the Miocene.

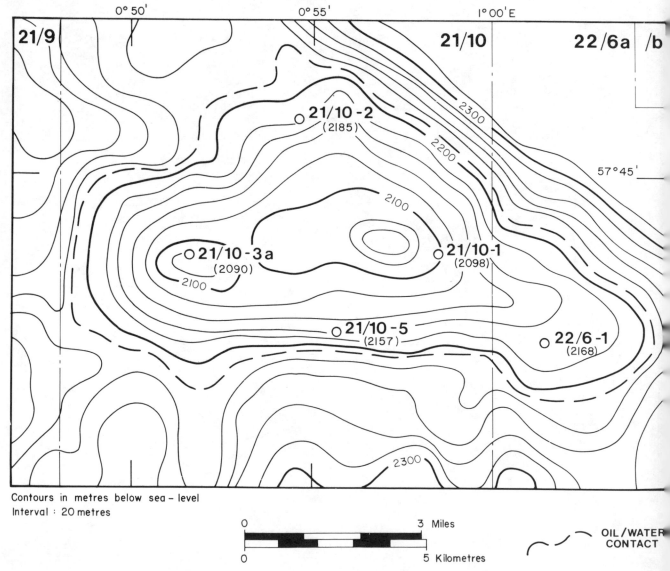

Contours in metres below sea-level
Interval : 20 metres

0   3 Miles

0   5 Kilometres

OIL/WATER CONTACT

FIG 6.   Structural contours on the top of the Palaeocene reservoir.

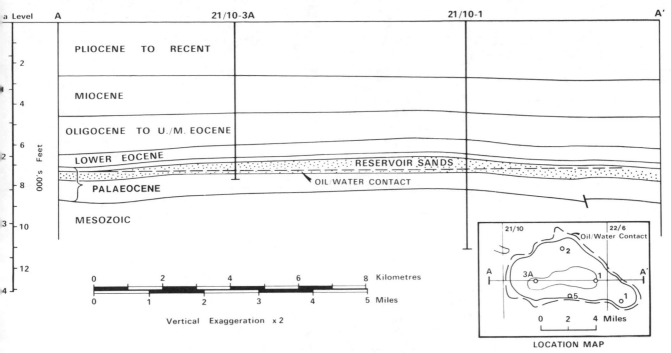

FIG 7. Structural section east—west across the Forties Field.

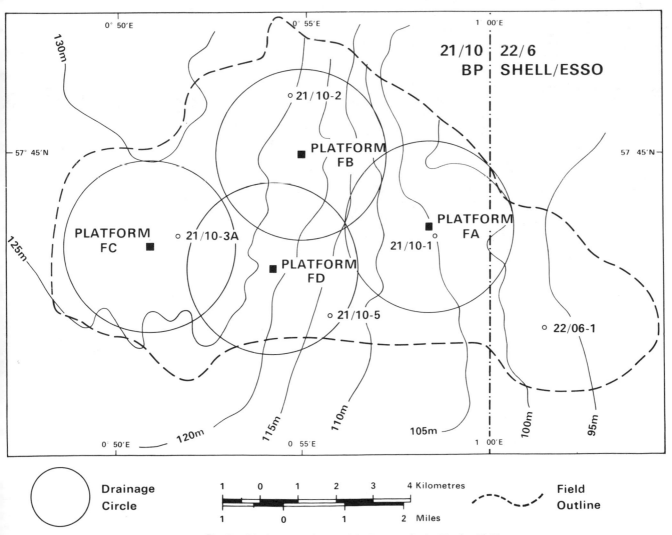

FIG 8. Platform locations and bathymetry in the Forties Field.

RESERVOIR CHARACTERISTICS, RESERVES AND PRODUCTION PLANS

BP 21/10-1 tested the interval 2108.5–2113m subsea and produced 37° API oil at a rate of 4730bbl/day on a 54/64-in surface choke with an estimated gas–oil ratio of 250scf/bbl. The crude was low-sulphur (0.3 percent) and medium-wax (8.5 percent). BP 21/10-3A tested the interval 2092–2137.5m subsea at rates up to 3260bbl/day, production being limited by test equipment restrictions. Bottom-hole PVT samples during the test established the gas-oil ratio at 330scf/bbl. Analysis of flow tests indicates that wells are capable of producing in excess of 15 000bbl/day in the crestal area. For productivity estimates an average rate of 8000bbl/day for development wells has been assumed. The initial reservoir pressure is about 3200psi, the oil being undersaturated and without a gas cap. Poroperm characteristics are good with core porosities ranging from 25–30 percent and permeabilities varying up to 3900md.

Oil in place and recoverable oil calculations are based on the whole accumulation including that part which falls in Shell/Esso Block 22/6. The calculations suggest an average oil-in-place figure of about 1400bbl/acre-ft, which in turn leads to a figure of about 4.4×10^9bbl stock tank oil initially in place. A recovery of 40 percent would yield about 1.8×10^9bbl recoverable oil.

Development drilling is to take place from four fixed platforms (Fig 8). Up to twenty-seven wells can be completed from each at a spacing of 120 acres, this having been selected as the maximum required to ensure full recovery of the field over a 20–25 year period for an initial well-rate of 8000bbl/day. In all 106 wells are planned.

By October 1974 the first two platforms had been emplaced and both the 32-in submarine pipeline to Cruden Bay and the 36-in land-line to Kerse of Kinneil near Grangemouth had been laid. Drilling is due to start from Platform FA in January 1975 and from Platform FC later that year.

It is estimated that the 54 wells to be drilled from these first two platforms will achieve a production rate of 250 000bbl/day by 1976. Following the emplacement of the second two platforms in 1975 and the completion of a further 52 wells the peak production of 400 000bbl/day is expected to be achieved by 1978 and will be maintained for three to four years before decline sets in.

Pressure will be maintained at about 2500psi and will be commenced early in the life of the field. Initially oil and gas will be separated at a pressure of 125psi and about a third of the solution gas piped ashore dissolved in oil. 25 percent of the remainder will be required as fuel on the platforms leaving about $15–20 \times 10^6$cu ft/day for disposal from each platform. During the drilling phase this must be flared, but subsequently it will be refrigerated for the recovery of natural gas liquids which will be shipped to shore mixed with the crude. Of the total gas recovered about half will be in the form of dry gas and half as condensed fluids in the C_3–C_5 range.

The discovery of the Forties Field, construction of the drilling platforms and their successful emplacement must be ranked as a major British achievement. By 1978 only 8 years from the date of discovery, it will be producing about one-fifth of Britain's total consumption of petroleum.

ACKNOWLEDGEMENTS

The author is indebted to the American Association of Petroleum Geologists for permission to reproduce parts of the text and diagrams from their Bulletin and to his colleagues who have contributed towards this account. I would like to thank the Chairman and directors of The British Petroleum Company Limited for permission to publish this paper.

REFERENCES

Howitt, F., Aston, E. R. and Jacqué, M. 1975. Occurrence of Jurassic volcanics in the North Sea. *Paper 28, this volume.*

Parker, J. R. 1975. Lower Tertiary sand development in the central North Sea. *Paper 34, this volume.*

Thomas, A. N., Walmsley, P. J. and Jenkins, D. A. L. 1974. Forties Field, North Sea. *Bull. Am. Ass. Petrol. Geol.*, **58**, 396–406.

DISCUSSION

R. C. Selley (Conoco Europe Ltd): It is apparent from the logs of the Forties sands that there is considerable variation of sand content. Clean sand units represent the submarine channel deposits of J. R. Parker's model (Paper 34) (with which I concur), while the "ratty" shaly sands correspond with turbidite fans.

Could Mr Walmsley say whether the development wells in the Forties Field would be located by engineers, or whether a geological facies model would be used to site wells in the better sand reservoir units? If so what model would be used.

P. J. Walmsley: The five wells so far drilled are widely spaced, lying four to five kilometres apart. Correlation of the various reservoir units between wells cannot be established. At this stage, therefore, it is not possible to forecast the extent and thickness of the different facies, so that the location of development wells must be based on a regular drainage pattern. Given detail from a large number of development wells it may be possible to confirm a geological facies-model and site some wells in the best reservoir units during the latter stages of development. However, I doubt if this will happen. The "ratty" shaly sands also contain oil and have reasonably good production characteristics.

38

Palaeotemperatures in the Northern North Sea Basin

By B. S. COOPER, S. H. COLEMAN, P. C. BARNARD and J. S. BUTTERWORTH

(Robertson Research International Limited, Ty'n-y-coed, Llandudno)

SUMMARY

Methods of palaeotemperature determination are discussed and causes of non-linear palaeotemperature gradients described. Maximum palaeotemperatures by the ESR method for North Sea wells have been integrated and a vertical profile over the Northern North Sea Basin, on which palaeotemperature isotherms have been plotted, show that there are generally high gradients in the uppermost 2000ft of section but the isotherms as a whole do not show much parallelism with basin structure. Usually isotherms and gradients appear to cross stratigraphic boundaries without discordance indicating that the maximum palaeotemperatures were reached in the Tertiary.

INTRODUCTION

A knowledge of past geothermal regimes and palaeotemperature gradients is important in assessing the history of hydrocarbon generation from sediments for the following reasons:

1. Oil and gas are released from the organic matter of sediments by geothermal heating at temperatures of 150°–350° F (60°–120°C). The general relationships between temperature, other indices of maturity and hydrocarbon product are shown in Fig 1.

2. During the process of oil accumulation, the precipitation of hydrocarbon fluids in the reservoir rock depends upon the temperature gradient in the vicinity of the reservoir (Klemme, 1972). It may be assumed that during basin development, sediment and its kerogen pass into zones of increased temperature whereas the expelled fluids, that is the compaction waters and released hydrocarbons retreat into zones of cooler temperatures.

3. After accumulation of hydrocarbons, the temperature regime further determines the rate at which the hydrocarbon deposit matures and changes its chemical composition, and also the rate at which it may be dissipated by diffusion and water washing.

It is obviously then of paramount interest when evaluating the results of a surface geological survey, when extracting the maximum amount of information from a dry well or when exploring the possibility of deeper pay zones in a producer, that a record of past temperature fluctuations which have affected the rock section through geological time be assessed.

MEASUREMENT OF PALAEOTEMPERATURES

Palaeotemperatures are determined by a number of methods:

1. The temperatures of the fluids at time of deposition may be derived from the physical and chemical properties of fluid-inclusions in minerals which have crystallized or been recrystallized in the sediment. The technique is more applicable to mineral veins and is most useful when there is complementary age determination of the thermal event.

2. The reflectivities of vitrinites contained in coals and shales give a measure of maturity of the sediments, which is the integral of all the thermal events which have affected the sediment. However, if the details of geological history are known, such as rates of sedimentation, etc., then estimates of palaeotemperatures may be attempted, using the data of Karweil (1956), as discussed by Teichmüller and Teichmüller (1968) and Bostick (1973).

Temperatures which had been reached at particular moments in time can be realized when vitrinites are found close to an igneous dyke or sill and their reflectivities compared with those produced by heating the unaffected coal or shale in the laboratory and applying the calculations of Jaeger (1964) with respect to temperature distributions in the vicinity of igneous intrusions.

3. Maximum palaeotemperatures are given by electron spin resonance spectra of kerogens after their isolation from their mineral matrices. This method, developed by Pusey (1973) measures the concentration and distribution of free electrons distributed through the chemical framework of the kerogen molecules. These free electrons are particularly associated with benzenoid structures and, to some extent, with organically complexed transition elements. Their electronic response is conditioned by the presence of heteroatoms such as oxygen, sulphur and nitrogen. It is assumed, in this method that kerogen is predominantly an amorphous lipo-humic complex which undergoes diagenetic change only in response to temperature change and not, to any great extent, to the duration

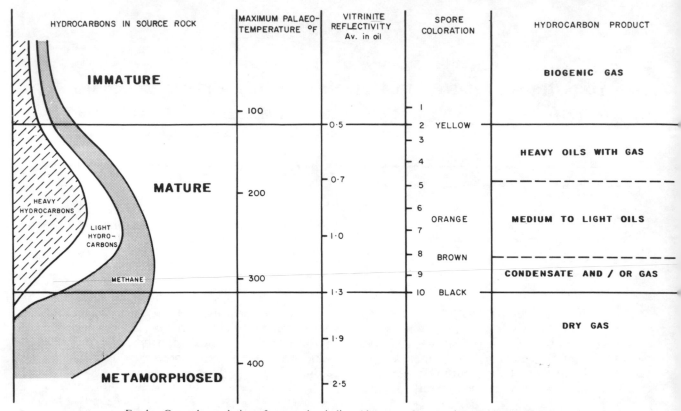

FIG 1.   General correlation of maturation indices (there may be some inter-basinal variation).

of heating. This property is comparable with the colour changes which occur in palynomorphs. It is in contrast to the changes which occur during the maturation of vitrinite, which is particularly susceptible to the actual duration of heating. The temperature-induced changes in kerogen are the loss of heteroatoms and the development of aromatic structures causing an increase in number but decrease in diversity of associations of free electrons in the kerogen structure. It is these parameters of quantity and quality of free electrons which are used to estimate maximum palaeotemperatures by using them as a calibration for well samples where present day temperatures are known to be maximal.

DISTRIBUTION OF TEMPERATURE

The variability of present-day temperature-gradients in the North Sea as demonstrated by Evans and Coleman (1974) must be a typical feature of sedimentary basins, both present and past. Their data show a generally diminishing heat flow from west to east, with higher geothermal gradients over the Tertiary and Carboniferous coal-bearing basins and low gradients over structural highs. This variability depends for the most part on the contrasting thermal conductivities and thicknesses of basement, sub-basement and basin fill (Fig 2).

From time to time, geothermal events in the upper mantle and lower crust also give rise to sudden short periods of heating and the progression of the heat front upwards depends upon the thermal diffusivities of the rock units (Table I) and may give rise to transient temperature gradients, different from those obtaining under normal heat flow conditions.

However, during sedimentation, within the basin and around its margins localized hot and cold spots occur because of the migration of connate waters, activated by the release of fluids during compaction (Fig 3) The direction of migration is controlled by the disposition of aquifers. There will be a general tendency for migration and upwards movement towards the edges of basins, and the process will be promoted if basement highs are stripped of their cover of soft sealing sediments, allowing the rapid release of connate waters through open joints and fractures (Fig 4). In this situation, deposition of both minerals and mineral hydrocarbons occurs.

In the basin itself, migration of connate waters takes place upwards along faults and, more rarely, downwards along faults so long as a suitable hydrodynamic gradient occurs. In each of these cases anomalous temperature distributions result and hot spots and cold spots occur.

PALAEOTEMPERATURES FROM ELECTRON SPIN RESONANCE

The determinations of palaeotemperatures for a typical well-section (Fig 5) are based on the

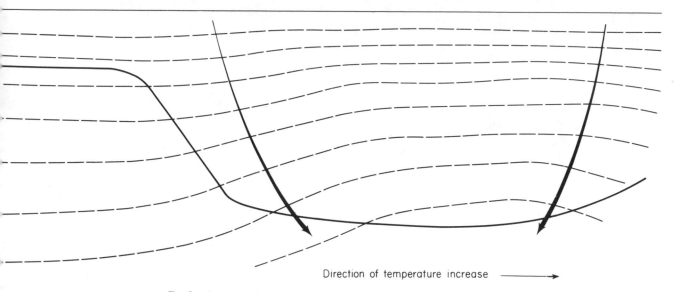

Direction of temperature increase ⟶

FIG 2. Isotherms in a homogeneous basin-fill overlying basement.

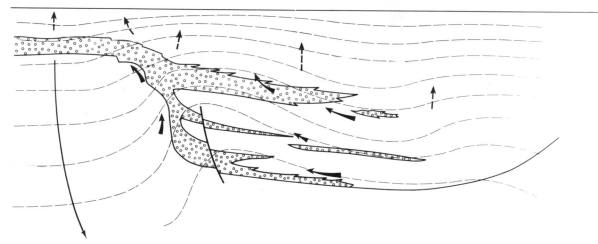

FIG 3. Isotherms of basin affected by lateral movement of waters through porous strata.

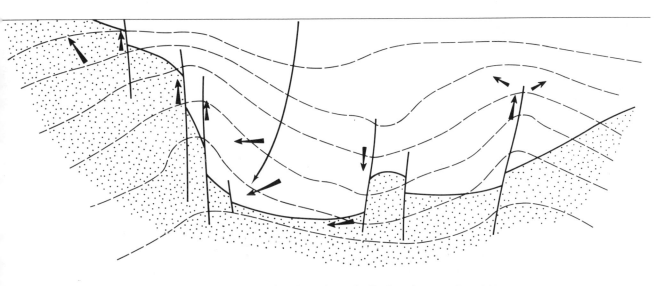

FIG 4. Isotherms of basin affected by redistribution of waters along faults.

TABLE I
**Comparison of conductivities and diffusivities
for various rock types**
(from Lovering, 1936)

	Conductivity	Diffusivity
Quartzite	0.012 8	0.031
Shale	0.001 9	0.004
Coal	0.000 8	0.002
Gabbro	0.004 3	0.008
Granite	0.005 7	0.009

measurement of certain electronic parameters which have been matched against values obtained from wells in which present-day temperatures are known and believed to be at a maximum.

In this typical example, the data lie as a scatter within the envelope, the scatter reflecting the non-homogeneity of kerogen. Kerogen is a mixture of components of diverse origin and chemistry, derived from algal bodies, spores, woody and carbonized tissue, all of which may be seen under the microscope. The usually dominant component is an amorphous lipo-humic component, residual after bacterial activity in the sediment. The influence of other components is to give a slightly elevated or slightly attenuated palaeotemperature value hence the scatter of data. The palaeotemperature gradient is obtained by drawing a median line within the envelope.

This example shows two features, firstly an increased temperature gradient at the top of the section in the youngest unconsolidated samples where the thermal conductivities are lowest and secondly, that there are no breaks at the unconformities, showing that the maximum palaeotemperatures were reached during the final stage of maturation.

PALAEOTEMPERATURES IN VERTICAL TRANSECT

In this study, palaeotemperature data for 15 wells were plotted on a vertical section across the Northern North Sea Basin from the Halibut Horst across the Fladden Ground Spur of the East Shetland Platform and along the Viking Graben (Fig 6).

The temperature distributions (Fig 7) may be compared with a sketch of the structure and stratigraphy (Fig 8). Several features stand out in the zone of isotherms between 200°F (93°C) and 300°F (149°C) at which temperatures the main events leading to the formation of oil take place.

The highest temperature gradients are in the uppermost least compacted sediments with values up to 4°F/100ft (7.3°C/100m), with maximum values at each end of the section over basement highs. At deeper levels, the palaeotemperature gradients have values of 1°–2°F/100ft (1.8°–3.6°C/100m) and over the middle part of the

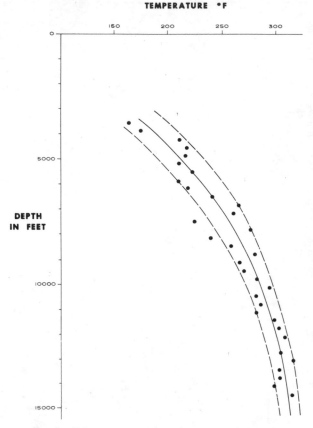

FIG 5.   Palaeotemperature data for a North Sea well.

section (where the Tertiary sediments are at their deepest) the isotherms are convex upwards as in the simplest model system (Fig 2). However, towards the north end of the section, although the isotherms at deep levels are regularly disposed through the section interface, a hiatus develops in the data in which both rapid lateral and vertical changes in palaeotemperature gradient occur. Here particularly there seems to be evidence that heat transfer had taken place under the influence of migrant fluids. They appear to have originated at a deep level and migrated upwards before being dissipated in the Lower Tertiary strata. It would appear unlikely that the fluids could break through the Upper Tertiary blanket of uncompacted sediments, but will be dispersed into adjacent coarse-grained sediments. During dispersion, these fluids lose heat to their surroundings, may react chemically with their host rocks and, more importantly, may precipitate hydrocarbons because of their lowered solubilizing power and by the effects of osmotic filtering as they pass from coarse-grained into fine-grained sediment.

Possibly a similar, but simpler, process occurs in the southern part of the section, close to the graben boundary fault. On the northern side of the fault there are rather lower temperatures than normal; on the southern side temperatures are about normal, except in the basal Tertiary where there is a steep palaeogeothermal gradient. It seems

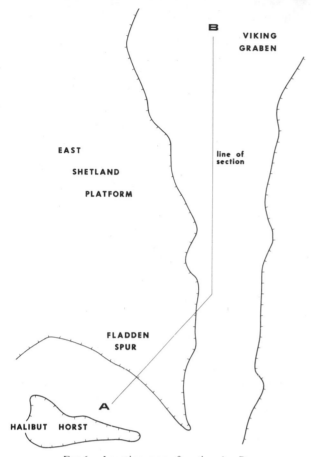

FIG 6.   Location map of section A—B.

reasonable to conjecture that fluids migrating from the graben area are ascending in the fault zone, then passing laterally through upper Mesozoic strata towards the Halibut Horst. A rather more sharply delineated flow appears to be associated with the edge of the Fladden Ground Spur.

Significantly, there is a general conformity in the palaeotemperature profiles; that is, there appears to be no sudden increases from low temperatures to higher temperatures, particularly at unconformities, as a well-section is traversed. Occasionally there are rapid changes in gradient but these are from high values to low values going down the well-section. For these reasons we believe that the palaeotemperatures recorded all belong to one geothermal event, which is, the development and filling of the Tertiary basins of the North Sea.

However, the possibility must be considered that some of the palaeotemperatures were impressed upon the sediments during a restricted episode of block-faulting which facilitated escape of connate waters, and that subsequent sedimentation, with increase and re-equilibration of temperatures, has not damped out that temporary event. In fact the palaeotemperatures and palaeotemperature-gradients often have values higher than those of the present day, particularly at each end of the section. The data seem to point to a past heating event over the area. It also seems probable that oil generation and entrapment are products of this Tertiary geothermal event with concomitant enhancement of its effect by redistribution of waters along fault

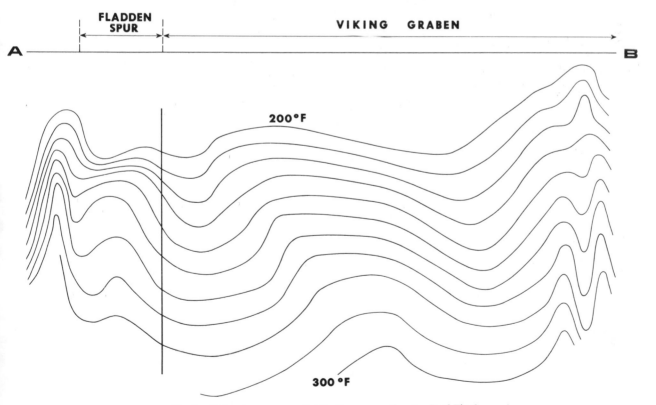

FIG 7.   Palaeotemperature distributions on section A—B of Fig 6.

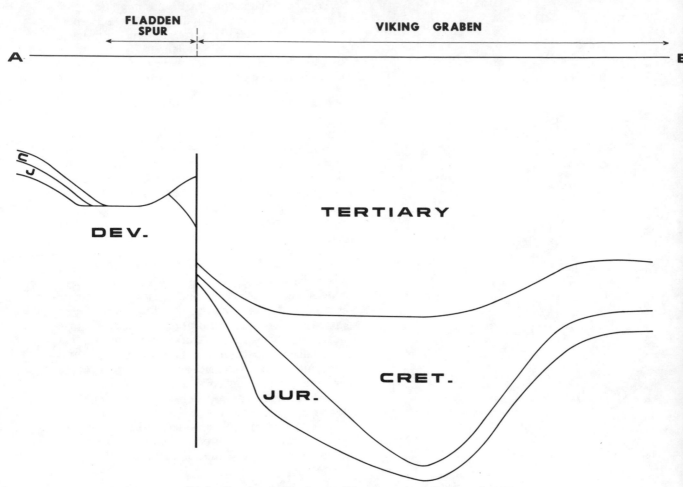

FIG 8.   Stratigraphic section A—B based mostly on Dunn *et al.* (1973).

channels. The extensive, though localized, vertical movement of waters suggests that, on occasion, oil may have migrated substantial vertical distances.

Although we are suggesting a fairly recent oil-generating event, the presence of previously generated oils is not precluded. They, too, will have been affected by redistribution and thermal alteration so that their presence may only be indicated by residual bitumens in reservoir rocks.

Finally, very high temperatures have not been recorded and there seems every indication that there are prospects for future finds of hydrocarbons derived from source-rocks older than the Jurassic, particularly if Devonian bituminous flagstones, such as those of Caithness, are present in the area.

ACKNOWLEDGEMENTS

We wish to express our thanks to the numerous oil companies who allowed us access to their data during the preparation of this paper.

REFERENCES

Bostick, N. H. 1973. Time as a factor in thermal metamorphism of phytoclasts (coaly particles). *C.r. 7ième congr. Int. Stratigr. Geol. Carb., Krefeld,* **2,** 183–93.

Dunn, W. W., Eha, S. and Heikkila, H. H. 1973. North Sea is a tough theater for the oil-hungry industry to explore. Pt. 1. *Oil and Gas Jl.,* **71,** (2), 122–8.

Evans, T. R. and Coleman. N. C. 1974. North Sea geothermal gradients. *Nature, Lond.,* **247,** 20–30.

Jaeger, J. C. 1964. Thermal effects of intrusion. *Rev. Geophys.,* **2,** 443–65.

Karweil, J. 1956. Die metamorphose der Kohlen vom Standpunkt der physikalischen chemie. *Z. dt. geol. Ges.,* **107,** 132–9.

Lovering, T. S. 1936. Heat conduction in dissimilar rocks and the use of thermal models. *Bull. geol. Soc. Am.,* **47,** 87–100.

Pusey, W. 1973. The E.S.R.-kerogen method, a new technique of estimating the organic maturity of sedimentary rocks. *Petr. Times,* **77,** 1952.

Teichmüller, M. and Teichmüller, R. 1968. Geological aspects of coal metamorphism: In *D. G. Murchison and T. S. Westoll (Eds.) Coal and Coal-bearing Strata.* Oliver and Boyd, Edinburgh.

Closing Address

By SIR PETER KENT, F.R.S.

(Chairman, Natural Environment Research Council)

My final task at this Conference is to sum up a most notable week's activities.

First I should refer to the three days of *Geology* which began the meeting. Most of the 900 geologists who attended returned to their desks or drilling rigs on Thursday morning, but they had taken part in a remarkable unfolding of the complex geology of the 200 000 sq miles of the North Sea, an occasion which was probably without parallel for the great quantity of critical and unpublished data released in a single conference. One may hazard the statement that never again will such a mass of new information be available to be released at such a meeting.

The North Sea information came from the oil companies themselves, with supporting contributions on other areas from non-industrial sources. Commercial secrecy is important—it means power to compete, it means (perhaps most importantly) ability to exchange information and to build up a more complete geological picture. It took an occasion like this to ensure that not only the two or three traditionally helpful major companies but also many of the others were willing to join in mutually unveiling their well-kept secrets. It is greatly to the credit of the Organizing Committee, and of Myles Bowen in particular, that so much was released; that the companies were persuaded to provide such an even spread of new data on the regional geology and on the detailed structure, stratigraphy and development of the complex areas which are the sites of our new oilfields.

The emphasis has inevitably been mainly on the North Sea; elsewhere the evaluation of knowledge and the internal exchange of data has not yet provided a basis for any well grounded account of the regional geology, nor has the competitive situation yet eased to a point where companies can be expected to release data of the few wells which have so far been drilled. So the westerly and southerly areas still remain available for the more speculative reconstructions of our academic friends.

In this respect—in the matter of the variation of attitude—we have been informed and indeed entertained by two contrasting approaches. We have had on the one hand the very broad brush painting of northwestern Europe in relation to the rest of the continent and in its setting in the global tectonic scheme by the brothers Ziegler. They were by no means in full agreement with one another or (as regards details) with many members of the audience, but it was a useful and thought-provoking exercise based on a phenomenal amount of research into literature and many unpublished data. On the other side we had the highly informed accounts of the regional stratigraphy as it is now known from the six hundred North Sea wells against the background of the surrounding lands, and the detailed description of the oil-field areas with their fascinating histories of intra-Mesozoic development, already based on years of meticulous research with seismic and well data, supported by all the expertise in palaeontology, petrography and geochemistry which the companies can muster in their sophisticated studies.

Some of the speakers, and in particular many of the academic speakers who asked questions, showed a tendency to regimentation of data on rigid lines which is open to some question. The assumption of a *universal* stratigraphic break—whether it is called Cimmerian, Hardegsen, Asturian or whatever—is a geological contradiction. One cannot have highs without lows, swells without basins, and there must be continuous deposition over some part of any area. We are of course brainwashed by the generous flow of information from the well-drilled positive areas, but it has to be remembered that they are atypical. Furthermore it is not at all a useful contribution in the long term to use a standard term, carrying the assumption of strict contemporaneity, if the stratigraphical information does not provide an exact dating.

In the tectonic field, it could perhaps be commented that problems are not *solved* by using such a term as "triple junction". It may or may not be a satisfactory description. Its use does not imply that we know what happened to the different plate elements, or that their movements were necessarily contemporaneous. The feed of knowledge must be the other way—if the North Sea is associated with a triple junction this is the first opportunity, anywhere in the world, to analyse in detail how such a structure developed. The concept poses a question; it does not supply an answer.

Other problems were more down to earth (literally) and of more direct importance in the location of economic oil sands. Notably among the matters still arousing controversy is the deposition of the Palaeocene sands, whether of turbidite origin or not; this bears on their productivity, extent and the prospects of

493

uncontrolled areas. It was clear that a major body of data on such matters is now available, and this Conference was a step forward in putting it all together. And mention should be made of the discoveries of volcanic rocks in the post-Permian, with a well-documented ash and lava pile of Middle Jurassic age east of the Moray Firth, a smaller Upper Jurassic centre off southern Norway, and a Lower Cretaceous volcano on the northern Netherlands and ash intercalations in the North Sea Eocene.

The three days have provided us with intellectual meals for years to come, and we must pay a tribute to the hundreds of geologists and geophysicists whose work was summarized at the symposium; industrial scientists whose expertise is unsurpassed in any other field, a body of highly trained men without parallel in either government research or in academic life.

On Thursday the scene changed, and so did much of the audience, as consideration swung from the geology and the oilfields to the two-day discussion of *Environmental problems.*

The first environmental day was largely devoted to the work of the companies in meeting their moral and statutory obligations to avoid the contingency of pollution and to deal with it as it occurs, under the eagle eye of government inspectors whose duty it is to ensure that high standards are maintained in every operational field. The Conference was told of the proposals for sea-floor completions and sea-bed storage, of pipelines and their valve systems, of the plans for dealing with oil-spill if a serious accident occurs. It was provided with factual data on the toxicity to fish of the material which might get into the sea in normal or abnormal circumstances.

The theme was continued on the final day by biologists concerned with the investigation of the effects of pollution of all kinds on the animal and vegetable life of the seas and their margins. It was a balanced and thoughtful presentation; the emphasis remained on the appraisal of the effects of the occurrence of hydrocarbons (both naturally occurring and introduced) on the environment, with recognition of the relatively large quantities which could be "assimilated" by the seas, and also repeated warnings that the damage due to the use of emulsifying detergents in the interests of short-term amenities was likely to be considerably greater than that of normal crude oil. It was clear that of the organisms associated with the seas it is the diving sea birds which are at greatest risk in the event of an oil spillage, and particularly the slowly reproducing species which can recover only very slowly from a disaster.

Perhaps the most important point arising from the day's discussions was the emphasis on high speed in dealing with a spillage—that action on the spot within two hours of an accident is by far the most effective (and least expensive) remedial action. The corollary was that present government-sponsored emergency facilities are too widely spaced and too far from potential danger points for maximum efficiency, particularly if they are to be deployed by surface transport.

So in this five days' Conference the geologists produced a mass of new information which must greatly improve understanding of the problems of the North Sea and their resolution, while the operational speakers and the biologists exchanged information on the hazards of spillage, the safeguards which exist, on the different aspects of trouble when it occurs and on the desiderata for remedial action. It has been a most notable Conference which must contribute to progress in discovery and in ensuring that the necessary economic progress takes place with minimum disturbance of the environment.

Throughout the text of this volume the recommendations of the joint Oil Industry–Institute of Geological Sciences Committee on North Sea Nomenclature have been followed as far as practical. Because of the difficulties of emending diagrams at short notice before publication certain variations between text and figures may be noticed—the spelling of "palaeo-" and "paleo-" and "Cimmerian" and "Kimmerian" may be instanced. The indulgence of the reader is sought in regard to such inconsistencies.

Index